DIFFUSE SCATTERING AND THE FUNDAMENTAL PROPERTIES OF MATERIALS

DIFFUSE SCATTERING AND THE FUNDAMENTAL PROPERTIES OF MATERIALS

Edited by

R. I. BARABASH
G. E. ICE
P. E. A. TURCHI

MOMENTUM PRESS

MOMENTUM PRESS, LLC, NEW JERSEY

Diffuse Scattering and the Fundamental Properties of Materials

First published in 2009 by
Momentum Press®, LLC
222 East 46th Street, New York, N.Y. 10017
www.momentumpress.net

ISBN-13: 978-1-60650-000-2 (hard back, case bound)
ISBN-10: 1-60650-000-7 (hard back, case bound)

ISBN-13: 978-1-60650-002-6 (e-book)
ISBN-10: 1-60650-002-3 (e-book)

DOI forthcoming

Cover Design by Jonathan Pennell
Interior Design by Scribe, Inc.

First Edition March 2009

10 9 8 7 6 5 4 3 2 1

Printed in the United States of America.

Contents

Part III: Scattering from Complex and Disordered Materials: Experiments and Theory

Part IV: Scattering from Distorted Crystals

Part V: Final Remarks

PREFACE

This book highlights emerging research areas that exploit the ability of diffuse scattering to characterize local structures and defects in materials. An emphasis is placed on the coming renaissance in diffuse scattering driven by new sources, better instrumentation, novel new materials, and advanced theories and methods. The book provides an overview of some of the most exciting recent advances in diffuse scattering and provides guidance for students and researchers interested in new methods to characterize their materials.

One transparent way to distinguish between conventional Bragg scattering and diffuse scattering is the access to the associated correlations. Bragg scattering measures single point (atom) positions in the *average lattice*, including the mean square thermal or static displacements. With diffuse scattering, however, we gain access to the pair correlations between atoms on separate sites; that is, we measure two-point correlations, including both compositions and displacements and their cross-terms (often referred to as the size-effect scattering). We thus extend the characterization of atomic structure to the two-point correlations, which are manifested in the scattering distribution throughout the Brillouin zone. Attendant on this is the availability of mean (molecular) field treatments of the susceptibility (intensity) that can extract the pairwise interactions, which is analogous to the exchange interaction of magnetism but usually extends to distant neighbors due to both long-range electronic and elastic effects. Finally, by using coherent diffraction, and/or first-principles-based models and higher order expansions in the displacements, one has, in principle, access to higher order concentration correlations. With the new X-ray and neutron sources, such effects may be explored.

In developing this overview of current and future trends in diffuse scattering research, we have solicited monographs from the world's leading experts. From these monographs it is clear that diffuse scattering is being revolutionized by intense new X-ray and neutron sources, by the development of massively parallel area and energy-dispersive detectors, and by the unprecedented availability of powerful and inexpensive computational resources. These advances have opened up new research opportunities that are being exploited through novel new techniques and theoretical developments. Together, these developments have transformed diffuse scattering from a difficult signal-limited research area into a mainstream materials analysis tool with exquisitely detailed characterization of defects and short-range structure and with a growing capacity for combinatorial, real-time and *in situ* studies. Furthermore, the rise of nanomaterials, the ability to fabricate low-dimensional structures, and the need to understand entropy and the role of disorder in materials properties are all important drivers for emerging diffuse X-ray scattering research and impact our understanding of the properties of materials.

During the last four decades, technical developments have greatly extended the range of experiments that are now practical. In the case of X-rays, third-generation synchrotron radiation sources now have about fourteen orders of magnitude higher brilliance (photons/s/mm^2/mrad^2/eV) than traditional laboratory sources. New third-generation sources are coming on-line around the world, and fourth-generation sources are under development, which will allow for completely new levels of materials characterization that will naturally extend diffuse scattering analysis of materials. New massively parallel area detectors with and without energy resolution will also extend studies of diffuse scattering and make possible materials characterization that until recently would simply not have been practical. In addition, the unprecedented power of computational resources allows for new approaches to data analysis in a timely fashion.

In the case of neutrons, the development of powerful spallation sources and the embrace of massive area detector arrays make routine the collection of three-dimensional reciprocal-space neutron-based diffraction maps. Structures with important low-Z components can particularly benefit from the strong neutron scattering of some low-Z isotopes. New beamlines optimized for diffuse scattering research will separate dynamic and static structure scattering for unprecedented detailed information about local structures. New neutron sources can also exploit neutron scattering sensitivity to magnetic structure to help us understand magnetic structure defects.

In this context, the book is organized into four main parts with twenty one chapters that emphasize diffuse scattering research applied to important classes of materials—namely, low-dimensional materials, bulk alloys, complex and disordered systems, and distorted crystals—with a fifth part that serves as a conclusion and explores the promises of the coming renaissance in diffuse scattering. Although there is a level of arbitrariness in sectioning a topic like this, the vastly disparate classes of materials discussed emphasize how the common denominator of diffuse scattering can provide deep insights into the fundamental properties of materials. We hope that this selection will invigorate debates on some of the pending challenging issues and provide inspiration for solving them.

Low-dimensional systems, such as surfaces, interfaces and thin films, nanostructures, and antiphase and orientation boundaries, play a critical role in modern technology. Among the six contributions of Part I, it is shown that interface roughness, capillary wave fluctuations, and nanoparticle inclusions, through their spatial correlation properties, shape, elastic strain, and local composition, give rise to diffuse scattering. Hence, anisotropic organization and incipient crystallinity in amorphous networks can be revealed by X-ray grazing-incidence scattering, and ordered and disordered thin films and the transition among them can be studied by a combination of surface X-ray diffraction and low-energy electron microscopy. It is also shown that X-ray diffuse scattering can provide model validation for the case of self-organized and composition-modulated growth of nearly lattice-matched superlattices.

In Part II, among the four contributions, the issue of atom size and how it affects scattering is discussed, and it is shown how the size factor impacts the modern understanding of electronic interaction in materials. X-ray scattering brings decisive answers to bulk alloy microstructure by elastic diffuse X-ray or neutron scattering, and this is illustrated by two examples that deal with large atomic size difference and species-dependent static atomic displacements. Fundamental aspects of diffuse scattering from lattice defects within the coherent wave theory and the use of Huang and asymptotic diffuse scattering for determining size, type, and internal structures of clustered defects is discussed. To close this chapter, thermally populated phonons in crystals give rise to thermal diffuse scattering, and this so-called TDS provides fundamental information on bonding, including phonon dispersion relations and interatomic forces.

Seven contributions in Part III deal with complex and disordered materials, mostly molecular crystals, quasicrystals, and materials with large unit cells. Once again, a quantitative investigation of complex disorder phenomena, displacement correlations, short-range order and phason spectrum in quasicrystals, or internal flexibility in molecular crystals can take advantage of scattering techniques and, at the same time, pair-density functions obtained from neutron scattering can be advantageously used in models to study structural defects, mesoscopic modulations, and other phenomena as they occur in complex systems.

The three contributions in Part IV show with ample illustrations how defects in irradiated materials, dislocation structures, and short-range ordered modulations caused by correlated atomic displacements can benefit from diffuse scattering studies.

Finally, Part V includes some concluding remarks on the role of dynamical effects in Bragg and diffuse scattering and describes strategies to exploit modern diffractometry to explore challenging defect structures in materials that cannot be addressed within the kinematic approximation.

In closing, we, as editors, wish to express our gratitude to all the contributors who worked diligently to bring this project to fruition. We also thank Joel Stein and his staff at Momentum Press, LLC, for their guidance and their patience and understanding during the time that it took to assemble the present volume. It is our fervent hope that this book will both inform and inspire further progress in this area of science that is very much alive and will convince the reader that diffuse scattering not only is and remains a pertinent tool to probe nature but also can advantageously be used to verify, validate, and challenge theory.

Rozaliya Barabash
Gene Ice
Patrice Turchi
June 2008

Part I

Scattering in Low Dimensions

1

Diffuse Scattering from Surfaces, Interfaces, and Thin Films

Sunil K. Sinha

Physics Department, University of California, San Diego, CA 92093, USA

Abstract

We discuss the formalism for diffuse scattering of X-rays or neutrons from surfaces, interfaces, thin films, and multilayers for nonmagnetic systems. Mechanisms that give rise to diffuse scattering include surface or interface roughness, capillary wave fluctuations, and inclusion of nanoparticles on the surface or in the layers. We discuss self-affine fractal models for surface roughness and models for roughness due to capillary waves on liquid surfaces and to what extent these models are successful in explaining experimental data. We also discuss scattering in both the in-plane off-specular and grazing-incidence small-angle X-ray or neutron scattering (GISAXS or GISANS) geometries and illustrate with a few examples of what can be learned from these types of studies.

1 Introduction

The field of diffuse or off-specular scattering of X-rays and neutrons from surfaces and interfaces has undergone explosive growth in the last decade as researchers have begun to realize its potential and as the advent of higher brightness neutron and X-ray sources have made these types of experiments more practicable. Studies of this type of scattering are now carried out by increasing numbers of researchers at neutron and synchrotron facilities worldwide, as well as with in-house X-ray sources. This article is not intended to be a comprehensive review of the field, but rather an outline of the basic principles, together with a few illustrative examples.

Specular reflectivity from surfaces is a well-established technique that has been widely employed for decades to get structural information normal to the surface or, more precisely, to model the scattering length density profile normal to the surface. The theory of specular reflectivity from single or multiple interfaces can be formulated more or less rigorously and can be calculated with high accuracy using numerical recipes that are well known [1,2] and for which numerical codes are widely available. In order to obtain the true specular reflectivity, the off-specular scattering under it must be estimated and subtracted [3]. As fields of investigation in themselves, however, *off-specular diffuse scattering* or *grazing-incidence small-angle scattering (GISAS)* are methods for studying lateral structural inhomogeneities in the plane of the surface or interface. As in all conventional scattering experiments, these experiments do not provide images of the type obtainable from the various scanning probe surface microscopies, but they have the advantage of providing global statistical information efficiently, as well as the ability to probe structural information at buried interfaces, besides being applicable in a variety of environments, such as low or high temperatures, large magnetic fields, and so on. They therefore constitute an extremely valuable complementary probe to the scanning probe microscopies.

Surface diffuse scattering and GISAS have been applied to study a variety of surface and interface phenomena: the morphology of the roughness found on surfaces [4–16], film-growth exponents [15,17–24], pitting corrosion and oxidation [25], wetting phenomena [26–30], capillary wave fluctuations on liquid surfaces [31–38], the morphology of block copolymer films [39–42], magnetic domain distributions [43–46], magnetic roughness in thin films [47–49], and nanodot arrays on surfaces [50–53]. (*Note*: These are meant to be representative rather than comprehensive references.)

In this article, we shall restrict ourselves to scattering at grazing incidence, where the component Q_z of the wave vector transfer **Q** defined by (see Figure 1.1)

$$\mathbf{Q} = \mathbf{k}_f - \mathbf{k}_i \, , \tag{1.1}$$

($\mathbf{k}_f$, $\mathbf{k}_i$ being the outgoing and incident wave vectors, respectively, of the neutron or X-ray beam) is small compared to the inverse of the interatomic spacing. Let k_0 represent the magnitude of these vectors in free space.

At this stage, we also introduce the notion of the coherence lengths of the incident beam. These are given by $\xi_t = \lambda R/s$ for the transverse coherence length, where λ is the wavelength of the radiation, R is the distance from the source to the sample, and s is the source size in the direction transverse to the beam, and $\xi_l = \lambda^2/\Delta\lambda$ for the longitudinal coherence length, where $\Delta\lambda$ is the wavelength spread of the beam. In favorable cases, these are typically ~1 to 10 microns for current synchrotron X-ray sources and ~100 to 500 nanometers for neutrons. At a grazing angle of incidence α_i, the projection of the

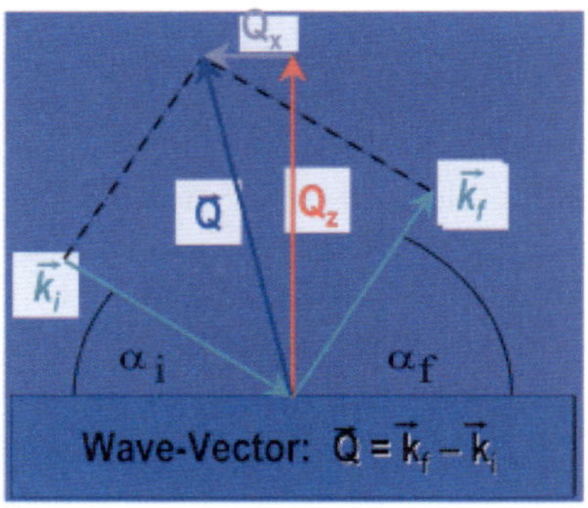

Figure 1.1. *Schematic of the scattering geometry in the DS mode. The red arrow denotes the specular ridge along Q_z.*

coherence lengths along the surface in the direction of the beam has the effect of multiplying them by the factor $[\sin(\alpha_i)]^{-1}$ so that the effective coherence length in this direction can be as large as tens of microns, whereas in the perpendicular direction, in the plane of the surface, there is no such multiplication. The *coherence area* of the beam on the surface, defined as the area from which the scattering is coherent with itself, is thus in general a small anisotropic area within the illuminated area of the surface. All measured surface scattering intensities (specular or diffuse) are the intensities of the scattering from within these coherence areas averaged over all such areas within the illuminated surface area.

The dimensions of the coherence volume (including the third direction normal to the surface) are, roughly speaking, proportional to the inverses of the corresponding dimensions of the instrumental resolution function in **Q** space. In this chapter, we shall use $S(\mathbf{Q})$ to denote the differential scattering cross section for the scattering.

For measurements in which the plane of scattering defined by the vectors $\mathbf{k}_f$, $\mathbf{k}_i$ contains the vector normal $\mathbf{n}$ to the surface, which we shall refer to as off-specular DS (as mentioned above), the projection of the coherence length on the surface is large and can extend to tens of microns, so longer length scale structures can be studied, but the reachable in-plane wave vector transfer is severely limited due to the limited angular range of the scattered beam. For shorter length scale in-plane structures, it is often more convenient to move the detector out of the plane containing $\mathbf{k}_i$ and $\mathbf{n}$ in the so-called GISAS (referred to as GISAXS or GISANS in the case of X-rays or neutrons, respectively) configuration. This is often achieved by having a two-dimensional detector (Figure 1.2).

The most common type of lateral inhomogeneity found on a surface or interface is lateral roughness. At reasonably long length scales, this is often statistically well described by random Gaussian height fluctuations about an average surface in terms of what has been called random Brownian motion by Mandelbrodt [54]. (This approximation breaks down in certain cases, which will be discussed later). The standard deviation of these Gaussian height fluctuations is the root mean square roughness σ. A height-height correlation function for points separated by a distance **R** parallel to the xy plane (i.e., the plane of the average surface) may be defined by the function

$$C(\mathbf{R}) = \langle z(\mathbf{r})z(\mathbf{r}+\mathbf{R})\rangle, \tag{1.2}$$

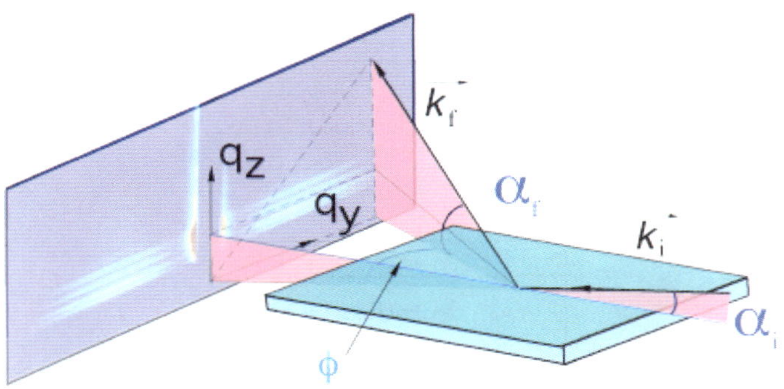

Figure 1.2. *Schematic of scattering geometry in the GISAS mode.*

where $z(\mathbf{r})$ denotes the height fluctuation at in-plane position $\mathbf{r}$. The angle brackets denote an average over the surface. The two-dimensional Fourier transform (FT) of $C(\mathbf{R})$ is the power spectral density function of the height fluctuations $|z(q)|^2$. We assume that, in such an average, the correlation function depends only on the separation between the points and not on the positions of each point. Related to this correlation function is the mean square height deviation function, $g(\mathbf{R}) = <[z(\mathbf{R}) - z(0)]^2> = 2[\sigma^2 - C(\mathbf{R})]$. For a large number of types of solid surfaces or interfaces (and sometimes for liquid surfaces, too), the correlation function $C(\mathbf{R})$ can be represented analytically by the self-affine fractal form [4]

$$C(\mathbf{R}) = \sigma^2 e^{-(R/\xi)^{2h}}, \tag{1.3}$$

where $0 < h < 1$, where h is the roughness or Hurst exponent for the surface, and ξ is the roughness correlation length. This yields a $g(\mathbf{r})$ function for the surface that is of the general form predicted by scaling theories of film growth. This has been discussed by Salditt et al. [55]. Thus the morphology of a particular surface can be expressed in terms of the three parameters: σ, ξ, and h.

For multiple surfaces (as in thin films or multilayers), we choose the notation shown in Figure 1.3. For these systems we also need to define a height-height correlation function for height fluctuations on different interfaces i and j; that is,

$$C_{ij}(\mathbf{R}) = \langle z_i(\mathbf{r}) z_j(\mathbf{r}+\mathbf{R}) \rangle, \tag{1.4}$$

where $z_i(\mathbf{r})$ is the height fluctuation at position $\mathbf{r}$ in the xy plane on surface i, whose average height is given by z_{0i}. If the interfaces i and j were perfectly conformal, then the correlation function $C_{ij}(\mathbf{R})$ would be identical to that for each interface with itself, that is, $C_{ii}(\mathbf{R})$. In general, there is only partial correlation between interfaces, which decreases as the separation between them increases.

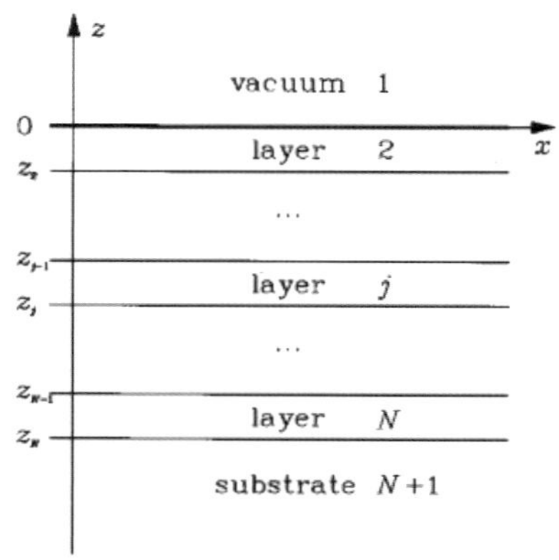

Figure 1.3. *Diagram indicating notation for layers and interfaces referred to in text. Note that the interfaces shown here are the average (smooth) interfaces.*

The issue of how the roughness propagates from one interface i to a subsequently deposited interface j above it depends on growth conditions and has been theoretically addressed by several authors [56,59]. This was done in the context of the Edwards-Wilkinson growth model by Spiller, Stearns, and Krumrey [56] by solving the linearized Langevin equation describing the evolution of the film surface with film thickness. However, this is a restrictive case, and for more general applications to multilayers, one may represent C_{ij} by the self-affine fractal form [5–10,14,57,58]

$$C_{ij}(\mathbf{R}) = \sigma_i \sigma_j e^{-(R/\xi_{ij})^{2h}} e^{-|Z_{0i}-Z_{0j}|/\xi_\perp}, \tag{1.5}$$

with h taken to be the same for both interfaces (for simplicity) and $\xi_\perp$ representing the length scale over which the interface roughnesses are correlated along the z direction. In order to reduce the number of parameters, ξ_{ij} is often taken to be the geometric mean $\sqrt{\xi_i \xi_j}$ of the in-plane roughness correlation length of the two surfaces. This assumption has been given theoretical justification by Stettner et al. [14]. With the above model for surface and interface roughness, one is now in a position to calculate the scattering from such surfaces. (In Equation 1.5, we have assumed isotropic roughness. The anisotropic case, as in the case of stepped surfaces, will be discussed later.)

There are other kinds of roughnesses found on surfaces that do not conform to the self-affine form. These include capillary wave roughness on liquid surfaces, surfaces with corrosion-induced pitting or decorated with islands or particles, and miscut single-crystal surfaces with semiperiodic arrays of steps. These will be discussed later.

1.1 Scattering Theory for Rough Surfaces

We first consider the scattering in the kinematic or Born approximation (BA). Although this approximation fails in the regions where the scattering is strong—that is, in the vicinity of or below the critical angle of total reflection, for example—it is often very useful in fitting the data over large regions of **Q** space and also illustrates several important qualitative features of the scattering. In the following, we use the notation $S(\mathbf{Q})$ for the differential scattering cross section of a system for neutrons or X-rays.

Consider a bulk material with a single rough surface. If we assume that **Q** is small compared to the interatomic spacing in the material, then we may replace the medium below the surface with a smeared-out scattering length density (SLD), $\rho(\tau)$, as is commonly done for small-angle scattering. Here τ is a position vector in three-dimensional space. The SLD for a uniform medium may also be expressed in terms of the refractive index $n = 1 - \delta + i\beta$ by

$$k_0^2(1-n^2) = 4\pi\rho_0. \tag{1.6}$$

If we assume that the SLD is constant in the bulk and equal to ρ_0, then the scattering from the bulk,

$$S(\mathbf{Q}) = \iint d\tau\, d\tau'\, \rho(\tau)\rho(\tau')\, e^{-i\mathbf{Q}\cdot(\tau-\tau')}, \tag{1.7}$$

may be transformed into an integral over the top surface (the bottom surface for a bulk material can be neglected by formally assuming that the beam is attenuated to negligible proportions by the time it reaches the surface):

$$S(\mathbf{Q}) = \frac{\rho_0^2}{Q_z^2} \int_{S_0} \int_{S_0} d\mathbf{r}\, d\mathbf{r}'\, e^{-iQ_z[z(\mathbf{r})-z(\mathbf{r}')]} e^{-i\mathbf{Q}_{\|}\cdot(\mathbf{r}-\mathbf{r}')} \tag{1.8}$$

Here **r**, **r**′ denote vectors in the plane of the surface S_0 (the *xy* plane). $\mathbf{Q}_{\|}$ is the projection of **Q** in the *xy* plane. We note that this scattering, in principle, depends on the mutual interference of all the height fluctuations across the surface. If they are random and static and the beam is coherent across the surface, this will give rise to a speckle distribution pattern in the scattered beam. At this stage, we may assume that the beam is incoherent in the sense that the illuminated surface area comprises a large number of projected coherence areas or else that the height fluctuations are dynamic in time (as for capillary waves on a liquid surface, for example) so that we may carry out a statistical average of the expression in Equation 1.8, which we may write as

$$S(\mathbf{Q}) = \frac{\rho_0^2 A}{Q_z^2} \int_{S_0} d\mathbf{R} \left\langle e^{-iQ_z[z(\mathbf{r})-z(\mathbf{r}+\mathbf{R})]} \right\rangle e^{-i\mathbf{Q}_{\|}\cdot\mathbf{R}} \tag{1.9}$$

We have assumed that the statistical average depends only on the two-dimensional vector separation **R** separating the points **r**, **r**′ in the plane.

A is the illuminated surface area. Using well-known theorems for averages over random Gaussian variables, a little algebra shows that Equation 1.9 may be written in the form

$$S(Q) = \frac{\rho_0^2 A}{Q_z^2} e^{-Q_z^2\sigma^2} \int_{S_0} d\mathbf{R}\, e^{Q_z^2 C(\mathbf{R})} e^{-i\mathbf{Q}_{\|}\cdot\mathbf{R}}, \tag{1.10}$$

which expresses *S*(**Q**) in terms of a two-dimensional FT over the surface. Here *C*(**R**) is the height-height correlation function introduced in Equation 1.2. Since, in general, $C(\mathbf{R}) \to 0$ as $\mathbf{R} \to \infty$, *S*(**Q**) will contain a delta function that may be explicitly separated by adding and subtracting unity from $\exp[Q_z^2 C(\mathbf{R})]$ in Equation 1.10. The result yields

$$S(\mathbf{Q}) = S(\mathbf{Q})_s + S(\mathbf{Q})_d, \tag{1.11}$$

where

$$S(\mathbf{Q})_s = \frac{4\pi\rho_0 A}{Q_z^2} e^{-Q_z^2\sigma^2} \delta(Q_x)\delta(Q_y), \tag{1.12a}$$

and

$$S(\mathbf{Q})_d = \frac{\rho_0 A}{Q_z^2} e^{-Q_z^2\sigma^2} \int_{S_0} d\mathbf{R} \left[e^{Q_z^2 C(\mathbf{R})} - 1 \right] e^{-i\mathbf{Q}_{\|}\cdot\mathbf{R}} \quad , \tag{1.12b}$$

where $S_s(\mathbf{Q})$ represents the specular part of the scattering and $S_d(\mathbf{Q})$ represents the off-specular or diffuse part. We shall focus on the diffuse part here.

The general properties of $S_d(\mathbf{Q})$ have been discussed in Sinha et al. [4]. For small Q_z, we may expand the exponential and, keeping the leading term, we then obtain the simple result that $S(\mathbf{Q})_d$ is proportional to the FT of $C(\mathbf{R})$:

$$S(\mathbf{Q})_d = \rho_0^2 A e^{-Q_z^2\sigma^2} \int_{S_0} d\mathbf{R}\, C(\mathbf{R}) e^{-i\mathbf{Q}_{\|}\cdot\mathbf{R}} \quad . \tag{1.13}$$

In the more general case, if $C(\mathbf{R})$ is isotropic, the two-dimensional integral can be replaced by a one-dimensional integral:

$$S(Q)_d = \frac{2\pi\rho_0^2 A}{Q_z^2} e^{-Q_z^2\sigma^2} \int_0^\infty dR\, R \left[e^{Q_z^2 C(R)} - 1 \right] J_0(Q_{\|} R) \quad . \tag{1.14}$$

In many cases, the scattering experiments for DS are carried out with slits that are wide open in the direction perpendicular to the plane of scattering. This corresponds to the y direction in the plane of the surface (see Figure 1.1). We shall refer to this case as scattering with slit (rather than point) geometry. In such a case, we may integrate over Q_y to obtain $S(Q_x, Q_z)$. The integration over Q_y yields a delta function in y, reducing $S(\mathbf{Q})$ again to the form of a one-dimensional integral:

$$S'(\mathbf{Q})_d = \frac{\rho_0^2 A}{Q_z^2 k_0} e^{-Q_z^2\sigma^2} \int dx \left[e^{Q_z^2 C(x)} - 1 \right] e^{-iQ_x x} \quad . \tag{1.15}$$

(*Note*: We have replaced $S(\mathbf{Q})$ with $S'(\mathbf{Q})$ to denote that, because of the integration over the out-of-plane angle in the scattered beam, this differs from $S(\mathbf{Q})$ by a constant of proportionality.)

Unfortunately, analytic forms of such integrals (especially if the correlation function is of the self-affine fractal form) cannot be obtained in general, and one is reduced to calculating them numerically. An exception is the case of liquid capillary wave fluctuations, which will be discussed later. A further complication is that the observed scattering represents $S(\mathbf{Q})$ folded with the instrumental resolution function (in the above case, this is due to beam divergences in the plane of scattering and nonmonochromaticity of the beam). This is sometimes done directly numerically, or it may be approximated by multiplying the integrand of Equation 1.14 by a damping factor $\exp(-x^2/L_{res}^2)$, where L_{res} is taken to be the inverse of the resolution function width in Q_x, or the effective coherence length along the x direction.

Diffuse scattering is often measured as a series of transverse scans (i.e., along Q_x) for different fixed values of Q_z on either side of the specular reflection at $Q_x = 0$. Such scans are approximated by rocking curves about the specular reflection at a fixed 2θ value. ($2\theta = \alpha_i + \alpha_f$ is the total scattering angle.) The general form of the diffuse scattering is illustrated in Figure 1.4. The specular reflection is represented by a (resolution-broadened) delta function with the diffuse scattering as a broad peak centered on it. (The more accurate approximation used to fit the DS, namely the distorted-wave Born approximation [DWBA], will be discussed below.) In general, larger roughness correlation lengths will lead to narrower DS distributions, but one should note that it is not correct simply to take the width of the DS curve as inversely proportional to the correlation length as is done in the case of critical scattering.

In some instances, DS is represented in terms of intensity contour plots, where one axis is α_i and the other axis is α_f (α_i and α_f being the grazing angles of incidence and scattering, respectively), or where one axis is $(\alpha_i + \alpha_f)$ and the other is $(\alpha_i - \alpha_f)$. For neutron time-of-flight experiments, data are often exhibited as intensity contours on a neutron wavelength vs. detector position plot. Each of these has certain representational advantages with respect to identifying special features in the DS. We note that all of these plots may be mapped onto (Q_x, Q_z) space in the case of DS geometry.

Now consider a system of layered films on a substrate. Let us assume that we have constant scattering length density between interfaces. The scattering length density for layer i above interface i (see Figure 1.3) is defined as

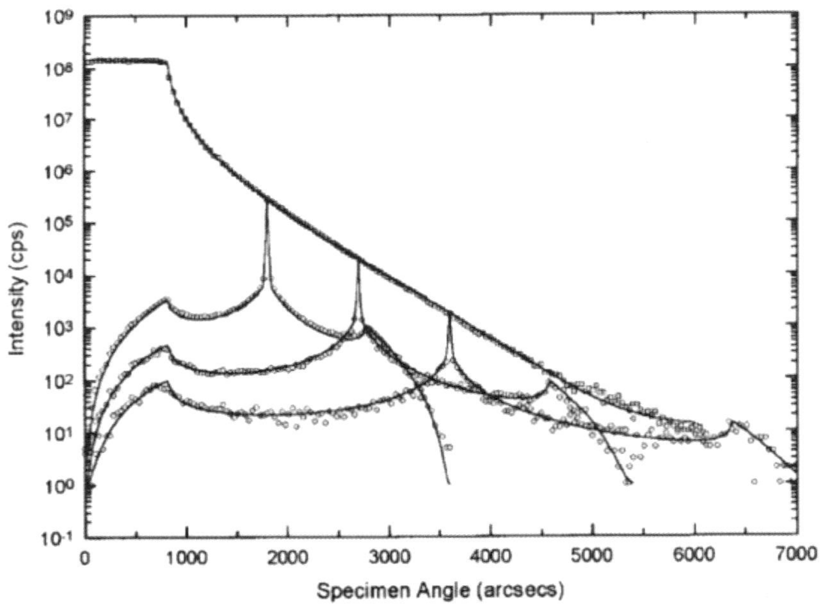

Figure 1.4. *Specular reflectivity and diffuse scattering as measured in rocking curve scans from a polished sample of Zerodur glass. The rocking curves are shown as measured about the points on the specular reflectivity, where they intersect the specular curve. Note the Yoneda peaks on either side of the specular peak. The solid lines represent an overall DWBA fit with a graded density near the surface. (From Wormington [60].)*

$$\rho_i = \sum_m b_m n_m \quad . \tag{1.16}$$

For neutrons, b_m denotes the scattering length and n_m is the number density for the mth type of nucleus in the layer. For X-rays, b_m stands for the Thomson scattering length r_0 (which is equal to e^2/mc^2 or 0.27 10^{-12} cm) times f_m, that is, the scattering factor for the mth type of atom, whereas n_m is the number density for that type of atom in the layer.

The calculation of the DS from multilayer systems in the BA was first carried out by Lagally and coworkers [6–8]. In the BA, the scattering from the system of layers may similarly be transformed into the scattering from the set of interfaces $i = 1$ to N, and we will obtain as the generalization of Equation 1.11

$$S(\mathbf{Q})_s = \frac{4\pi A}{Q_z^2} \sum_{i,j=1}^{N} e^{-iQ_z(z_{0i}-z_{0j})} (\rho_{i+1}-\rho_i)(\rho_{j+1}-\rho_j) e^{-\frac{1}{2}Q_z^2(\sigma_i^2+\sigma_j^2)} \delta(Q_x)\delta(Q_y) \tag{1.17a}$$

and

$$S(\mathbf{Q})_d = \frac{A}{Q_z^2} \sum_{i,j=1}^{N} e^{-iQ_z(z_{0i}-z_{0j})} (\rho_{i+1}-\rho_i)(\rho_{j+1}-\rho_j) e^{-\frac{1}{2}Q_z^2(\sigma_i^2+\sigma_j^2)} \int_{S_0} d\mathbf{R} \left[e^{Q_z^2 C_{ij}(\mathbf{R})} - 1 \right] e^{-i\mathbf{Q}_{\|}\cdot\mathbf{R}} \quad , \tag{1.17b}$$

where the notation for the symbols is as defined previously. We note that Equations 1.17a and 1.17b formulate the resultant scattering in terms of a sum of the scattering amplitudes from the individual interfaces, modulated by the SLD contrast across these interfaces. As mentioned previously, there are well-established methods for evaluating the specular reflectivity from multilayer systems without making the BA, and we shall concentrate here on the diffuse scattering component. As before, for scattering experiment geometries performed in slit geometry, the integral over Q_y can be implicitly performed in Equation 1.17b, leading to one-dimensional integrals of the form of Equation 1.15, and in-plane instrumental resolution effects must be allowed for, as in the discussion above for single surfaces.

For a periodically repeated set of layers along the z axis with periodicity Λ, Equation 1.17a shows that the specular scattering will peak strongly at values of Q_z given by

$$Q_z = n\left(\frac{2\pi}{\Lambda}\right) \qquad (n = \text{integer}) \quad . \tag{1.18}$$

These are the well-known "multilayer Bragg peaks" for a periodic multilayer. (Refraction effects will shift the Q_z positions of these slightly, as will be discussed later when we go beyond the BA.) A comparison of Equation 1.17a and 1.17b shows that the intensity variation of the DS along Q_z for a film with multiple interfaces is governed by the same interference function as the specular if the roughness between layers is conformal, that is, highly correlated. Consider the extreme case in which all the C_{ij} are identical for all pairs of interfaces. Then the integral may be factored out, leaving the same interference factor in Q_z in front as the one that determines the maxima in Equation 1.17a for the specular. For a single film on a surface (two interfaces), the interference function along Q_z leads to the well-known Kiessig fringes or oscillations in the specular reflectivity with period $2\pi/d$, with d being the film thickness. If there is a degree of conformality between the top and bottom film surfaces, which is usually the case, the longitudinal DS similarly will exhibit Kiessig-like fringes along the Q_z direction. (Longitudinal DS scans are often made by slightly offsetting the sample orientation by an amount sufficient to move off the specular peak and doing the usual $\theta - 2\theta$ scan [3].) This circumstance makes it possible to obtain the thickness of a film even if it is not possible to obtain a good specular reflectivity curve due to nonflatness or excessive roughness of the surfaces. In the case of a multilayer, the appearance of "Bragg ridges" in the DS centered on the same Q_z values as the multilayer Bragg peaks is the telltale sign of conformal roughness between the interfaces. (The opposite extreme of totally uncorrelated interfaces corresponds to C_{ij} being zero if $i \neq j$, in which case this interference factor becomes unity and the DS is simply the sum of the DS from each of the interfaces.) The width along Q_z of the ridges in the DS is inversely proportional to the correlation length $\xi_\perp$ of the roughness between interfaces along the z direction (see Figure 1.5).

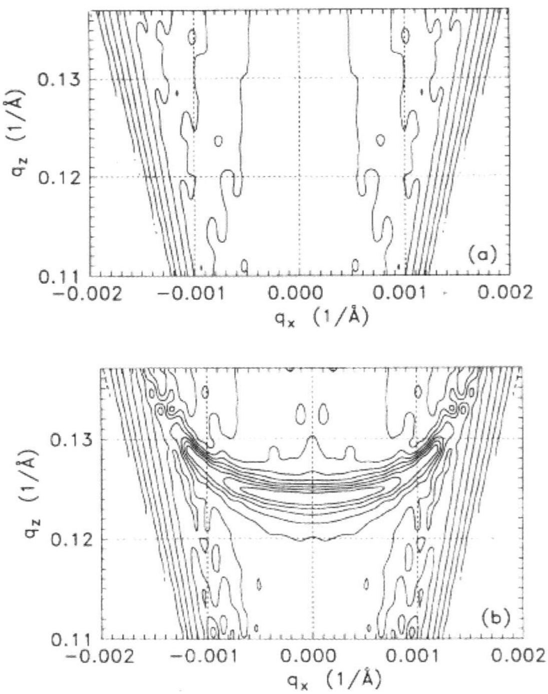

Figure 1.5. *Plot of intensity distribution of DS for a multilayer with (a) no correlated roughness between layers and (b) a degree of correlated roughness, as calculated in the DWBA by Holy and Baumbach [10]. Notice how the intensity is bunched into a ridge going through the multilayer Bragg peak on the specular ridge ($Q_x = 0$). In the DWBA, these ridges are curved due to refraction effects.*

As mentioned above, for Q_z values close to or below the values for total reflection, and particularly for multilayer systems in the vicinity of the Bragg peaks, the BA fails. This is because the reflected wave is strong, and its effect on the scattering cannot be neglected. An approximation that is widely used for such cases and often works reasonably well, except in cases of extreme surface roughness, is the DWBA [4]. The idea is to calculate the wave functions for the perfect system, that is, without roughness (which can be done using standard methods), and then to use the roughness as a perturbation and calculate the scattering using the eigenstates of the perfect system (including the reflected wave) instead of plane waves as in the BA. We do not have space to go into the detailed derivation of this approximation here, but we refer the reader to the literature [4,61–63] and simply quote the results here.

For a single surface, the result in Equation 1.12b is replaced by

$$S(\mathbf{Q})_d = \rho_0^2 A |T(\alpha_i)|^2 |T(\alpha_s)|^2 \frac{e^{-\frac{1}{2}(\tilde{Q}_z + \tilde{Q}_z^{*2})\sigma^2}}{|\tilde{Q}_z|^2} \int d\mathbf{R} \left[e^{|\tilde{Q}_z|^2 C(\mathbf{R})} - 1 \right] e^{-i\mathbf{Q}_\parallel \cdot \mathbf{R}} \quad , \qquad (1.19a)$$

or its one-dimensional equivalent in the case of slit geometry,

$$S'(\mathbf{Q})_d = (1/k_0) \rho_0^2 A |T(\alpha_i)|^2 |T(\alpha_f)|^2 e^{-\frac{1}{2}(\tilde{Q}_z^2 + \tilde{Q}_z^{*2})\sigma^2} \int dx \left[e^{|\tilde{Q}_z|^2 C(x)} - 1 \right] e^{-iQ_x x} \quad , \qquad (1.19b)$$

where $\tilde{Q}_z$ is the value of Q_z in the medium and $T(\alpha_i)$, $T(\alpha_f)$ are the Fresnel transmission coefficients for the corresponding smooth surface for grazing angles of incidence α_i, α_f, respectively. $\tilde{Q}_z$ differs from Q_z in free space by a refraction effect related to the index of refraction in the medium. This can be obtained from the expression for the transmitted z component of the wave vector k_z in the medium in terms of that in free space k_{0z}:

$$k_z^2 = k_{0z} - 4\pi\rho_0 \, . \qquad (1.20)$$

$T(\alpha)$ has a peak at $\alpha = \alpha_c$, the critical angle of total reflection given by $\alpha_c = \sqrt{2\delta}$, where δ is the real part of $1 - n$, and this results in peaks in the DS on either side of the specular peak, when the surface makes an angle with either the incident or outgoing beam equal to α_c. These are the well-known Yoneda wings in the DS [64]. Physically, they are due to the fact that, at the critical angle, the incident and reflected beams are in phase so that the wave amplitude at the surface is twice that of the incident wave. The principle of optical reciprocity requires that this be true of the scattered wave as well.

Sometimes the assumption that the density of the material under the surface is constant under the surface is not a good one, even for the **Q** values of interest. Thus the material may have a graded density in the vicinity of the surface, represented by a factor $f(z)$, which multiplies the SLD. In such a case, the expression for $S(\mathbf{Q})$ is multiplied by $|F(Q_z)|^2$, the one-dimensional Fourier transform of $f(z)$. This was done by Wormington [60] in a fitting of the DS from a sample of Zerodur glass using the DWBA. His results are shown in Figure 1.4. The Yoneda wings in the transverse scans are clearly visible and are fitted well using the DWBA.

TABLE 1.1

Table of expressions for the symbols used in Equation 1.21

$\mathbf{Q}_j^0 = \mathbf{k}_{f,j} - \mathbf{k}_{i,j}$	$\mathbf{G}_j^0 = T_{i,j} T_{f,j}$
$\mathbf{Q}_j^1 = \mathbf{k}_{f,j}^R - \mathbf{k}_{i,j}$	$\mathbf{G}_j^1 = T_{i,j} R_{f,j}$
$\mathbf{Q}_j^2 = \mathbf{k}_{f,j} - \mathbf{k}_{i,j}^R$	$\mathbf{G}_j^2 = R_{i,j} T_{f,j}$
$\mathbf{Q}_j^3 = \mathbf{k}_{f,j}^R - \mathbf{k}_{i,j}^R$	$\mathbf{G}_j^3 = R_{i,j} R_{f,j}$

Note: $\mathbf{k}_{i,j}$, $\mathbf{k}_{i,j}^R$ REPRESENT THE WAVE VECTORS OF THE WAVE PROPAGATING IN THE "DOWN" DIRECTION AND ITS SPECULARLY REFLECTED WAVE PROPAGATING IN THE "UP" DIRECTION (SEE FIGURE 1.3) IN MEDIUM j FOR THE BEAM INCIDENT ON THE MULTILAYER AT GRAZING ANGLES OF INCIDENCE A_I. $\mathbf{k}_{f,j}^R$, $\mathbf{k}_{f,j}$ ARE THE WAVE VECTORS OF THE CORRESPONDING WAVES IN MEDIUM j PROPAGATING IN THE "DOWN" AND "UP" DIRECTIONS, RESPECTIVELY, FOR THE BEAM INCIDENT AT GRAZING ANGLES OF INCIDENCE A_F.

Holy and coworkers were the first to apply the DWBA to the case of DS from multiple interfaces [9,10]. The formulation may be found in several papers [5,14,65].

We must consider the incident (i.e., traveling in the same direction as the incident beam) and reflected waves in each layer, and so we must define four possible $\mathbf{Q}$ values for each layer. These are defined in Table 1.1 as $\mathbf{Q}_j^m$ (m = 0 to 3), where j refers to the layer j, bounded below by interface j and on top by interface (j – 1) (see Figure 1.3). $T_{i,j}$, $T_{f,j}$ refer to the amplitudes of the transmitted beams in medium j for beams incident on the multilayer at grazing angles of incidence α_i and α_f, respectively, and $R_{i,j}$ and $R_{f,j}$ refer to the amplitudes of the corresponding reflected beams. These amplitudes can be calculated, for example, using the Parratt formalism [1].

The DWBA then provides a recipe for calculating the DS from this system of layers. One must make the further assumption that the wave functions in the layer j can be analytically continued into layer (j – 1) within the height of the roughness fluctuations. For the case of slit geometry scattering experiments, the scattering cross section may be written as

$$S'(\mathbf{Q}) = \frac{Ak_0^3}{8\pi^2} \sum_{j,k=1}^{N} \left(n_j^2 - n_{j+1}^2\right)\left(n_k^2 - n_{k+1}^2\right)^* \times \sum_{m,n=0}^{3} e^{-i\left(Q_{z,j}^m z_{0,j} - Q_{z,k}^{n*} z_{0,k}\right)} G_j^m G_k^{m*} e^{-\frac{1}{2}\left[\left(Q_{z,j}^m \sigma_j\right)^2 + \left(Q_{z,k}^{n*} \sigma_k\right)^2\right]} \times \frac{1}{Q_{z,j}^m Q_{z,k}^{n*}} \int_0^\infty dx \left[e^{Q_{z,j}^m Q_{z,k}^{n*} C_{jk}(x)} - 1\right] \cos(Q_x x) \quad , \tag{1.21}$$

where the G_j^m (m = 0, ..., 3) refers to the appropriate combinations of the T and R coefficients given in Table 1.1.

With the above formalism, and the self-affine expression for the height-height correlation functions given by Equation 1.5, the data for DS from multilayers can usually be fitted fairly well. The transverse DS scans from multilayer systems usually exhibit fine structure manifested as a set of abrupt peaks and dips, which are also reproduced by the DWBA calculations (see Figure 1.6). Physically, these effects arise as follows [66]: Whenever the angle of incidence α_i is close to a multilayer Bragg reflection, standing waves are set up within the layers due to interference from the incident and reflected beams. According to the dynamical theory of diffraction, as α_i is swept through the Bragg condition, these standing waves change from having nodes at the interfaces of the repeating layer units to having antinodes at those interfaces, thereby significantly changing the relative amounts of scattering from the roughness at those interfaces. By the optical reciprocity principle, the same will be true for α_f. They are thus the generalizations to multilayers of the Yoneda peaks discussed above in conjunction with a single surface.

A further effect of the use of the DWBA in place of the BA is that refractive index differences between the layers can shift the positions of the ridges in the DS that arise from conformal roughness between the interfaces. In the BA, as discussed above, these ridges are at constant Q_z through the multilayer Bragg peaks. In the DWBA, they are bent toward smaller Q_z at large $|Q_x|$ values due to the refraction effect, forming banana-shaped ridges, as shown first by Holy et al. [10] (see Figure 1.5).

1.2 Diffuse Scattering Studies of Rough Surfaces

As mentioned above, it is now possible to use X-ray and neutron DS as a detailed probe of the roughness morphology of surfaces and interfaces, and many such studies have been carried out. One of the most detailed studies is that of Stettner et al. [14], who made careful measurements of the specular and diffuse scattering from a molecular beam epitaxy (MBE) grown $CoSi_2$/Co/$CoSi_2$ layer system deposited on Si(111) and analyzed the results using the DWBA (see Figure 1.7).

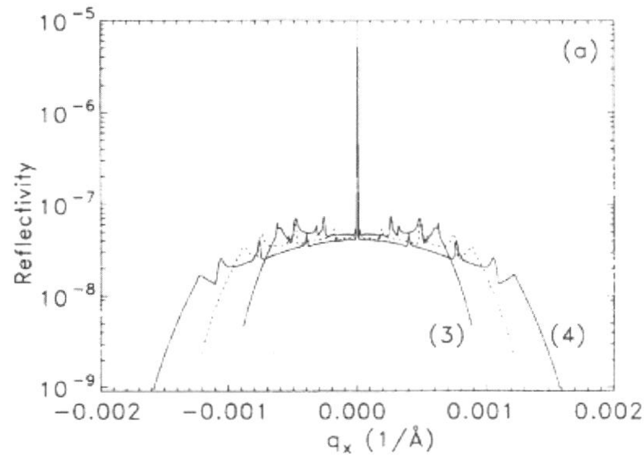

Figure 1.6. *A DWBA calculation of transverse DS scans through the third (line 3) and fourth (line 4) Bragg peaks of a multilayer together with a transverse DS scan between the peaks (dotted line). (From Holy and Baumbach [10].)*

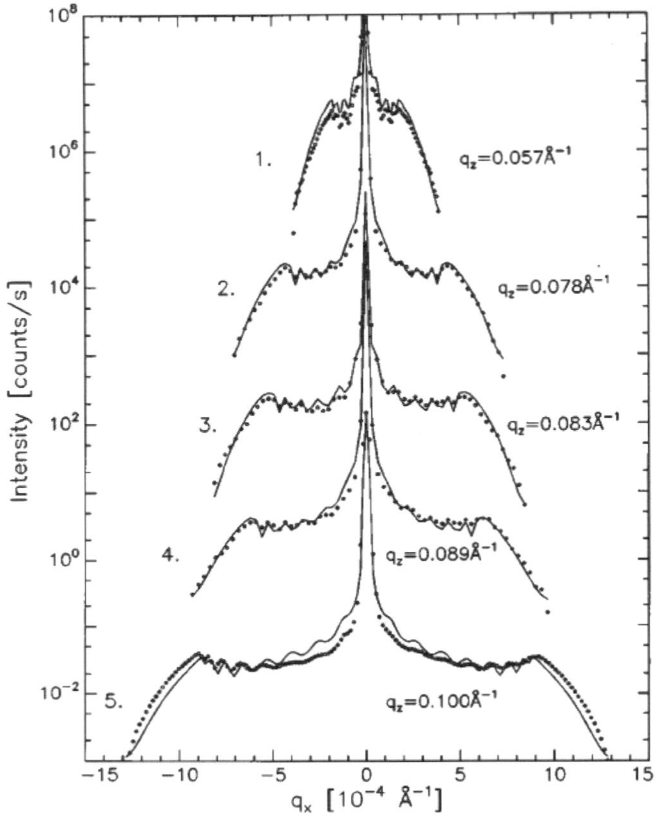

Figure 1.7. *Transverse DS scans for various Q_z values of an MBE grown $CoSi_2$/Co/$CoSi_2$ film on Si(111) together with fits using the DWBA. (From Stettner et al. [14].)*

They also compared the parameters resulting from their analysis with those obtained from other techniques, such as Rutherford backscattering, transmission electron microscopy, and scanning tunneling microscopy (STM). Reasonably good agreement was found with the parameters from the X-ray analysis. The Si substrate had a slight miscut angle, resulting in a stepped surface with the steps normal to the true (111) axis of the Si crystal. The results showed a significant degree of conformality between the interfaces (see Figure 1.8).

The results also showed that, for large enough length scales, the self-affine fractal model is still a good representation for the height-height correlation function of the interfaces and agreed with the STM measurements (with slight periodic deviations due to the step separations) (see Figure 1.9), although the roughness correlation length depended on the angle made by the separation **R** and the step direction, which was at maximum when **R** was parallel to the step direction. Similar conclusions were reached by Sanyal et al. [57] in a study of DS from a GaAs/AlAs multilayer [58], in which, however, only the simple BA was used to analyze the data. Holy and Baumbach [10] published the

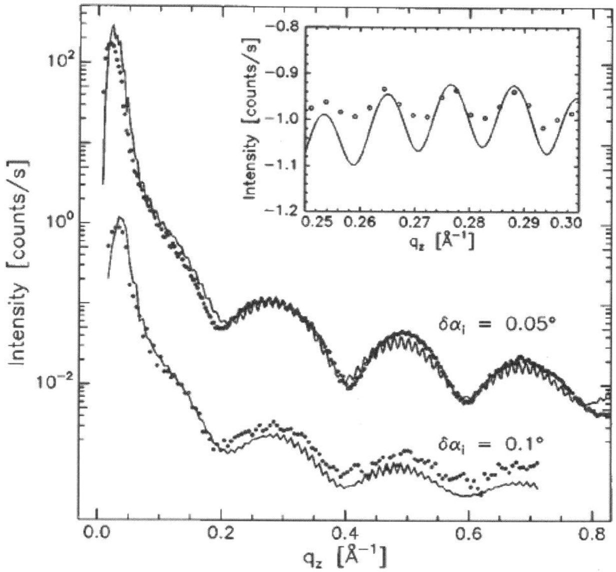

Figure 1.8. *Longitudinal diffuse scans for the same sample as in Figure 1.7, carried out for two different offset angles of the sample position from the specular condition, namely, 0.05° and 0.1°. The solid lines denote the best fit with the DWBA, and the open symbols denote the measurements. The inset shows an enlargement of the upper curves within the region $0.25 < Q_z < 0.30$ Å^{-1}. The curves have been shifted for clarity. (From Stettner et al. [14]).*

first DWBA study of the DS from a GaAs/AlAs multilayer and obtained good agreement using the self-affine fractal model and demonstrated "bunching" of the DS into ridges through the multilayer Bragg peaks, including the bending in $Q_x - Q_z$ space due to refraction effects. Salditt et al. [67] have also carried out a direct comparison between STM and DS measurements of a Zr(0.35)Co(0.65) film sputtered onto a Si susbstrate. They found good agreement with the self-affine fractal form for the mean square height deviation function for the surface obtained from the STM measurements. Earlier measurements by Weber and Lengeler [59] showed similarly good agreement between the parameters obtained from fitting the self-affine fractal model to the DS and the results of atomic force microscopy measurements. This form has also been verified for Fe/Cr multilayers using transmission electron microscopy by Gomez et al. [68] and for electrodeposited Cu films using atomic force microscopy (AFM) by Lafouresse et al. [69]. Fe/Cr/Fe trilayers are also of importance as examples of systems exhibiting giant magnetoresistance, but the contrast for X-rays between the SLD of Fe and that of Cr is very small, so Feygenson et al. [70] employed anomalous X-ray DS, using X-ray photons with energies close to that of the K absorption edge of Cr to enhance the contrast, and obtained good fits with the self-affine fractal model using the DWBA. Interestingly enough, in their analysis of the DS from a magnetron sputtered W/Si multilayer, Salditt et al. [21] found the height-height correlation function could not be fitted with the self-affine fractal form of Equation 1.3, but rather with a logarithmic form as predicted by the Edwards-Wilkinson Langevin equation, which, as we shall see below, also holds for liquid surfaces.

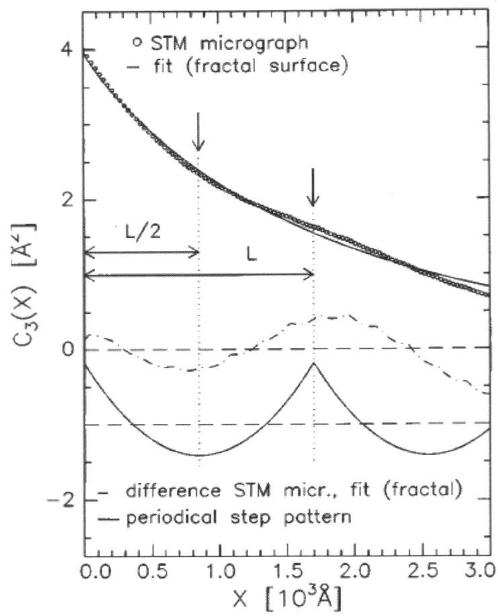

Figure 1.9. *Height-height correlation function $C_3(X)$ of the upper $CoSi_2$ layer determined by the statistical analysis of the STM micrograph (open symbols). The fit based on the self-affine fractal model is denoted by the upper solid line. Additionally, the difference between the upper curves (dashed-dotted line) and the correlation function of a step structure (lowest curve, solid line) with ideal periodicity (~ average terrace width L) is shown. The difference curve is enlarged by a factor of 5, and the correlation function for the ideal periodical step pattern has been shifted for clarity. The arrows mark the minima and maxima of the latter curve. (From Stettner et al. [14].)*

As stated above, there are far fewer neutron scattering studies of the DS from rough surfaces, but they have yielded very similar results. Thus Singh and Basu [71] have recently published a neutron DS measurement from a Ni film and also have determined the morphology of the surface roughness using AFM. They found excellent quantitative agreement with the DS based on the self-affine fractal model and also good agreement between the AFM measurements for the quantity $g(\mathbf{R}) = \langle[z(\mathbf{r}) - z(0)]^2\rangle$ and the calculated function based on parameters extracted from the neutron DS data.

The predictions of various models of film growth such as the Kardar-Parisi-Zhang model [72] for quantities that are essentially related to the roughness exponent h in the self-affine fractal model (often referred to as α in these theories) and the time rate of increase of σ and ξ have been checked by DS, electron diffraction, and AFM. Various exponents have been obtained, depending on growth conditions, indicating that it is difficult to identify universality associated with these exponents.

It should be noted that even if the self-affine fractal model for the correlation does not hold, as long as the height fluctuations are Gaussian, the scattering theory for the DS outlined in the previous section may still be applied and, in fact, inverted to obtain the function $C(\mathbf{R})$. This may be seen for a single surface by applying the inverse FT to Equation 1.19 (after correcting for the transmission factors) or its one-dimensional analog for slit geometry. This has been carried out by Noh et al. [73] and more recently by Smigiel et al. [74].

1.3 Liquid Surfaces

The conventional theory of capillary waves on a bulk liquid surface or an interface between bulk liquids can be used to show [31] that the height-height correlation function, which is isotropic, can be written as

$$C(\mathbf{R}) = \tfrac{1}{2} BK_0(\kappa\mathbf{R}) , \tag{1.22}$$

where $B = k_B T/\pi\gamma$ (γ being the interface tension) and κ is the inverse of the capillary length and is equal to $(\Delta\rho g/\gamma)^{1/2}$, where $\Delta\rho$ is the difference in mass density across the interface. This is known as the gravitational cutoff at small-wave vectors and is negligible compared to the instrumental resolution, so it cannot be observed in scattering experiments. However, in the case of a liquid film with a van der Waals interaction with the substrate, κ can be appreciably larger and is given by

$$q_{\text{vdw}}(d) = \left(A_{\text{eff}}/2\pi\gamma\right)^{\frac{1}{2}}/d^2 , \tag{1.23}$$

where d is the film thickness and A_{eff} is the effective Hamaker constant. This can result in a cutoff for observable wave vectors, showing up as a plateau or shoulder in the transverse DS at $Q_x = q_{\text{vdw}}$ if the film is sufficiently thin. The evidence of these for molecular liquids will be discussed below.

As can be seen from its definition at $\mathbf{R} = 0$, $C(\mathbf{R})$ should be equal to the mean square roughness, σ^2. However, Equation 1.22 is not valid for small distances as $\mathbf{R}$ approaches the intermolecular spacing. Sometimes this is corrected for by replacing the lower limit in the integral in Equation 1.9 by the intermolecular distance d_m or by replacing $\mathbf{R}$ by $[\mathbf{R}^2 + d_m^{\,2}]^{1/2}$ [75].

The expression for the mean square roughness given directly by capillary wave theory is

$$\sigma^2 = \frac{1}{4} B \ln\left(q_u^2 + k^2/\kappa^2\right) \approx \frac{1}{2} B \ln\left(q_u/\kappa\right) , \tag{1.24}$$

where q_u is an upper wave vector cutoff, which is $\sim 2\pi/d_m$, and is typically >> κ. This cuts off the capillary waves shorter than a wavelength d_m. We note that by considering higher order terms in the expansion of the free energy, a natural cutoff can be obtained without recourse to q_u. This yields a short wavelength cutoff for the capillary waves equal to $(3kT/8\pi\gamma)^{1/2}$ and a slightly modified expression for $C(\mathbf{R})$ and for σ. Typical values of σ for liquids range from 3 Å to 8 Å.

For $\kappa R << 1$, we can approximate $K_0(\kappa R)$ as

$$K_0(\kappa R) = -\gamma_E - \ln\left(\kappa R/2\right), \tag{1.25}$$

where γ_E is the constant 0.5772. This turns out to be a good approximation because at large values of R the integrand in Equation 1.15 is effectively cut off by the coherence length factor exp $(-x_2/L_{res}^{\ 2})$ discussed in connection with Equation 1.15. Without separating the scattering into specular and diffuse, we may insert the expression from Equation 1.25 into Equation 1.22 and then into the DWBA expression for a single surface (i.e., Equation 1.19 without the 1 subtracted from the exponential) to obtain

$$S(\mathbf{Q})_{TOT} = \rho_0^2 A \left|T(\alpha_i)\right|^2 \left|T(\alpha_f)\right|^2 \frac{e^{-\frac{1}{2}\left[\tilde{Q}_z^2 + \tilde{Q}_z^{*2}\right]\sigma^2}}{\left|\tilde{Q}_z\right|^2} e^{-\left|\tilde{Q}_z\right|^2\left[\frac{1}{2}B\gamma_E\right]} \times \int_{S_0,\, R>d_m} d\mathbf{R} \left(\frac{2}{\kappa R}\right)^{\eta} e^{-R^2/L_{res}^2} e^{-i\mathbf{Q}_{\parallel}\cdot\mathbf{R}} , \quad (1.26)$$

where we have introduced the finite coherence length or resolution cutoff. η is the quantity $({}^1/_2 BQ_z^{\ 2})$. If the lower length cutoff d_m is taken as zero, the expression from Equation 1.26 may be evaluated analytically, and this works reasonably well unless η approaches the value unity when singularities occur. This can happen when Q_z becomes sufficiently large. (This is due to the use of the approximate form of $C(\mathbf{R})$ rather than the rigorous expression.) Evaluating Equation 1.26 in slit geometry (Q_y integrated over), we obtain the expression for the intensity in the detector,

$$I = I_0 \frac{16\pi^2 \rho_0^2 \left|T(\alpha_i)\right|^2 \left|T(\alpha_f)\right|^2}{Q_z^3} \left[\frac{1}{2k_0 \sin\alpha_i}\right] e^{-Q_z^2\sigma_{\text{eff}}^2} \frac{1}{\sqrt{\pi}} \Gamma\left\{\frac{1-\eta}{2}\right\} \times {}_1F_1\left\{\frac{1-\eta}{2};\frac{1}{2};\frac{Q_x^2 L_{\text{res}}^2}{4\pi^2}\right\}, \quad (1.27)$$

where I_0 is the intensity of the incident beam illuminating the liquid surface, $\Gamma(x)$ is the gamma function, and ${}_1F_1(x;y;z)$ is the Kummer function [76], and where we have neglected the difference between Q_z in the liquid and Q_z in free space.

The asymptotic form of this expression for large Q_x is proportional to $(Q_x)^{\eta-1}$. (For point geometry, this tail would be of the form $(Q_x)^{\eta-2}$.) Such power law tails are typical of scattering from many two-dimensional systems. As $Q_x \rightarrow 0$, the scattering rises smoothly to saturate on the specular ridge. This expression also yields for the scattering at the specular ridge ($Q_x = 0$), the "effective specular reflectivity" (since it inevitably includes the diffuse scattering that is not easily separable from the true specular reflectivity as it is in the case of solid surfaces) given by

$$R = R_F e^{-Q_z^2\sigma_{\mathit{eff}}^2} \frac{1}{\sqrt{\pi}} \Gamma\left\{\frac{1-\eta}{2}\right\}, \quad (1.28)$$

where σ_{eff} is the effective mean square roughness related to the true surface roughness by

$$\sigma_{\mathit{eff}}^2 = \sigma^2 + \frac{1}{2}B\gamma_E - \frac{1}{2}B\ln\left(2\pi/\kappa L_{res}\right) . \quad (1.29)$$

It is a property of the scattering from liquid surfaces that the value for the surface roughness has to be corrected for the instrumental resolution. This was pointed out first by Braslau et al. [78] and Schwartz et al. [32].

From Equations 1.29 and 1.24, we may relate σ_{eff} to q_u:

$$\sigma_{eff}^2 = \frac{1}{2}B\gamma_E + \frac{1}{2}B\ln\left(q_u L_{res}/2\pi\right) \quad (1.30)$$

Ghaderi [63] has shown that calculations of the specular reflectivity using Equation 1.28 can result in artificial singularities for sufficiently large values of Q_z, owing to using the expression given in Equation 1.22, but using a rigorous (although complicated) analytical expression that takes into account the lower distance cutoff at d_m, prevents this.

Studies of the specular and diffuse scattering by Sanyal et al. [31] on liquid ethanol at room temperature showed that the capillary wave theory represented by Equation 1.29 was able to give an excellent fit to the data with parameters that were in good agreement with bulk values for the surface tension and mass density of ethanol (see Figure 1.10). Equations 1.30 and 1.31 yielded values for the

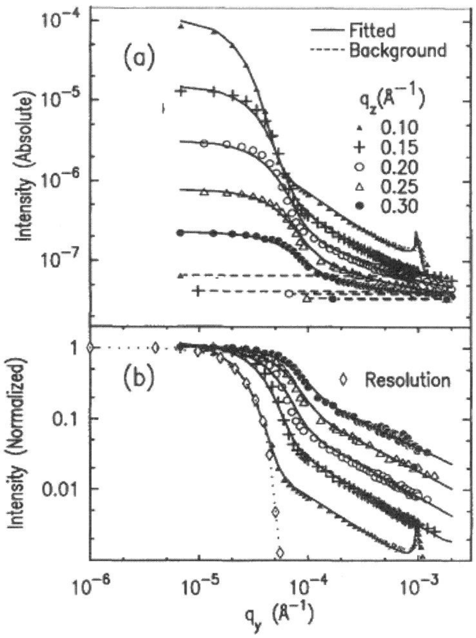

Figure 1.10. *(a) Raw transverse DS scans from the surface of liquid ethanol at room temperature for various Q_z values, together with the fitted Q_z dependent background. (b) Background subtracted transverse DS scans on a semilog plot showing the asymptotic power law tails from capillary wave scattering. The diamond symbols indicate the width of the specular as determined from the instrumental resolution function. (From Sanyal et al. [31].)*

true capillary wave roughness of 6.9 Å and a value for q_u of 0.5 Å^{-1} that agreed well with the intermolecular distance.

From the above arguments, it may be seen that DS from a liquid surface can never be properly subtracted experimentally to obtain the true specular reflectivity but must be allowed for theoretically when interpreting the reflectivity curve. This is particularly important if the liquid surface density profile shows variations with z apart from the surface roughness due to capillary waves. An example is the case of liquid metal surfaces, which often exhibit layering of their density at the free liquid surface [77].

In order to extract the true density profile for this layering, Tostmann et al. [34] and DiMasi et al. [35] allowed for this by extracting the factor $\exp(-Q_z^2\sigma_{eff}^2)$ from the measured reflectivity profile (although they used the expression for σ_{eff}^2 without the factor $\exp(\gamma_E)$ as given by Equation 1.30 and without the factor $(1/\sqrt{\pi})\,\Gamma[(1-\eta)/2]$ in Equation 1.28).

Studies of diffuse scattering on liquid surfaces have now been carried out by several groups. These liquid surfaces include water, organic liquids, molten polymers, polymer brushes, surfactants, membranes and liquid metals, and alloys at liquid-vapor and also at liquid-liquid interfaces [31–38,78–80]. In many cases, good agreement is obtained using the liquid capillary wave theory, but in some cases the data do not fit this theory well.

Uniform films of liquids such as alkanes and ethanol on smooth Si substrates have been studied for various film thicknesses by Doerr et al. [33]. Good agreement was found with the predictions of the capillary wave scattering theory outlined above, although with a lower value for the surface tension γ than the bulk value γ_0 (providing support for the predictions of a q-dependent decrease in γ by the density function-based theory of Dietrich and coworkers [80,82] to be discussed below). They also found shoulders in the transverse DS indicating the presence of cutoffs, which is in good agreement with the predictions of Equation 1.23 for cutoffs arising from the van der Waals interaction with the substrate (see Figure 1.11).

However, care should be exercised in extracting exponents from the asymptotic form of the transverse DS for thin films, especially if there is significant scattering from the substrate. Calculations by Lippmann et al. [83] using the DWBA show that spurious power law exponents can be obtained in the total scattering from a film whose top surface has capillary waves on a substrate with fractal roughness. Thus care must be exercised in interpreting the asymptotic slope of the scattering with Q_x as directly related to η and hence the surface tension.

An interesting case is that of the interface between two liquids close to their consolute critical point. Since the interface tension goes to zero, the capillary waves develop very large amplitudes. On the other hand, the contrast between the SLDs on both sides goes to zero, thus the scattering does not diverge. McClain et al. [84] measured both the specular reflectivity and the DS from the capillary waves. They found that an intrinsic width appears at the interface close to the critical point, as expected from phase transition theory, but a much larger width at the interface is induced by capillary wave fluctuations, which are well fitted by the capillary wave model.

Thin polymer films whose thicknesses are comparable to the radius of gyration of the macromolecule exhibit viscoelastic effects that cause a deviation from the predictions of capillary wave theory. Such scattering is better represented in terms of an expression first given by Fredrickson et al. [85] for the FT of the height-height correlation function, that is, the power spectral density function

$$\left|z(\mathbf{q})\right|^2 = \frac{kT}{\pi}\left[\gamma\left(q^2+\kappa^2\right)+2\mu_0 q+3\mu_0 d^{-3}q^{-2}\right]^{-1}, \qquad (1.31)$$

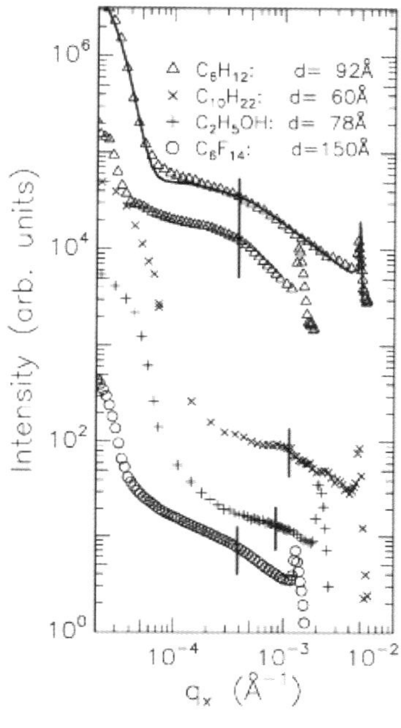

Figure 1.11. *Transverse DS scans from liquid films of various thicknesses showing the shoulders indicating the van der Waals cutoffs ~a/d² (solid vertical lines). The top two curves (triangles) are for two different Q_z values for a 92 Å cyclohexane film together with the theoretical capillary wave model (solid lines) discussed in the text. (From Doerr et al. [33].)*

where μ_0 is the shear modulus of the film and d is the film thickness.

When integrated over the Q_y direction, the DS given by this expression possesses a shoulder at a cutoff wave vector given by

$$q_{l,c} = (3\mu_0 / \gamma)^{1/4} h_0^{-3/4} \quad . \tag{1.32}$$

Since a cutoff can arise both from a van der Waals interaction with the substrate (Equation 1.25) in the case of pure capillary waves (as discussed above) or via the occurrence of a shear modulus in the case of viscoelastic films, it is important to distinguish these cases. We note that the dependence of the cutoff on film thickness is different in the two cases (see Equations 1.23 and 1.32). Thus Wang et al. [86] concluded from the thickness dependence of the observed cutoff for thin films of polystyrene that the films were behaving in a viscoelastic manner rather than as pure capillary waves with a van der Waals cutoff (see Figure 1.12). This has been confirmed in more detail by dynamic measurements [87].

Lippmann et al. [83] carried out studies of the DS from polystyrene films of various thicknesses and molecular weights. They concluded that, for thin films or high molecular weights, the DS was fitted better with a self-affine fractal model than with a capillary wave model. It is possible that this is due to the fact that thermal equilibrium is not reached for these films or that the relaxation times

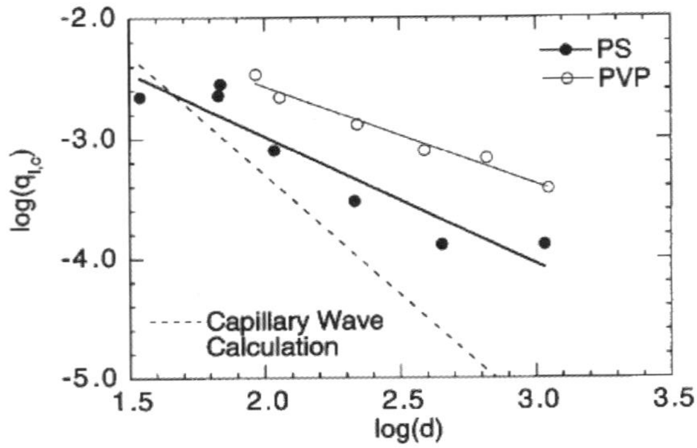

Figure 1.12. *Log-log plot of the low wave-number cutoff ql,c as a function of the film thickness for a series of polystyrene (PS) samples (solid circles) and a polyvinyl-pyridene (PVP) series (open circles). The solid lines are power-law fits according to $q_{l,c} = b/d^m$ yielding m = 1.0 and m = 0.8 for PS and PVP, respectively. The dashed line corresponds to the case of free capillary waves with m = 2 and b = 5 Å. (From Wang et al. [86].)*

are longer than the measuring times. Akgun et al. [88] also found in their study of DS from polymer brushes that a self-affine fractal model fits the data, whereas capillary wave models do not. Thus capillary wave models do not work for all types of liquidlike surfaces.

A liquid film on a substrate may not be uniform due to dewetting effects, which have the effect of decorating the surface with droplets or islands. An example is the case of a polyethylene-polypropylene (PEP) film on a Si substrate [26]. An AFM image of the film revealed a distribution of islands on the surface. The height-height correlation function of the islands $C_{isl}(\mathbf{R})$ can be represented well by the expression

$$C_{isl}\left(\mathbf{R}\right) = \sigma_{isl}^2 \exp\left(-\Delta q R\right) \mathrm{Cos}\left(q_d R\right), \tag{1.33}$$

where Δq is the width of the island size distribution in reciprocal space and q_d is $2\pi/D$, D being the mean spacing between the islands. If we assume that there are capillary wave height fluctuations superimposed on the island height fluctuations and that they are statistically independent, then we may write

$$C(\mathbf{R}) = C_{isl}(\mathbf{R}) + C_{cap}(\mathbf{R}), \tag{1.34}$$

where $C_{cap}(\mathbf{R})$ is the usual correlation function for capillary wave fluctuations, as given by Equation 1.22. Consider the expression for scattering in the BA in slit geometry without separating the specular and diffuse, as given by Equation 1.10. Since this expresses the scattering in terms of the FT of $\exp[Q_z^2 C(\mathbf{R})]$, we may use the convolution theorem for FTs to write the scattering function as

$$S'(Q_x, Q_z) = \frac{\rho_0^2 k_0^2 A}{Q_z^2} e^{-Q_z^2(\sigma_{cap}^2 + \sigma_{isl}^2)} S'_{cap}(Q_x, Q_z) * S'_{isl}(Q_x, Q_z), \tag{1.35}$$

where $S'_{cap}(Q_x, Q_z) = \int dX \exp[Q_z^2 C_{cap}(X)] \exp[-iQ_xX]$ and $S'_{isl}(Q_x, Q_z) = \int dX \exp[Q_z^2 C_{isl}(X)] \exp[-iQ_xX]$ are, separately, the scattering functions from the capillary waves and the islands alone, respectively, and where * indicates a convolution over Q_x.

Equation 1.35 can be shown to yield asymptotic power law expressions for $S'(Q_x, Q_z)$ as a function of Q_x with exponent $(\eta - 2)$ (instead of the usual $(\eta - 1)$) for capillary wave scattering in slit geometry for Q_x values greater than a cutoff wave vector Q_{eff}, which is given by $Q_{eff} = [\kappa^2 + \eta^2 (q_d^2 - \Delta q^2)]^{1/2}$. Thus both the cutoff and the exponent are dependent on Q_z. This may be seen from Figure 1.13, where the transverse DS is plotted, together with fits using the above model. Good agreement was found with the model, and the value of κ from the fits was found to scale as a/d^2, as predicted from the van der Waals theory, with a reasonable value for the effective Hamaker constant.

The capillary wave theory, as usually derived, is somewhat phenomenological. Napiorkowski and Dietrich [81] and Mecke and Dietrich [82] derived capillary wave theory from a more fundamental standpoint, namely, the density functional formalism. An interesting consequence of their calculations is that the surface tension γ is actually q-dependent and that $\gamma(q)$ has a minimum at large q. For water, this turns out to be near the first peak in the bulk liquid structure factor and should be manifested by an increase in the DS at those values of in-plane wave vector. This was observed by Daillant et al. [89] in a study of DS from water. Subsequently, similar effects were observed in liquid metals and alloys by Li et al. [90]. Although the results were interpreted in terms of the capillary wave scattering theory outlined above, it should be noted that the assumption underlying equations such as Equation 1.28 is that the liquid under the surface scatters as a continuum, based on the assumption that $Q <<$ interatomic

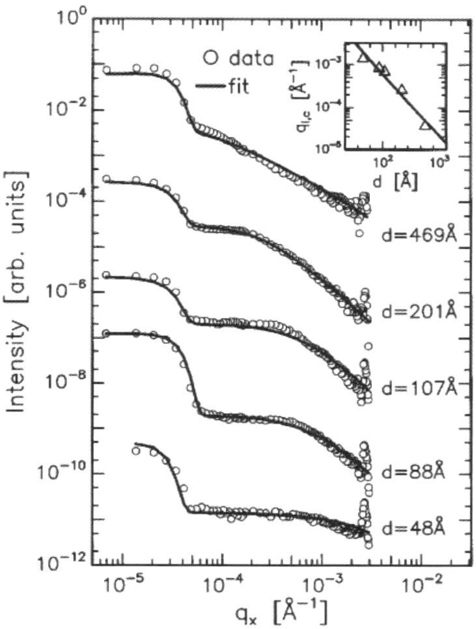

Figure 1.13. *Transverse DS for $Q_z = 0.2$ Å^{-1} from PEP films on Si substrates with various film thicknesses. The solid lines are fits to the data. The inset depicts a log-log plot of the lower capillary wave cutoff $q_{l,c}$ vs d (open triangles $q_{l,c}$ obtained from the diffuse scattering fits; line: linear fit). The Yoneda peak at $Q_x = 3.3\ 10^{-3}$ Å^{-1} was excluded from the data analysis. (From Tolan et al. [26].)*

spacing. This is no longer true when the in-plane Q is of the order of the first peak in the bulk liquid structure factor. Thus the use of the above scattering theory (although not of the capillary wave model per se) needs to be more carefully justified.

1.4 Grazing-Incidence Small-Angle Scattering (GISAXS and GISANS)

Grazing-incidence small-angle X-ray scattering is an increasingly popular method at synchrotron sources for studying decorated surfaces or configurations of nanoparticles, nanodots, nanorods, block copolymers, and so on arranged on surfaces. The potential of this technique was first realized and popularized by Naudon and coworkers [50]. It is particularly useful if these objects are buried under a protective cover layer so that it is not possible to image them using one of the surface-scanning probe microscopies. In this technique, the incident beam is made to fall at grazing angles of incidence (near or below the critical angle for total reflection) and the scattering is often measured by a two-dimensional detector or a linear position-sensitive detector that can measure scattering both in and out of the plane of incidence. GISANS is the equivalent technique with neutrons but is not as common because it usually suffers from intensity limitations. It has, however, been put to effective use and will undoubtedly be increasingly popular at the newer high-intensity neutron sources.

Grazing-incidence small-angle scattering (GISAS) experiments measure the same thing as in the DS geometry, but with a couple of differences. By going out of the plane of incidence (in the Q_y direction), it is possible to go to large in-plane wave vector transfers, but the instrumental resolution width in Q in this direction is not nearly as good as in the Q_x direction in the plane of incidence. Another way of stating this is that in the Q_y direction the projected coherence length of the incident beam is much smaller because it cannot take advantage of the grazing angle of incidence.

The main difference between GISAS experiments and conventional small-angle scattering experiments is that we cannot neglect the effects of the reflected beam. Let us consider a set of fairly smooth layers and calculate the scattering from an array of particles situated in the layer j using the DWBA. We shall neglect the diffuse scattering from the roughness of the layers and consider only the incident and specularly reflected beams in the layer j and the scattering from the particles. As before, we define 4 $\mathbf{Q}$ vectors $\mathbf{Q}_j^m$ ($m = 0, \ldots, 3$) in layer j and coefficients $T_{i,j}$, $T_{f,j}$ for the transmitted wave amplitudes in layer j for beams incident on the system at grazing angles of incidence α_i and α_f, respectively, and $R_{i,j}$ and $R_{f,j}$ refer to the amplitudes of the corresponding reflected beams (see Table 1.1). Then the DWBA calculation yields

$$S(\mathbf{Q})_d = \frac{Ak_0^4}{8\pi^2} \sum_{m,n=0}^{3} G_j^m G_j^{n*} \left(n_\mu^2 - n_j^2\right)\left(n_\nu^2 - n_j^2\right)\left\langle f(\mathbf{Q}_j^m)\right\rangle\left\langle f^*(\mathbf{Q}_j^n)\right\rangle e^{-i(Q_{z,j}^m - Q_{z,j}^n)R_z} s(\mathbf{Q}_{\|}) \quad .(1.36)$$

In the above equation, μ,ν index the particles in the layer. The particle μ is assumed to have a form factor $f(\mathbf{Q})$ and a refractive index n_μ, and its center is located at lateral position $\mathbf{R}_\mu$. n_j is the refractive index of layer j. We assume all the particles are centered at the same average height R_z. We also assume that the sizes and shapes of the particles—that is, their form factors—are uncorrelated with their positions. $s(\mathbf{Q}_{||})$ represents the in-plane structure factor of the nanoparticles, given by

$$s(\mathbf{Q}_{\|}) = \left\langle \sum_{\mu,\nu} e^{-i\mathbf{Q}_{\|}\cdot(\mathbf{R}_{\mu}-\mathbf{R}_{\nu})} \right\rangle . \tag{1.37}$$

If the particles are in air above a single surface with a bulk medium below, then the layer $j = 0$, the T_i and $T_f = 1$, and the R_i and R_f represent the reflection coefficients from the surface at angles of incidence α_i and α_f, respectively. For particles on the surface of a fixed height d (e.g., islands), $\langle f(\mathbf{Q})\rangle$ will produce fringes along Q_z. However, this fringe pattern may be complex because there are two distinct combinations of wave vectors along the z axis involved (Table 1.1). In the case of pitting corrosion on a single surface, we can think of the pits as regions of negative scattering density (which cancel the scattering density of the solid) sitting just below the surface. We choose $j = 1$, the R_i, R_f in the medium below the surface are zero, and the T_i, T_f are the transmission coefficients for angles of incidence α_i and α_f, respectively. Then Equation 1.36 simplifies to

$$S(\mathbf{Q})_d = \frac{Ak_0^4}{8\pi^2}\left(1-n_j^2\right)^2 \left\langle f(\mathbf{Q})\right\rangle^2 s(\mathbf{Q}_{\|}) \left|T(\alpha_i)\right|^2 \left|T(\alpha_f)\right|^2 , \tag{1.38}$$

where $f(\mathbf{Q})$ is the form factor of the pit and $s(\mathbf{Q}_{\|})$ is the in-plane structure factor of the pit distribution on the surface.

Grazing-incidence small-angle X-ray scattering measurements have been carried out on a variety of nanoparticle systems, including gold nanoparticles deposited on a polymer film [91], CdS nanoparticles implanted in a SiO2 matrix [92], Ge nanowires grown on a Si(111) surface [94], Ge nanoparticles on Si substrates [95,96], Si/Ge islands grown on a vicinal Si(001) single crystal surface [97], and a variety of other systems in order to study the size, shape, orientations, and spacings of the nanoparticles. Very often AFM and other scanning probe microscopies could not be used for such studies due to the fact that the nanoparticles were buried. If the islands possess a particular shape and orientation (as is sometimes the case) with respect to the crystalline substrate, this may be derived by a careful mapping of the GISAXS patterns versus azimuthal orientation of the beam with respect to the substrate and the positions of the crystal truncation rods from the substrate. Rauscher et al. [96] argue that shapes of nanoparticles not obtainable uniquely using transmission SAXS experiments often can be sorted out by GISAXS owing to the lower symmetry provided by the substrate reference. Islands of uniform height, such as block copolymer islands on a multilamellar block copolymer film, have also been studied [99], as well as pitting of a Cu surface in an electrochemical cell [100].

The in-plane structure factor and its modeling are often an issue. Sometimes, rather crude models for the interparticle correlation function, such as that given by Equation 1.33 or an $s(\mathbf{Q}_{\|})$ given by Lazzari et al. [98], Hosemann and Bagchi [101], and Narayanan et al. [102], of the form

$$s(\mathbf{Q}_{\|}) = \frac{1-\exp(-2Q_{\|}^2\sigma_d^2)}{1-2\exp(-Q_{\|}^2\sigma_d^2)\cos(Q_{\|}d)+\exp(-2Q_{\|}^2\sigma_d^2)} \tag{1.39}$$

appear to work reasonably well. In other cases, particularly for dense depositions of nanoparticles, one needs more sophisticated models where the correlations between the sizes, shapes, and interparticle distances need to be taken into account by appealing to growth models. In addition, the layer of nanoparticles itself forms its own layer and must be treated self-consistently in the DWBA. Such refinements have been formulated recently in models by Lazzari et al. [65].

Time-dependent GISAXS measurements have been used to monitor the growth of nanoparticles on a surface, for example, on a Si surface subjected to Ar ion bombardment [51]. GISAXS measurements have also been used to study the growth of Co nanoparticles on the dislocation network produced in a thin-strained Al film on a single crystal MgO(001) surface [93] and the growth of Au nanoparticles on TiO_2(110) [98], which is a model catalyst system. They also have been carried out on lamellar structures of block copolymer films on substrates [103]. Modeling and numerical simulation of GISAXS patterns from such systems have been presented by Busch et al. [104]. Another important area of investigation comprises GISAXS studies of nanoparticles on liquid surfaces [105].

We discuss one particularly interesting case where the medium containing the particles acts as a thin film waveguide [106]. The rigorous condition for this is that the scattering length density of this layer must be smaller than the layers above and below it so that total reflection occurs at both interfaces, and the beam must be introduced into the guide via an evanescent wave from the top layer (which itself should be thin) by adjusting α_i to be smaller than the critical angle for the top layer. In such a circumstance, at particular values of α_i, a resonant guided mode can be excited in the guide with amplitude that can be considerably larger than the incident wave amplitude and thus can enhance the GISAXS from the particles in the guide layer. By the optical reciprocity theorem, similar enhancement occurs when α_f also takes these values. Even if the condition that the overlayer have a scattering length density greater than that of the guide is relaxed, one can still obtain "pseudo-resonances" in the film with corresponding gain [107]. These results are correctly reproduced by the GISAS theory given by Equation 1.33. This method of enhancing GISAXS was first demonstrated by Salditt et al. [108]. In a recent GISAXS experiment carried out by Narayanan et al. [102] on gold nanoparticles sandwiched between two poly(tert-butyl acrylate) (PtBA) polymer layers on a Si substrate, one may see in the two-dimensional detector pattern a set of bright streaks corresponding to this large enhancement of the GISAXS at the waveguide resonances for α_f (Figures 1.15a and 1.15b). The GISAXS pattern was analyzed using the general DWBA formalism given in Equation 1.36 using the in-plane structure factor represented by Equation 1.39 and yielded information about the spatial distribution of the nanoparticles in the plane of the film (Figure 1.14b). By comparing the changes in this distribution at high annealing temperatures to the changes in distribution normal to the film, it was concluded that the diffusion of the nanoparticles in-plane was about five times more rapid than in the perpendicular direction. There have been relatively fewer neutron scattering studies of nanoparticles with such techniques. As an example, Lauter-Pasyuk et al. [109] have studied with neutron specular and off-specular scattering the conformation and distribution of nanoparticles in block-copolymer films. For a review see the article by Hamilton [110].

2 Summary and Conclusions

We have derived the basic formalism of diffuse scattering from surfaces in both the kinematic (Born) approximation and the DWBA. We have shown how this formalism can be applied to a wide variety of systems if one can model the height-height correlation functions well. For solid surfaces, thin

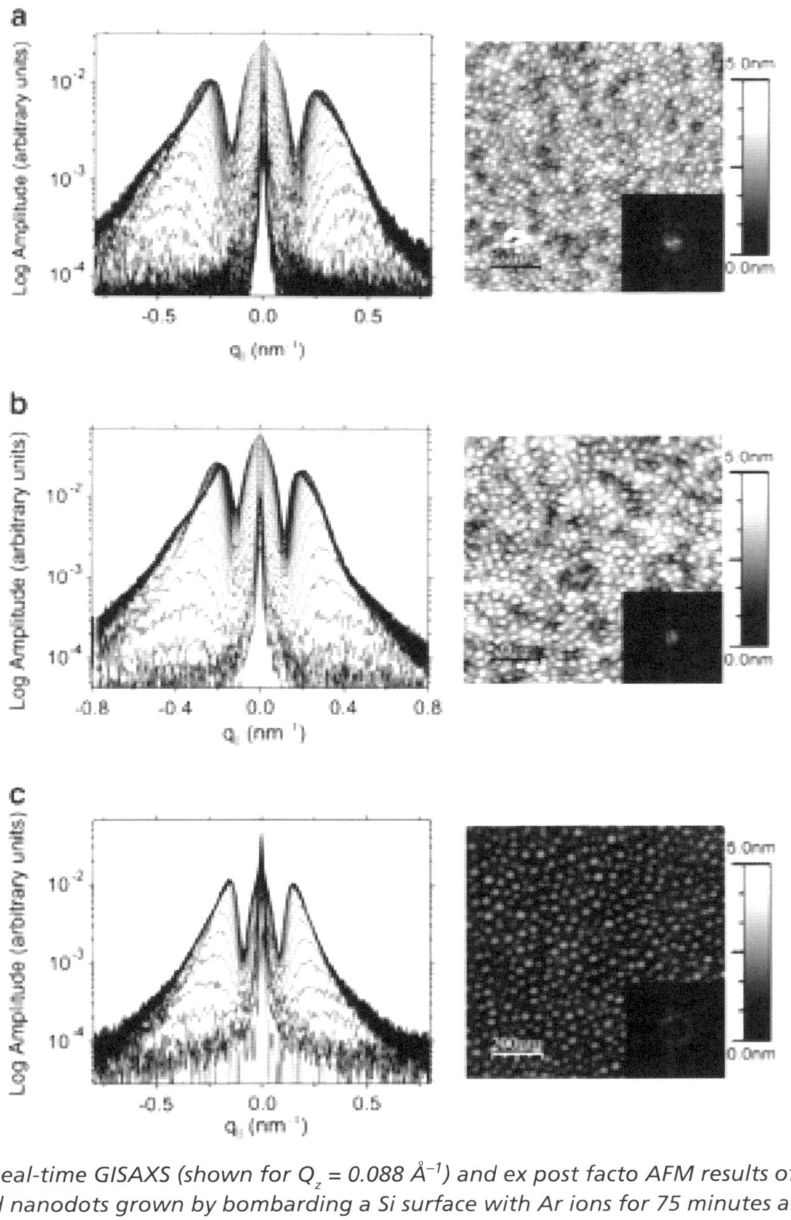

Figure 1.14. *Real-time GISAXS (shown for $Q_z = 0.088$ Å^{-1}) and ex post facto AFM results of samples of Mo-seeded nanodots grown by bombarding a Si surface with Ar ions for 75 minutes at (a) 300 eV, (b) 500 eV, and (c) 1000 eV. Each curve was recorded at 140-second intervals during the growth process. The kinetic roughening and the dot correlation peaks are shown. (From Ozaydin et al. [51].)*

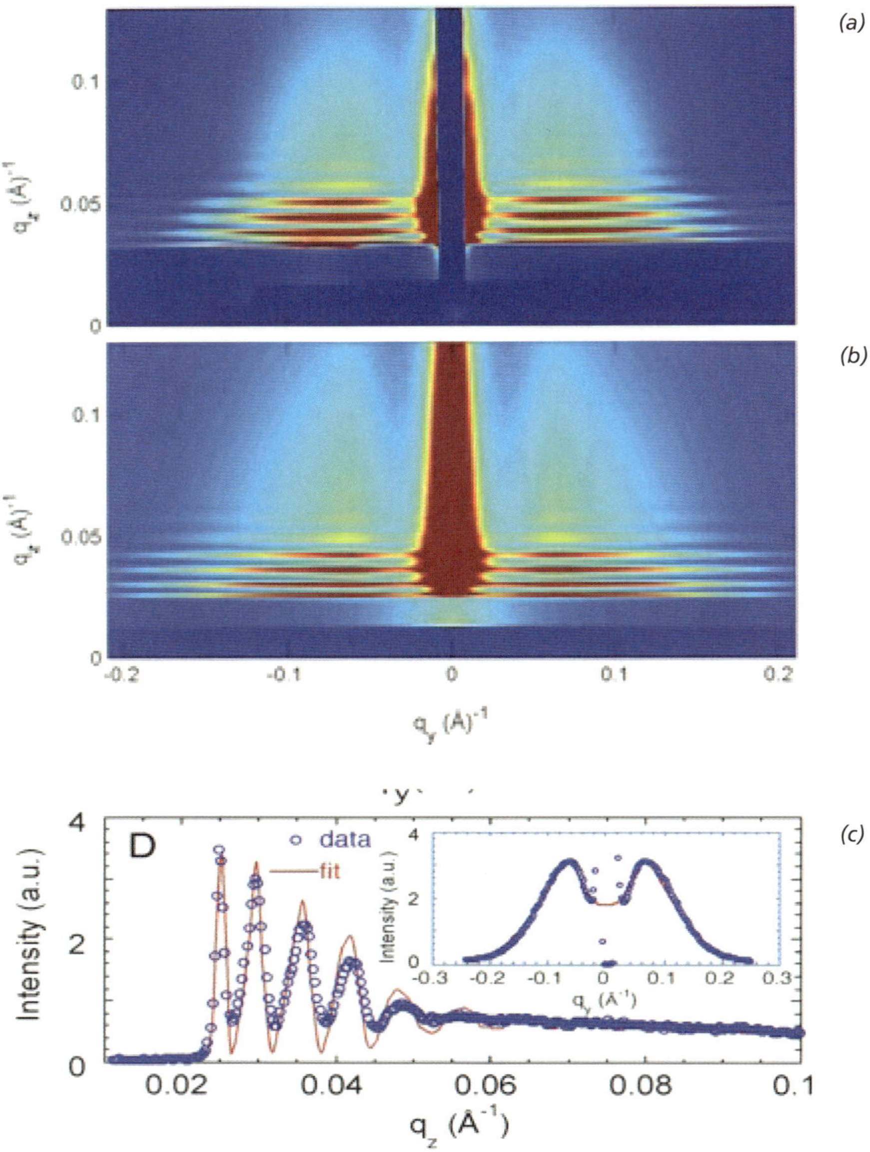

Figure 1.15. *(a) Measured and (b) simulated reciprocal space GISAXS patterns of an as-deposited Au nanoparticle/PtBA sandwich sample. The strong specular streak in (a) is blocked with a beam-stop. (c) Measured (open circles) and DWBA fitted (line) intensity cross sections drawn along the Q_z direction from (a) and (b), with inset showing the measured (open circles) and DWBA fitted (line) intensity cross section along the Q_y direction. The in-plane structure factor for the nanoparticle correlations was taken as given by Equation 1.39. (From Narayanan et al. [102].)*

solid films, and multilayers (and even for some complex fluid surfaces), the surface roughness and the corresponding correlation functions can often be successfully modeled in terms of the self-affine fractal model. Where complementary microscopy studies have been carried out on the surface, the parameters obtained from the microscopy and X-ray diffuse studies are often in good agreement. For liquid surfaces, we have discussed the conventional capillary wave predictions for the height-height correlation function and have shown that it can lead to power law tails in the diffuse scattering from a liquid surface. Modifications due to van der Waals interactions with the substrate in the case of thin films or due to viscoelastic effects in the fluid can terminate these power laws due to low wave-vector cutoffs. Effects due to the wave-vector dependence of the surface tension may also change the shape of the scattering at large **Q** values.

There are two important related topics that are beyond the scope of this article and that have not been discussed. The time autocorrelation function of the diffuse scattering from a liquid surface can yield important information about the slow *dynamics* of the capillary waves, and this can be obtained by using a *coherent* X-ray beam to study the time dependence of the speckles that take the place of the diffuse scattering patterns discussed here, but it is not our purpose to discuss this topic in this article. Also, magnetic diffuse scattering from thin films is beginning to play an important role in studying magnetic domain patterns at surfaces and interfaces, magnetic roughness, and so on, and has been studied using both neutrons and resonant X-ray scattering at the L or M edges of magnetic atoms. The DWBA approach has been formulated for both these cases, but the situation is more complicated than for pure charge scattering because both up- and down-spin neutron beams (or, in the case of X-rays, both photon polarizations) must be considered. We refer the interested reader to the literature [111–113]. It is clear that diffuse scattering will continue to play a major role in the investigation of the properties of surfaces, interfaces, thin films, and multilayers.

Acknowledgments

I am grateful to my many collaborators in the area of research discussed here and to Dr. G. Chen for help with the manuscript. I also wish to acknowledge support from Basic Energy Sciences, the U.S. Department of Energy, and the National Science Foundation.

References

1. L. G. Parratt, *Phys. Rev.* 95 (1954): 359.
2. M. Born and E. Wolf, *Principles of Optics,* 7th ed. (Cambridge: Cambridge University Press, 1999).
3. A. Gibaud, G. Vignaud, and S. K. Sinha, *Acta Crystallogr. A* 49 (1993): 642.
4. S. K. Sinha, E. B. Sirota, S. Garoff, and H. B. Stanley, *Phys. Rev. B* 38 (1988): 2297.
5. S. K. Sinha, *J. de Physique III, France* 4 (1994): 1543.
6. D. E. Savage, J. Kleiner, N. Schimke, Y. H. Phang, T. Jankowski, J. Jacobs, R. Kariotis, and M. G. Lagally, *J. Appl. Phys.* 69 (1991): 1411.
7. C. Teichert, J. F. Mackay, D. E. Savage, M. G. Lagally, M. Brohl, and P. Wagner, *Appl. Phys. Lett.* 66 (1995): 2346.
8. Y. H. Phang, D. E. Savage, R. Kariotis, and M. G. Lagally, *J. Appl. Phys.* 74 (1993): 3181.
9. V. Holy, J. Kubena, I. Ohlidal, K. Lischka, and W. Plotz, *Phys. Rev. B* 47 (1993): 15896.
10. V. Holy and T. Baumbach, *Phys. Rev. B* 49 (1994): 10668.
11. D. K. G. de Boer, *Physica B* 221 (1996): 18.
12. D. K. G. de Boer, *Phys. Rev. B* 53 (1996): 6048.

13. T. Salditt, T. H. Metzger, J. Peisl, G. Vignaud, P. Hoghoj, O. Scharpf, P. Hinze, and R. Lauer, *Phys. Rev. B* 54 (1996): 5860.
14. J. Stettner, L. Schwalowsky, O. H. Seeck, M. Tolan, W. Press, C. Schwarz, and H. V. Kanel, *Phys. Rev. B* 53 (1996): 1398.
15. J. M. Freitag and B. M. Clemens, *J. Appl. Phys.* 89 (2001): 1101.
16. S. A. Stepanov, E. A. Kondrashkina, M. Schmidbauer, R. Kohler, J. U. Pfeiffer, T. Jach, and A. Y. Souvorov, *Phys. Rev. B* 54 (1996): 8150.
17. C. Thompson, G. Palasantzas, Y. P. Feng, S. K. Sinha, and J. Krim, *Phys. Rev. B* 49 (1994): 4902.
18. C. V. Dharamadikari, A. O. Ali, N. Suresh, D. M. Phase, S. M. Chaudhuri, A. Gupta, and B. A. Dasannacharya, *Matls. Science Eng. B-Solid State Materials for Advanced Technology* 75 (2000): 29.
19. L. Peverini, E. Ziegler, and I. Kozhevnikov, *Appl. Phys. Lett.* 91 (2007): 053121.
20. A. C. Dürr, F. Schreiber, K. A. Ritley, V. Kruppa, J. Krug, H. Dosch, and B. Struth, *Phys. Rev. Lett.* 90 (2003): 016104.
21. T. Salditt, T. H. Metzger, and J. Peisl, *Phys. Rev. Lett.* 73 (1997): 4091.
22. M. Lütt, J. P. Schlomka, M. Tolan, J. Stettner, O. H. Seeck, and W. Press, *Phys. Rev. B* 56 (1997): 4085–4091.
23. J. K. Basu, S. Hazra, and M. K. Sanyal, *Phys. Rev. Lett.* 82 (1999): 4675.
24. J. Krim and G. Palazantsas, *Int. J. Mod. Phys. B* 9 (1995): 599.
25. Y. P. Feng, S. K. Sinha, C. A. Melendres, and D. D. Lee, *Physica B* 221 (1996): 251.
26. M. Tolan, O. H. Seeck, J.-P. Schlomka, W. Press, J. Wang, S. K. Sinha, Z. Li, M. H. Rafailovich, and J. Sokolov, *Phys. Rev. Lett.* 81 (1998): 2731.
27. W. Zhao, M. H. Rafailovich, J. Sokolov, L. J. Fetters, R. Plano, M. K. Sanyal, S. K. Sinha, and B. B. Sauer, *Phys. Rev. Lett.* 70 (1993): 1453.
28. I. M. Tidswell, T. A. Rabedeau, P. S. Pershan, and S. D. Kosowsky, *Phys. Rev. Lett.* 66 (1991): 2108.
29. O. H. Seeck, P. Muller-Buschbaum, M. Tolan, and W. Press, *Europhys. Lett.* 29 (1995): 699.
30. P. Müller-Buschbaum and M. Stamm, *Physica B* 248 (1998): 229.
31. M. K. Sanyal, S. K. Sinha, K. G. Huang, and B. M. Ocko, *Phys. Rev. Lett.* 66 (1991): 628.
32. D. K. Schwartz, M. L. Schlossmann, E. H. Kawamoto, G. J. Kellogg, P. S. Pershan, and B. M. Ocko, *Phys. Rev. A* 41 (1990): 5687.
33. A. K. Doerr, M. Tolan, W. Prange, J.-P. Schlomka, T. Seydel, and W. Press, *Phys. Rev. Lett.* 83 (1999): 3470.
34. H. Tostmann, E. DiMasi, P. S. Pershan, B. M. Ocko, O. G. Shpyrko, and M. Deutsch, *Phys. Rev. B* 59 (1999): 783.
35. E. DiMasi, H. Tostmann, B. M. Ocko, P. S. Pershan, and M. Deutsch, *Phys. Rev. B* 58 (1998): 13419.
36. D. X. Li, B. Yang, B. H. Lin, M. Meron, J. Gebhardt, T. Graber, and S. A. Rice, *Phys. Rev. Lett.* 92 (2004): 136102.
37. A. M. Tikhonov, M. Li, and M. L. Schlossman, *J. Phys. Chem. B* 105 (2001): 8065.
38. J. Penfold, *Rep. Progr. Phys.* 64 (2001): 777.
39. P. Busch, M. Rauscher, D. M. Smilgies, D. Posselt, and C. M. Papadakis, *J. Appl. Cryst.* 39 (2006): 433.
40. B. Lee, I. Park, J.yoon, S. Park, J. Kim, K. W. Kim, T. Chang, and M. Ree, *Macromolecules* 38 (2005): 4311.
41. P. Muller-Buschbaum, *Anal. Bioanal. Chem.* 376 (2003): 3.
42. Z. Sun, M. Wolkenhauer, G.-G. Bumbu, D. H. Kim, and J. S. Gutmann, *Physica B* 357 (2005): 141–143.
43. S. Langridge, J. Schmalian, C. H. Marrows, D. T. Dekadjevi, B. J. Hickey, *Phys. Rev. Lett.* 85 (2000): 4964.
44. F. Radu, M. Etzkom, R. Siebrecht, T. Schmitte, K. Westerholt, and H. Zabel, *Phys. Rev. B* 67 (2003): 134409.
45. S. G. E. te Velthuis, A. Berger, G. P. Felcher, B. K. Hilland, D. Dahlberg, *J. Appl. Phys.* 87 (2000): 5046.
46. J. B. Kortright, S.-K. Kim, G. P. Denbeaux, G. Zeltzer, K. Takano, and E. E. Fullerton, *Phys. Rev. B* 64 (2001): 092401.
47. R. M. Osgood, S. K. Sinha, J. W. Freeland, Y. U. Idzerda, and S. D. Bader, *J. Magn. Magnetic Materials* 198 (1999): 698.
48. C. S. Nelson, G. Srajer, J. C. Lang, C. T. Venkataraman, S. K. Sinha, H. Hashizume, N. Ishimatsu, and N. Hosoito, *Phys. Rev. B* 60 (1999): 12234.
49. J. F. Mackay, C. Teichert, D. E. Savage, and M. G. Lagally, *Phys. Rev. Lett.* 77 (1996): 3925.
50. A. Naudon and D. Thiaudiere, *Surf. Coating Tech.* 79 (1996): 103.
51. G. Ozaydin, A. S. Ozcan, Y. Y. Wang, K. F. Ludwig, H. Zhou, and R. L. Headrick, *Nucl. Instr. Methods in Phys. Res. B-Beam Interactions with Matls. & Atoms* 264 (2007): 47.

52. D. R. Lee, Y. S. Chu, Y. Choi, J. C. Lang, G. Srajer, S. K. Sinha, V. Metlushko, B. Ilic, *Appl. Phys. Lett.* 82 (2003): 982.
53. I. A. Vartanyants, I. K. Robinson, J. D. Onken, M. A. Pfeifer, G. J. Williams, F. Pfeiffer, H. Metzger, Z. Zhong, and G. Bauer, *Phys. Rev. B* 71 (2005): 245302.
54. B. Mandelbrodt, *The Fractal Geometry of Nature* (New York: Freeman, 1982); R. F. Voss, in *Fundamental Algorithms for Computer Graphics*, ed. R. A. Earnshaw (Berlin: Springer-Verlag, 1985), 805; R. F. Voss, in *Scaling Phenomena in Disordered Systems*, ed. R. Pynn and A. Skjeltorp (New York: Plenum, 1985), 1.
55. T. Salditt, T. H. Metzger, C. Brandt, U. Klemradt, and J. Peisl, *Phys. Rev. B* 51 (1995): 5617.
56. E. Spiller, D. Stearns, and M. Krumrey, *J. Appl. Phys.* 74 (1993): 107.
57. M. K. Sanyal, S. K. Sinha, A. Gibaud, S. K. Satija, C. F. Majkrzak, and H. Homa, in *Surface X-ray and Neutron Scattering*, ed. H. Zabel and I. K. Robinson, Springer Proceedings in Physics, vol. 61 (Berlin: Springer-Verlag, 1992), 91–94.
58. S. K. Sinha, M. K. Sanyal, S. K. Satija, C. F. Majkrzak, D. A. Neumann, H. Homma, S. Szpala, A. Gibaud, and H. Morkoc, *Physica B* 198 (1994): 72.
59. W. Weber and B. Lengeler, *Phys. Rev. B* 46 (1992): 7953.
60. M. Wormington, I. Pape, T. P. A. Hase, B. K. Tanner, and D. K. Bowen, *Phil. Mag. Letters* 74 (1996): 211.
61. L. I. Schiff, *Quantum Mechanics* (New York: McGraw-Hill, 1968).
62. G. H. Vineyard, *Phys. Rev. B* 50 (1982): 4146.
63. T. Ghaderi, "X-ray Intensity Correlation Spectroscopy from Fluid Surfaces—Theory and Experiment" (Thesis, U. Dortmund, 2006).
64. Y. Yoneda, *Phys. Rev.* 131 (1963): 2010.
65. R. Lazzari, F. Leroy, and G. Renaud, *Phys. Rev. B* 76 (2007): 125411.
66. S. K. Sinha, *MRS Symposium Proc.,* ed. D. A. Neumann, T. P. Russell, and B. J. Wuensch, 376 (1995): 175.
67. T. Salditt, T. H. Metzger, J. Peisl, B. Reinker, M. Moske, and K. Samwer, *Europhys. Lett.* 32 (1995): 331.
68. M. E. Gomez, J. Santamaria, M. C. Cyrille, E. C. Nelson, K. M. Krishnan, and I. K. Schuller, *Eur. Phys. J. B* 30 (2002): 17.
69. M. C. Lafouresse, P. J. Heard, and W. Schwarzacher, *Phys. Rev. Lett.* 98 (2007): 236101.
70. M. Feygenson, E. Kentzinger, N. Ziegenhagen, U. Rucker, G. Goerigk, Y. Wang, and T. Bruckel, *J. Appl. Cryst.* 40 (2007): 532.
71. S. Singh, and S. Basu, *Surf. Sci.* 600 (2006): 493.
72. M. Kardar, G. Parisi, and Y. C. Zhang, *Phys. Rev. Lett.* 56 (1986): 889.
73. D. Y. Noh, Y. Hwu, H. K. Kim, and M. Hong, *Phys. Rev. B* 51 (1995): 4441.
74. E. Smigiel, A. Knoll, N. Broll, and A. Cornet, *J. Phys. IV France* 10 (2000): Pr 10–237.
75. B. R. McClain, D. D. Lee, B. L. Carvalho, S. G. J. Mochrie, S. H. Chen, and J. D. Litster, *Phys. Rev. Lett.* 72 (1994): 246.
76. *Handbook of Mathematical Functions*, ed. M. Abramowitz and I. A. Stegun (Washington, DC: Natl. Bureau of Standards, 1972).
77. O. M. Magnussen, B. M. Ocko, M. J. Regan, K. Penanen, P. S. Pershan, and M. Deutsch, *Phys. Rev. Lett.* 74 (1995): 4444.
78. A. Braslau, P. S. Pershan, G. Swislow, B. M. Ocko, and J. Als-Nielsen, *Phys. Rev. A* 38 (1988): 2457.
79. L. Lurio, H. J. Kim, A. Ruhm, J. K. Basu, J. Lal, S. K. Sinha, and S. G. J. Mochrie, *Macromolecules* 36 (2003): 5704.
80. Y. S. Seo, T. Koga, M. H. Rafailovich, M. Tolan, and S. K. Sinha, *Phys. Rev. Lett.* 94 (2005): 157802.
81. M. Napiorkowski and S. Dietrich, *Phys. Rev.* E 47 (1993): 1836.
82. K. R. Mecke and S. Dietrich, *Phys. Rev.* E 59 (1999): 6766.
83. M. D. Lippmann, O. Seeck, H. Kim, and S. K. Sinha, article in *Macromolecules* (forthcoming 2008).
84. B. R. McClain, M. Yoon, J. D. Litster, and S. G. J. Mochrie, *Eur. Phys. J. B* 10 (1999): 45.
85. G. H. Fredrickson, A. Ajdari, L. Leibler, and J.-P. Carton, *Macromolecules* 25 (1992): 2882.
86. J. Wang, M. Tolan, O. H. Seeck, S. K. Sinha, O. Bahr, M. H. Rafailovich, and J. Sokolov, *Phys. Rev. Lett.* 83 (1999): 564.
87. Z. Jiang, H. Kim, X. Jiao, H. Lee, Y. J. Lee, Y. Byun, S. Song, D. Eom, C. Li, M. H. Rafailovich et al., *Phys. Rev. Lett.* 98 (2007): 227801.

88. B. Akgun, D. R. Lee, H. Kim, H. Zhang, O. Prucker, J. Wang, J. Ruhe, and M. D. Foster, *Macromolecules* 40 (2007): 6361.
89. S. Mora, J. Daillant, K. Mecke, D. Luzet, A. Braslau, M. Alba, and B. Struth, *Phys. Rev. Lett.* 90 (2003): 216101.
90. D. X. Li, X. Jiang, B. H. Lin, M. Meron, and S. A. Rice, *Phys. Rev. B* 72 (2005): 235426.
91. M. Kimura, A. Acosta, H. Fujioka, and M. Oshima, *J. Appl. Phys.* 93 (2003): 2034.
92. M. Buljan, K. Salamon, P. Dubcek, S. Bernstorff, I. D. Desnica-Francovic, O. Milat, and U. V. Desnica, *Vacuum* 71 (2003): 65.
93. F. Leroy, G. Renaud, A. Letoublon, R. Lazzari, C. Mottet, and G. Goniakowski, *Phys. Rev. Lett.* 95 (2005): 185501.
94. Z. X. Xu, Y. T. Zhang, R. L. Headrick, H. Zhou, L. Zhou, and T. Fukamachi, *Phys. Rev. B* 75 (2007): 233310.
95. M. I. Richard, T. U. Schulli, E. Wintersberger, G. Renaud, and G. Bauer, *Thin Solid Films* 508 (2006): 213.
96. M. Rauscher, R. Paniago, H. Metzger, Z. Kovats, J. Domke, J. Peisl, H.-D. Pfannes, J. Schulze, and I. Eisele, *J. Appl. Phys.* 86 (1999): 6763.
97. T. Roch, V. Holy, J. Stangl, E. Hoflinger, A. Daniel, G. Bauer, I. Kegel, H. Metzger, J. Zhu, K. Brunner et al., *Phys. Stat. Sol. (b)* 224 (2001): 241.
98. R. Lazzari, G. Renaud, J. Jupille, and F. Leroy, *Phys. Rev. B* 76 (2007): 125412.
99. G. Vignaud, A. Gibaud, J. Wang, S. K. Sinha, J. Daillant, G. Grubel, and Y. Gallot, *J. Phys. Cond. Mat.* 9 (1997): L125.
100. C. A. Melendres, Y. P. Feng, D. D. Lee, and S. K. Sinha, *J. Electrochem Soc.* 142 (1995): L19.
101. R. Hosemann and S. N. Bagchi, *Direct Analysis of Diffraction by Matter* (Amsterdam: North-Holland, 1962).
102. S. Narayanan, D. R. Lee, R. S. Guico, S. K. Sinha, and J. Wang, *Phys. Rev. Lett.* 94 (2005): 145504.
103. D.-M. Smilgies, P. Busch, D. Posselt, and C. M. Papadakis, *Synchrotron Radiation News* 15 (2002): 35.
104. P. Busch, M. Rauscher, D.-M. Smilgies, D. Posselt, and C. M. Papadakis, *J. Appl. Cryst.* 39 (2006): 433.
105. M. K. Bera, M. K. Sanyal, S. Pal, J. Daillant, A. Datta, G. U. Kulkarni, D. Luzet, and O. Konovalov, *Europhys. Lett.* 78 (2007): 56003.
106. Y. P. Feng, S. K. Sinha, H. W. Deckman, J. B. Hastings, and D. P. Siddons, *Phys. Rev.* Lett 71 (1993): 537.
107. J. Wang, M. J. Bedzyk, and M. Caffrey, *Science* 258 (1992): 775.
108. T. Salditt, F. Pfeiffer, H. Perzl, A. Vix, U. Mennicke, A. Jarre, A. Mazuelas, and T. H. Metzger, *Physica B* 336 (2003): 181.
109. V. Lauter-Pasyuk, H. J. Lauter, G. P. Gordeev, P. Müller-Buschbaum, B. P. Toperverg, M. Jernekov, and W. Petry, *Langmuir* 19 (2003): 7783–7788.
110. W. A. Hamilton, *Current Opinion in Colloid and Interface Science* 9 (2005): 390.
111. V. G. Syromyatnikov, A. Menelle, B. P. Toperverg, Z. N. Soroko, and A. F. Schebetov, *Physica B* 268 (1999): 190.
112. R. M. Osgood, S. K. Sinha, J. W. Freeland, Y. U. Idzerda, and S. D. Bader, *J. Magn. Magnetic Materials* 198–199 (1999): 698.
113. D. R. Lee, S. K. Sinha, C. S. Nelson, J. C. Lang, C. T. Venkataraman, G. Srajer, and R. M. Osgood III, *Phys. Rev. B* 68 (2003): 22410.

2

Diffuse X-ray Scattering from Semiconductor Nanostructures

Václav Holý[1]

Julian Stangl[2]

Günther Bauer[2]

1. Faculty of Mathematics and Physics, Charles University, Prague, Czech Republic

2. Institute of Semiconductor Physics, Kepler University, Linz, Austria

Abstract

X-ray scattering methods are frequently used for the investigation of semiconductor nanostructures. From the distribution reciprocal space of diffusely scattered intensity, it is possible to determine the spatial correlation properties of the nano-objects, their shapes, and the elastic strains in and around the nano-objects as well as their local chemical compositions. In this chapter, we present basic theoretical approaches for the description of diffuse X-ray scattering from nanostructures and present some experimental examples.

1 Introduction

In the last two decades, X-ray scattering methods for structure investigations of nanostructures have been facilitated by powerful third-generation synchrotron sources with many orders of magnitude more brilliance than conventional X-ray sources. X-ray scattering methods became broadly accepted due to their relative simplicity and nondestructivity. In many studies on nanostructures, they are used as a routine tool for the determination of basic structural properties, such as the size of the nano-objects and their mean lattice parameters.

In this chapter we present basic theoretical approaches for the description of diffuse X-ray scattering from ensembles of nanostructures as well as some experimental results. These examples demonstrate that sophisticated X-ray scattering experiments can bring a deeper insight into the structure of nano-objects. The aim of this chapter is to present a few *typical* experimental examples that demonstrate the strengths (and weaknesses) of X-ray scattering methods. For a more comprehensive bibliography, see the specialized review papers or monographs on this subject [1,2,3,4].

The chapter is organized as follows. In the next section, we briefly summarize the basic approaches used in the theoretical description of X-ray diffuse scattering from nanostructures and then discuss kinematic approximation and the distorted-wave Born approximation (DWBA) method in detail. Then models of the random arrangement of nanostructures will be introduced, including short-range and long-range ordering. In section 4 a simple numerical example will be presented showing the characteristic features of the reciprocal-space distribution of scattered intensity. Experimental examples in section 5 will be restricted only to semiconductor nanostructures in epitaxial layers and multilayers.

2 Basic Scattering Theory

Elastic scattering of X-rays can be described by the wave equation for the electric intensity **E**,

$$(\nabla^2 + K^2)\mathbf{E}(\mathbf{r}) = \widehat{\mathbf{V}}(\mathbf{r})\mathbf{E}(\mathbf{r}), \tag{2.1}$$

where $K = 2\pi / \lambda$ is the wave vector length in a vacuum and $\widehat{\mathbf{V}}$ is the scattering potential. The form of $\widehat{\mathbf{V}}$ depends on the scattering mechanism. If we restrict ourselves to charge scattering (i.e., if we neglected the magnetic scattering of X-rays from magnetic moments),

$$\widehat{\mathbf{V}}(\mathbf{r}) = \mathrm{graddiv} - K^2\chi(\mathbf{r}) \tag{2.2}$$

holds, where $\chi(\mathbf{r})$ is the crystal polarizability proportional to the electron density $\rho(r)$,

$$\chi(\mathbf{r}) = -\frac{r_{\mathrm{el}}\lambda^2}{\pi} C\rho(\mathbf{r}),$$

where $r_{\mathrm{el}} \approx 2.82\times10^{-15}$ m is the classical electron radius and C is the linear polarization factor, which is assumed to be constant in further considerations.

A frequent description of the scattered wave is based on the differential scattering cross section $d\sigma/d\Omega$ determining the flux of the scattered radiation into an elementary solid angle $d\Omega$ pointing to the detector. This description can be used in the far-field limit, that is, if the sample is much smaller than the first Fresnel zone, the radius of which is $\sqrt{\lambda R}$ (R is the sample-detector distance), that is, a few micrometers in a usual case. On the other hand, the concept of a differential cross section is also applicable to large, statistically homogeneous samples. In this case, however, the meaning of the differential cross section is different; it determines the flux of the photons scattered into an elementary solid angle around an independently chosen wave vector **K** of the scattered radiation. In the following sections, we restrict ourselves to this kind of sample.

A direct calculation yields for the differential scattering cross section the formula [2]

$$\frac{d\sigma}{d\Omega}=\frac{1}{16\pi^2}\left\langle|\mathcal{T}|^2\right\rangle,\ \mathcal{T}=\langle\mathbf{K}|\widehat{\mathbf{T}}|\mathbf{K}_i\rangle\ , \tag{2.3}$$

where $\mathbf{K}_i$ is the wave vector of the primary radiation (assumed here ideally plane and monochromatic) and the scattering operator $\widehat{\mathbf{T}}$ can be expressed symbolically as [2]

$$\widehat{\mathbf{T}}=\widehat{\mathbf{V}}+\widehat{\mathbf{V}}G_0\widehat{\mathbf{V}}+\widehat{\mathbf{V}}G_0\widehat{\mathbf{V}}G_0\widehat{\mathbf{V}}+\cdots, \tag{2.4}$$

where G_0 is the Green function of a free particle. In the kinematic approximation (first Born approximation), we put $\widehat{\mathbf{T}}\approx\widehat{\mathbf{V}}$.

We will investigate scattering from randomly disordered samples. We will assume that the irradiated sample volume W is very large so that the measured signal is averaged over a statistical ensemble of microscopic configurations of the nanostructures present in W. This ensemble averaging is denoted by $\langle\ \rangle$ in Equation 2.3. The differential cross section can be divided into coherent and diffuse parts:

$$\left(\frac{d\sigma}{d\Omega}\right)_{\mathrm{coh}}=\frac{1}{16\pi^2}|\langle\mathcal{T}\rangle|^2,\left(\frac{d\sigma}{d\Omega}\right)_{\mathrm{diff}}=\frac{1}{16\pi^2}\left\langle(\mathcal{T}-\langle\mathcal{T}\rangle)(\mathcal{T}-\langle\mathcal{T}\rangle)^*\right\rangle. \tag{2.5}$$

In the following, we investigate only the diffuse part of the differential scattering cross section.

The kinematic approximation is applicable if the multiple scattering of the same photon can be neglected. In the case of nanostructures, this approximation can be used if the wave irradiating a nano-sized object (quantum dot, for instance) can be replaced by the incident wave; that is, if we can neglect diffraction and absorption in the substrate as well as refraction and specular reflection at the sample surface. Therefore the kinematic approximation can be used for the description of diffraction from a sample with nanostructures if the angle of incidence α_i and/or the angle of exit α_f of the scattered wave are much larger than the critical angle α_c of total external reflection at the sample surface. Since the waves in vacuum are transversal with zero divergence, the kinematic approximation of the matrix element $\mathcal{T}$ is proportional to the Fourier transformation of the electron density:

$$\mathcal{T}\approx\mathcal{V}=-K^2\int_W d^3\mathbf{r}\chi(\mathbf{r})e^{-i\mathbf{Q}.r},\mathbf{Q}=\mathbf{K}-\mathbf{K}_i\,. \tag{2.6}$$

Therefore, in the kinematic approximation, the scattered intensity is a function of the scattering vector $\mathbf{Q}$, and it can be represented by a distribution in reciprocal space (*reciprocal-space map*).

If the incidence and/or exit angle is close to the critical angle α_c or if diffraction in the substrate cannot be neglected, then the kinematic approximation must be replaced by a more exact approach. A full dynamical description of the scattering process in a sample with nanostructures is usually not feasible, and a semikinematic method can be used instead [2]. In this method, the scattering potential $\widehat{\mathbf{V}}$ is divided into two parts, $\widehat{\mathbf{V}}=\widehat{\mathbf{V}}_A+\widehat{\mathbf{V}}_B$, where $\widehat{\mathbf{V}}_A$ is the scattering potential of a nondisturbed system (assumed nonrandom) and $\widehat{\mathbf{V}}_B$ is a random disturbance. In the DWBA, the wave equation (Equation 2.1) for $\widehat{\mathbf{V}}_A$ is solved *exactly*, finding two independent solutions: $E_1(\mathbf{r})$ and $E_2^{(-)}(\mathbf{r})$; the latter is time inverted [5,2]. The scattering due to $\widehat{\mathbf{V}}_B$ is then assumed to be kinematic, and the differential cross section of diffuse scattering is

$$\left(\frac{d\sigma}{d\Omega}\right)_{\text{diff}}^{\text{DWBA}} = \frac{1}{16\pi^2}\left\langle (\mathcal{T}_B - \langle \mathcal{T}_B \rangle)(\mathcal{T}_B - \langle \mathcal{T}_B \rangle)^* \right\rangle, \text{ where } \mathcal{T}_B = \langle E_2^{(-)} \mid \widehat{\mathbf{V}}_B \mid E_1 \rangle. \tag{2.7}$$

The main challenge in this approach is the right choice of the nondisturbed system. If, for instance, the diffraction of the primary or scattered waves can be neglected (the substrate is amorphous or we investigate the intensity scattered with such a scattering vector **Q** for which the diffraction condition in substrate is not fulfilled), a semi-infinite *amorphous* substrate is the nondisturbed system and the nanostructures represent the disturbance. The solution of the wave equation in the nondisturbed system is simple—it can be expressed by the Fresnel coefficients $r_{1,2}$ and $t_{1,2}$ of reflection and transmission [2,3,6]:

$$E_1(\mathbf{r}) = \begin{cases} e^{i\mathbf{K}_{1i}\cdot\mathbf{r}} + r_1 e^{i\mathbf{K}_{1R}\cdot\mathbf{r}} & \text{for } z > 0 \\ t_1 e^{i\mathbf{k}_{1T}\cdot\mathbf{r}} & \text{for } z < 0 \end{cases}, E_2^{(-)}(\mathbf{r}) = \begin{cases} e^{i\mathbf{K}_{2i}\cdot\mathbf{r}} + r_2^* e^{i\mathbf{K}_{2R}\cdot\mathbf{r}} & \text{for } z > 0 \\ t_2^* e^{i\mathbf{k}_{2T}^*\cdot\mathbf{r}} & \text{for } z < 0 \end{cases}. \tag{2.8}$$

The solutions $E_1, E_2^{(-)}$ are chosen so that $\mathbf{K}_{1i} \equiv \mathbf{K}_i$ is the wave vector of the actual incidence wave and $\mathbf{K}_{2i} \equiv \mathbf{K}_i$ is the actual emitted (measured) wave.

If the nanostructures are buried below the sample surface, the matrix element $\mathcal{T}_B$ is

$$\mathcal{T}_B = -K^2 t_1 t_2 \int_W d^3\mathbf{r}\Delta\chi(\mathbf{r})e^{-i\mathbf{Q}_T\cdot\mathbf{r}}, \mathbf{Q}_T = \mathbf{k}_{2T} - \mathbf{k}_{1T}, \tag{2.9}$$

where $\Delta\chi(\mathbf{r})$ is the difference in the polarizabilities of the whole system and the amorphous substrate and $\mathbf{Q}_T$ is the complex scattering vector corrected to refraction and absorption. The scattered intensity is therefore proportional to the factor $|t_1 t_2|^2$, giving rise to a maximum if $\alpha_i = \alpha_c$ or $\alpha_f = \alpha_c$ (the Yoneda wings). The intensity is usually expressed as a function of the scattering vector **Q**; however, it depends on the incidence angle α_i in addition because this angle enters the Yoneda term $|t_1 t_2|^2$ as well as $\mathbf{Q}_T$.

In the case of nanostructures deposited on the substrate surface, the formula for $\mathcal{T}_B$ is more complicated—it can be expressed as a *coherent superposition* of four scattering processes [2,3,6],

$$\mathcal{T}_B = -K^2 \sum_{\alpha=1,\dots,4} A_\alpha \int_W d^3\mathbf{r}\Delta\chi(\mathbf{r})e^{-i\mathbf{Q}_\alpha\cdot\mathbf{r}}, \tag{2.10}$$

where

$$A_1 = 1, A_2 = r_1, A_3 = r_2, A_4 = r_1 r_2$$

are the amplitudes of the individual scattering processes and

$$\mathbf{Q}_1 = \mathbf{K}_{2i} - \mathbf{K}_{1i}, \mathbf{Q}_2 = \mathbf{K}_{2i} - \mathbf{K}_{1R}, \mathbf{Q}_3 = \mathbf{K}_{2R} - \mathbf{K}_{1i}, \mathbf{Q}_4 = \mathbf{K}_{2R} - \mathbf{K}_{1R}$$

are their scattering vectors. The physical meaning of Equation 2.10 is illustrated in Figure 2.1; both the primary (A_1) and specularly reflected (A_2) beams are scattered from the nanostructures, and the scattered wave can also be specularly reflected from the substrate surface (A_3, A_4). The scattering processes included in the DWBA simulations are schematically sketched in Figure 2.1.

In both the kinematic approximation and the DWBA method, the main step in the simulation of the diffusely scattered intensity is to calculate the Fourier transformation

$$C(\mathbf{Q}) = \int_W d^3\mathbf{r}\Delta\chi(\mathbf{r})e^{-i\mathbf{Q}.\mathbf{r}}$$

of $\Delta\chi(\mathbf{r})$ occurring in the matrix element $\mathcal{T}_B$ (or $\mathcal{V}$ in the kinematic approximation) and the covariance

$$\mathrm{cov}(\mathcal{T}_B) = \left\langle (\mathcal{T}_B - \langle\mathcal{T}_B\rangle)(\mathcal{T}_B - \langle\mathcal{T}_B\rangle)^* \right\rangle = K^4 \sum_{\alpha,\beta} A_\alpha A_\beta^* C_{\alpha\beta}, C_{\alpha\beta} = \mathrm{cov}(C(\mathbf{Q}_\alpha), C(\mathbf{Q}_\beta)).$$

The explicit formulas for this covariance depend on the structure model of the nanostructures. In the following sections we deal with two basic experimental arrangements corresponding to two model classes: small-angle X-ray scattering (SAXS) and X-ray diffraction (XRD). In the remaining descriptions, we will always assume the DWBA method, since the kinematic approximation can be treated as DWBA, in which the nondisturbed system is vacuum.

In a SAXS experiment, we investigate the neighborhood near the reciprocal-space origin. Here the scattered intensity depends on the electron density averaged over the crystal unit cell (or over a region much larger than the distance of neighboring atoms in the case of an amorphous structure). Therefore, the scattered intensity is not affected by deformation in and around the nanostructures and depends only on the contrast of the refraction index $n = 1 - \delta = 1 + \chi_0/2$, where $\chi_0 = \langle\chi(\mathbf{r})\rangle$ is the averaged value of the polarizability.

If we assume, for the sake of simplicity, that the nanostructures (quantum wires, for instance) are chemically homogeneous, the disturbance $\Delta\chi(\mathbf{r})$ is

$$\Delta\chi(\mathbf{r}) = \Delta\chi_0 \sum_j \Omega_j(\mathbf{r} - \mathbf{R}_j), \tag{2.11}$$

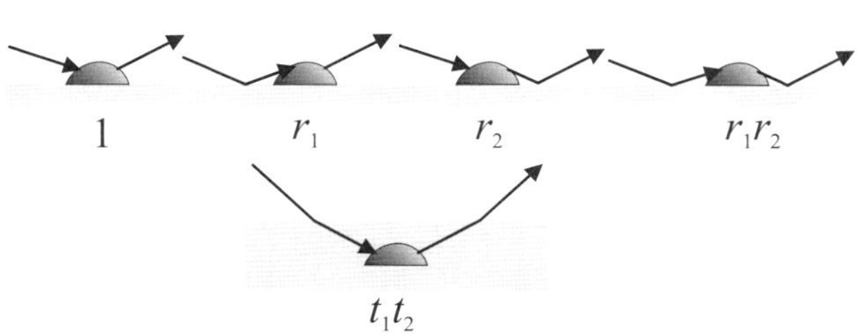

Figure 2.1. *Sketch of four scattering processes included in the DWBA simulation of diffuse scattering from freestanding objects (upper row) and a single scattering process in the case of buried objects (lower row).*

where $\Omega_j(\mathbf{r})$ is the shape function of the jth nano-object (unity in its volume and zero outside it), and $\mathbf{R}_j$ is its random position vector. If we assume that the size of a given nano-object is not statistically correlated with its position, we obtain [7]

$$C_{\alpha\beta}=\left[\left\langle F_0(\mathbf{Q}_\alpha)F_0(\mathbf{Q}_\beta)^*\right\rangle-\left\langle F_0(\mathbf{Q}_\alpha)\right\rangle\left\langle F_0(\mathbf{Q}_\beta)\right\rangle^*\right]G^{(1)}_{\alpha\beta}+\left\langle F_0(\mathbf{Q}_\alpha)\right\rangle\left\langle F_0(\mathbf{Q}_\beta)\right\rangle^* G^{(2)}_{\alpha\beta}, \quad (2.12)$$

where

$$G^{(1)}_{\alpha\beta}=\left\langle\sum_j \mathrm{e}^{-\mathrm{i}\mathbf{R}_j(\mathbf{Q}_\alpha-\mathbf{Q}^*_\beta)}\right\rangle, G^{(2)}_{\alpha\beta}=\left\langle\sum_{j,k} \mathrm{e}^{-\mathrm{i}(\mathbf{Q}_\alpha.\mathbf{R}_j-\mathbf{Q}^*_\beta.\mathbf{R}_k)}\right\rangle$$

are the correlation functions of the object positions and

$$F_0(\mathbf{Q})=\int d^3\mathbf{r}\,\Omega(\mathbf{r})\mathrm{e}^{-\mathrm{i}\mathbf{Q}.\mathbf{r}} \quad (2.13)$$

is the structure factor of a single nano-object. In Equation 2.12 the averaging is performed over random object sizes and random positions. In the kinematic approximation, $\mathbf{Q}_\alpha$ is the vacuum scattering vector and the first correlation function $G^{(1)}$ equals the number N of the irradiated objects.

In the diffraction arrangement—that is, if the scattering vector $\mathbf{Q}$ is close to a chosen nonzero reciprocal-lattice vector $\mathbf{h}$ (diffraction vector)—the description of X-ray scattering is complicated by the fact that the scattered wave is affected both by the contrast in the electron density (the chemical contrast) and by the elastic strain field around the nano-objects (the strain contrast). The main problem of the interpretation of experimental data is to distinguish these two contributions.

The polarizability of a deformed crystal can be expressed by means of a distorted Fourier series,

$$\chi(\mathbf{r})=\sum_{\mathbf{g}}\chi_{\mathbf{g}}(\mathbf{r})\mathrm{e}^{\mathrm{i}\mathbf{g}.(\mathbf{r}-\mathbf{u}(\mathbf{r}))}, \quad (2.14)$$

where $\mathbf{g}$ are the vectors of the lattice reciprocal to the nondisturbed lattice, $\mathbf{u}(\mathbf{r})$ is the displacement of the atom in position $\mathbf{r}$, and $\Delta\chi_{\mathbf{g}}(\mathbf{r})=\chi_{\mathbf{g}}(\mathbf{r})-\chi^{\mathrm{id}}_{\mathbf{g}}$ is the disturbance of the $\mathbf{g}$th Fourier coefficient of the crystal polarizability due to the chemical contrast in the nanostructures.

Since the scattering vector $\mathbf{Q}$ as well as all the scattering vectors $\mathbf{Q}_{\alpha}$, $\mathbf{Q}_T$ are close to the diffraction vector $\mathbf{h}$, the terms in the Fourier series in Equation 2.14 yield negligibly small contributions to $C(\mathbf{Q})$, except for the $\mathbf{h}$th term. Then the disturbance of the polarizability is

$$\Delta\chi(\mathbf{r})=\left[\chi^{\mathrm{id}}_{\mathbf{h}}\left(\mathrm{e}^{-\mathrm{i}\mathbf{h}.\mathbf{u}(\mathbf{r})}-1\right)+\Delta\chi_{\mathbf{h}}\sum_j\Omega(\mathbf{r}-\mathbf{R}_j)\mathrm{e}^{-\mathrm{i}\mathbf{h}.\mathbf{u}(\mathbf{r})}\right]\mathrm{e}^{\mathrm{i}\mathbf{h}.\mathbf{r}}. \quad (2.15)$$

An analytic calculation of the covariances $C_{\alpha\beta}$ is possible only for completely disordered object positions. If we assume that the displacement field at point $\mathbf{r}$ caused by a single nano-object placed in point $\mathbf{r}'$ depends only on $\mathbf{r}-\mathbf{r}'$ (this is valid for an object buried in a sufficient depth below the sample surface), the covariance $C_{\alpha\beta}$ is [8,2]

$$C_{\alpha\beta} = nS\left\langle F_{\mathbf{h}}^{\mathrm{FT}}(\mathbf{Q}_{\alpha} - \mathbf{h})\left[F_{\mathbf{h}}^{\mathrm{FT}}(\mathbf{Q}_{\beta} - \mathbf{h})\right]^{*}\right\rangle \int_{W} dz\mathrm{e}^{-\mathrm{i}z.(Q_{\alpha z} - Q_{\beta z})}. \qquad (2.16)$$

Here n denotes the density of nano-objects, the averaging $\langle\,\rangle$ is performed over the object sizes, S is the irradiated sample surface, the superscript FT means the Fourier transformation

$$F_{\mathbf{h}}^{\mathrm{FT}}(\mathbf{q}) = \int d^{3}(\mathbf{r} - \mathbf{r}')F_{\mathbf{h}}(\mathbf{r} - \mathbf{r}')\mathrm{e}^{-\mathrm{i}\mathbf{q}.(\mathbf{r}-\mathbf{r}')},$$

and [8]

$$F_{\mathbf{h}}(\mathbf{r} - \mathbf{r}') = \chi_{\mathbf{h}}^{\mathrm{id}}\left(\mathrm{e}^{-\mathrm{i}\mathbf{h}.\mathbf{v}(\mathbf{r}-\mathbf{r}')} - 1\right) + \Delta\chi_{\mathbf{h}}\Omega(\mathbf{r} - \mathbf{r}')\mathrm{e}^{-\mathrm{i}\mathbf{h}.\mathbf{v}(\mathbf{r}-\mathbf{r}')} \qquad (2.17)$$

is the structure factor of a single nano-object in diffraction $\mathbf{h}$. The vector $\mathbf{v}(\mathbf{r} - \mathbf{r}')$ is the displacement in point $\mathbf{r}$ due to an object in point $\mathbf{r}'$.

For nano-objects at the surface or close to it, the displacement at point $\mathbf{r}$ due to a single object at point $\mathbf{r}'$ may not depend on $\mathbf{r} - \mathbf{r}'$ but on $\mathbf{r}$ and $\mathbf{r}'$ independently. In particular, this effect is caused by the relaxation of internal stresses at the sample surface. Then the expression for $C_{\alpha\beta}$ is more complicated (see Revenant et al. [8] and the references therein):

$$C_{\alpha\beta} = n\int_{W} d^{3}\mathbf{r}'\left\langle \Phi_{\mathbf{h}}(\mathbf{Q}_{\alpha} - \mathbf{h};\mathbf{r}')\left[\Phi_{\mathbf{h}}(\mathbf{Q}_{\beta} - \mathbf{h};\mathbf{r}')\right]^{*}\right\rangle, \Phi_{\mathbf{h}}(\mathbf{q};\mathbf{r}') = \int_{W} d^{3}\mathbf{r}F_{\mathbf{h}}(\mathbf{r},\mathbf{r}')\mathrm{e}^{-\mathrm{i}\mathbf{q}.\mathbf{r}}. \qquad (2.18)$$

Equations 2.16 and 2.18 are valid if the positions of the nano-objects are completely random. For partially correlated positions, we can approximate the covariance $C_{\alpha\beta}$ by these formulas as well, replacing the density n of the objects by the expression $G_{\alpha\beta}^{(2)}/W$.

3 Models of Nanostructures

Models for samples with nanostructures should include a description of the shapes of the nano-objects, a statistical distribution of the shapes, and the correlation function of the positions of the nano-objects. In the case of XRD, the model should also include the displacement field around each nano-object.

The shape of the nano-object is described by the shape function $\Omega(\mathbf{r})$ introduced in the previous section. Calculating the differential cross section of small-angle scattering, we have to calculate the averages $\langle \Omega^{\mathrm{FT}}(\mathbf{Q})\rangle$ and $\langle |\Omega^{\mathrm{FT}}(\mathbf{Q})|^{2}\rangle$. This averaging can be performed directly assuming a simple object shape (an ellipsoid, for instance) and the gamma distribution of the object sizes. The resulting formulas are rather cumbersome; in Figure 2.2 we present the averages $|\langle \Omega^{\mathrm{FT}}(Q)\rangle|^{2}, \langle |\Omega^{\mathrm{FT}}(Q)|^{2}\rangle$ and their difference calculated for spherical nano-objects; their sizes are distributed according to the gamma distribution with the order $m = 30$. The root mean square (RMS) dispersion of the object radii R is $\langle R\rangle/m$.

For XRD, the calculation of the size averages is rather complicated because the object size affects the displacement field $\mathbf{v}(\mathbf{r}-\mathbf{r}')$ that has to be included in the averaging procedure, and the averaging can be carried out only numerically.

A short-range order (SRO) model is well suited for modeling the positions of nanostructures. In order to explain the important properties of the SRO model, we consider a simple situation in which the nano-objects are arranged along the x axis parallel to the sample surface. In this model we assume that the distance L_j between the objects j and $j-1$ is random with the mean value $\langle L\rangle$ and RMS dispersion $\sigma_{\|}$. For $j \neq k$, the distances L_j and L_k are statistically independent. Since the x coordinates of all scattering vectors $\mathbf{Q}$ and $\mathbf{Q}_\alpha$ are the same, the correlation function $G^{(2)}_{\alpha\beta}$ is

$$G^{(2)}_{\alpha\beta} \equiv G_x(Q_x) = \left\langle \sum_{j,k} \mathrm{e}^{-\mathrm{i}Q_x(X_j - X_k)} \right\rangle.$$

For a sufficiently large number N_x of irradiated objects along the x axis, this correlation function is [9,10]

$$G_x(Q_x) = N_x\left[1 + 2\,\mathrm{Re}\left(\frac{\xi_x}{1-\xi_x}\right)\right],\ \xi_x = \left\langle \mathrm{e}^{-\mathrm{i}Q_x L}\right\rangle = \mathrm{e}^{-\mathrm{i}Q_x\langle L\rangle}\mathrm{e}^{-(\sigma_{\|}Q_x)^2/2}. \tag{2.19}$$

Along the Q_x axis, this correlation function exhibits an almost periodical sequence of maxima. If the dispersion $\sigma_{\|}$ is sufficiently small, the position of the mth maximum is $Q_{xm} \approx 2\pi m/\langle L\rangle$, the width of the maxima is proportional to m^2, and its integrated intensity does not depend on m. This is

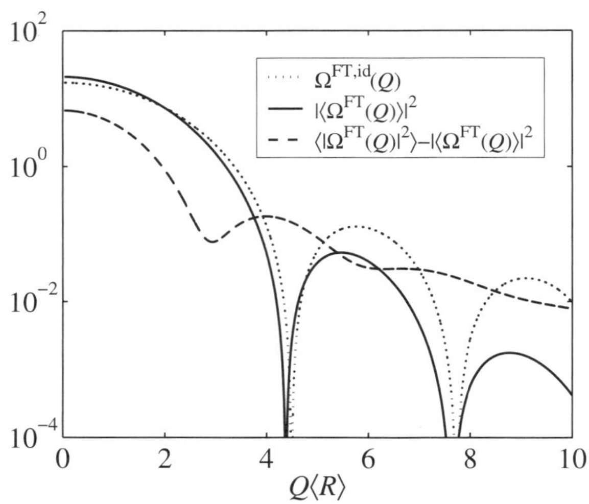

Figure 2.2. *The averages of the Fourier transformation of the shape function of spherical nano-objects. The sizes of the objects are distributed according to the gamma distribution with the order m = 30.*

demonstrated in Figure 2.3, which shows the one-dimensional correlation functions calculated within the SRO model with various values of $\sigma_{\|}$. The δ-like central maximum ($m = 0$) is omitted.

The one-dimensional SRO model can easily be extended to two dimensions, assuming that the X_{1j} (X_{2j}) coordinates of the object j defined with respect to a pair of (not necessarily orthogonal) axes $x_{1,2}$ obey the one-dimensional SRO model. Then the lateral correlation function of the object positions is

$$G_{\|}(\mathbf{Q}_{\|}) = \left\langle \sum_{j,k} \mathrm{e}^{-\mathrm{i}\mathbf{Q}_{\|}\cdot(\mathbf{X}_j - \mathbf{X}_k)} \right\rangle = N_1 N_2 \left[1 + 2\,\mathrm{Re}\left(\frac{\xi_1}{1-\xi_1}\right)\right]\left[1 + 2\,\mathrm{Re}\left(\frac{\xi_2}{1-\xi_2}\right)\right], \tag{2.20}$$

where $\mathbf{Q}_{\|} = (Q_1, Q_2)$ are the orthogonal projections of $\mathbf{Q}$ to these two axes and

$$\xi_p = \left\langle \mathrm{e}^{-\mathrm{i}Q_p L_p} \right\rangle = \mathrm{e}^{-\mathrm{i}Q_p \langle L_p \rangle} \mathrm{e}^{-(\sigma_{\|} Q_p)^2/2}, \; p = 1, 2\,.$$

In the case of a three-dimensional distribution of nano-objects, we restrict ourselves to nano-objects distributed at the interfaces of a periodic multilayer with the period T so that the Z coordinates of the objects are not random. We assume that the two-dimensional object arrays at different interfaces of the multilayer are randomly shifted laterally; the mutual shift of neighboring arrays has the mean value $\langle \mathbf{d} \rangle$ and the RMS dispersion $\sigma_{\|\perp}$. The correlation functions are

$$G^{(1)}(Q_z) = \frac{\mathrm{e}^{-2N_3 T \mathrm{Im} Q_z} - 1}{\mathrm{e}^{-2T\mathrm{Im}Q_z} - 1},\; G^{(2)}(\mathbf{Q}) = G_{\|}(\mathbf{Q}_{\|})\left\{G^{(1)}(Q_z) + 2\,\mathrm{Re}\left[G_{\|\perp}(\mathbf{Q})\right]\right\}, \tag{2.21}$$

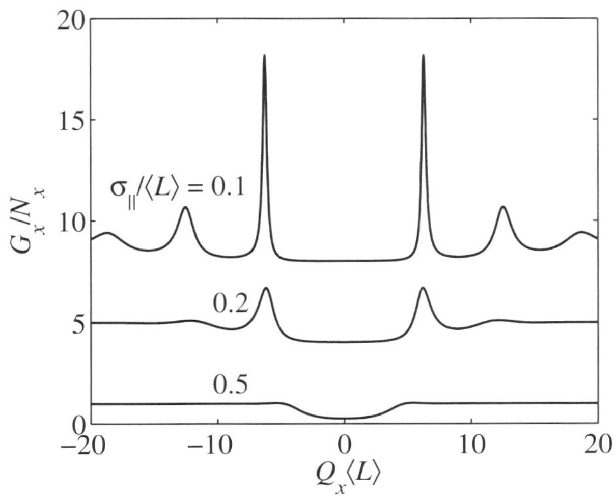

Figure 2.3. *One-dimensional correlation function G_x calculated within the SRO model for various RMS dispersions $\sigma_{\|}$ of the object distances. The curves are shifted vertically for clarity.*

where N_3 is the number of multilayer periods,

$$G_{\|\perp}(\mathbf{Q}) = \frac{1}{\left(e^{iTQ_z}\xi_3\right)^* - 1}\left[\frac{|e^{iTQ_z}\xi_3|^{2N_3} - |e^{iTQ_z}\xi_3|^2}{|e^{iTQ_z}\xi_3|^2 - 1} - \frac{\left(e^{iTQ_z}\xi_3\right)^{N_3} - e^{iTQ_z}\xi_3}{e^{iTQ_z}\xi_3 - 1}\right], \quad (2.22)$$

and

$$\xi_3 = e^{-i\mathbf{Q}_\| \langle \mathbf{d} \rangle} e^{-(\sigma_{\|\perp}|\mathbf{Q}_\||)^2/2}.$$

The SRO model cannot be used if the positions of the nano-object are predefined, for instance, by a patterning of the substrate surface before depositing the nano-objects. In this case, the surface pattern induces a long-range order (LRO) of the object positions [11]. Here we describe the one-dimensional LRO model; its generalization to two or three dimensions is analogous to the SRO model. We assume that the position X_j of the jth object deviates from its ideal position X_j^{id} by a random displacement U_j with a zero mean and RMS dispersion $\sigma_\|$; displacements $U_{j,k}$ are not correlated for $j \neq k$. The one-dimensional correlation function is

$$G_x(Q_x) = N_x\left(1 - |\xi|^2\right) + |\xi|^2 \left|\sum_j e^{-iQ_x X_j^{\mathrm{id}}}\right|^2, \; \xi = \left\langle e^{-iQ_x U}\right\rangle = e^{-\sigma_\|^2 Q_x^2/2}. \quad (2.23)$$

If the ideal positions of the objects along the x axis are periodic with period a, the sum Σ_j of the phase factor in the last equation gives rise to a periodic sequence of maxima in positions $2\pi m/a$ (m is an arbitrary integer) and equal widths $\Delta Q = 2\pi/(N_x a)$. The height of the maxima (as well as their integrated intensity) decreases with increasing $\sigma_\|$. From the comparison of the correlation functions of the SRO and LRO models, it follows that the SRO model can be recognized from the m^2 dependence of the maxima width. However, in the LRO model, the width of the maxima does not depend on the maximum order m, and it is inversely proportional to the (coherently irradiated) sample size or to a size of the domain, where the nano-objects obey the LRO model. Figure 2.4 shows examples of the calculated correlation function G_x in the LRO model for various values of $\sigma_\|$.

The simulation of the diffusely diffracted intensity also includes calculation of the displacement field **v** around and inside a single nano-object. If the objects are buried below the sample surface, the displacement field can be calculated analytically assuming the validity of the continuum elasticity approach [2,12,13]. For nano-objects at the free substrate surface, a suitable numerical method must be used (the finite-element method, for instance).

4 Numerical Examples

In this section we demonstrate the theory described above by a series of numerical simulations. We restrict ourselves to the grazing-incidence small-angle X-ray scattering (GISAXS) or grazing-incidence diffraction (GID) geometries sketched in Figure 2.5. In both geometries, we calculate the reciprocal-space distributions of the scattered intensity, keeping constant the incidence angle α_i of the primary radiation (and consequently its penetration depth).

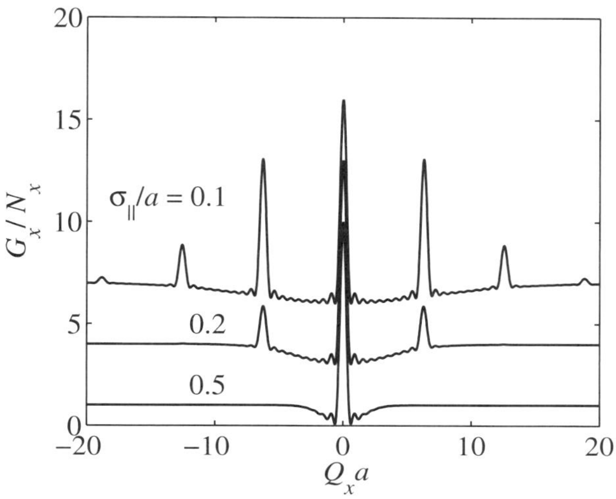

Figure 2.4. *One-dimensional correlation function G_x within the LRO model calculated for various RMS dispersions $\sigma_{||}$. The curves are shifted vertically for clarity.*

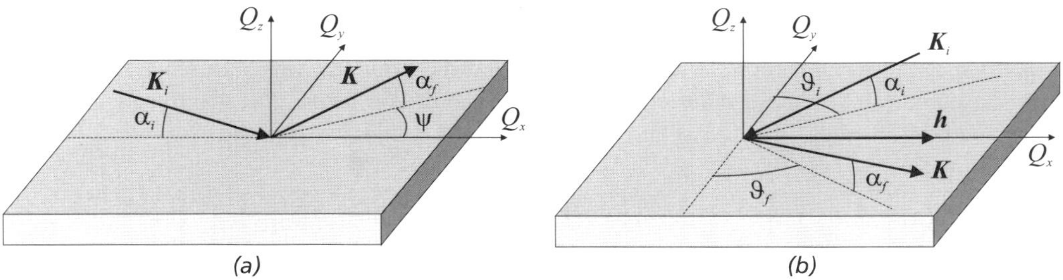

Figure 2.5. *GISAXS (left) and GID (right) geometries.*

The simulation of the GISAXS reciprocal-space maps will be performed on a system of spheres creating a three-dimensional disordered cubic lattice with the lattice parameter $a = 10$ nm. The spheres, with a mean radius of 3 nm, are embedded in a semi-infinite (amorphous) SiO_2 matrix, and the lateral distances of the spheres obey the SRO model. The contrast in the mean electron density of spheres and the host matrix (i.e., in the index of refraction) causes small-angle diffuse scattering that is simulated in the GISAXS geometry. Figure 2.6 presents the distribution of the intensity scattered in the Q_yQ_z plane calculated for various values of the RMS dispersions $\sigma_{||}$ and $\sigma_{||\perp}$ as well as for two values of the order m of gamma distribution of the sizes of the nanospheres.

From the figure, it follows that the widths of the satellite maxima in horizontal and vertical directions depend on $\sigma_{||}$ and $\sigma_{||\perp}$, respectively. The vertical widths are proportional to the square of the *lateral* order of the satellites. If we decrease the incidence angle below the critical angle of total external reflection, the primary radiation irradiates only a few vertical periods of the lattice, which results in the vertical broadening of the satellite maxima as well (see Figure 2.6f). This type of broadening differs from that caused by $\sigma_{\|\perp}$; the limited penetration depth of the primary radiation vertically broadens all satellite maxima in the same way. Decreasing the order m of the statistical distribution of the sphere sizes (i.e., increasing the RMS dispersions of the sizes), the diffuse background increases and the relative intensity of the satellite maxima decreases (Figure 2.6e).

For the simulation of GID, we used a system of randomly placed Ge pyramids; all pyramids are buried in the same depth of 30 nm in a semi-infinite Si(001) substrate. All the pyramids have the same sizes; their square bases have the edges of 50 nm length along ⟨100⟩, and the pyramid height was 25 nm. The side facets of the pyramids were {101}. Since we do not include any ordering of the pyramid positions, the scattered intensity is proportional only to the square of the structure factor $F_{\mathbf{h}}^{\mathrm{FT}}$ of a single pyramid.

In order to obtain the GID intensity, first we have to calculate the displacement field around a single pyramid. Using the continuum-mechanics approach described elsewhere [2,12,13] we obtain the lateral and vertical components of the displacement vector **v**; in Figure 2.7 we plotted the components of **v** in the xz plane (010). In the figure, the influence of the free sample surface is obvious, causing the surface relaxation of internal stresses. In a near-surface layer, both the lateral and vertical displacements are enhanced.

Using this displacement field, we have calculated the distribution of the scattered intensity in $q_x q_z$ plane (radial plane parallel to the diffraction vector **h**) and in the angular plane $q_y q_z$ (perpendicular to **h**). We used in-plane diffraction **h** = 400 and assumed a constant incidence angle of α_i = 0.3 deg; the results are plotted in Figure 2.8. In this figure, $q_{x,y,z}$ are the reciprocal-space coordinates with respect to the reciprocal lattice point **h**, that is, $\mathbf{q} = \mathbf{Q} - \mathbf{h}$.

The lattice mismatch compresses the Si lattice around the pyramids, which causes the shift of the intensity maximum in the radial q_x direction (i.e., parallel to **h**) (see Figure 2.8a). In the angular plane

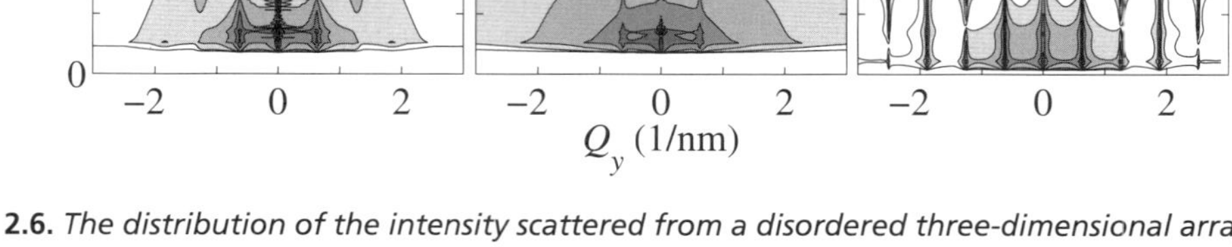

Figure 2.6. *The distribution of the intensity scattered from a disordered three-dimensional array of spheres embedded in a semi-infinite matrix. The calculations were performed for various RMS dispersions $\sigma_{\|}$, $\sigma_{\|\perp}$ for two orders m of the gamma distribution of the sizes and for two incidence angles α_i (a): $\sigma_{\|}$ = 0.01 nm, $\sigma_{\|\perp}$ = 0.01 nm, m = 100, α_i = 0.5 deg; (b): $\sigma_{\|}$ = 1 nm, $\sigma_{\|\perp}$ = 0.01 nm, m = 100, α_i = 0.5 deg; (c): $\sigma_{\|}$= 0.01 nm, $\sigma_{\|\perp}$ = 1 nm, m = 100, α_i = 0.5 deg; (d): $\sigma_{\|}$ = 1 nm, $\sigma_{\|\perp}$ = 1 nm, m = 100, α_i = 0.5 deg; (e): $\sigma_{\|}$ = 0.01 nm, $\sigma_{\|\perp}$ = 0.01 nm, m = 2, α_i = 0.5 deg; (f): $\sigma_{\|}$= 0.01 nm, $\sigma_{\|\perp}$ = 0.01 nm, m = 100, α_i = 0.1 deg. See the text for other parameters. The step of the iso-intensity contours is 10^1.*

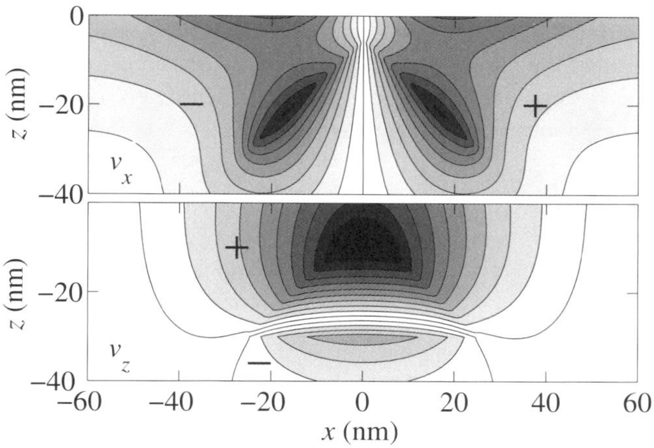

Figure 2.7. *The horizontal (upper panel) and vertical (lower panel) components of the elastic displacement vector around a single Ge pyramid in a semi-infinite Si substrate. The coordinate z = 0 corresponds to the free surface, the sign of the displacement is denoted by + and – signs, and the contour step is 0.02 nm and 0.04 nm for v_x and v_z, respectively.*

(Figure 2.8b), the intensity pattern is almost not affected by the displacement field, and it remains symmetric with respect to the q_z axis. This intensity distribution is very similar to a GISAXS pattern in the Q_yQ_z plane. In both planes, the intensity distribution exhibits distinct streaks perpendicular to the side facets of the pyramids. From the streak width, the size of the facets can be estimated. The period of the intensity fringes along q_z is inversely proportional to the depth of the pyramid base below the surface.

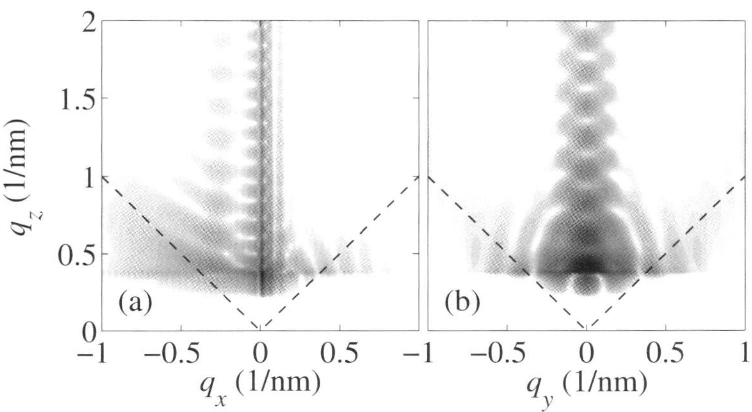

Figure 2.8. *The reciprocal-space distribution of the intensity scattered in the GID geometry from the Ge pyramids around the reciprocal lattice point* **h** *= 400 in plane (a) q_xq_z and (b) q_yq_z. The dashed lines are parallel to the {101} side facets of the pyramids.*

The intensity distribution along q_z exhibits a maximum at $q_z \approx 0.4$ nm^{-1}. This maximum is caused by the maximum of the Fresnel transmission coefficient t_2 occurring if the exit angle α_f equals α_c (the so-called Yoneda peak). For smaller q_z, the scattered intensity is very weak due to the absorption of the scattered wave in the substrate with the wave vector $\mathbf{k}_{2T}$.

5 Experimental Examples

In this section we present typical experimental results of X-ray scattering on nanostructures in order to demonstrate the capability of X-ray scattering for characterization of nanostructures.

5.1 *Small-Angle X-ray Scattering*

In the previous section we showed that the GISAXS method is sensitive to the positions and shapes of nanostructures. Since the scattered intensity is independent of elastic deformation, the chemical composition of the nanostructures can be deduced from the contrast in the mean electron density (i.e., from the index of refraction). However, such a determination is typically unreliable because it requires an absolute measurement of the scattered intensity.

Here we illustrate the strength of the GISAXS method with two examples: (1) an almost perfectly ordered three-dimensional lattice of PbSe quantum dots in PbSe/PbEuTe superlattices and (2) a fairly disordered three-dimensional distribution of Ge nanoclusters in an amorphous SiO_2 substrate.

5.1.1 *PbSe Quantum Dots*

We investigated a series of three periodic PbSe/PbEuTe superlattices grown by molecular beam epitaxy (MBE) on a thick (111)PbTe buffer deposited on a cleaved (111)BaF_2 substrate. In all samples there were sixty periods, and the thicknesses of the PbSe layers were nominally five monolayers. The thicknesses of the PbEuTe spacer layers were 35.5 nm in sample #1, 47.0 nm in sample #2, and 66.0 nm in sample #3. The Eu concentration in the PbEuTe layers was adjusted in order to achieve complete strain symmetry in which the mean lattice parameter of the period was the same as that of the PbTe buffer underneath. The work of Springholz et al. [14] provides details of the growth procedure. Cross section transmission electron micrographs (TEM) revealed that the PbSe dots were ordered vertically in sample #1. In sample #2 they created a highly ordered trigonal lattice, and no ordering of the dot positions was observed in sample #3. In the most interesting trigonal case, the dots create hexagonal arrays at the PbSe interfaces, the axes of the arrays are along $[\bar{1}\bar{1}2]$ and $[\bar{1}2\bar{1}]$.

The GISAXS measurements were performed at the European Synchrotron Radiation Facility (ESRF) Grenoble TROÏKA II beamline using a wavelength of λ = 1.5545 Å and a linear position-sensitive detector (PSD) oriented parallel to the sample surface. One PSD scan measures the dependence of the scattered intensity on the in-plane angle ψ defined in Figure 2.5. For constant α_i and α_f we took a series of the PSD scans for various azimuthal directions of the plane of incidence from which we reconstructed the distribution of the scattered intensity in the Q_xQ_y plane parallel to the sample surface. Figure 2.9 presents examples of the resulting intensity maps. Lateral intensity maxima in the map of sample #2 demonstrate the lateral ordering of the sample positions.

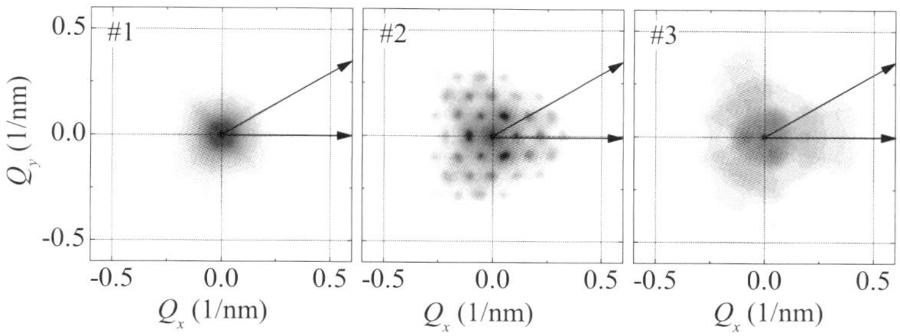

Figure 2.9. *GISAXS intensity distributions of PbSe/PbEuTe multilayer in reciprocal planes parallel to sample surface $Q_Z = 1\ nm^{-1}$ = constant . The arrows denote the directions in which the line scans have been extracted. The Q_x and Q_y axes are parallel to $[\bar{2}11]$ and $[\bar{1}01]$, respectively [15,16].*

From the measured maps we extracted linear scans depicted in Figure 2.10 and simulated these scans using the formulas in section 2. In agreement with TEM and atomic-force microscopy (AFM) observations, we assumed that a PbSe island has the form of a truncated pyramid with {100} side facets and a (111) top facet. The edges of the pyramid are rounded. The shape of the pyramid bases was determined from the fit of the experimental line scans to theory (see the insets in the figure).

The lateral size of the dot bases agrees qualitatively with AFM; for smaller PbEuTe thickness D the dots are largest, and for large D they are smallest. The rounded dot shape following from the GISAXS data indicates that the dot base has no sharp corners, probably the side "ribs" of the pyramids are truncated as well by tall {110} facets. These facets, however, could not be seen immediately in the X-ray data. Although the coincidence of fits and measurements is not perfect, the trends of the dot shapes are clearly visible [15,16].

The GISAXS method was not suitable for determination of the correlation of the vertical positions of the PbSe dots because the accessible range of Q_z was limited. To achieve a broader Q_z range, we measured reciprocal-space maps in the Q_xQ_z plane in symmetrical 111 diffraction (coplanar diffraction arrangement); the maps are depicted in Figure 2.11.

For smaller D, horizontal intensity sheets are visible corresponding to the vertical correlation of the dot positions. In the case of intermediate D, the intensity maxima correspond to a reciprocal trigonal "dot lattice"; for larger D, no sharp intensity maxima occur, indicating that the dot positions are uncorrelated [17].

From the widths of the intensity maxima, we determined the RMS dispersions $\sigma_{||}$ and $\sigma_{||\perp}$, assuming the validity of the SRO model (see section 3 for the definitions of $\sigma_{||}$ and $\sigma_{||\perp}$). From this model it follows that the lateral and vertical widths $\delta q_{||}$ and $\delta q_{\perp}$ of the satellites are

$$\delta q_{||} \approx \frac{(\sigma_{||} q_{||})^2}{\langle L \rangle}, \delta q_{\perp} \approx \frac{(\sigma_{||\perp} q_{||})^2}{D}, \tag{2.24}$$

where $\langle L \rangle$ is the mean lateral distance of the dots. In Figure 2.12 we plotted the dependences of the satellite widths on their positions $(q_{||}, q_{\perp})$. Since the lateral correlation of the dot positions in sample #1 is very weak, the lateral width of the satellites $\delta q_{||}$ is very large. On the other hand, the lateral dot

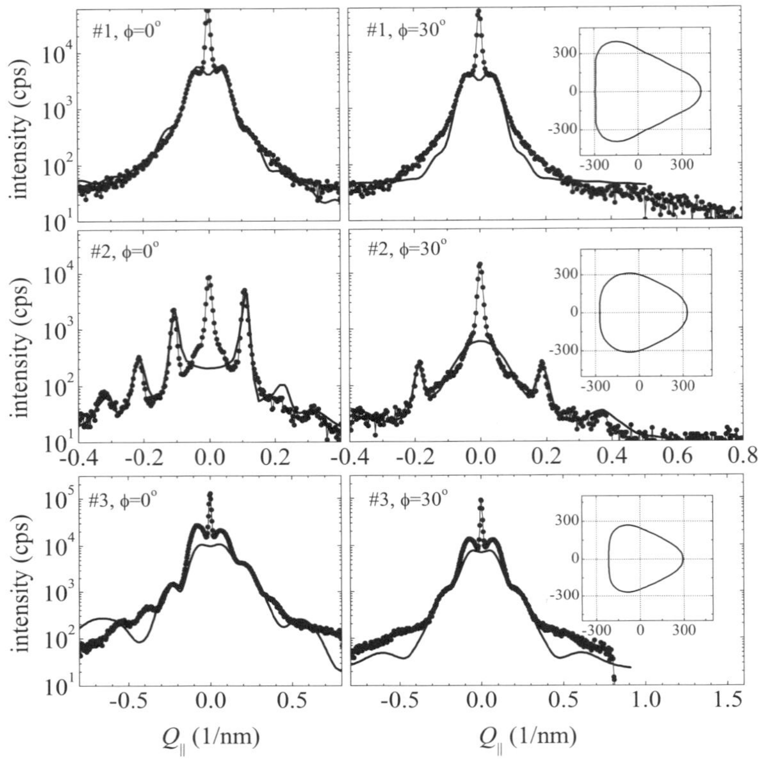

Figure 2.10. *The linear scans extracted from the GISAXS maps in Figure 2.9 along the arrows (points) and their simulations (lines). In the insets, the shapes of the pyramid bases are plotted following from the fit of the line scans to the theory [16].*

positions are correlated vertically, since $\delta q_{\perp}$ is proportional to $q_{\|}^{2}$; $\sigma_{\|\perp} = (5.2 \pm 0.5)$ nm. In sample #2 the lateral dot positions are correlated laterally with the RMS dispersion $\sigma_{\|\perp} = (6 \pm 2)$ nm. The vertical widths $\delta q_{\perp}$ of the maxima are independent of $q_{\|}$, and they are determined by the total thickness of the multilayer. In sample #3 the dot positions are completely noncorrelated so that the intensity map is determined only by the dot structure factor $F_{\mathbf{h}}^{\mathrm{FT}}(\mathbf{q})$ [16,17].

5.1.2 Ge Nanoclusters

Grazing-incidence small-angle scattering was used for the investigation of an ordered structure of Ge nanocrystals in an amorphous silica matrix prepared by magnetron sputtering deposition of twenty alternating (Ge + SiO_2)/SiO_2 bilayers with the period of 13 nm. In each Ge + SiO_2 layer, the Ge nanocrystals create a highly disordered two-dimensional hexagonal array; the lateral position of the arrays in different layers are correlated so that the nanocrystals are arranged in a trigonal lattice with the [111] direction perpendicular to the sample surface. Since the substrate is amorphous, the lattice of the nanocrystals consists of randomly rotated trigonal domains with common [111] axis;

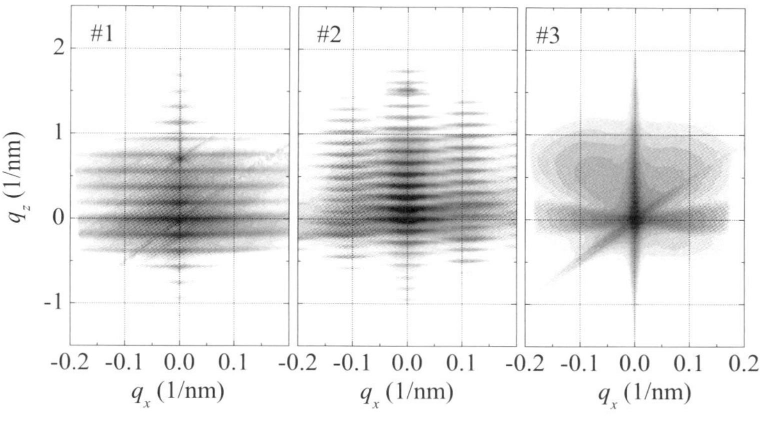

Figure 2.11. *The reciprocal-space maps of PbSe/PbEuTe multilayers taken in symmetric coplanar 111 diffraction; the lateral coordinate Q_x is parallel to $[\bar{2}11]$ [17].*

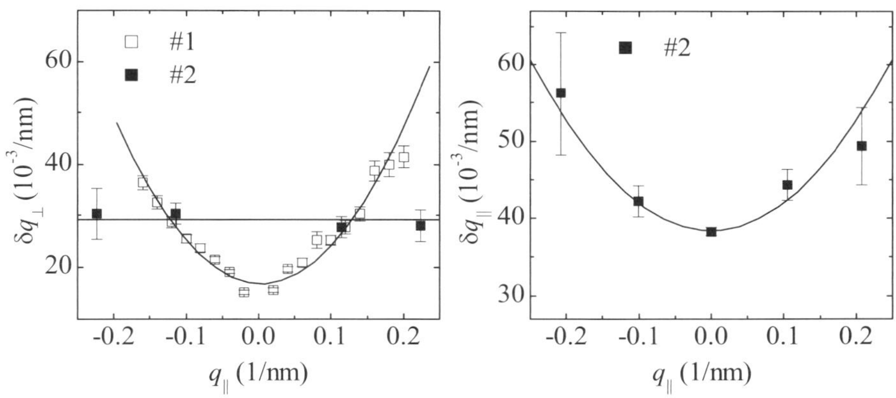

Figure 2.12. *The dependencies of the lateral ($\delta_\parallel$) and vertical ($\delta q_\perp$) widths of the satellite maxima of PbSe/PbEuTe superlattices on their lateral and vertical positions $q_\parallel$ and $q_\perp$ [17].*

within one domain, the SRO model can be applied. The GISAXS measurement has been carried out at ELETTRA Trieste (SAXS beamline) using the wavelength λ = 1.54 Å and the incidence angle α_i = 0.249 deg (i.e., close to the critical angle of total external reflection). The scattered intensity was measured by a two-dimensional CCD detector (1024 × 1024 pixels, pixel size 80 × 80 μm^2) parallel to the Q_yQ_z plane [18,19].

The detector signal was recalculated to reciprocal space (Figure 2.13, the left panel). The central vertical shadow in the figure is caused by a beamstop absorbing the specularly reflected beam. Two lateral maxima in the picture stem from the lateral ordering of the clusters, the vertical maxima correspond to the period of the multilayer, and the Yoneda peak is clearly visible at $Q_z \approx 0.35$ nm^{-1}.

From the position of the horizontal maxima, we estimated the mean distance between the clusters to $\langle L \rangle = (14.0 \pm 0.5)$ nm. We simulated the intensity pattern using the formulas presented above; the simulated image is shown in the right panel of Figure 2.13. In order to determine the parameters of the cluster array, we extracted horizontal and vertical linear scans at $Q_z = 0.5$ nm^{-1} and at $Q_y = 0.45$ nm^{-1}; the latter was chosen along the lateral intensity maximum (Figure 2.14). We fitted these scans to the theory, and from the fit we determined the mean cluster radius to be $\langle R \rangle = (4 \pm 1)$ nm and the RMS dispersions to be $\sigma_{\|} = (5 \pm 1)$ nm and $\sigma_{\|\perp} = (2.5 \pm 2.0)$ nm.

The value of $\sigma_{\|\perp}$ is statistically correlated to $\langle R \rangle$ so that both parameters can be simultaneously determined only with a larger statistical error. It was not possible to determine the RMS dispersion of the cluster sizes from the experimental data; in the simulations we assumed that all clusters have the same diameters. The correspondence of the measured and simulated scans is not perfect because diffuse scattering from the surface roughness was not included in the theoretical model.

5.2 X-ray Diffraction

In contrast to small-angle scattering (and to GISAXS, in particular), methods based on XRD are mainly sensitive to elastic strains in nanostructures and in the surrounding crystal matrix. In this section we present three representative applications of diffraction methods. In the first example, we show a "brute force" approach, in which the measured intensity distribution in reciprocal space is compared with numerical simulations based on a structure model. The numerical simulation consists of two steps. First, elastic strains in and around a nano-object are calculated from an assumed nano-object shape and chemical composition using a suitable numerical method (most frequently a finite-element method). Then these strains are used to calculate the scattered intensity; for this calculation either kinematic or DWBA formulas are used. The "brute force" method is time consuming and requires a preliminary characterization of the nano-object by TEM or AFM.

In further examples we present direct methods that allow (to some extent) determination of the object shape and chemical composition directly without a structure model. The iso-strain method uses the fact that for larger objects the intensity at a given point in reciprocal space stems from a given *iso-strain volume* with a given strain status. This method is suitable for the determination of the shape and strain of freestanding objects. The anomalous-scattering method uses two photon energies and a

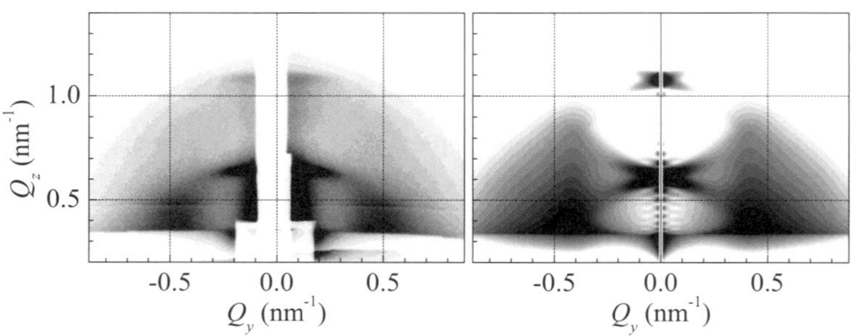

Figure 2.13. *Measured (left) and simulated (right) GISAXS intensity scattered from a system of Ge nanoclusters in the Q_yQ_z plane. (Courtesy of M. Buljan, Zagreb.)*

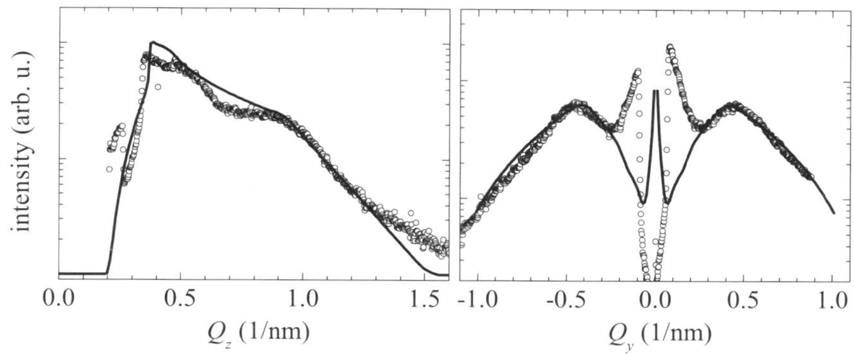

Figure 2.14. *Vertical (left) and horizontal (right) linear scans extracted from the intensity map in Figure 2.13 (points) and their theoretical fits (lines).*

steep energy dependence of the structure factor around a chosen absorption edge. This method can be used for the determination of the local chemical composition of nano-objects.

5.2.1 Brute Force Method

The "brute force" method is frequently used for the determination of nanostructure parameters in semiconductors. The method consists of the following steps:

1. Choose a suitable structure model based on TEM or AFM observations and on reasonable assumptions about the local chemical composition.
2. Simulate the elastic displacement field in the nanostructure volume and around it. Usually, numerical methods based on continuum elasticity are used, such as finite-element methods. For nanostructures buried below a flat surface, an analytical approach can be used.
3. Simulate the scattered intensity using kinematic or DWBA approaches according to the formulas in the previous section.
4. Compare the measured and simulated intensity distributions, refine or modify the structure model, and iterate.

Usually the method is very time consuming; however, it is rather general and can be used for any type of nanostructure.

As a state-of-the-art example of the brute force method, we show here measured and simulated reciprocal-space maps of intensity scattered from large SiGe islands deposited by liquid-phase epitaxy on a Si(001) surface [20]. From scanning electron micrographs (SEMs) the mean shape of the islands was determined; the islands have the form of a truncated pyramid with a square base (edges along [110] and $[\bar{1}10]$) and {111} side facets. The base width and island height were 130 and 65 nm, respectively; the nominal Ge content was 30%. The X-ray measurement was carried out at the BW2 beamline (HASYLAB) in standard coplanar diffraction geometry using symmetric 004 diffraction and a photon energy of 8 keV [21,22].

The measured intensity map is shown in Figure 2.15b. Several Ge distributions in the island volume were used for the simulations; the best comparison was achieved using the structure models in Figure 2.15a,c. For both structure models shown here, the correspondence of the measured and simulated maps is very good. From AFM observations it follows that the dispersion of the island sizes is very small, so no averaging over the island sizes was performed in the simulations. This is the reason why the simulated map well reproduces tiny fringes stemming from the side maxima of the Fourier transformation of the island structure factor $F_{\mathbf{h}}$. Any averaging over the island sizes would smear out these fringes. Narrow vertical stripes in Figure 2.15 stem from the correlation of the island sizes; these stripes correspond to the maxima of the correlation function.

5.2.2 Iso-Strain method

The methods described so far were based on a comparison of experimental data with numerical simulations of the scattered intensity based on a suitably chosen defect model. The iso-strain method introduced in this section is *direct*; that is, it can yield the mean size and strain in nanostructures without an a priori defined structure model [23,24]. As we show in this section, the method is applicable mainly for freestanding islands deposited onto a smooth substrate surface [25].

The main problem in a direct determination of the shape and strain profile in an island is to distinguish the contribution of the chemical contrast caused by $\chi_{\mathbf{h}}(\mathbf{r})$ from the strain contrast due to $\exp(-i\mathbf{h}.\mathbf{v}(\mathbf{r}))$. Both factors enter the Fourier transformation of the structure factor in Equation 2.17 in a product.

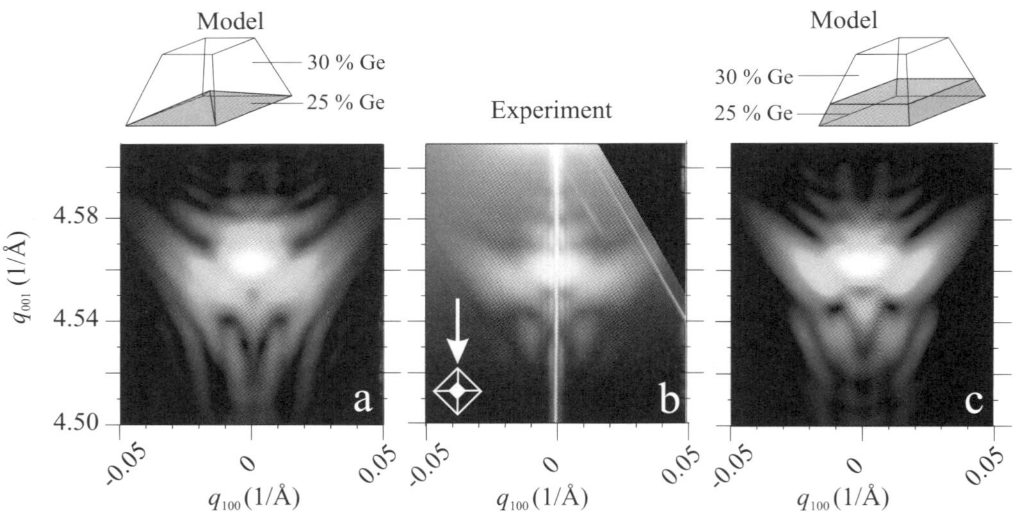

Figure 2.15. *Reciprocal-space maps from large SiGe islands deposited on a Si substrate by liquid-phase epitaxy. The map measured in symmetric 004 diffraction is in panel (b); panels (a) and (c) show the maps simulated using different structure models sketched above the maps [22]. (Courtesy of M. Hanke, Halle.)*

Let us define the *iso-strain volume* as the volume of the part of the island where the values ε_{jk} of the strain tensor can be assumed nearly constant. The whole island volume can be expressed as a set of these iso-strain volumes defined for various strains. Each iso-strain volume contributes to the total intensity distribution $I_{\mathbf{h}}(\mathbf{q})$ by a partial intensity distribution $I_{\mathbf{h}}^{(\varepsilon)}(\mathbf{q})$, with a maximum in the point $\mathbf{q}_{\max}^{(\varepsilon)} = -\hat{\varepsilon}\mathbf{h}$. For many semiconductor islands, the shape of the iso-strain volumes can be approximated by rather flat disks with diameters $L^{(\varepsilon)}$. Then the lateral width of the central peak in reciprocal space is inversely proportional to $L^{(\varepsilon)}: \Delta q^{(\varepsilon)} = 2\pi / L^{(\varepsilon)}$. The contributions of two iso-strain volumes with strains ε and $\varepsilon + \Delta\varepsilon$ can be distinguished if their contributions $I_{\mathbf{h}}^{(\varepsilon)}(\mathbf{q})$ and $I_{\mathbf{h}}^{(\varepsilon+\Delta\varepsilon)}(\mathbf{q})$ do not overlap, that is, if

$$\frac{2\pi}{L^{(\varepsilon)}} \ll h\Delta\varepsilon.$$

If, for instance, the strain gradient in the island is oriented mainly parallel to the vertical axis z, the shape contrast and the strain can be distinguished if $HLh\frac{d\varepsilon}{dz} \gg 1$, where H is the island height and L is the island width. If this condition is fulfilled, the integral in the Fourier transformation of the structure factor $F_{\mathbf{h}}(\mathbf{r}-\mathbf{r}')$ (Equation 2.17) can be calculated using the stationary-phase method. If we assume the GID geometry and neglect the shear terms of the strain tensor $\varepsilon_{jk}, j \neq k$ in the case of laterally homogeneous islands, the Fourier transformation of the structure factor can be approximated by

$$F_{\mathbf{h}}(\mathbf{q}) \approx \frac{2\pi}{h} L(x_0,z_0)\chi_{\mathbf{h}}(z_0)\mathrm{sinc}(L(x_0,z_0)q_y/2)\left[\left.\frac{d\varepsilon_{xx}}{dz}\right|_{x=x_0,z=z_0}\right]^{-1} \mathrm{e}^{-iq_z z_0}, \qquad (2.25)$$

Here sinc(x) = sin(x)/x, and we denoted (x_0,z_0) the coordinates of the stationary point in the xz plane

$$q_x + h\varepsilon_{xx}(x_0,z_0) = 0,\ q_z + hx_0 \left.\frac{d\varepsilon_{xx}}{dz}\right|_{x=x_0,z=z_0} = 0.$$

In Equation 2.25, $L(x_0,z_0)$ is the island width along y through the stationary point (x_0,z_0). The stationary phase method can be used only if *one* stationary point exists for a given $\mathbf{q}$. This is certainly the case if the in-plane strain varies monotonically with height.

Using Equation 2.25 we can reconstruct the mean shape of the islands from the measured distribution $I_{\mathbf{h}}(\mathbf{q})$. The contribution to the scattered intensity in a point $\mathbf{q} = (q_x,q_y,q_z)$ depends on the size of the iso-strain volume having the strain component $\varepsilon_{xx} = -q_x/h$. The intensity distribution along q_y is determined by the function $\mathrm{sinc}(L(x_0,z_0)q_y)$; from its width the size $L^{(\varepsilon)} \equiv L(x_0,z_0)$ can be determined.

The vertical position z_0 of the iso-strain volume affects the phase factor $\exp(-iq_z z_0)$. This factor affects the measured intensity for freestanding islands within DWBA. Since all four scattering processes included in Equation 2.10 are added coherently, the resulting scattered intensity depends on the phase $-iq_z z_0$. Thus from the intensity distribution $I_{\mathbf{h}}(\mathbf{q})$ measured in a three-dimensional region in reciprocal space, it is possible to determine the widths $L^{(\varepsilon)}$ of a certain iso-strain volume in y direction and the corresponding vertical position $z_0^{(\varepsilon)}$ of this volume. We can therefore reconstruct both the shape and the strain distribution in the island.

To demonstrate the iso-strain method, we show here GID measurements on a sample with free-standing SiGe islands grown by MBE on a Si(001) substrate. Six monolayers of Ge were deposited at a growth temperature of 600°C, resulting in dome-shaped islands with a density of about 4×10^9 cm^{-2} and a height and base diameter of about 13 nm and 110 nm, respectively. The measurements were carried out using the wavelength of 1.55 Å in two in-plane diffractions of 220 and 400. The scattered intensity was detected by a PSD perpendicular to the sample surface [25].

Figure 2.16 shows a sequence of angular scans along the Q_y axis perpendicular to the diffraction vector **h** collected in various radial positions Q_x (along **h**). In order to improve the counting statistics, the intensities have been integrated over the entrance PSD window, that is, along the exit angle α_f. From Figure 2.16 it is obvious that the diffraction maxima are broader in the angular direction for larger radial distance ΔQ_x; that is, the iso-strain volumes are laterally smaller for larger lateral lattice parameters. From the dependence of the angular width ΔQ_y of the intensity maxima in angular direction on the radial position Q_x in reciprocal space, we can reconstruct the dependence of the lateral size $L^{(\varepsilon)}$ of the iso-strain volume on the lateral strain ε_{xx} (i.e., on the degree of local elastic relaxation of the island lattice) (Figure 2.16, right panel). The dependencies determined for the 220 and 400 diffractions are almost identical.

The vertical position of the iso-strain area with a given lateral strain was determined from the α_f dependencies of the scattered intensity measured at $Q_y = 0$ for various Q_x (Figure 2.17). To improve the statistics, the intensity was integrated along Q_y. From the dependence of the intensity maxima on Q_x (denoted by empty circles), we determine the vertical position z_0 of the iso-strain area as a function of ε_{xx} (the right panel in Figure 2.17). Here the profiles resulting from the 220 and 400 diffractions are different above the relaxation degree of approximately 70%. Most likely, the relaxation degrees greater than 70% are not present in the island, and the intensities in these Q_x positions stem from the

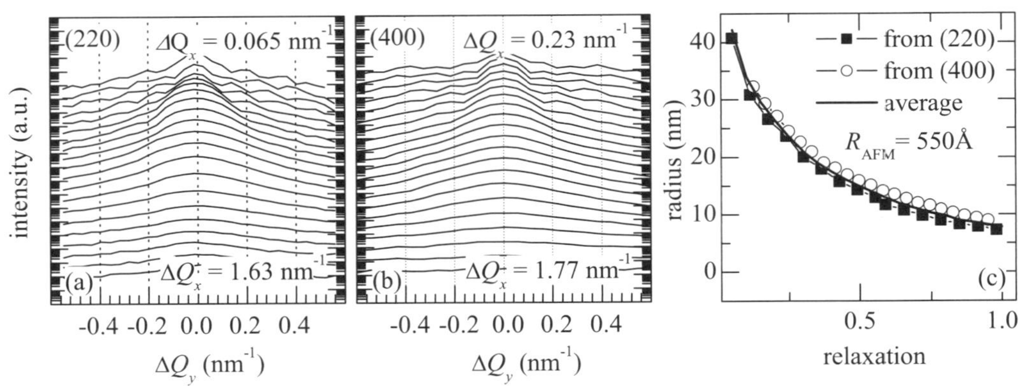

Figure 2.16. *A series of angular scans measured for various radial positions ΔQ_x in reciprocal space with respect to the substrate maximum using GID diffractions (a) 220 and (b) 400; freestanding SiGe islands on (001)Si. Panel (c) shows the dependence of the radius of the iso-strain area on the degree of elastic relaxation [25].*

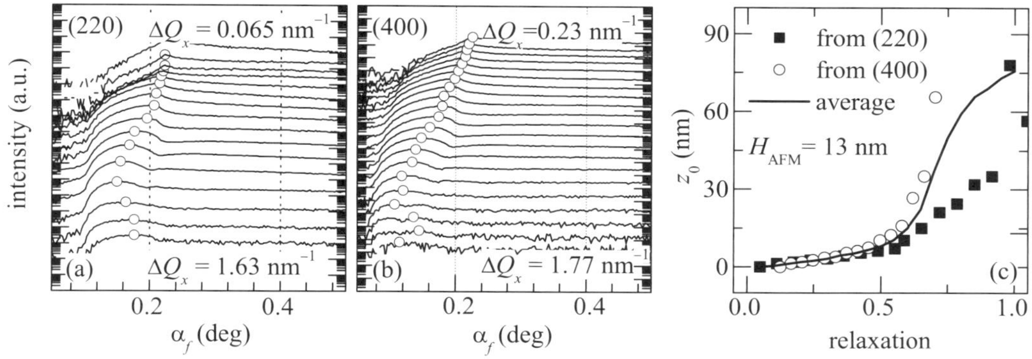

Figure 2.17. *A series of α_f scans measured for various radial positions Q_x in reciprocal space in GID diffractions (a) 220 and (b) 400; freestanding SiGe islands on (001)Si. Panel (c) shows the dependence of the vertical coordinate z_0 of the iso-strain area on the degree of elastic relaxation [25].*

iso-strain volumes with smaller relaxation. (The width of the intensity maximum of a circular iso-strain volume is the same in the Q_y and Q_z directions.) Taking this fact into account, we reconstructed the dependence of the in-plane strain $\varepsilon_{||}$ on z_0 and compared it with finite-element simulations based on the island shape determined by AFM. The resulting dependence is shown in Figure 2.18; both profiles compare very well.

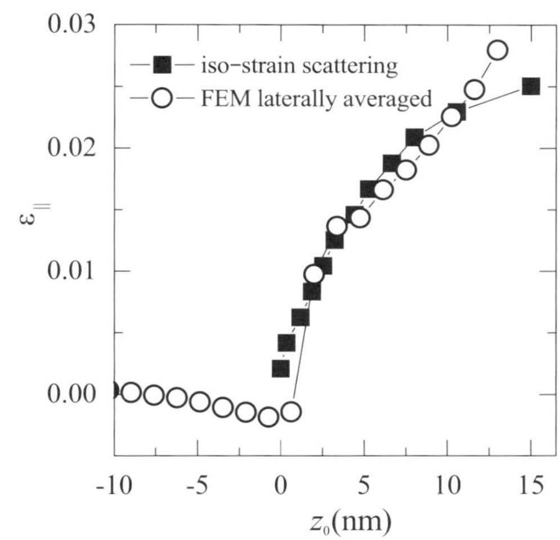

Figure 2.18. *Dependence of the lateral strain in the iso-strain volume on its vertical coordinate. The dot base is at $z_0 = 0$ [25].*

5.2.3 Anomalous Diffraction

Comparing the intensities of diffuse scattering for two photon energies close to and far away from the absorption edge of a given element, we can determine the local density of this element. This so-called *anomalous-diffraction method* is used in grazing-incidence geometry in combination with the iso-strain method [26–28]. We demonstrate this method on the example of freestanding SiGe quantum dots deposited on a Si(001) substrate; the growth parameters were similar to those of the SiGe quantum dots in the previous section.

The direct measurement of the Ge content in the SiGe quantum dots is based on the difference between the values $|\chi_{\mathbf{h}}|^2$ of Si and Ge; these values determine the diffracted intensity. Figure 2.19 shows the energy dependence of these values calculated at the 004 and 008 diffractions. The steep dependence of the polarizability around the GeK absorption edge is caused by the dispersion corrections $f'_{\mathrm{Ge}}(E)+if''_{\mathrm{Ge}}(E)$ that depend on the energy and (in first approximation) not on the scattering vector $\mathbf{Q}$. Increasing the length of the scattering vector and keeping E constant, we decrease the value of the classical Thomson scattering factor $f_{0\,\mathrm{Si,Ge}}(Q)$, whereas the dispersion correction remains constant. Therefore, for higher diffractions such as 800, the dispersion corrections become comparable to f_0 and therefore the sensitivity of the diffracted intensity to chemical composition is improved.

Figure 2.20 shows the intensity distribution along the Q_x axis parallel to the diffraction vector 800 (radial scan) measured for two energies: below the GeK absorption edge (11043 eV, crosses) and above it (11103 eV, dots). From the ratio of the intensities for a given radial position Q_x, the chemical composition in the iso-strain volume with a corresponding lateral lattice parameter is determined (squares). These data were combined with the iso-strain analysis explained in the previous section. From the angular Q_y scans measured for various radial positions Q_x, the lateral size of the iso-strain area was determined as a function of its vertical position (i.e., the mean shape of the dots). The vertical profile of the Ge concentration was obtained from the dependence of the vertical position z_0 of the iso-strain volume on the lateral lattice parameter. The shape of the quantum dots as well as the vertical profiles of the Ge concentration and the lateral lattice parameter are plotted in Figure 2.21.

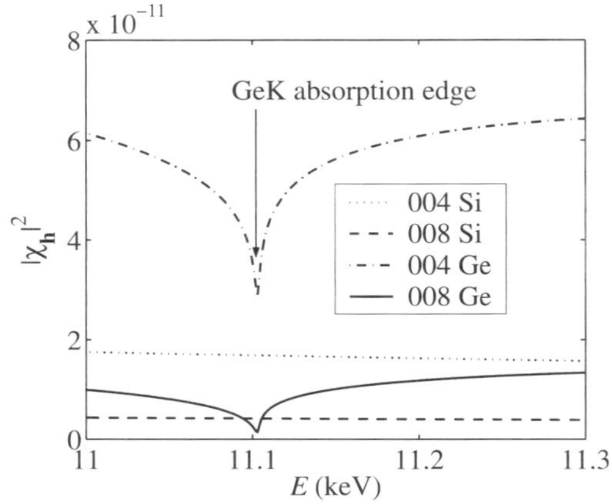

Figure 2.19. *The energy dependence of the squares of the polarizability coefficients (proportional to the diffracted intensity) calculated for pure Si and Ge and diffractions 004 and 008.*

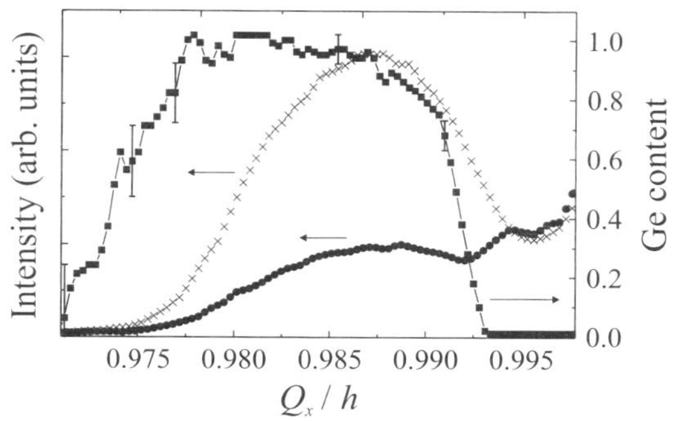

Figure 2.20. *The radial scans measured in GID diffraction 800 on a system of freestanding SiGe quantum dots using energies 11043 eV (crosses) and 11103 eV (dots), as well as the Ge concentration (squares) determined from the ration of these intensities [28]. (Courtesy of T. Schuelli, Grenoble.)*

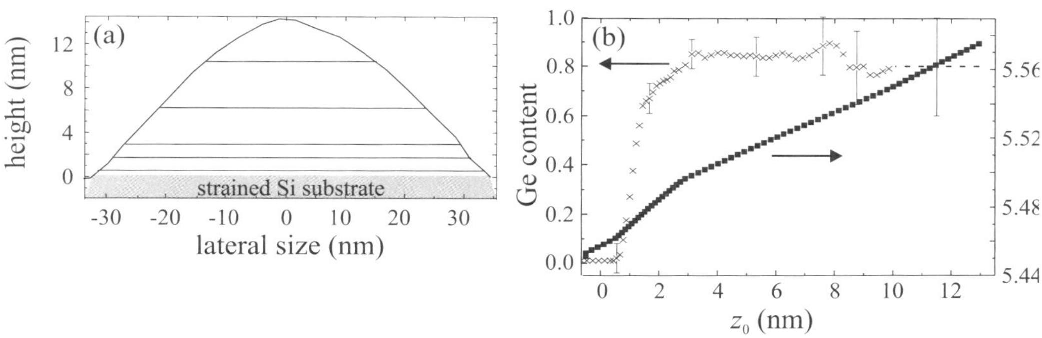

Figure 2.21. *The mean shape of the dot cross section determined by (a) the iso-strain method, the z profiles of the lateral lattice parameter (squares), and (b) the Ge concentration (crosses) [28]. (Courtesy of T. Schuelli, Grenoble.)*

6 New Developments

In the last decade, experimental methods outlined in the previous sections have been used for investigations of various types of semiconductor and metallic nanostructures, and they have been used almost routinely for the determination of shapes, strains, and local chemical composition. More recently, substantial progress has been achieved in X-ray instrumentation. In particular, new X-ray optical systems such as X-ray waveguides and X-ray lenses can produce very narrow X-ray beams (with the diameter below 100 nm), which makes it possible to investigate the structure of a *single* nano-object. If the coherence volume of the incoming X-ray beam is larger than the object size, the primary beam can be considered as fully coherent. Then the phase of the scattered beam can be recovered numerically using an iterative procedure, which makes it possible to fully reconstruct the shape and strains of a single nano-object [29,30]. The coherent-scattering phenomenon in the GISAXS arrangement has been successively applied for the investigation of metallic nanoparticle shapes [30] and semiconductor quantum dots [31]. Coherent XRD has also been used for the investigation of strains in metallic nanocrystals [32]. Strictly speaking, scattering from a single nano-object cannot be considered as diffuse scattering because the diffusely scattered intensity is usually defined by an average over a statistical ensemble. However, this method can yield much deeper insight into the object structure because the ensemble averaging smears out details in the reciprocal-space map.

Increasing the brilliance of X-ray sources also makes it possible to investigate diffuse X-ray scattering from nanostructures in situ during their growth [33]. For this purpose, X-ray beamlines are equipped with growth chambers (ion-sputtering chambers on MBE systems) and two-dimensional X-ray detectors. An MBE system attached to a bending-magnet beamline at ESRF was used for an in situ investigation of the growth of Ge quantum dots on Si, both in GISAXS and GID geometries [34,35]. In the former geometry, an evolution of the size of crystallographic facets of the dots has been studied, revealing a qualitative change in the dot shape at a certain Ge coverage. Using the latter method, it was found that this change in the dot shape is connected with an abrupt change in the elastic strain relaxation in the dot volumes.

Important progress has also been achieved in the combination of a "standard" diffuse scattering measurement with X-ray absorption spectroscopy. In this diffraction anomalous fine structure (DAFS) method, the intensity of diffuse scattering is measured at a given point **Q** in reciprocal space as a function of the energy of the primary photons around a suitably chosen absorption edge. From a series of energy scans performed at various points **Q**, it is possible to determine the local neighborhood of a given atom *in a given iso-strain area*. This method was successfully applied for the investigation of the local atomic structure of semiconductor quantum dots [36,37].

The theoretical description and experimental examples presented above consider only ideal nanostructures without any structural defects. Creation and kinetics of defects in quantum dots during their growth and evolution have been studied theoretically, and only a few works have been devoted to their experimental investigation. In order to study the defects in nanostructures by XRD, it is necessary to suppress the signal from the perfect nano-object lattice measuring diffuse scattering close to a forbidden reciprocal lattice point (200 in Si or Ge, for instance). These studies revealed the presence of atomically ordered domains in $Si_{0.5}Ge_{0.5}$ quantum dots [38,39] and small stacking faults or dislocation loops in large SiGe islands [40].

Substantial progress has been achieved in the last few years in the technology of nanostructure growth. Several emerging types of nano-objects include semiconductor "quantum sticks" grown by a vapor-liquid-solid epitaxy process on semiconductor substrates [41] and "rolled-up nanotubes" prepared by rolling an epitaxial bilayer sandwich [42]. These and other new nano-objects represent a challenge for characterization methods and further development of X-ray methods and instrumentation.

Acknowledgment

This work is a part of the research program MSM 0021620834 that is financed by the Ministry of Education of the Czech Republic.

References

1. I. K. Robinson and D. Tweet, *Rep. Prog. Phys.* 55 (1992): 599.
2. U. Pietsch, V. Holý, and T. Baumbach, *High-Resolution X-ray Scattering from Thin Films to Lateral Nanostructures* (Berlin, Heidelberg: Springer-Verlag, 2004).
3. J. Stangl, V. Holý, and G. Bauer, *Rev. Mod. Phys.* 76 (2004): 725.
4. M. Schmidbauer, *X-ray Diffuse Scattering from Self-Organized Mesoscopic Semiconductor Structures* (Berlin, Heidelberg: Springer-Verlag, 2004).
5. S. K. Sinha, E. B. Sirota, S. Garoff, and H. B. Stanley, *Phys. Rev. B* 38 (1988): 2297.
6. M. Rauscher, R. Paniago, T. H. Metzger, Z. Kovats, J. Domke, J. Peisl, H.-D. Pfannes, J. Schultze, and I. Eisele, *J. Appl. Phys.* 86 (1999): 6763.
7. C. Revenant, F. Leroy, R. Lazzari, G. Renaud, and C. R. Henry, *Phys. Rev. B* 69 (2004): 035411.
8. M. A. Krivoglaz, *X-ray and Neutron Diffraction in Nonideal Crystals* (Berlin, Heidelberg: Springer-Verlag, 1996).
9. P. R. Pukite, C. S. Lent, and P. I. Cohen, *Surf. Sci.* 161 (1985): 39.
10. F. Leroy, R. Lazzari, and G. Renaud, *Surf. Sci.* 601 (2007): 1915.
11. J. Wollschlaeger and M. Larsson, *Phys. Rev. B* 57 (1998): 14937.
12. V. Holý, G. Springholz, M. Pinczolits, and G. Bauer, *Phys. Rev. Lett.* 83 (1999): 356.
13. J. Novák, V. Holý, J. Stangl, T. Fromherz, Z. Zhong, G. Chen, G. Bauer, and B. Struth, *J. Appl. Phys.* 98 (2005): 073517.
14. G. Springholz, V. Holý, M. Pinczolits, and G. Bauer, *Science* 282 (1998): 734.
15. V. Holý, J. Stangl, G. Springholz, M. Pinczolits, G. Bauer, I. Kegel, and T. H. Metzger, *Physica B* 283 (2000): 65.
16. V. Holý, J. Stangl, G. Springholz, M. Pinczolits, and G. Bauer, "Processes Of Self-Organization During Epitaxial Growth of Semiconductor Superlattices—An X-ray Scattering Study" in *Morphological and Compositional Evolution of Heteroepitaxial Semiconductor Thin Films*, ed. J. Mirecki Millunchick, A. L. Barabasi, N. A Modine, and E. D. Jones, Symposium proceedings of Material Res. Society, vol. 618 (2001): 161.
17. V. Holý, J. Stangl, G. Springholz, M. Pinczolits, and G. Bauer, *J. Phys. D* 34 (2001): A1.
18. U. V. Desnica, P. Dubcek, K. Salamon, I. D. Desnica-Frankovic, M. Buljan, S. Bernstorff, U. Serincan, and R. Turan, *Nucl. Inst. and Meth. in Phys. Res. B* 238 (2005): 272.
19. U. V. Desnica, M. Buljan, P. Dubcek, Z. Siketic, I. Bogdanovic Radovic, S. Bernstorff, U. Serincan, and R. Turan, *Nucl. Inst. and Meth. in Phys. Res. B* 249 (2006): 843.
20. W. Dorsch, H. P. Strunk, H. Wawra, and G. Wagner, *Appl. Phys. Lett.* 72 (1998): 179.
21. T. Wiebach, M. Schmidbauer, M. Hanke, H. Raidt, and R. Köhler, *Phys. Rev. B* 61 (2000): 5571.
22. M. Hanke, M. Schmidbauer, D. Grigoriev, H. Raidt, P. Schäfer, R. Köhler, A.-K. Gerlitzke, and H. Wawra, *Phys. Rev. B* 69 (2004): 075317.
23. I. Kegel, T. H. Metzger, A. Lorke, J. Peisl, J. Stangl, G. Bauer, J. M. Garcìa, and P. M. Petroff, *Phys. Rev. Lett.* 85 (2000): 1694.
24. I. Kegel, T. H. Metzger, A. Lorke, J. Peisl, J. Stangl, G. Bauer, K. Nordlund, W. V. Schoenfeld, and P. M. Petroff, *Phys. Rev. B* 63 (2001): 035318.
25. J. Stangl, T. U. Schülli, A. Hesse, G. Bauer, and V. Holý,, "Indirect and Direct X-ray Scattering Methods for the Study of Epitaxial Self-assembled Quantum Dots in *Quantum Dots: Fundamentals, Applications and Frontiers*, ed. B. A Joyce, P. C. Kelires, A. G. Naumovets, and D. D. Vvedensky, NATO Sci. Series II, vol. 190 (Dordrecht: Springer, 2005), 183.
26. R. Magalhaes-Paniago, G. Medeiros-Ribeiro, A. Malachias, S. Kycia, T. I. Kamins, and R. Stan Williams, *Phys. Rev. B* 66 (2002): 245312.
27. T. U. Schülli, M. Sztucki, V. Chamard, and T. H. Metzger, *Appl. Phys. Lett.* 81 (2002): 448.
28. T. U. Schülli, J. Stangl, Z. Zhong, R. T. Lechner, M. Sztucki, T. H. Metzger, and G. Bauer, *Phys. Rev. Lett.* 90 (2003): 066105.
29. J. R. Fienup, *Opt. Lett.* 3 (1982): 27.

30. I. K. Robinson, I. A. Vartanyants, G. J. Williams, M. A. Pfeifer, and J. A. Pitney, *Phys. Rev. Lett.* 87 (2001): 195505.
31. I. A. Vartanyants, I. K. Robinson, J. D. Onken, M. A. Pfeifer, G. J. Williams, F. Pfeiffer, T. H. Metzger, Z. Zhong, and G. Bauer, *Phys. Rev. B* 71 (2005): 245302.
32. M. A. Pfeifer, G. J. Williams, I. A. Vartanyants, R. Harder, and I. K. Robinson, *Nature* 442 (2006): 63.
33. G. Renaud, R. Lazzari, C. Revenant, A. Barbier, M. Noblet, O. Ulrich, F. Leroy, J. Jupille, Y. Borensztein, C. R. Henry et al., *Science* 300 (2003): 1416.
34. M.-I. Richard, T. U. Schülli, E. Wintersberger, G. Renaud, and G. Bauer, *Thin Solid Films* 508 (2006): 213.
35. T. U. Schülli, M.-I. Richard, G. Renaud, V. Favre-Nicolin, E. Wintersberger, and G. Bauer, *Appl. Phys. Lett.* 89 (2006): 143114.
36. M. G. Proietti, H. Revevier, J. L. Hodeau, J. Garcìa, J. F. Bérar, and P. Wolfers, *Phys. Rev. B* 59 (1999): 5479.
37. S. Grenier, M. G. Proietti, H. Renevier, L. González, J. M. Garcìa, and J. Garcìa, *Europhys. Lett.* 57 (2002): 499.
38. A. Malachias, T. U. Schülli, G. Medeiros-Ribeiro, L. G. Cancado, M. Stoffel, O. G. Schmidt, T. H. Metzger, and R. Magalhaes-Paniago, *Phys. Rev. B* 72 (2005): 165315.
39. A. Malachias, T. H. Metzger, M. Stoffel, O. G. Schmidt, and V. Holý, *Thin Solid Films* 515 (2007): 5587.
40. M.-I. Richard (forthcoming 2007).
41. J. Johansson, L. S. Karlsson, C. P. T. Svensson, T. Martensson, B. A. Wacaser, K. Deppert, L. Samuelson, and W. Seifert, *Nature Materials* 5 (2006): 574.
42. O. G. Schmidt and K. Eberl, *Nature* 410 (2001): 168.

3

X-ray Scattering from Antiphase and Orientation Boundaries in III–V Semiconductor Ternary Alloy Films

Jianhua Li

Rigaku Americas Corporation, 9009 New Trails Dr., The Woodlands, TX 77381, USA

Abstract

This chapter presents a review of our recent X-ray scattering study of partially ordered III–V semiconductor ternary alloy films. X-ray scattering of the semiconductor alloy films often exhibits extended zigzag or streaky patterns in reciprocal space. Random size distributions of both orientation and antiphase domain boundaries, along with atomic displacements due to bond length difference between the constitutive binaries (III_1–V and III_2–V or III–V_1 and III–V_2), are taken into account to interpret the observed scattering phenomena of these alloy films.

1 Introduction

Spontaneous partial atomic ordering in epitaxial semiconductor alloy films is generally believed to result from surface reconstruction during film growth [1–4]. Change of crystal symmetry, as well as the formation of defects (such as orientation and antiphase boundaries and atomic displacements) associated with the partial atomic ordering, have a strong impact on the electronic and optical properties of the alloy semiconductors [5–8]. Understanding the atomic structure and defects of the epitaxial semiconductor alloy films is thus of great interest from both a technological and a fundamental point of view.

X-ray scattering has long been used for quantitative analysis of uniformly ordered bulk metallic materials [9]. However, these approaches cannot be applied to address the ordering of epitaxial alloy films due to phenomena such as the occurrence of preferably oriented ordering domains and the coexistence of both long-range and short-range ordering. In addition, unlike metallic alloys, the atoms in a semiconductor alloy are covalently bonded. Different bond lengths in the alloy films, along with the constraint exerted by the substrate lattices, can result in significant deviation of the local atomic structure from that of an ideal crystal [10,11].

In the sections below, X-ray scattering results from some partially ordered III–V semiconductor alloy films with either mixed group III elements or mixed group V elements grown by metal-organic vapor-phase epitaxy (MOVPE) and molecular-beam epitaxy (MBE) techniques will be discussed.

2 X-ray Scattering from $Ga_{0.5}In_{0.5}P$ Alloy Films on GaAs (001) Substrates

An important type of long-range ordering observed in $Ga_{0.5}In_{0.5}P$ alloy films on (001) substrates is the so-called CuPt type; in this case, the mixed group III elements, Ga and In, prefer alternate {111} lattice planes, as shown in Figure 3.1. Because the ordering is not caused by bulk diffusion (as a matter of fact, such an ordered structure is thermodynamically unstable in bulk alloys [12]), but rather by stress induced by the surface dimmers that form reconstructed surfaces, ordering can only occur in two of the four equivalent {111} lattice planes [3]. Under normal growth conditions of $Ga_xIn_{1-x}P$, that is, medium growth temperature (~510°C) and medium phosphorus partial pressure, the P-terminated surface always forms (2 × 1) reconstruction with the P-dimmers oriented along the $[\bar{1}10]$ direction, as shown in Figure 3.2. These dimmers exert stress on the underlying atomic layers, leading to subsurface diffusion. Consequently, ordering occurs only on $(\bar{1}11)$ and $(1\bar{1}1)$ lattice planes, forming the so-called CuPt-B ordering. The ordering on (111) and $(\bar{1}\bar{1}\bar{1})$ lattice planes, called CuPt-A ordering, occurs only if the films are grown under unusual conditions, as will be discussed later.

X-ray scattering intensities in reciprocal space along both the $[\bar{1}11]$ and $[1\bar{1}1]$ directions were measured on a standard four-circle diffractometer on the beamline X14A of the National Synchrotron

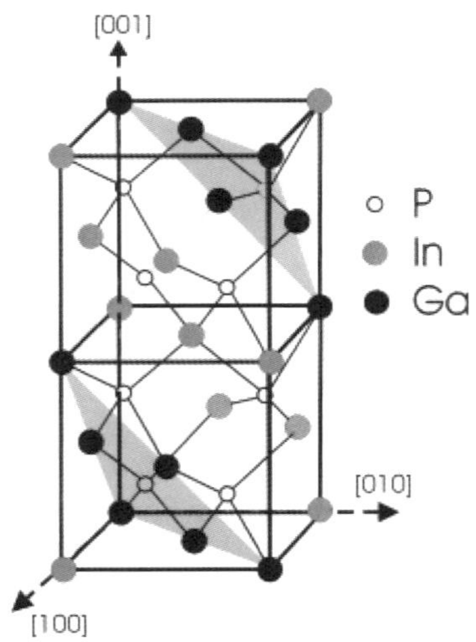

Figure 3.1. *CuPt-type atomic ordering in $Ga_{0.5}In_{0.5}P$ alloy films.*

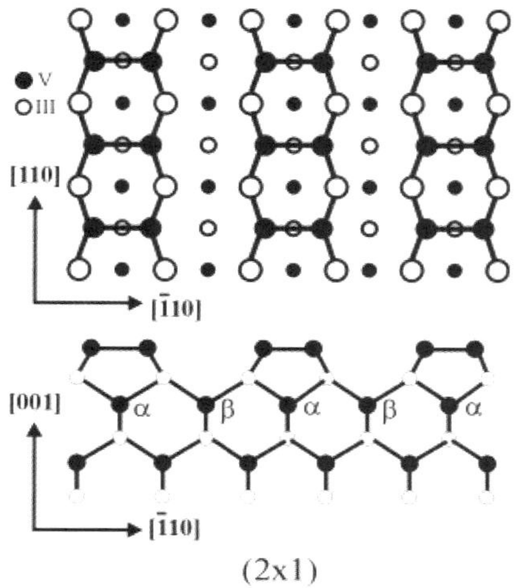

Figure 3.2. *(2 × 1) surface reconstruction of GaInP film during growth.*

Light Source (NSLS) at the Brookhaven National Laboratory (BNL) with an X-ray energy of 8.0478 keV. Most of our measurements were done in the reciprocal space plane sketched in Figure 3.3. Special care was taken to eliminate the higher harmonics that may interfere with the scattering of the ordered alloy structures [13].

A typical scattering pattern of CuPt-B ordered $Ga_{0.5}In_{0.5}P$ alloy films is shown in Figure 3.4a. Diffuse scattered intensities spanning $l = 1$ to $l = 3$ form inclined streaks passing through the $(\frac{\bar{h}}{2},\frac{k}{2},\frac{l}{2})$ and $(\frac{h}{2},\frac{\bar{k}}{2},\frac{l}{2})$ (h,k,l = 3 and 5) reciprocal points, where sharp long-range CuPt-B ordering peaks would

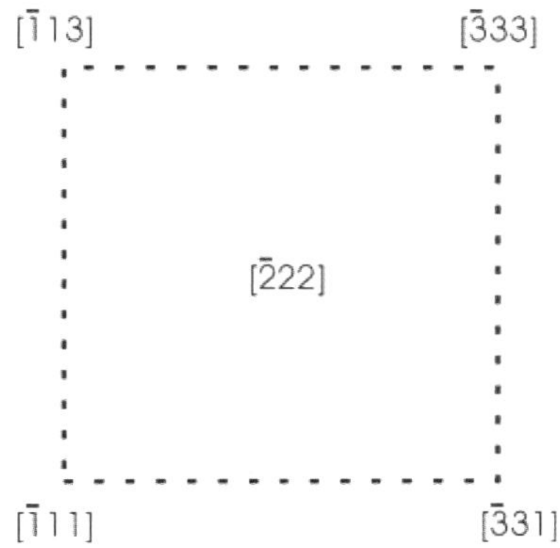

Figure 3.3. *Sketch of the reciprocal plane in which most of our measurements were done.*

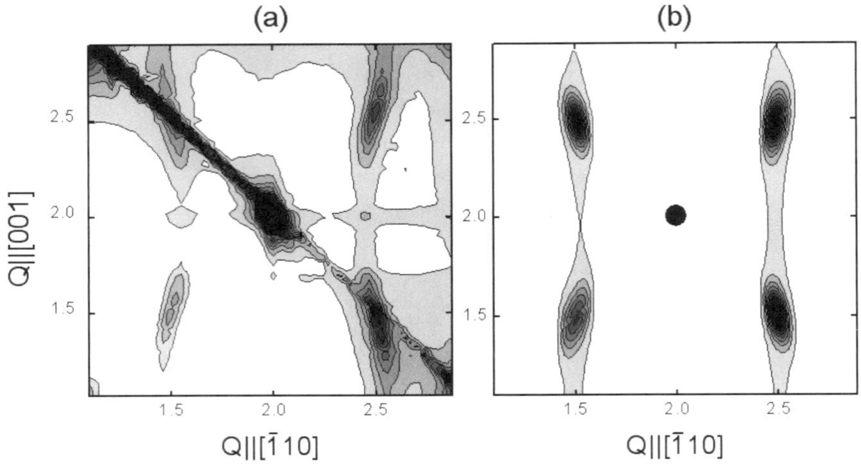

Figure 3.4. *A typical X-ray scattering pattern of partially CuPt-B ordered $Ga_{0.5}In_{0.5}P$ alloy films. (a) and (b) are the measured and simulated data, respectively. (With kind permission of Springer Science and Business Media.)*

appear had such a structure formed during growth. The streaks around $(\frac{\bar{3}}{2},\frac{3}{2},\frac{3}{2})$ and $(\frac{\bar{5}}{2},\frac{5}{2},\frac{5}{2})$ correspond to the $(\bar{1}11)$ lattice planes, whereas the streaks around $(\frac{3}{2},\frac{\bar{3}}{2},\frac{5}{2})$ and $(\frac{5}{2},\frac{\bar{5}}{2},\frac{3}{2})$ correspond to the $(1\bar{1}1)$ lattice planes. This clearly indicates that two orientation domains coexist in the film. Notice that the inclination of the intensity streaks arising from different orientation domains is quite regular, forming a pronounced zigzag pattern.

Moreover, it is clear that the intensity of the diffuse streak around $(\frac{\bar{5}}{2},\frac{5}{2},\frac{5}{2})$ is substantially higher than that of the streak around $(\frac{\bar{3}}{2},\frac{3}{2},\frac{3}{2})$. This is quite unusual because both the decreasing atomic form factor at a higher scattering vector and the thermal effect normally result in a weaker scattering intensity at a higher scattering vector.

Besides these interesting scattering features related to partial atomic ordering, a diagonal streak running from $(\bar{1}13)$ to $(\bar{3}31)$, due to stacking faults [14], is also observed. However, discussion of the stacking faults is beyond our scope here and thus will be ignored.

To account for the scattering phenomenon shown in Figure 3.4, we consider three types of possible defects in the film, namely, antiphase boundaries, orientation boundaries, and atomic displacements due to different Ga-P and In-P bond lengths. Figure 3.5a shows schematically an in-plane antiphase boundary in each orientation domain. In Figure 3.5b, out-of-plane orientation boundaries are shown (dashed lines). Notice that the distance between the two orientation boundaries affects the relative phase of the domains at the bottom and on the top; they could be either in phase or out of phase, as indicated by the inclined dotted lines.

With this in mind, the structure factor of a partially ordered structure with N orientation domains can be written as

$$F(\mathbf{Q}) = \sum_{j=1}^{N} \left[F_A(\mathbf{Q}) \frac{\alpha^{n_{Aj}} - 1}{\alpha - 1} + F_B(\mathbf{Q}) \beta \alpha^{n_{Aj}-1} \frac{\beta^{n_{Bj}-1}}{\beta - 1} \right] \cdot (\alpha\beta)^{j-1} \prod_{k=1}^{j-1} \alpha^{n_{Ak}-1} \beta^{n_{Bk}-1}, \tag{3.1}$$

where F_A and F_B are the structure factors of the two possible orientation domains, each consisting of n_{Aj} and n_{Bj} atomic layers. α and β are the phase shifts of the neighboring atomic layers in each domain.

As seen in Figure 3.5, the antiphase boundaries result in a phase shift. The collective effect of M antiphase boundaries on the structure factor F_S ($S = A,B$) can be written as

$$F_S(\mathbf{Q}) = \sum_{p=1}^{M} F_{0S}^{(p)} \phi^{p-1}, S = A, B, \tag{3.2}$$

where ϕ represents the phase shift at an antiphase boundary and F_{0S} the structure factor of a domain between two consecutive boundaries.

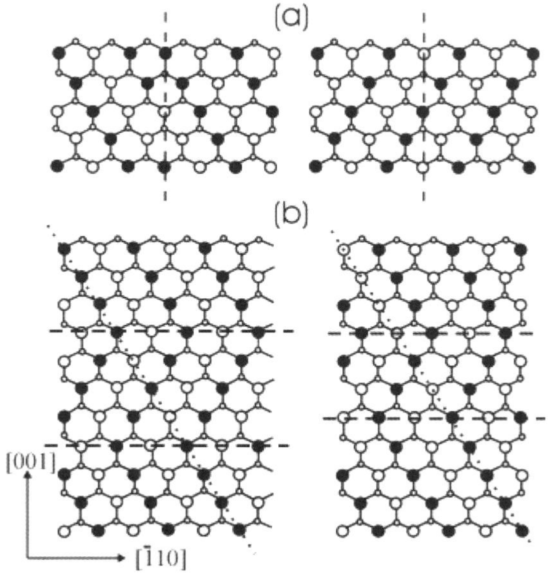

Figure 3.5. *(a) Antiphase boundaries, marked by the dashed lines, in two differently oriented domains. (b) Orientation domain boundaries (marked by horizontal dashed lines) and the effect of the domain thickness on the relative phase of the domains of the same orientation: they could be either in phase (left) or out of phase (right).*

To calculate F_{0S}, atomic displacements due to mixed bond lengths in the alloy film need to be considered. Bulk semiconductor alloys $A_xB_{1-x}C$ are known to obey Vegard's law quite accurately: the overall lattice constant is the average of the bulk AC and BC lattice constants, weighted by mole fraction. If we imagine the alloy to be a virtual crystal (VC), where each atom sits on geometrically precise zincblende lattice sites, then its lattice constant can be expressed as $a_{ABC} = xa_{AC} + (1-x)a_{BC}$. Indeed, diffraction anomalous fine structure (DAFS) measurements have shown that the second-nearest-neighbor distances between group III atoms (or between group V atoms) in the lattice are very nearly those—$a_{ABC} / \sqrt{2}$ —expected for such a VC approximation.

In contrast, however, first-nearest-neighbor distances between group III and group V atoms deviate significantly from $\sqrt{3}a_{ABC} / 4$, expected from VC approximation [10,11]; the A-C (or B-C) bond length is always shorter (or longer) than the VC bonds. Therefore, one can imagine that, instead of occupying the VC lattice sites, the atoms will sit off the VC sites, resulting in local strain in the alloy. Such local strain can be neatly treated by valence force field (VFF) theory [15,16]. The structure factor, F_{0S}, including displacements, can be written in as

$$F_{0S} = \sum_i f_i(\mathbf{r_i} + \delta_\mathbf{i})e^{2\pi i Q\cdot(\mathbf{r_i}+\delta_\mathbf{i})} . \tag{3.3}$$

Since the displacement, $\delta_\mathbf{i}$, affects only the nearest neighbors, Equation 3.3 can be approximated as

$$\begin{aligned} F_{cS} &= \sum_i \left\{ \bar{f}_i + \langle f(\mathbf{r}_i)\rangle + 2\pi i \mathbf{Q}\cdot\langle \delta_i \rangle \left(\bar{f}_i + \langle f(\mathbf{r}_i)\rangle\right)\right\}\cdot e^{2\pi i \mathbf{Q}\cdot\mathbf{r}_i} \\ &= F_0 + F_s + F_d, \end{aligned} \tag{3.4}$$

where

$$\begin{aligned} F_0 &= \sum_i \bar{f}_i e^{2\pi i Q\cdot\mathbf{r}_i} \\ F_s &= \sum_i \langle f(r_i)\rangle\, e^{2\pi i Q\cdot\mathbf{r}_i} \\ F_d &= 2\pi i \sum_i Q\cdot\langle\delta_i\rangle\left(\bar{f}_i + \langle f(\mathbf{r}_i)\rangle\right)\, e^{2\pi i Q\cdot\mathbf{r}_i} . \end{aligned}$$

Here F_s and F_d are the contributions to the total structure factor from a perfect VC, from atomic ordering, and from atomic displacements, respectively. F_0 is thus only responsible for the fundamental Bragg peaks and can be ignored in our discussion.

The calculation of the average atomic displacements is more complicated. In an $A_xB_{1-x}C$ alloy, VFF model calculations indicate that mainly the group V atoms are displaced to accommodate bond length difference. Different orientation domains thus result in different displacement directions, but the magnitude can be written as

$$|\delta| = \left[\left(\frac{1+s_\tau}{2}\right)^4 - \left(\frac{1-s_\tau}{2}\right)^2\right] d_t + \left(\frac{1+s_\tau}{2}\right)\left(\frac{1-s_\tau}{2}\right)\left[\left(\frac{1+s_\tau}{2}\right)^2\left(\frac{1-s_\tau}{2}\right)^2\right] d_s, \tag{3.5}$$

where $\tau = 1,2$, corresponding to the orientation domains A and B, respectively, and d_t and d_s are parameters determined from VFF model calculations. For different materials, they have different values. Since this calculation is beyond the scope of this chapter, readers are directed to references 10 and 11 for details.

Using Equations 3.1 through 3.5, we can simulate the scattering pattern in Figure 3.4a. The result is shown in Figure 3.4b. Since we have ignored the Bragg diffraction, only the diffuse streaks are calculated. The simulated pattern agrees quite well with the measurement. In particular, notice that, just like the measurement results, the calculated $(\frac{\bar{5}}{2},\frac{5}{2},\frac{5}{2})$ streak is stronger than the $(\frac{\bar{3}}{2},\frac{3}{2},\frac{3}{2})$ streak. This intensity modulation is caused by the term F_d in Equation 3.4 relating to the atomic displacements due to bond length difference in the alloy. As an example, Figure 3.6 shows a calculated cross section of the streaks along the $(\bar{1}11)$ direction. If there were no bond length difference, that is, no term F_d in Equation 3.4, the relative intensities of the two streaks would look like the one in Figure 3.6a, with the $(\frac{\bar{3}}{2},\frac{3}{2},\frac{3}{2})$ streak clearly more intense than the $(\frac{\bar{5}}{2},\frac{5}{2},\frac{5}{2})$ streak. The effect of F_d on the total scattering intensity is an atomic-displacement-induced modulation, as shown in Figure 3.6b, which causes the $(\frac{\bar{5}}{2},\frac{5}{2},\frac{5}{2})$ streak to be more intense than the $(\frac{\bar{3}}{2},\frac{3}{2},\frac{3}{2})$ streak, as shown in Figure 3.6c.

As mentioned earlier in this section, growth conditions can alter the surface reconstruction and hence the ordering in the GaInP alloy films. For example, low growth temperature reduces the evaporation rate of the phosphorus atoms and consequently results in the existence of more than one atomic

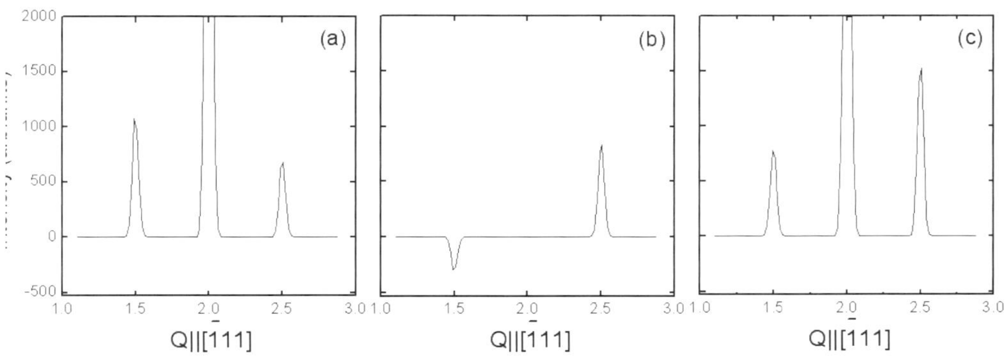

Figure 3.6. *Relative intensity of the* $(\frac{\bar{3}}{2},\frac{3}{2},\frac{3}{2})$ *and* $(\frac{\bar{5}}{2},\frac{5}{2},\frac{5}{2})$ *peaks (a) without atomic displacements, (b) with modulation by atomic displacement, and (c) with atomic displacements. (With kind permission of Springer Science and Business Media.)*

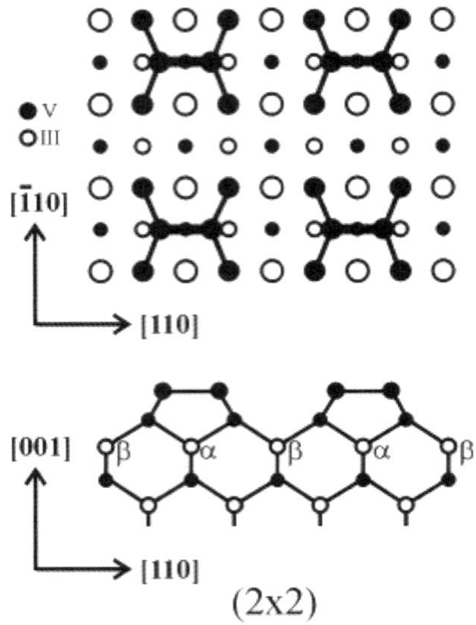

Figure 3.7. *(2 × 2) surface reconstruction of GaInP alloy films during low-temperature growth. Surface is terminated with more than one atomic layer of phosphorus.*

layer of phosphorus atoms at the terminal surface during growth. Figure 3.7 shows a case in which the surface is terminated by one and one-half layers of phosphorus atoms. The P-dimmers in such a surface are oriented along the [110] direction. Consequently, CuPt-A ordering occurs. This has been observed in our experiments. For example, Figure 3.8 shows both the [110] and [$\bar{1}$10] zone X-ray scattering patterns from a film grown at 480°C. Diffuse streaks are only seen in the [110] zone pattern, indicating that only CuPt-A type ordering exists in this sample.

An interesting feature in Figure 3.8 is that all streaks are inclined inward instead of outward, as in Figure 3.4. We believe that the inclination of the diffuse streaks is related to the average domain shape bounded by the antiphase and orientation boundaries. For example, Figure 3.9 shows two calculated streaks assuming two different domain shapes. As expected, one shape results in an inward inclination, and the other results in an outward inclination.

3 X-ray Scattering from $GaAs_{0.87}Sb_{0.13}$ Alloy Films on GaAs (001) Substrates

Epitaxial $GaAs_{0.87}Sb_{0.13}$ alloy films on GaAs (001) substrates exhibit a different type of ordering along the crystallographic [110] direction with a long-range periodicity that quadruples the (110) lattice plane spacing—an ordering called the quadruple-period ordering (QPO). This type of ordering is directly related to the (2 × 4) surface reconstruction with missing dimmer rows, as shown in Figure 3.10. This creates alternative local compressive and tensile strain in the subsurface atomic layers. Consequently, the larger GaSb molecules migrate to the site with tensile strain and the smaller GaAs molecules migrate to sites with compressive strain, forming the observed QPO along the [110] direction.

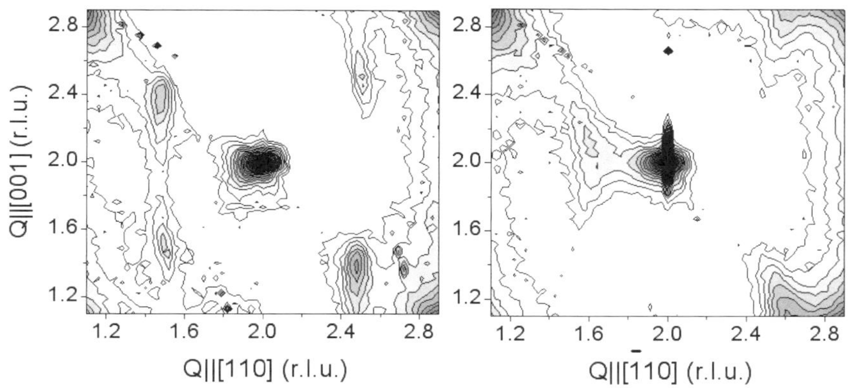

Figure 3.8. *[110] (left) and [$\bar{1}$10] (right) zone X-ray scattering patterns of a low-temperature grown GaInP alloy film.*

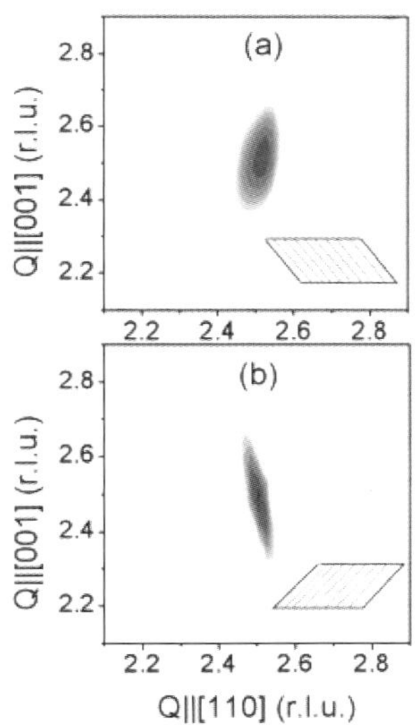

Figure 3.9. *Calculated scattering streaks considering different domain shapes. The direction of inclination of the streak is determined by the average shape of the domains.*

As in the GaInP alloy films, antiphase boundaries are almost inevitable. They appear in both [110] and [$\bar{1}$10] directions. Along the [001] growth direction, phase-shifting boundaries exist that separate the antiphase domains above and below them.

Figure 3.11 shows a scattering pattern of epitaxial $GaAs_{0.87}Sb_{0.13}$ alloy films on GaAs (001) in reciprocal space. The measurement was done on a beamline X14A of the NSLS at the BNL with an X-ray energy of 8.0478 keV. Diffuse streaks spanning through the neighboring reciprocal (00*l*) lattice planes are clearly observed.

With the in-plane antiphase boundaries and out-of-plane phase shifts in mind, we can write the structure factor of the unit cell at position $\mathbf{R} = n_1\mathbf{a} + n_2\mathbf{b} + n_3\mathbf{c}$ as

$$F(n_1, n_2, n_3) = F(\mathbf{Q})\exp[i\phi(n_1, n_2, n_3)]\exp\left[i2\pi\mathbf{Q}\cdot(n_1\mathbf{a} + n_2\mathbf{b} + n_3\mathbf{c})\right], \tag{3.6}$$

where $\phi(n_1, n_2, n_3)$ represents the phase shift of the unit cell (n_1, n_2, n_3). The total structure factor of the structure is thus

$$F_{total} = \sum_{n1,n2,n3} F(\mathbf{Q})\exp[i\phi(n_1, n_2, n_3)]\exp\left[i2\pi\mathbf{Q}\cdot(n_1\mathbf{a} + n_2\mathbf{b} + n_3\mathbf{c})\right]. \tag{3.7}$$

F(**Q**) in Equation 3.7 considers the possible phase shift along the [110] and [$\bar{1}10$] directions between the two atomic monolayers.

In Equation 3.8, $F_a(\mathbf{Q})$ and $F_b(\mathbf{Q})$ are the structure factors of the two adjacent monolayers, in which the III-V covalent bonds are oriented differently:

$$F(\mathbf{Q}) = F_a(\mathbf{Q}) + F_b(\mathbf{Q})\exp(i\phi_c). \tag{3.8}$$

They include the effect of atomic displacements due to bond length difference, which is similar to that discussed in the previous section [17]. ϕ_c is the phase shift between the two monolayers.

Adopting the method described by Warren [9], which treats the antiphase boundaries using a statistic method, we can now simulate the observed streaky diffuse pattern using Equation 3.7. In this method, the average phase factor is related to the probabilities of finding a random antiphase boundary along each direction **a**, **b**, and **c**, which in turn relates to the average size of the antiphase domains and ultimately determines the shape of the observed streaky pattern. As an example, Figure 3.11b shows

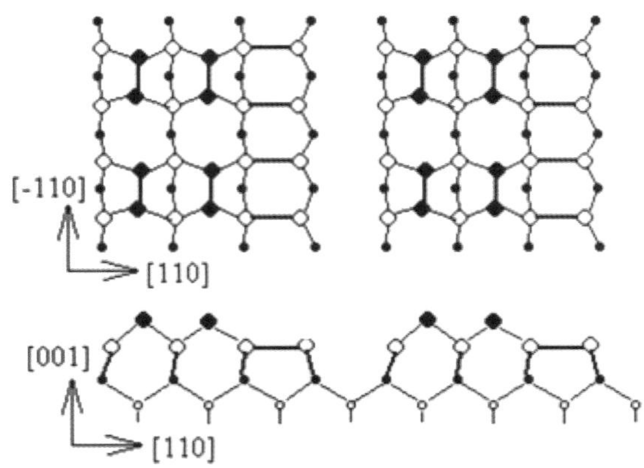

Figure 3.10. *(2 × 4) surface reconstruction of GaAsSb alloy films with missing dimmer rows during growth.*

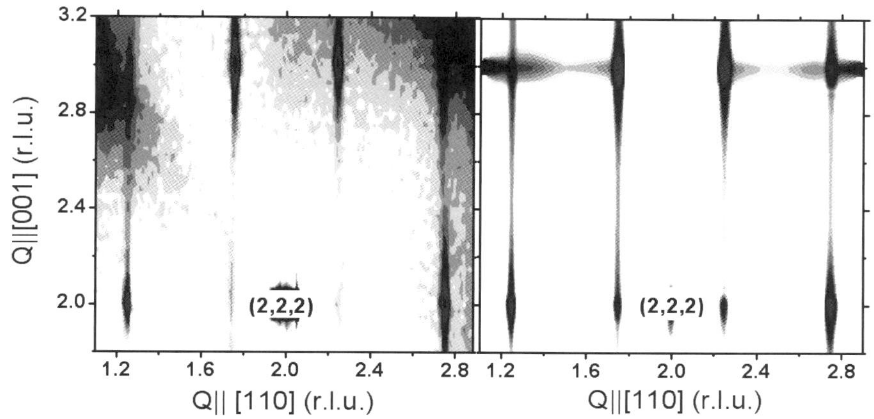

Figure 3.11. *The measured (left) and calculated (right) scattering pattern of an GaAsSb alloy films.*

a simulated scattering pattern for the $GaAs_{0.87}Sb_{0.13}$ film sample, which is in good agreement with the experimental result.

4 Summary

The surface structure of epitaxial semiconductor alloy films on GaAs substrates results in partially ordered structures that contain local lattice strain, atomic ordering, and antiphase, orientation-phase, and phase-shifting boundaries. These phase boundaries cause distinct diffuse scattering patterns. By analyzing these patterns, detailed information of the film structural defects can be derived, which may help us to better understand the physical properties of the semiconductor thin films.

Acknowledgment

I am grateful to my former colleagues, Professor Simon Moss and his graduate student, Zhenyang Zhong, at the University of Houston for their valuable contributions to the work summarized in this chapter.

References

1. A. Gomyo, K. Matika, I. Hino, and T. Suzuki, *Phys. Rev. Lett.* 72 (1994): 673.
2. B. A. Philips, I. Kamiya, K. Hingerl, L. T. Florez, D. E. Aspnes, S. Mahajan, and J. P. Harbison, *Phys. Rev. Lett.* 74 (1995): 3640.
3. S. B. Zhang, S. Froyen, and A. Zunger, *Appl. Phys. Lett.* 67 (1995): 3141.
4. S. Froyen and A. Zunger, *Phys. Rev. B* 53 (1996): 4570.
5. S. H. Wei and A. Zunger, *Phys. Rev. B* 57 (1998): 8983.
6. Y. Zhang, A. Mascarenhas, and L.-W. Wang, *Phys. Rev. B* 63 (2001): R201312.
7. T. Kanata, M. Nishimoto, H. Nakayama, and T. Nishino, *Phys. Rev. B* 45 (1992): 6637.
8. D. J. Mobrary, R. A. Hogg, M. S. Skolnick, M. C. DeLong, S. R. Kurtz, and J. M. Olson, *Phys. Rev. B* 46 (1992): 7232.

9. B. E. Warren, *X-ray Diffraction* (New York: Dover, 1990).
10. D. C. Meyer, K. Richter, and P. Paufler, *Phys. Rev. B* 59 (1999): 15253.
11. L. Alagna, T. Properi, S. Turchini, C. Ferrari, L. Francesio, and P. Franzosi, *J. Appl. Phys.* 83 (1998): 3552.
12. P. Gavrilovic, F. P. Dakowiski, K. Meehan, J. E. Williams, W. Stutius, K. C. Hsieh, N. Holonyak, M. A. Shahid, and S. Mahajan, *J. Cryst. Growth* 93 (1988): 426.
13. J. H. Li, J. Kulik, V. Holy, Z. Zhong, S. C. Moss, Y. Zhang, A. Mascarenhas, and J. Bai, *Phys. Rev. B* 63 (2001): 155310.
14. R. L. Forrest, J. Kulik, T. D. Golding, and S. C. Moss, *J. Mater. Res.* 15 (2000): 45.
15. Y. Cai and M. F. Thorpe, *Phys. Rev. B* 46 (1992): 15872; *Phys. Rev. B* 46 (1992): 15879.
16. J. S. Chung and M. F. Thorpe, *Phys. Rev. B* 55 (1997): 1545.
17. Z. Zhong, V. Holy, J. H. Li, J. Kulik, J. Bai, T. D. Golding, and S. C. Moss, *J. Appl. Phys.* 90 (2001): 244.

4

X-RAY STUDIES OF THERMALLY GROWN THIN VITREOUS SiO_2 OVER Si(001)

M. Castro-Colin[1]

W. Donner[2]

S. C. Moss[2]

Z. Islam[3]

S. K. Sinha[4]

R. Nemanich[5]

H. T. Metzger[6]

T. Shülli[6]

P. Bösecke[6]

1. Department of Physics, University of Texas El Paso, El Paso, TX 79968, USA

2. Department of Physics, University of Houston, Houston, TX 77204, USA

3. Advanced Photon Source, Argonne National Laboratory, Argonne, IL, 60439, USA

4. Department of Physics, University of California San Diego, San Diego, CA, 92093, USA

5. Department of Physics, North Carolina State University, Raleigh, NC, 27695, USA

6. E. Synchrotron Radiation Facility, BP 220; 38043 Grenoble CEDEX 9, France

Abstract

We have explored the thin gate oxide formed when SiO_2 is grown through thermal oxidation of Si(001). X-rays in grazing-incidence mode from a synchrotron source were used to explore the thin (500 and 100 Å) amorphous layer, whose first sharp diffraction peak exhibited a fourfold modulation. Such modulation is more stressed close to the Si(001) substrate and exponentially decays toward the film surface, thus being present throughout the entire film. This indicates that the building blocks of the amorphous network anisotropically organize themselves, in this case along the [110] directions, revealing at the same time an incipient crystalline structure within the amorphous matrix.

1 Introduction

Thermally grown amorphous SiO_2 on crystalline Si has served as the gate oxide in complementary metal-oxide semiconductor (CMOS) processing due to several key features that include stability, high SiO_2-Si interface quality, and electrical isolation properties. These properties may be adversely affected by the continuous thinning of the gate oxide. One example is the considerably increased leakage current in 10 Å SiO_2 [1]. Thus the system is technologically of great interest but also of interest from a more fundamental point of view as has been recently more thoroughly exposed elsewhere [2–7].

Studies carried out on this SiO_2-Si system have used techniques such as electron energy loss spectroscopy (EELS) [8], X-ray-photoelectron spectroscopy (XPS) [4,9], transmission electron microscopy [10], X-ray reflectivity [11,12], ion scattering [13], and neutron reflectivity [14]. Theoretical studies [13,15–17] have suggested interesting structural models to describe some of the experimental observations. In the work presented here and in regard to some of those models [15,16], we note that no evidence for cell doubling or other periodic interfacial layer was found.

In the present document, we exhibit results about the behavior, as a function of thickness, of the amorphous structure factor of thin SiO_2 via observation of the first sharp diffraction peak (FSDP). A prominent feature of the well-known bulk structure factor [18] is the presence of an FSDP, as displayed in Figure 4.1, whose interpretation has received great attention [19–25]. In the present work we are interested in the structural correlations associated with the FSDP that yield information about the intermediate-range order in vitreous solids, particularly in SiO_2, the distances well beyond the first near Si-O neighbors [19] (Figure 4.1, inset). The FSDP thus indicates the degree of order among structural tetrahedra in the film.

The samples consisting of thin SiO_2 were prepared at North Carolina State University with an RCA clean that leaves a thin native oxide on the p-type Si(001)-oriented, high-quality wafers from Okmetic, which showed ≈0.2° miscut at ≈5° off the <100> (a vicinality that is not responsible for the fourfold modulation discussed below). The samples of 100 and 500 Å SiO_2 were created through thermal oxidation of Si wafers at 850°C and 950°C, respectively.

2 Experimental Methods and Results

We resorted to grazing incidence diffraction (GID) to measure the lateral (in-plane) structure factor in the thin SiO_2 films. The high flux available at the Advanced Photon Source (beamline 4ID, APS) allowed us to study the surface structure by limiting the X-ray penetration depth while working close to the critical angle for total external reflection. This technique reduces the substrate scattering signal,

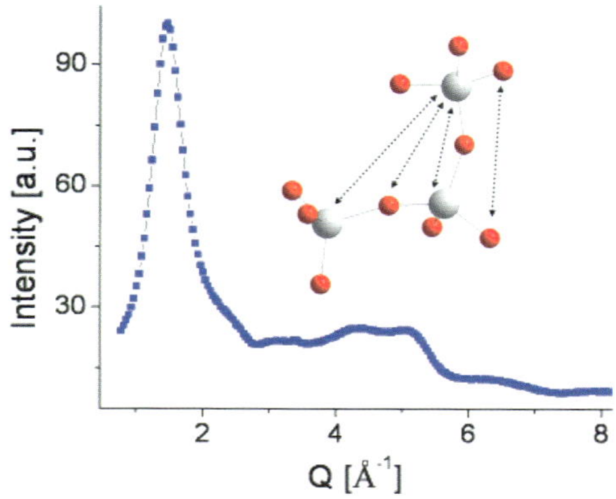

Figure 4.1. *X-ray scattering from vitreous SiO_2 at a labsource. Inset: SiO_2 structural units indicating distances that contribute to the FSDP, including the intertetrahedral Si-O-Si distance, which defines the bond angle.The small dark spheres are oxygen; the larger spheres are silicon. (Reprinted with permission from Castro-Colin et al. [5].)*

although substrate Si thermal diffuse scattering (TDS) will always be present, which must be either avoided or accounted for.

As noted earlier, SiO_4 tetrahedra, corner-shared through O atoms, give rise to the FSDP at a momentum transfer Q ≈ 1.5 Å^{-1}. There are three SiO_4 units with the distances d(Si – O_{2nd}), d(Si – Si_{2nd}), d(O – O_{2nd}), highlighted in Figure 4.1 (inset), which have been associated with the FSDP [19,21]. The familiar profile (Figure 4.1) shows the X-ray intensity from a standard bulk SiO_2 glass measured with monochromatic Mo-Kα radiation. Using the expression $2\pi/d \approx Q$, we can estimate [19] that d(Si – O) =1.617 Å, d(Si – Si) = 3.10 Å, and d(O – O) = 2.626 Å [18,26–28] cannot produce momentum transfers at the FSDP position. Strictly, however, the FSDP appears at a maximum of $\sin(Qr_{ij})/(Qr_{ij})$, where r_{ij} refers to a particular distance; the maximum then appears at $Q = 2\pi\,(0.715)/d$, which perhaps may permit the intertetrahedral distance, d(Si – Si), to contribute to the FSDP (1.45 Å^{-1} vs. 1.5 Å^{-1}, an aspect that could be further explored through computer modeling).

The medium-range correlations responsible for the FSDP are present throughout the glass and are normally of an isotropic nature. These correlations have associated partial structure factors, $S^{\text{Si-Si}}(Q)$, $S^{\text{Si-O}}(Q)$, $S^{\text{O-O}}(Q)$, that contribute to the intensity of the FSDP [21]. The present work shows that the familiar bulk isotropy is altered by the fourfold Si(001) template. The anisotropy is measured (in Q space and intensity) as the samples are rotated in ϕ about the Si<001> surface normal (Figure 4.2).

An X-ray energy of 20 keV was used to perform the GID measurements on the 100 Å sample, which was enclosed in a beryllium dome under helium atmosphere. In a 4S + 2D six-circle diffractometer [29], the sample was vertically aligned, and scattered photons were collected with a scintillation detector after passing through a vertically open slit oriented normal to the sample surface. The slit had an aperture width of 0.75 mm at 740 mm away from the sample. At this energy (20 keV), the critical angle for total external reflection is $\alpha_c = 0.08°$. The incident X-ray beam impinged on the sample at an angle, $\alpha_i = 0.05°$, such that α_f could probe the entire SiO_2 film thickness.

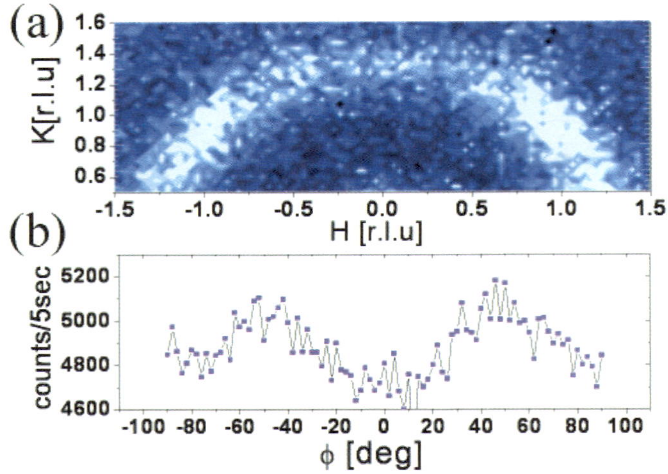

Figure 4.2. *(a) The HK0-map in the Si crystal notation, which clearly shows that the FSDP modulation peaks in the Si[110] directions. (b) A ϕ scan of the 100 Å SiO_2 FSDP at a radial position corresponding roughly to the Si(110) distance in reciprocal space. (Reprinted with permission from Castro-Colin et al. [5].)*

The HK0-map (Figure 4.2a) shows the variation in reciprocal space of the FSDP. It should be noted that the FSDP describes an intensity variation that radially expands to 1.636 $Å^{-1}$ at the positions of the Si(110) reflections, which are forbidden by symmetry. The intensity is reduced by 20 counts per second at positions away from the Si(110) reflections, and the FSDP position changes radially to 1.4 $Å^{-1}$. For the sake of clarity, a set of data points were selected to highlight the intensity trend of the HK0-plane (Figure 4.2b).

The coincidence of the FSDP with the Si(110) reflections provides a reference frame to scale the degree of structural compression of the tetrahedral units. There is a structural compression of 8% along <110>, with a corresponding expansion along <100>. It should also be pointed out that the enhanced intensity does not contain harmonic contamination because it was investigated with attenuators that would eliminate any $\lambda/2$ contribution.

In another experiment (beamline X22C, National Synchrotron Light Source [NSLS]), we used X-ray radiation at 11 keV, in a bisecting reflectivity mode, to determine the thickness, roughness, and electron density profile of the sample. Figure 4.3 displays the intensity plotted against the perpendicular momentum transfer. As is common to reflectivity measurements, multiplying the intensity by Q^4 (Fresnel normalized) enhances the inherent characteristics disclosed by the oscillations. This normalization permits a more accurate fit. The medium is modeled assuming that is composed by three strata and a substrate. The uppermost layer (12 Å) is most probably due to organic material [14] denatured by the synchrotron radiation during measurements. The prominent oscillation period corresponding to 74 Å represents the amorphous SiO_2. Lastly, we have the interfacial layer of 20 Å, identified through modeling.

The inset in Figure 4.3 shows the variation with thickness of $\delta = \lambda^2 N r_0 Z / 2\pi$ and $\beta = \lambda N \sigma_0 / 4\pi$ (λ = wavelength, N = electron density, r_0 = Thompson scattering length, and σ_0 = absorption

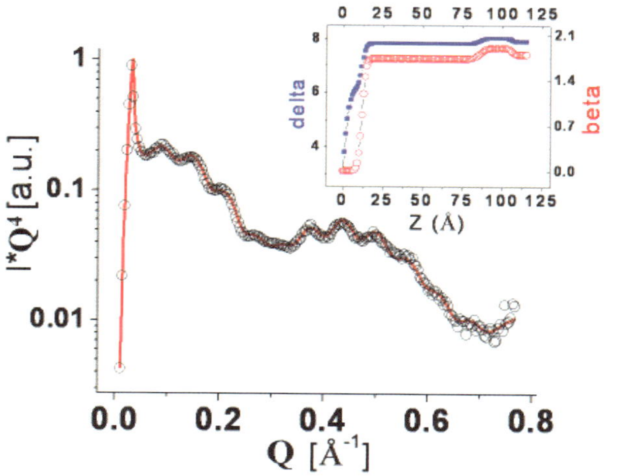

Figure 4.3. *Fresnel normalized reflectivity of 100 Å SiO_2 film; line = fit (see text), open circles = data. Inset: Calculated electron density (≈ delta, in units of 10^{-6}) and absorption (≈ beta, in units of 10^{-7}); surface at z = 0. Note that while the absorption at the surface drops abruptly, the electron density falls more gradually, indicating some low Z (carbon compound) contamination, which has no adverse influence in our analyses. (Reprinted with permission from Castro-Colin et al. [5].)*

coefficient). These parameters are associated with the complex refractive index, where $n = 1 - \delta + i\beta$. The film surface is located on the left-hand side (at $z = 0$). The top layer has a low electron density and the interfacial layer has a density 3% larger than the remaining 74 Å. This 3% density increase at the interface confirms earlier results [12], but we cover twice the Q range and ten times the intensity range (rotating anode vs. synchrotron). An independent study by another group [14] did not find any density increase in neutron reflectivity fits. Our fitting values were obtained from the Parrat formalism [30] and an implementation of least-squares fitting. These results show a root mean square interface roughness of ≈5 Å.

To explore the amorphous film in more detail, depth-dependent measurements were also carried out at the APS on a 500 Å SiO_2 film at 10 keV (λ = 1.239 Å). The sample was placed inside a cryostat at 10 K to suppress the bulk Si crystal TDS contribution. This sample was cleaned with acetone, followed by ethanol and deionized water. Flowing nitrogen dried the sample surface. The overall intention was to rid the sample of possible organic contamination [14] observed on the 100 Å thin film. Due to technical constraints, the sample was measured while set horizontally. We collected the scattered photons with a Braun position-sensitive detector (PSD) and a slit perpendicular to the sample surface, as in the 100 Å GID measurement. With this geometry the PSD resolution of 100 μm corresponds to an angular step size of 0.004° in the final angle, α_f. Thus α_f was recorded between 0.176° below and 0.848° above the critical angle of 0.174° for two different incident angles of 0.15° and 0.25°. Successive radial scans, rotating the sample about its normal at intervals of 10° , collected both surface and bulk information within a 180° range. Scattering depths between 40 and 130 Å can be explored with the 0.15° = α_i, while scattering depths for the 0.25 Å = α_i range from 60 to 8000 Å

[31] (a scattering depth is essentially the $1/e$ penetration of the evanescent wave). Each PSD data set was separated into two regions of interest (ROI), one below the critical angle (at which the intensity vs. α_f peaks) and another above the critical angle; the intention is to quantify the FSDP depth-dependence intensity variation as a function of azimuthal position. The ROI above the critical angle includes information down to the bulk (film ROI, 0.25° = α_i, Figure 4.4a), whereas the ROI below the critical angle is mostly surface sensitive (surface ROI, 0.15° = α_i, Figure 4.4b).

We confirmed the depth-dependence modulation of the 500 Å film through an additional experiment at ID-1, European Synchrotron Radiation Facility (ESRF) (Figure 4.5), using λ = 1.722 Å X-rays. Again, two prominent peaks are observed at ϕ = –75° and ϕ =15° along Si<$\bar{1}$10> and Si<110>, respectively. The film ROI shows strong modulation (Figure 4.5a), whereas the surface ROI does not (Figure 4.5b).

As others did earlier [32,33], we also investigated the (1 1 L) crystal truncation rod (CTR) of the 100 and 500 Å films; L represents a Miller index. We show the measurement 100 Å film within 0.1 and 0.8 reciprocal lattice units (RLUs) using λ = 0.6199 Å (Figure 4.6). Successive ϕ scans at every L position exhibited a full width at half-maximum (FWHM) that indicated a coherent domain size of 9800 Å, corresponding to a crystalline component distributed throughout the film thickness also responsible for the Laue oscillations (Figure 4.6). Later we will comment on this aspect.

The 500 Å film (Figure 4.7) at λ = 1.239 Å, with better resolution than in the 100 Å film, allowed the identification of two kinds of coherent domains, one of 1500 Å and another one ten times larger. The results shown were measured at H = 1 RLU, K = 0.9939 to 1.0045 RLU, and L = 0.3 to 0.6 RLU In both samples, 100 and 500 Å, the CTR shows a crystalline reflection at about Si(1 1 0.485) and Si(1 1 0.455), respectively, while no additional reflections are found along the rod. Laue oscillations are again observed.

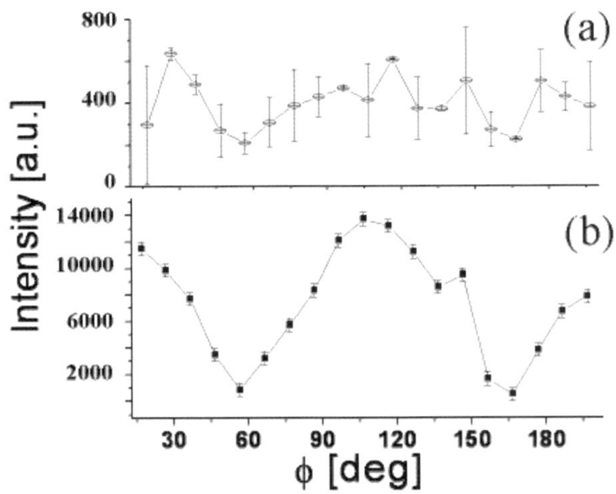

Figure 4.4. *FSDP values at their corresponding ϕ positions for a 500 Å SiO_2 film; the plotting procedure shows the intensity variation about an (arbitrary) average set to zero: (a) the thin film bulk modulation is clearly expressed, while in (b) the film surface modulation has virtually vanished. (Reprinted with permission from Castro-Colin et al. [5].)*

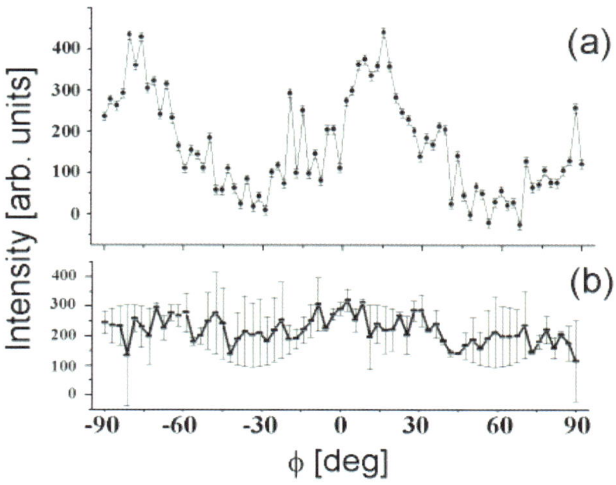

Figure 4.5. *Azimuthal scans of the 500 Å film at the FSDP position: (a) surface ROI and (b) film ROI. (Reprinted with permission from Castro-Colin et al. [5].)*

The CTRs were fitted using a proposed model of a high-pressure phase of SiO_2 that here we call pseudo-coesite. The least-squares method helped in identifying adequate parameters of fractional occupancy of the crystallites embedded in the amorphous SiO_2 matrix (as a function of the electron density), bond-length variation of the tetrahedral structural units, and Si/SiO_2 interface roughness.

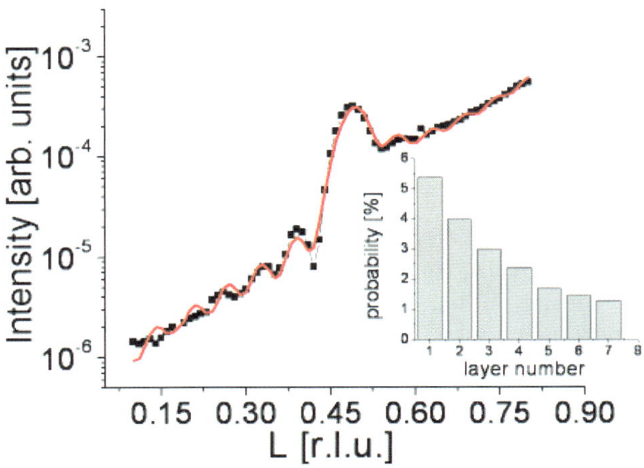

Figure 4.6. *Si (1 1 L) rod (black circles) of 100 Å film with its fitted profile (line). Inset: fitted occupational probability of crystallites; the numbers indicate the layers in units of the lattice size of the crystalline component. The occupational probability decays toward the surface (Reprinted with permission from Castro-Colin et al. [6]).*

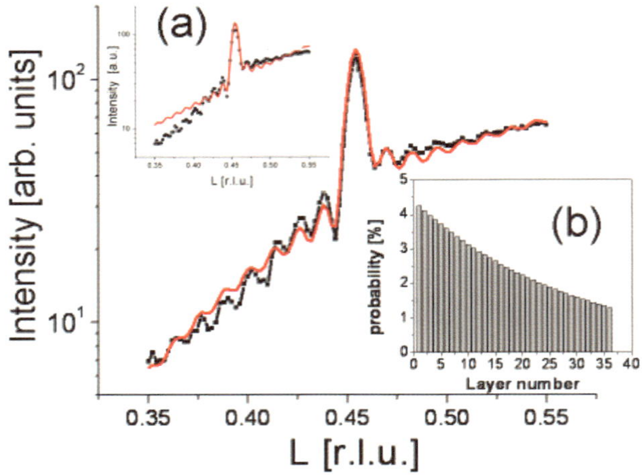

Figure 4.7. *Si (1 1 L) rod (circles) of 500 Å film with its fitted profile (line). (a) Fit (line) obtained when no interfacial layer is included in the model. (b) Fitted occupational probability of crystallites. The numbers indicate the layers in units of the lattice size of the crystalline component. The occupational probability decays toward the surface. (Reprinted with permission from Castro-Colin et al. [6].)*

3 Discussion and Conclusions

Since the FSDP reflects network connectivity of structural units in SiO_2, here it has been used to disclose information about the modulation imprinted by a crystalline substrate onto an amorphous film. The azimuthal depth dependence correspondence between the structural variations and the film thickness shows consistency with previously observed refractive index variations [34,35]. The difference in position of the FSDP for the 500 Å film in reciprocal space is smaller than in the 100 Å film, which can be attributed to a relaxation of the structural units, as expected for a film (thicker) grown at higher temperature [34,35]. To the former we add that FSDP intensity variations are known to have their origins in pressure changes [36–39] that arise in a system like the present one. Thin films grown at lower temperatures would be under a higher compression that induces a higher electron density, as exhibited in the reflectivity measurements. The reflectivity measurements clearly probe the density enhancement by averaging over the entire interfacial transition layer, measured normal to the (001) plane, including both compressive and tensile stress, to yield an overall density increase. Regarding the denser interfacial layer, we would agree [13] that it is more disordered than previously considered [15,16]—a conclusion supported by our modeling of the CTRs.

The concomitant result obtained is that of the crystalline component present throughout the entire film, whose fractional occupation probability exponentially decays [32] toward the surface of the film (Figures 4.6 and 4.7). As a precursor of the SiO_2 crystallite, we use the high-pressure phase coesite-like structure. Coesite, in a distorted fashion, fulfills the epitaxial match, combined with high electron density, to satisfy the interfacial value reported by our reflectivity measurements, FSDP reciprocal space variations, and crystalline reflection observed along the Si(1 1 L) rod (Figures 4.6 and 4.7). More quantitative details of the pseudo-coesite structure proposed can be found in Castro-Colin et al. [6]. During modeling of the Si(1 1 L) rod, pseudo-coesite reflections correspond to those measured experimentally at Si(1 1 0.485) and Si(1 1 0.455) in 100 and 500 Å films, respectively. The model [6] suggested a density of about 2.31 g/cm^3, which is about 2.2% above that found in the reflectivity measurements. Unlike pseudo-cristobalite [2,32] or pseudo-trydimite [10] models, pseudo-coesite does not require large distortions of the bond lengths (<2%). By virtue of the geometry used to measure the rods, the distortion in the crystalline component precludes that a reflection at Si(1 1 1.455) and Si(1 1 1.365) for 100 and 500 Å, respectively, is experimentally observed. The aspect of structural distortion in the cristalline component within the amorphous thin film has been most recently studied by other groups [2,3,7]. We share [6] their opinion about the large degree of static disorder that would suppress the appearance of any higher order reflections along the Si(1 1 L) rod.

In conclusion, we have shown evidence of a fourfold modulation of thermal SiO_2 over Si(001), which suggests the existence of structural correlations with preferential orientation along the Si[1 1 0] directions. This anisotropy is observed throughout the amorphous thin film, more prominently close to the substrate and less so toward the film surface, and is definitely associated to a crystalline component clearly observed along the Si(1 1 L) rod. Whether such a crystalline component actually follows a coesite-like structure or a different one is a matter of further study. Some results suggest that the crystalline peaks observed along the Si(1 1 L) rod are produced by residual order of the Si(1 1 1) planes [3,7]. To such an assertion we would like to add another instance of residual order [40], where coesite as well as quartz were irradiated with neutrons. Quartz was completely amorphized while coesite maintained some stable regions, indicating that silicon can sustain preferentially some crystalline phase embedded in an amorphous surrounding. In the case of a thin film, the substrate during oxidation is able to sustain order imprinted as a crystalline structure with an occupational probability decaying toward the surface.

Acknowledgments

We would like to acknowledge the assistance of H. Zajonz (X22C, NSLS) and J. C. Lang (1-ID, APS). M. C-C would like to acknowledge J. H. Li (University of Houston, now at Rigaku Am. Co.) for insightful comments. This work was supported by the U.S. Department of Energy (DOE) under contract DE-FG03-01ER45880. Use of the APS was supported by the U.S. DOE, Office of Science, Office of Basic Energy Sciences, under contract W-31-109-Eng-38. Research at the NSLS, Brookhaven National Laboratory, was supported by the U.S. DOE, Division of Materials Sciences and Division of Chemical Sciences, under contract DE-AC02-98CH10886. We appreciate the permission granted by the American Physical Society, copyright 2005, to reprint some figures that appeared in their *Physical Review B* [5,6].

References

1. R. M. Wallace, and G. Wilk, *MRS Bulletin* 27 (2002): 192.
2. K. Tatsumura, T. Watanabe, D. Yamasaki, T. Shimura, T. Umeno, and I. Ohdomari, *Phys. Rev. B* 69 (2004): 085212.
3. A. Munkholm and S. Brennan, *Phys. Rev. Lett.* 93 (2004): 036106-1.
4. S. Dreiner, M. Shürmann, and C. Westphal, *Phys. Rev. Lett.* 93 (2004): 106101-1.

5. M. Castro-Colin, W. Donner, S. C. Moss, Z. Islam, S. K. Sinha, R. Nemanich, H. T. Metzger, T. Shülli, and P. Bösecke, *Phys. Rev. B* 71 (2005): 045310.
6. M. Castro-Colin, W. Donner, S. C. Moss, Z. Islam, S. K. Sinha, and R. Nemanich, *Phys. Rev. B* (2005): 71, 045311.
7. K. Tatsumura, T. Shimura, E. Mishima, K. Kawamura, D. Yamasaki, H. Yamamoto, T. Watanabe, M. Umeno, and I. Ohdomari, *Phys. Rev. Lett.*, 72 (2005): 045205.
8. D. A. Muller, T. Sorsch, S. Moccio, F. H. Baumann, E. Evans-Lutterodt, and G. Timp, *Nature* 399 (1999): 758.
9. K. Hirose, H. Nohira, T. Koike, K. Sakano, and T. Hattori, *Phys. Rev. B* 59 (1999): 5617.
10. A. Ourmazd, D. W. Taylor, J. A. Rentschler, and J. Bevk, *Phys. Rev. Lett.* 59 (1987): 213.
11. Y. Sugita, S.Watanabe, N.Awaji, and S.Komiya, *Appl. Surf. Sci.* 101 (1996): 268.
12. S. D. Kosowsky, P. S. Pershan, K. S. Krisch, J. Bevk, M. L. Green, D. Brasen, L. C. Feldman, and P. K. Roy, *Appl. Phys. Lett.* 70 (1997): 3119.
13. A. Bongiorno, A. Pasquarello, M. S. Hybertsen, and L. C. Feldman, *Phys. Rev. Lett.* 90 (2003): 186101.
14. J. A. Dura, C. A. Richter, C. F. Majkrzak, and N. V. Nguyen, *Appl. Phys. Lett.* 73 (1998): 2131.
15. Y. Tu and J. Tersoff, *Phys. Rev. Lett.* 89 (2002): 086102.
16. A. Demkov and O. F. Sankey, *Phys. Rev. Lett.* 83 (1999): 2038.
17. A. Pasquarello, M. S. Hybertsen, and R. Car, *Phys. Rev. Lett.* 74 (1999): 1024.
18. B. E. Warren, H. Krutter, and O. Morningstar, *J. Am. Ceram. Soc.* 32 (1936): 614.
19. S. C. Moss, *Amorphous and Liquid Semiconductors*, ed. J. Stuke and W. Brening (London: Taylor & Francis, 1973), 17.
20. S. C. Moss and D. L. Price, *Physics of Disordered Materials*, ed. D. Adler, H. Fritzsche, and S. R. Ovshinsky (New York: Plenum Press, 1985), 77.
21. J. L. Robertson and S. C. Moss, *J. Non-cryst. Solids* 106 (1988): 330.
22. S. R. Elliott, *Phys. Rev. Lett.* 67 (1991): 711.
23. J. Swenson and L. Borgesson, *J. Non-cryst. Solids* 223 (1998): 223.
24. C. Z. Tan and J. Arndt, , *J. Non-cryst. Solids* 249 (1999): 47.
25. C. Massobrio and A. Pasquarello, *J. Chem. Phys.* 114 (2001): 7976.
26. R. L. Mozzi and B. E. Warren, *J. Appl. Cryst* 2 (1969): 164.
27. J. Neuefeind and K. D. Liss, *Ber. Bunsen Phys. Chem.* 100 (1996): 1341. This is probably the most definitive of the studies, particularly regarding the bond angle distribution.
28. D. A. Keen and M. Dove, *J. Phys.: Condens. Matter* 11 (1999): 9263.
29. H. You, *J. Appl. Cryst.* 32 (1999): 614.
30. L. G. Parrat, *Phys. Rev.*, 95 (1954): 359.
31. H. Dosch, *Phys. Rev. Lett.* 35 (1987): 2137.
32. I. Takahashi, T. Shimura, and J. Harada, *J. Phys.: Condens. Matter* 5 (1993): 6525.
33. A. Munkholm, S. Brennan, F. Comin, and L. Ortega, *Phys. Rev. Lett.* 75 (1995): 4254.
34. A. G. Revesz, B. J. Mrstik, and H. L. Hughes, *The Physics and Technology of Amorphous* SiO_2, ed. R. A. B. Devine (New York: Plenum Press, 1987), 299.
35. A. Szekeres, K. Christova, and A. Paneva, *Philos. Mag. B* 65 (1992): 961.
36. H. Tsutsu, K. Tamura, and H. Endo, *Solid State Comm.* 52 (1984): 877.
37. S. Tanaka, *J. Non-Cryst. Sol.* 90 (1987): 363.
38. C. Meade, R. J. Hemley, and H. K. Mao, *Phys. Rev. Lett.* 69 (1992): 1387.
39. Y. Inamura, Y. Katayama, W. Utsumi, and K. Funakoshi, *Phys. Rev. Lett.* 93 (2004): 015501-1.
40. M. C. Wittels, *Philos. Mag.* 2 (1957): 1445.

5

Ordered and Disordered Surface Phases of Bi on Cu(111)

Elias Vlieg[1]

Daniel Kaminski[1]

Raoul van Gastel[2]

Bene Poelsema[2]

1. *IMM Solid State Chemistry, Radboud University Nijmegen, Toernooiveld 1, 6525 ED Nijmegen, The Netherlands*
2. *University of Twente, Solid State Physics, MESA+ Institute for Nanotechnology, PO Box 217, 7500 AE Enschede, The Netherlands*

Abstract

Using a combination of surface X-ray diffraction and low-energy electron microscopy, we have determined the structure of several ordered and disordered phases of Bi on Cu(111) and the transitions between these structures. At low coverage, a surface alloy is formed that is initially disordered but that orders into $\sqrt{3}$ domains up to a coverage of 1/3 ML. For increasing coverage, de-alloying occurs, leading to an overlayer structure with a coverage of 0.50 ML and with a [2012] unit cell. Finally, at 0.53 ML, a uniaxial incommensurate overlayer is formed. At temperatures around 250°C and a coverage near 0.50 ML, the Bi layer melts, but the local structure in the liquid closely resembles that of the two overlayer phases at room temperature. This results in a liquid with an orthorhombic orientational order.

1 Introduction

Surfaces provide a rich environment for the investigation of material properties that frequently differ from the bulk. Examples are (two-dimensional) melting [1] and surface alloy formation [2]. We present a detailed study of the behavior of submonolayer Bi phases on a Cu(111) substrate as a function of both temperature and coverage. We find that the fact that this is not a genuine two-dimensional system because of the presence of the substrate has significant consequences for the behavior [3,4].

The first, and for many years only, study of the Bi-Cu(111) system was performed in 1973 by Delamare and Rhead [5]. Using low-energy electron diffraction (LEED), they identified a number of room-temperature phases. We have extended these results using surface X-ray diffraction (SXRD) supplemented by low-energy electron microscopy (LEEM) in order to see the phase behavior in real space.

2 Experimental Section

Surface X-ray diffraction is very suitable for the determination of both surface structure and surface phase transitions [6]. The technique is similar to bulk diffraction but avoids the bulk Bragg peaks and uses instead the rodlike scattering from the surface in order to obtain surface specific information. These diffraction rods can either originate from the surface supercell in case of a reconstructed surface (fractional-order rods) or contain contributions from both bulk and surface in so-called crystal truncation rods (integer-order rods) [7]. Due to the weak surface signal, the strong X-ray beams from a synchrotron-radiation source are required in order to obtain a strong-enough signal. Here we used the DUBBLE beamline at the European Synchrotron Radiation Facility in Grenoble. The experimental setup consists of an ultrahigh vacuum chamber with deposition facilities [8] coupled to a (2 + 3) surface diffractometer [9]. From measured structure factors, the structure of the various phases was determined using the ROD surface crystallographic software [10].

We use a (111) surface unit cell to describe our SXRD results. The lattice vectors are expressed in terms of the conventional cubic lattice vectors by $\mathbf{a}_1 = \frac{1}{2}[10\bar{1}]_{\text{cubic}}$, $\mathbf{a}_2 = \frac{1}{2}[\bar{1}10]_{\text{cubic}}$, and $\mathbf{a}_3 = [110]_{\text{cubic}}$. With this choice, the diffraction index l is along the perpendicular direction, while all rods are indexed by (hk).

Low-energy electron microscopy data can complement the accurate but average structural information from SXRD by providing real-space images of the surface. The technique uses the surface sensitivity of low-energy electrons and employs diffraction contrast in order to obtain video-rate images of a surface over an area that is typically several micrometers wide [11]. A selected area in the image can be identified by obtaining a local diffraction pattern.

3 Results

3.1 Phase Diagram

We first determined the phase diagram over a wide coverage and temperature range by observing the X-ray scattering during deposition for a range of temperatures. One such deposition curve is shown in Figure 5.1, where (using other diffraction data as well) a number of phases can be identified. The phase diagram we derive from these data is shown in Figure 5.2. Three ordered phases can be identified: a $(\sqrt{3} \times \sqrt{3})R30°$ reconstruction (called $\sqrt{3}$ from now on) at an ideal coverage of 1/3 monolayer (ML), a

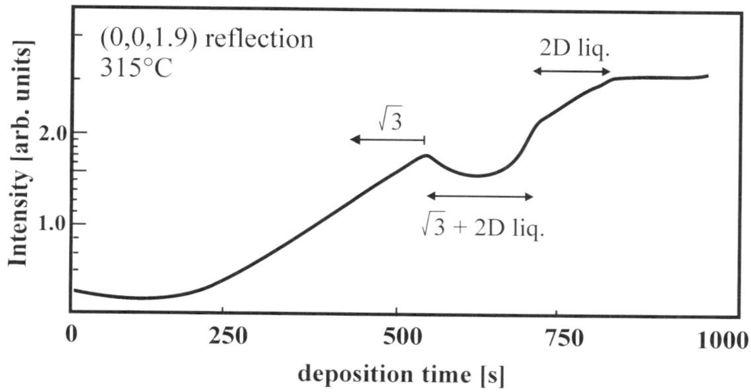

Figure 5.1. *The intensity of the (0,0,1.9) X-ray reflection during Bi deposition on Cu(111). Several phases can be identified.*

unit cell with matrix notation $\begin{pmatrix} 2 & 0 \\ 1 & 2 \end{pmatrix}$ (written as [2012] in the following) at 0.5 ML, and a uniaxial incommensurate (UIC) phase at 0.53 ML. For coverages in between the ideal values, typically two-phase coexistence is observed, but the [2012] phase is found to be stable over a coverage range from 0.47 to 0.50 ML. The phases melt at a temperature that depends strongly on coverage and that can be much lower than the bulk melting point of Bi of 271°C. The determination of the phase diagram is complicated by the desorption of Bi. For the typical time scale of an SXRD experiment, the dotted curve in Figure 5.2 indicates the temperature at which desorption becomes a problem. The melting of the √3 phase could therefore not reliably be observed using this technique. LEEM is much faster, and from LEEM data we estimate that this phase melts around 520°C.

3.2 Room-Temperature Structures

We determined the room-temperature structure of all three phases by acquiring sufficiently large data sets, including integer-order rods and (if applicable) fractional-order rods. In all cases, the analysis was straightforward and unambiguous [12]. Typical examples of measured rod data and the corresponding fits are shown in Figure 5.3. As we will show, knowing the room-temperature structures is important for understanding the high-temperature behavior.

The √3 phase is found to be a genuine surface alloy; that is, the 1/3 ML Bi is embedded in the first layer, while these materials do not mix in the bulk. One of every three Cu atoms in the top layer is substituted by a Bi atom (see Figure 5.4a). The driving force for this reconstruction is the tensile stress of the bare Cu(111) surface [13]. By embedding the larger Bi atoms, the electron density in the surface is increased. Several similar √3 structures have been found [2]; in particular, Pb on Ni(111) shows great similarity [14].

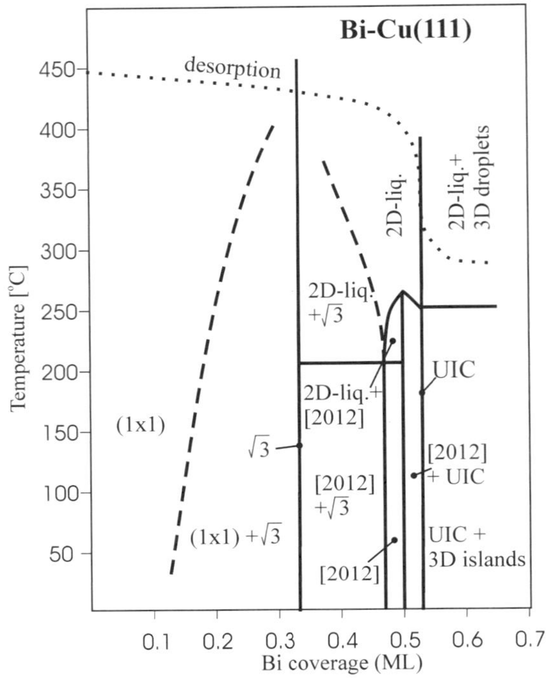

Figure 5.2. *Phase diagram of submonolayer Bi on a Cu(111) surface. For coverages above 0.53 ML, three-dimensional islands (solid or liquid) appear.*

When the Bi coverage exceeds 1/3 ML, it is not favorable to incorporate the extra Bi atoms in the surface layer. Instead, an overlayer with a [2012] unit cell is formed. Such an overlayer is an alternative means to provide the extra charge density for the substrate. The bonding of the Bi atoms to the substrate is not enough to reach a fully stable structure, because then one would expect the Bi atoms to fit exactly in the hollow sites of the substrate. The structure refinement shows that the Bi atoms are

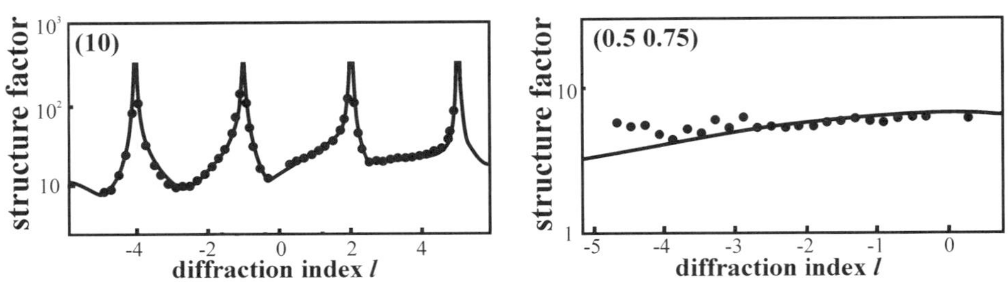

Figure 5.3. *The (10) and (0.5,0.75) rods of the [2012] phase. The circles are the data points, the solid curves the corresponding fit. The fact that the fractional-order rod has a nearly constant amplitude, shows that the reconstruction is confined to the single Bi layer.*

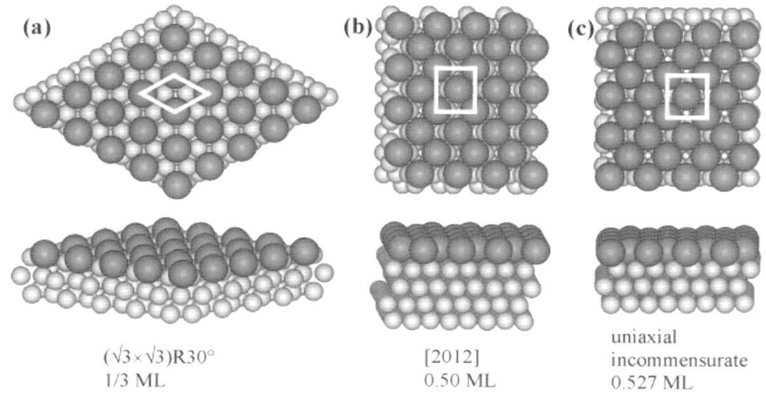

Figure 5.4. *The three stable room-temperature phases of Bi on Cu(111). The √3 phase is a surface alloy, whereas in the other two cases the Bi forms an overlayer. The surface unit cells are indicated in white.*

slightly displaced from this (by 0.13 Å), leading to a chainlike structure of the Bi (see Figure 5.4b). The rectangular lattice of the [2012] phase has a different symmetry than the p3 symmetry of the underlying substrate, and thus three rotational domains occur. One specific fractional-order reflection arises from one of these domains only, and thus the domains can be analyzed separately.

The [2012] phase has some room in the top layer. When more Bi is added, it turns out that this room is filled up. The uniaxial incommensurate phase that is formed can be considered as a [2012] phase squeezed in one direction (the vertical direction in Figure 5.4c). The lattice in the other direction remains the same; thus this direction is commensurate. For the UIC phase, the Bi-Bi interactions become more important than the Bi-Cu interactions, and the Bi forms its own lattice. From the location of the diffraction spots, the unit cell is found to be a centered rectangle. The structure determination is thus simple because the unit cell contains only one asymmetric atom. The height of the Bi layer can be determined because the commensurate direction of the overlayer will lead to interference with the substrate rods. Like in the [2012] phase, three rotational domains occur.

The rotational domains of the [2012] and the UIC phases can be observed directly using LEEM. Since each domain has unique reflections, diffraction contrast between them is possible. Figure 5.5a shows the three domains for the UIC phase.

From the deposition curve in Figure 5.1, one may get the impression that the √3 is formed immediately. The observed (0,0,1.9) reflection is not sensitive, however, to the lateral ordering of the deposited material; it only shows that the vertical arrangement of the Bi does not change until a coverage of 1/3 ML is reached. Repeating the experiment but then observing a fractional-order reflection from the √3 phase does allow the onset of the √3 domains to be detected. The coverage at which this occurs depends on temperature and is determined not only by thermodynamics but also by kinetics, particularly at low temperatures. The dashed curve in Figure 5.2 is only a rough indication of at what coverage the √3 phase occurs. LEEM allows a direct observation of the √3 domain formation. Figure 5.6 shows this at a temperature of 128°C, where the mobility of the Bi is limited and kinetics is very important.

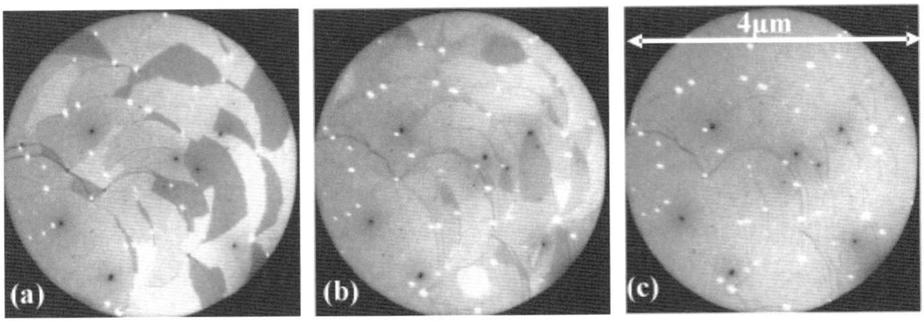

Figure 5.5. *LEEM images of the melting of the UIC phase (a) just below, (b) at, and (c) just above the melting point at 250°C. The different shades of gray in (a) and (b) are the different rotational domains of the UIC phase.*

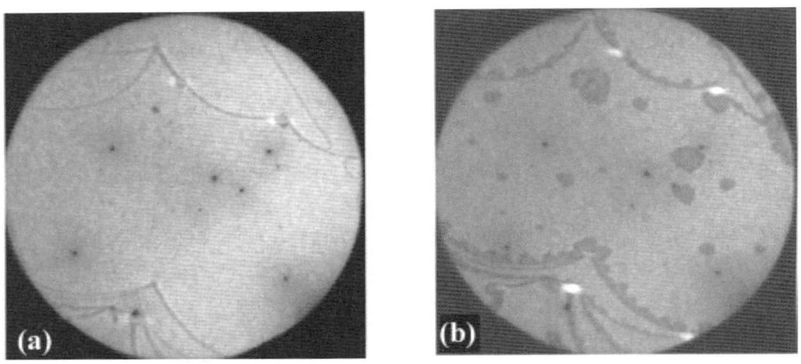

Figure 5.6. *LEEM images of the deposition of Bi on Cu(111) at a temperature of 128°C. (a) The initial surface, with a number of steps present. (b) After deposition of 0.08 ML Bi, showing the formation of √3 domains at both the steps and on the terraces.*

Combining the SXRD and LEEM results, we conclude that, at very low coverages, the Bi forms a disordered surface alloy. When the coverage reaches some specific value, ordered √3 domains are formed. At low temperatures, when the mobility is limited, the local coverage may be much higher than the average value, and thus √3 domains can form kinetically. Because of the kinetics, we are unable to determine the thermodynamic coverage value at which the transition from a Bi lattice gas to the √3 phase occurs. A simple description of this kind of ordering is given by the hard-hexagon model in statistical mechanics [15]. This model predicts a critical coverage of 0.28 ML but assumes no long-range interaction between the Bi atoms. Similar transitions from a lattice gas to a √3 phase have been observed for Sb on both Cu(111) and Ag(111) [16].

3.3 Liquid Phase

Now we turn our attention to the structure of the Bi layer above the melting transition [17]. We restrict ourselves to the coverage regime around 0.50 ML because then the melting temperatures are so low that Bi desorption is nearly negligible. The melting of the [2012] phase can easily be detected by observing the disappearance of a fractional-order spot as a function of temperature; for the UIC phase, an incommensurate reflection can play the same role. Here we will consider not the melting transition itself, but the structure of the liquid monolayer above the melting point. We have two options for determining the properties of this layer using diffraction. The first one is to detect the cylinder-like scattering expected from a two-dimensional liquid ("liquid ring") [3,18]. This diffuse scattering arises from the pair correlation in the liquid and will be azimuthally isotropic in a true two-dimensional liquid. Any deviation from this means the liquid is nonisotropic. The second option is to measure a substrate rod. If the liquid is partly commensurate with the substrate, one will see interference effects in these rods. In order to get as complete a picture as possible, we acquired both types of data.

Using a radial scan, we detected the presence of a liquid ring above the melting point. By performing such scans for different l values, we established that the diffuse scattering has a cylindrical shape as expected for a two-dimensional liquid and not a spherical shape as would be the case for a three-dimensional liquid. By fixing the detector and scanning the sample azimuth, the properties of the liquid ring can be found. For a temperature of 280°C and a coverage of 0.47 ML, an isotropic ring is determined (see Figure 5.7a). From this, however, we may *not* conclude that this layer closely resembles a two-dimensional liquid. Quite the contrary turns out to be the case, since an analysis of the substrate rods shows a large contribution from the Bi layer [17]. Thus this layer is found to have both solid and liquid properties. The behavior for increasing coverage while keeping the temperature fixed at 280°C is very interesting. For 0.50 ML, the liquid ring becomes anisotropic, and this effect becomes stronger at 0.53 ML (see Figure 5.7). Azimuthal anisotropy has been predicted [19] and observed [1] in two-dimensional melting as the occurrence of a hexatic phase close to the melting point. This, however, is not what is happening for the Bi layer. Not only is the anisotropy observed quite far from the melting point, but, more importantly, the number of azimuthal maxima is twelve instead of the six for a hexatic phase.

The origin of the twelve peaks can be deduced when the locations of the maxima are plotted in reciprocal space, together with those from the room-temperature [2012] and UIC phases. The liquid peak is, of course, much broader than the highly ordered room-temperature phases, but the peak positions can be determined quite accurately. Figure 5.8 shows the results. It is obvious that the peaks from the different phases are very close in position in reciprocal space, and thus the local structures in real space must be similar as well. Since the twelve peaks from the room-temperature phases arise from four peaks from three rotational domains, the liquid peaks must also come from three rotational domains. We thus find that the liquid has locally orthorhombic orientational order and that this occurs in three rotational domains. From the peak widths the liquid domain size is determined to be around 15 Å.

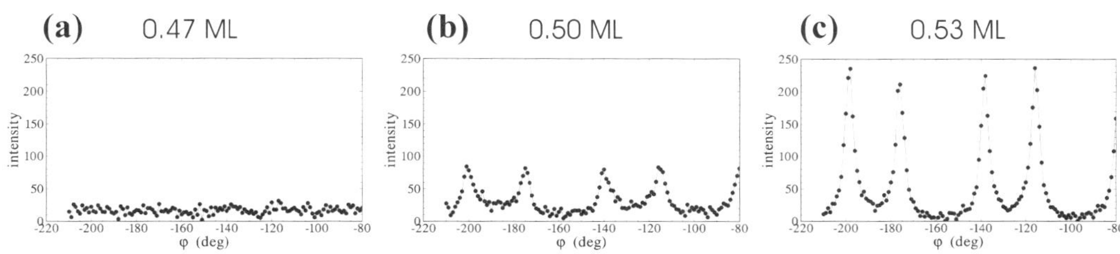

Figure 5.7. *Azimuthal scans (with background subtracted) through the diffuse scattering rings at Bi coverages of 0.47, 0.50, and 0.53 ML, and at a temperature of 280°C.*

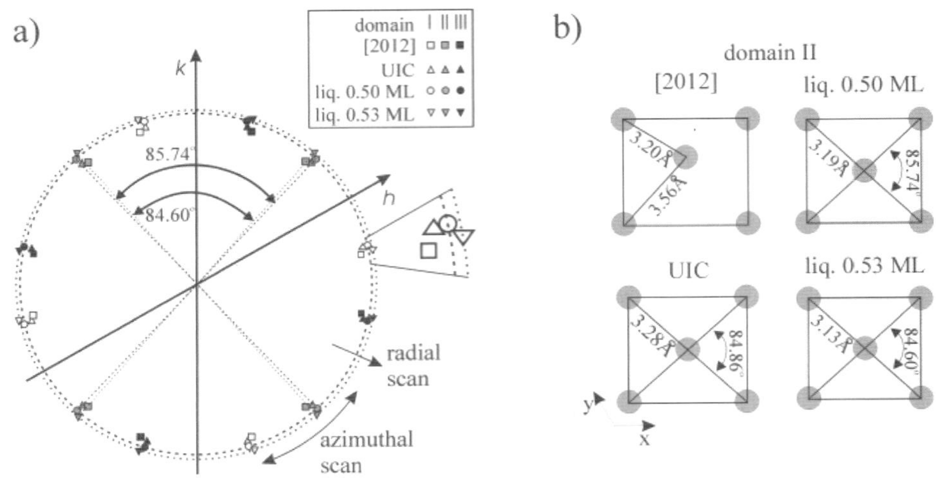

Figure 5.8. *(a) The positions in reciprocal space for selected reflections of the [2012] and UIC phases at room temperature, and for the maxima in the liquid ring at 280°C for coverages of 0.50 and 0.53 ML. (b) The local Bi structure in real space of the four structures shown in (a) for the domains with orientation II.*

For both 0.50 and 0.53 ML, we find from the substrate rods that this liquid phase continues to exhibit significant solid properties, although less than for 0.47 ML.

Figure 5.5 shows some LEEM images of the melting of the UIC phase. The LEEM data show that the domain boundaries of the UIC phase are quite mobile near the melting point. Above the melting point, the LEEM image shows no contrast between the domains because the domain size of 15 Å is below the resolution.

From the combined data of diffuse scattering and the substrate rods, the following picture emerges of the liquid structure at 280°C. For all three coverages, a significant fraction of the molten Bi atoms is located near the two different hollow sites of the substrate. At the lowest coverage (0.47 ML), there is little steric hindrance, and the azimuthal distribution is isotropic. For higher coverages, the Bi atoms start to interact more strongly and locally adapt a fourfold coordinated structure that closely resembles the ordered phases at a low temperature. The interaction with the substrate gives rise to three orientational domains, explaining the twelve observed peaks in the diffuse scattering. The increased steric hindrance causes a smaller fraction of the Bi atoms to be near the hollow substrate sites and leads to a larger fraction with liquid behavior. This is also reflected in the increased total intensity in the liquid rings (see Figure 5.7). The high mobility in the layer means that there is no strong distinction between "solid" and "liquid" Bi atoms: the measured solid fraction only indicates the time fraction that an atom can be associated with a hollow substrate site.

4 Conclusions

Bismuth forms three highly ordered phases at room temperature on Cu(111) whose structure can be fully determined using SXRD. The combination with LEEM data allows the formation of these phases and the transitions between them to be characterized in detail. We also identified two disordered phases. At low coverage and low temperature, a disordered surface alloy is formed that only orders to the $\sqrt{3}$ phase above a critical coverage. The other disordered phase is due to melting of the [2012] and the uniaxial incommensurate phases. We find that the liquid structure locally resembles the structure of these room-temperature phases, giving rise to a peculiar liquid with orthorhombic order.

Acknowledgment

This work is supported by the Council for Chemical Sciences of the Netherlands Organization for Scientific Research (CW-NWO).

References

1. R. J. Birgeneau and P. M. Horn, "Two-dimensional rare-gas solids," *Science* 232 (1986): 329.
2. D. P. Woodruff, *Surface Alloys and Alloy Surfaces,* vol. 10, *The Chemical Physics of Solid Surfaces* (Amsterdam: Elsevier Science B. V., 2002).
3. G. Reiter and S. C. Moss, "X-ray scattering from a two-dimensional liquid modulated by its periodic host," *Phys. Rev. B* 33 (1986): 7209.
4. S. C. Moss, G. Reiter, J. L. Robertson, C. Thompson, and J. D. Fan, "X-ray determination of the substrate modulation potential for a two-dimensional Rb liquid in graphite," *Phys. Rev. Lett.* 57 (1986): 3191.
5. F. Delamare and G. E. Rhead, "First stages of the deposition of Bismuth on Copper examined by LEED, II the (111): substrate," *Surf. Sci.* 35 (1973): 185.
6. R. Feidenhans'l, "Surface structure determination by X-ray diffraction," *Surf. Sci.* Rep. 10 (1989): 105; I. K. Robinson and D. J. Tweet, "Surface X-ray diffraction," *Rep. Prog. Phys.* 55 (1992): 599.
7. I. K. Robinson, "Crystal truncation rods and surface roughness," *Phys. Rev. B* 33 (1986): 3830.
8. E. Vlieg, A. van 't Ent, A. P. de Jongh, H. Neerings, and J. F. van der Veen, "An ultrahigh-vacuum chamber for surface X-ray diffraction combined with MBE," *Nucl. Inst. Meth.* A262 (1987): 522.
9. E. Vlieg, "A (2+3)-type surface diffractometer: mergence of the *z*-axis and (2 + 2)-type geometries," *J. Appl. Cryst.* 31 (1998): 198.
10. E. Vlieg, "ROD, a program for surface crystallography," *J. Appl. Cryst.* 33 (2000): 401.
11. E. Bauer, "Low-energy-electron microscopy," *Rep. Prog. Phys.* 57 (1994): 895.
12. D. Kaminksi, P. Poodt, E. Aret, N. Radenovic, and E. Vlieg, "Surface alloys, overlayer and incommensurate structures of Bi on Cu(111)," *Surf. Sci.* 575 (2005): 233.
13. H. Ibach, "The role of surface stress in reconstruction, epitaxial growth and stabilization of mesoscopic structures," *Surf. Sci. Rep.* 29 (1997): 195.
14. D. Brown, P. D. Quinn, D. P. Woodruff, P. Bailey, and T. C. Q. Noakes, "Medium-energy ion scattering structural study of the Ni(111)($\sqrt{3} \times \sqrt{3}$)R30°-Pb surface phase," *Phys. Rev. B* 61 (2000): 7706.
15. R. J. Baxter, *Exactly Solved Models in Statistical Mechanics* (London: Academic Press, 1982).
16. S. A. de Vries, W. J. Huisman, P. Goedtkindt, M. J. Zwanenburg, S. L. Bennett, I. K. Robinson, and E. Vlieg, "Surface atomic structure of the ($\sqrt{3} \times \sqrt{3}$)R30°-Sb reconstructions of Ag(111): and Cu(111)," *Surf. Sci.* 414 (1998): 159.
17. D. Kaminksi, P. Poodt, E. Aret, N. Radenovic, and E. Vlieg, "Observation of a liquid phase with an orthorhombic orientational order," *Phys. Rev. Lett.* 96 (2006): 56102.
18. F. Grey, R. Feidenhans'l, J. S. Pedersen, M. Nielsen, and R. L. Johnson, "Pb/Ge(111)1x1: An anisotropic two-dimensional liquid," *Phys. Rev. B* 41 (1990): 9519.
19. J. M. Kosterlitz and D. J. Thouless, "Ordering, metastability and phase transitions in two-dimensional systems," *J. Phys. C: Solid State Phys.* 6 (1973): 1181; B. I. Halperin and D. R. Nelson, "Theory of two-dimensional melting," *Phys. Rev. Lett.* 41 (1978): 121; A. P. Young, "Melting and the vector Coulomb gas in two dimensions," *Phys. Rev. B* 19 (1979): 1855.

6

Nonlinear Evolution of Surface Morphology in Short-Period Superlattices

Kevin E. Bassler[1,2]

Ondrej Caha[3]

1. *Department of Physics, University of Houston, 617 Science and Research Blvd. I, Houston, TX 77204, USA*
2. *Texas Center for Superconductivity at the University of Houston, 202 Houston Science Center, Houston, TX 77204, USA*
3. *Institute of Condensed Matter Physics, Masaryk University, Kotlářská 2, 611 37 Brno, Czech Republic*

Abstract

Continuum simulations of the self-organized, lateral compositional modulated growth of nearly lattice-matched semiconductor superlattices are presented. The results of the simulations are compared to the results of synchrotron X-ray diffuse scattering experiments of InAs/AlAs superlattices. We find that the time evolution of the compositional modulation during epitaxial growth can be explained by including a nonlinear dependence of the elastic energy of the growing epitaxial layer on its thickness. From the fit of the experimental data to the growth simulations, the parameters of thřs nonlinear dependence are determined.

1 Introduction

Nanostructures spontaneously develop in epitaxially grown semiconductor layers when the associated reduction in their strain energy outweighs the corresponding increase of the surface energy. This morphological instability is known as Asaro-Tiller-Grinfeld (ATG) instability. The instability was originally discovered by a linear analysis of a strained semi-infinite layer [1–3]. The nonlinear surface evolution for the same system was studied by Yang and Srolovitz [4] and Spencer and Meiron [5]. A generalization of the linear analysis to multilayers has also been made that, through a perturbation analysis, allows for the identification of unstable growth modes [6,7]. However, from this approach, it is not possible to predict the final morphological structure of the multilayer. We have, for the first time, simulated the full nonlinear evolution of epitaxially grown multilayers and predicted their final morphological structure. To accomplish this we simulate continuum nonlinear equations that account for the difference of the physical properties of very thin layers (down to a few monolayers) from the properties of the bulk. This difference leads to the creation of a thin stable layer at the surface, a so-called wetting layer, in the first stage of the Stranski-Krastanov growth mode. The occurrence of this "wetting effect" can be explained by a nonlinear dependence of the elastic energy density on the layer thickness [8–10]. To validate our approach, we simulate the evolution of laterally modulated short-period InAs/AlAs superlattices grown on InP and quantitatively compare the results to those obtained by X-ray scattering measurements.

2 Theory

The surface evolution via the surface diffusion is described by the equation [11]

$$\frac{\partial h(x,t)}{\partial t} = \frac{D_S \theta \Omega}{k_B T} \nabla^2 \mu(x,t) + F + \eta(x,t), \tag{6.1}$$

where $h(x,t)$ is the local height of the growing surface, D_S is the surface diffusion coefficient, θ is number of atoms per unit area on the surface, Ω is atomic volume, k_B is the Boltzmann constant, T is the temperature, μ is the chemical potential on the surface, F is the deposition rate, and η is the deposition and diffusion random noise. This model is assumed to be independent on the third coordinate y and hence only one-dimensional surface modulation in a two-dimensional (x,z) space can be simulated. Thus this approach is well justified for quantum wires or for strongly elongated quantum dots, but it can give only limited information for symmetric quantum dots.

The chemical potential μ can be expressed as [10]

$$\mu(x,t) = \mu_0 + \gamma\kappa + \frac{1}{2} C_{jklm} \in_{jk} \in_{lm} \Big|_{z=h(x)} + \frac{\mathrm{d} f_{el}^{(0)}(h)}{\mathrm{d}h}, \tag{6.2}$$

where μ_0 is the chemical potential of an ideally flat unstrained surface, γ is the surface tension, κ is the surface curvature, C_{jklm} are the components of the elastic tensor of the material, and $\in_{jk}$ is the strain tensor. The function $\overset{(0)}{fel}(h)$ describes a nonlinear dependence of the elastic energy density on the layer thickness h giving rise to a wetting effect. We assume that this function can be approximated by the exponential function [10]

$$f_{el}^{(0)}(h) = E_W (1 - \exp(-h / h_W)), \tag{6.3}$$

where E_W and h_W are parameters depending on the lattice misfit and elastic constants of the layer. To estimate the values of these parameters we use the atomistic valence force field method with the Keating potential [25]. In this calculation we neglect the surface relaxation and reconstruction and obtain the dependence of the density of the elastic energy on the thickness of a layer with a flat (001) surface. Fitting the result with the exponential formula (Equation 6.3), we obtain $E_W = 0.10 \times E_S$ and $h_W = 0.8 \times h_{ml}$, where E_S is the elastic energy density in the thick layer and h_{ml} is the thickness of one monolayer.

The elastic strain and displacement field are determined by solving the elastic equilibrium equation

$$\frac{\partial \sigma_{jk}}{\partial x_k} = 0, \tag{6.4}$$

where σ_{jk} are the components of the stress tensor. This equation can be solved by transforming it into the boundary integral equation

$$c_{jk}(x_0)u_k(x_0) = \int_S \mathrm{d}x\left[U_{jk}(x_0,x)t_k(x) - T_{jk}(x_0,x)u_k(x)\right], \tag{6.5}$$

where the integration is provided along the continuous boundary of a chosen region, $t_j(x)$ is the jth component of the surface traction, n_k is the kth component of outward normal, $U_{jk}(x_0,x)$ and $T_{jk}(x_0,x)$ are the elastic Green functions, and $c_{jk}(x_0) = \frac{1}{2}\delta_{jk}$ for x_0 on the boundary of the integrating region.

The Green functions for the periodic two-dimensional structure are presented in Yang and Srolovitz [4]. Each layer was taken to be an independent region for the purposes of integration. The complete system of equations is obtained by the boundary conditions on the surface and the interfaces. The boundary conditions represent the continuity of the displacement on the interfaces $u_j^{(+)}(x) = u_j^{(-)}(x)$, where the indexes (+) and (–) denote the values obtained from the upper and lower layer, respectively. The other condition is that the traction acting on the boundary is in equilibrium $t_j^{(+)}(x) = -t_j^{(-)}(x)$, where the minus sign corresponds to the opposite direction of the outer normal [14]. On the stress-free top surface, the surface traction t equals zero. We assume that the substrate far below the multilayer is undeformed. Since the neighbor layers are related by the boundary conditions, the coupled set of the integral equations has to be solved numerically. The interfaces and the surface were divided into equidistant elements and, assuming that the displacement and traction are constant inside each element, the integral Equation 6.6.5 can be transformed into a set of the linear algebraic equations. The particular integrals of the Green functions over the single elements are then calculated by Gaussian quadrature. The main difficulty occurs when $x = x_0$; then the Green functions U_{jk} and T_{jk} become singular. In such a case we use the technique described in Luo, Liu, and Berger [13], which calculates the singular part of the integrals analytically and calculates the nonsingular part numerically with Gaussian quadrature. For example, the distribution of the strain tensor components $\in_{xx}$ and $\in_{zz}$ in an InAs/GaSb superlattice are shown in Figure 6.1 [15].

3 X-ray Scattering Experiment

The results of the simulations have been compared to the X-ray diffuse scattering experiments on a series of InAs/AlAs superlattices grown by molecular beam epitaxy on InP (001) substrates [16]. The

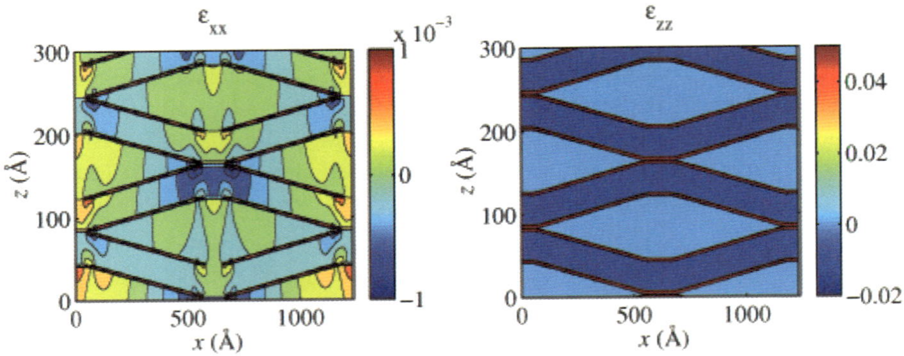

Figure 6.1. *The calculated components of the strain tensor inside the InAs/GaSb multilayer.*

samples were prepared in a series with two, five, ten, and twenty superlattice periods. The diffuse scattered intensity distributions around 400 and 040 reciprocal lattice points in the plane parallel to the sample surface were measured [17]. Two first-order lateral satellite maxima were observed at samples with five or more superlattice periods. The measured intensity distributions are shown in Figure 6.2. The distance of the lateral satellites from the reciprocal lattice point $q_{x,y} = 0$ is inversely proportional to the lateral modulation period; the period remains constant for all samples and was determined to be (267 ± 15) Å. From the position of the satellite maxima in the $q_x q_y$ plane, we determined the modulation

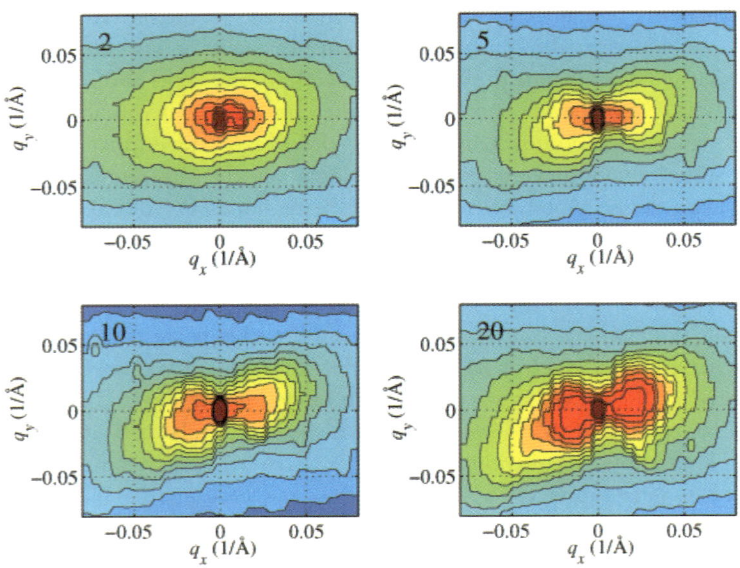

Figure 6.2. *Diffuse scattering intensity measured around the 400 reciprocal lattice point on the series of samples with two, five, ten, and twenty superlattice periods. The contour step is $10^{0.25}$. The diffraction vector is along the q_x axis.*

directions close to [310] and $[\bar{1}30]$. From the measured intensity, it is possible to extract the correlation function $\varepsilon(\mathbf{x} - \mathbf{x}') = \langle (c(x) - c_0)(c(x') - c_0) \rangle$, where $c(x)$ is the local InAs concentration averaged along the growth direction z and c_0 is the average InAs concentration [17,18]. The dependence of the first coefficient ε_1 of the Fourier series of ε obtained from the experimental data is plotted in Figure 6.3; this coefficient corresponds to the modulation amplitude.

4 Simulation of InAs/AlAs Multilayer Growth

Simulations of InAs/AlAs multilayer growth have been performed with known material parameters. The resulting structure of the interfaces inside the superlattice is shown in Figure 6.4. We find a modulation period of 300 Å, which is in reasonable agreement with the observed value $L_{exp} = (267 \pm 15)$ Å. It should be noted that the simulated modulation period is affected by the size of the simulated region because there can be only an integer number of the waves in the simulated system of a given size. To eliminate the influence of the system size, we simulated the growth of several systems of sizes 1500, 2250, 3000, and 4000 Å. For the various system sizes, the obtained modulation period was always in the interval (300 ± 20) Å, depending on the particular system size. We also found that the modulation amplitude is not affected by the system size.

During the growth of the first layers in the stack, the modulation amplitude grows exponentially as predicted by the linearized theory [6,7]. In the further growth stage, however, the rate of the growth of the modulation amplitude decreases (see Figure 6.3, where the experimentally obtained values of the first Fourier coefficient ε_1 of ε are compared with the simulation results).

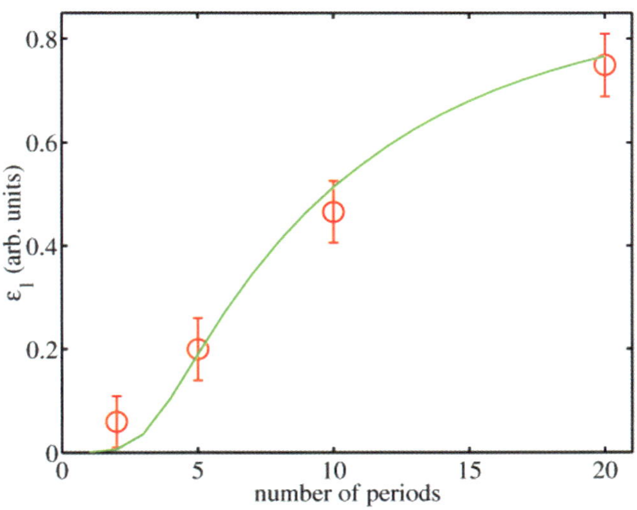

Figure 6.3. *The dependence of ε_1 on the number of superlattice periods; this dependence describes the time evolution of the modulation during the growth. The circles with error bars are the experimental points obtained from the X-ray data, and the full line represents the simulations.*

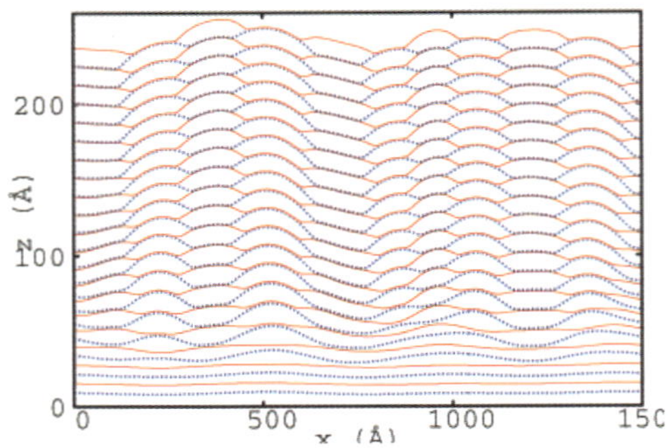

Figure 6.4. *Simulated interface profiles in a InAs/AlAs superlattice. Thick dashed lines denote the interfaces between InAs (under) and AlAs (above). Thin solid lines denote the interfaces between AlAs (under) and InAs (above).*

The simulations show a good agreement with the experimental results in spite of the simplified one-dimensional model of the surface used. Transmission electron microscopy (TEM) on similar samples [16] revealed that the modulation is indeed nearly one-dimensional, resulting in a quasi-periodic sequence of quantum wires; this fact explains why the one-dimensional model is sufficient for the simulation of the modulation kinetics. The continuum simulation also allows for the formation of nonphysical layers the thickness of which are fractional numbers of monolayers. However, our results based on a continuum approximation are qualitatively similar to those obtained using an atomistic model and a monolayer step corrugation [19].

According to Shitara et al. [20], the surface diffusion of In is about fifty times faster than Al, although an exact value of the surface diffusivity of In is not known. In Kasu and Kobayashi [21], the surface diffusion constant of Al at 530°C was found to be 1.5×10^{-7} cm^{2}s^{-1}. The deposition flux of As atoms is higher than the flux of In and Al atoms at usual molecular-beam epitaxy (MBE) conditions [22], therefore only diffusivities of Al and In atoms play a role.

On the other hand, our simulations have shown that the values of the diffusion constants have nearly no influence on the modulation amplitude because the diffusion process is sufficiently quick and the growing surface is nearly in an equilibrium state. The diffusion rate, however, does affect the modulation *period*. In the case of very slow diffusion (on the order of 10^{-10} cm^{2}s^{-1} for Al), the growth of the larger ripples at the expense of smaller ones (the Ostwald ripening) does not take place and the modulation remains constant during the growth of the whole superlattice stack. If the diffusion is very fast (on the order of 10^{-7} cm^{2}s^{-1} for Al), the Ostwald ripening takes place during the growth of the first layer, which leads to the creation of a smaller amount of larger, more distant dots separated by larger flat areas of a thin wetting layer. The nucleation of the ripples on the subsequent interfaces is affected by the local distribution of lateral strains originated from the large buried ripples. Due to the elastic anisotropy, this distribution gives rise to local minima of the chemical potential at the rims of the buried ripples (two local minima for each ripple) so that the number of the ripples is duplicated. After the deposition of several periods, the ripples cover the whole interface again and the flat areas between the ripples disappear. The resulting modulation period is again approaching the period obtained for a slow diffusion. In our simulations, we have achieved a good correspondence of both the modulation period and the time dependence of the modulation amplitude for any value of the diffusion constant of Al between 10^{-10} and 10^{-7} cm^2s^{-1}.

The resulting interface morphology is substantially affected by the wetting effect, that is, by a nonlinear dependence of the volume density of the elastic energy on the layer thickness. This dependence is approximated by Equation 6.6.3. The parameters E_W and h_W affect the modulation amplitude, and they have no influence on the modulation period. In the first stage of the multilayer growth, the modulation amplitude rapidly increases; this increase is slowed down after the growth of about ten superlattice periods. The parameter h_W affects mainly the rate of the initial amplitude growth; this rate increases with decreasing h_W. The parameter E_W determines the slowing-down process: for larger values of E_W, the slowdown of the amplitude growth is observed earlier than for smaller E_W. The best correspondence of the measured and simulated modulation amplitudes was obtained for the values $E_W = 0.15 \times E_S$ and $h_W = 0.6 \times h_{ml}$, which very well correspond to the values calculated from the valance force field method discussed above.

5 Conclusion

In conclusion, we have developed a novel method to simulate the multilayer growth of nearly lattice-matched lateral composition modulated semiconductors using a nonlinear continuum model. With this method we can for the first time predict the final morphology of nanostructured multilayers. The simulation results for InAs/AlAs multilayers agree very well with experimental data obtained by X-ray scattering. From the simulations performed for various values of material parameters, we have found that the wetting effect (the nonlinear dependence of the elastic energy density on the layer thickness) has a crucial influence on the resulting interface morphology; from the fit of the experimental data with the simulations, we have determined the parameters of this nonlinear dependence and we have compared these values with atomistic simulations. We also find that the diffusion rate of the atoms on the surface does not affect the modulation amplitude but does affect the modulation period.

6 Acknowledgment

The authors are grateful to S. C. Moss, who has helped this project in many ways both materially and intellectually. We are also grateful to A. Mascarenhas, A. G. Norman, and J. L. Reno for providing samples and to V. Holy, J. Li, D. Stokes, Z. F. Huang, and P. W. Voorhees for helpful discussions. One of us (O. C.) thanks the University of Houston for financial support during his stay in Houston. This work was supported by the project MSM 0021622410 of the Ministry of Education of the Czech Republic and by the National Science Foundation through grant DMR-0406323.

References

1. R. J. Asaro and W. A. Tiller, *Metall. Trans.* 3 (1972): 1789.
2. M. A. Grinfeld, *Sov. Phys. Dokl.* 31 (1986): 831.
3. D. J. Srolovitz, *Acta Metall.* 37 (1989): 621.
4. W. H. Yang and D. J. Srolovitz, *J. Mech. Phys. Solids* 42 (1994): 1551.
5. B. J. Spencer and D. I. Meiron, *Acta Metall. Mater.* 42 (1994): 3629.
6. Z.-F. Huang and R. C. Desai, *Phys. Rev. B* 67 (2003): 075416.
7. L. E. Shilkrot, D. J. Srolovitz, and J. Tersoff, *Phys. Rev. B* 62 (2000): 8397.

8. B. J. Spencer, *Phys. Rev. B* 59 (1999): 2011.
9. J. Tersoff, *Phys. Rev. B* 43 (1991): R9377.
10. H. R. Eisenberg and D. Kandel, *Phys. Rev. B* 66 (2002): 155429.
11. W. W. Mullins, *J. Appl. Phys.* 28 (1957): 333.
12. P. N. Keating, *Phys. Rev.* 145 (1966): 637.
13. J. F. Luo, Y. J. Liu, and E. J. Berger, *Comp. Mech.* 24 (2000): 448.
14. J. F. Luo, Y. J. Liu, and E. J. Berger, *Comp. Mech.* 22 (1998): 404.
15. J. H. Li, D. W. Stokes, O. Caha, S. L. Ammu, J. Bai, K. E. Bassler, and S. C. Moss, *Phys. Rev. Lett.* 95 (2005): 096104.
16. S. P. Ahrenkiel, A. G. Norman, M. M. Al-Jassim, A. Mascarenhas, J. Mirecki-Millunchick, R. D. Twesten, S. R. Lee, D. M. Follstaedt, and E. D. Jones, *J. Appl. Phys.* 84 (1998): 6088.
17. O. Caha, P. Mikulík, J. Novák, V. Holý, S. C. Moss, A. Norman, A. Mascarenhas, J. L. Reno, and B. Krause, *Phys. Rev. B* 72 (2005): 035313.
18. O. Caha, V. Holý, and K. E. Bassler, *Phys. Rev. Lett.* 96 (2006): 136102.
19. L. Bai, J. Tersoff, and F. Liu, *Phys. Rev. Lett.* 92 (2004): 225503.
20. T. Shitara, D. D. Vvedensky, J. H. Neave, and B. A. Joyce, *Mater. Res. Soc. Symp. Proc.* 312 (1993): 267.
21. M. Kasu and N. Kobayashi, *J. Appl. Phys.* 78 (1995): 3026.
22. M. A. Herman and H. Sitter, *Molecular Beam Epitaxy: Fundamentals and Current Status* (New York: Springer, 1996).

Part II

Elastic and Thermal Diffuse Scattering from Alloys

7

On the Sizes of Atoms

Simon C. Moss

Physics Department, University of Houston, Houston, TX 77204, USA

Abstract

In this selected guide to scattering studies of disordered alloys and related materials, the issue of atom size is prominent in the scattering data. A few examples will be covered that include cases of Bragg scattering strongly affected by a static Debye-Waller factor (C in Fe) [1] or a crystallographic determination of the Cr ion position (size) in ruby [2]. We have also collected lattice parameter and diffuse scattering data that required a reconsideration of the structure and diffusion in very dilute rapidly diffusing Au in Pb [3–5]. From our study of the disorder and static displacements in a nearly equiatomic bcc Fe-Cr crystal [6], in which, using anomalous X-ray scattering (AXS), we were able to recover the equilibrium pair correlations and the shell-by-shell species-dependent displacements of the Fe-Fe, Cr-Cr, and Fe-Cr distances from their average lattice spacing. These data revealed a quite anomalous value of the nearest-neighbor Cr-Cr displacement, which was negative and nearly an order of magnitude smaller in length than the negative Fe-Fe displacement, where a_0(Cr) > a_0(Fe). Finally, predictions exist for very dilute transition metal (TM)-Al liquids [7] whose dynamics we are currently studying using quasi-elastic neutron scattering to probe the remarkable influence of Fe and, especially, Ti on the viscosity (sluggish diffusion) of the liquid phases, which both show a large decrease in partial molar volume for the TM. The calculations by Turnbull [7] are for large clusters forming about a TM atom and thereby limiting the self-diffusion. Electronic structure plays a prominent role in all of these cases, as does elasticity theory in the solids. Clearly, the use of scattering data to access atomic size effects will continue as a very important factor in our contemporary theoretical understanding of electronic interactions in materials and the related issue of phase stability in alloys that was so brilliantly pioneered many years ago by Sir William Hume-Rothery.

1 Introduction

I begin with a quote from the late Sir Neville Mott in his introductory lecture to the book *Metallic Solid Solutions*, edited by J. Friedel and A. Gunier in 1963 [8]. It might be noted here that, while the European scientific community considered metallurgy and metal physics and chemistry as important theoretical topics, the U.S. physics community somewhat disdained it all in the rush to nuclear and high-energy physics. We thus found a real paucity of activity in most physics and metallurgy departments, the universities at Iowa State, Illinois, Chicago, Berkeley, Harvard, CalTech, Pennsylvania, and Cornell being among the notable early exceptions. (Once the Advanced Research Projects Agency [ARPA] formed the many Centers for Materials Science, all of that changed.) Mott wrote the following on page I-2 of the aforementioned book [8]:

> The first point that has to be emphasized about heats of solution is they are normally small compared with the heat of formation of the constituent metals. This is shown in Table 7.1. The second is that heats of solution certainly show a size effect, and often become large and positive if the difference in size between solute and solvent is too big. [William] Hume-Rothery and others [9] came to the conclusion that if the atomic radii differ by more than ca. 14% the solubility will normally be very small although there are of course exceptions, e.g., Zn in Cu. One has to ask therefore, how a metal atom manages to have a size.

TABLE 7.1

Heats of solution of certain metals (*italicized*) in matrix (*not italicized*)

	(KCAL/GRAM-ATOM)		
Cu-Ag	3480	81800	(Cu)
Ag-Cu	2940	69100	(Ag)
Cu-Au	–2900	111400	(Au)
Ag-Au	–3470		
Cu-*Zn*	–9180	31500	(Zn)
Ag-*Zn*	–1850		
Ag-*Cd*	–6200	26750	(Cd)
Au-*Cd*	–8500		
Ag-*Al*	–11900	76100	(Al)
Au-*Sn*	–7000		
Cr-Ni	–950		
Ni-Au	9600		

Aside from Cu-Zn, there is the case of Ni-Au, where the heat of the solution is large and positive, indicating a strong tendency toward phase separation, which is observed as a miscibility gap in the equilibrium phase diagram. In fact, early measurements of Golding et al. [2] on the pressure dependence of the elastic constants in Au-rich Au-Ni crystals could be simply interpreted using a

Born-Mayer potential and essentially assigning a vacancy to the Ni site. Our scattering data from that period also indicated a very prominent affect of atom size on the diffuse X-ray scattering that was large enough to be confused with short-range order as opposed to clustering [10]. The late J. B. Cohen and T. B. Wu did a much more detailed study, including measurement of the phonon dispersion curves, and found that there was a competition between ordering and clustering [11].

The much more recent high-energy synchrotron work of Reichert et al. [12] on Au-Ni alloys suggests a more subtle competition in the disordered state between clustering (phase separation) and local ordering—a very interesting result attributed to the dominance of size mismatch (strain induced) over purely chemical considerations (they are both electronic in origin, of course).

Indeed, this issue of atom size remains a principal topic in metal physics, especially on the theoretical side but also increasingly in scattering experiments where the details of the local and long-range species-dependent displacements can be measured. Both of these efforts are well represented in some very recent work by Rodriguez et al. [13] on Ni-Pt.

One of the most elegant early discussions of the conditions for alloy phase formation can be found in the Dover reprint of *Electrons, Atoms, Metals and Alloys* by Hume-Rothery [14]. This book is wonderfully written as a dialogue, in the Platonic fashion, between an "Older Metallurgist" (OM) and a "Young Scientist" (YS). Many materials scientists know it well, but I would recommend it for any student today wanting an elegant introduction to the subject of phase stability and phase diagrams. Later work, of course, has refined and replaced many of the details of what was written then, revised in 1955, and published with an epilogue from 1962. With a thorough grounding in modern quantum mechanics, some of the preliminary understanding must be relegated to history—the devil, once again, being in the details, which in turn can only be accessed using modern computers and advanced calculational algorithms as discussed in this book. Also discussed will be the seminal role played by atom size in semiconductor films and multilayer structures, including the fine-tuning of nanodots and nanowires, as elaborated by Jianhua Li in Chapter 3 and Kevin Bassler and Ondrej Caha in Chapter 6 of this book.

2 Use of Bragg Scattering Data

At this point, I would like to reiterate some work, which was reviewed in my von Hippel Lecture of 2001 and published in the *Materials Research Bulletin* [15]. (*Note*: In the *Materials Research Bulletin* paper, everywhere $<|c_q|>^2$ must be replaced by $<|c_q|^2>$—an unfortunate typographical error.)

Let us briefly explore defects and their static displacement fields. If we examine the scattering intensity from a crystal with either substitutional or interstitial defects that expand the lattice, three major effects obtain (actually, the defect could consist of a number of atoms as in irradiated materials or coherent precipitation in an alloy as with Guinier-Preston zones): (a) the lattice parameter will get larger and the Bragg peaks will shift to smaller values with increasing scattering wave vector **Q**; (b) there will be an attenuation of these peaks with increasing **Q**, often referred to as a static Debye-Waller factor; and (c) the intensity lost from the Bragg peaks appears as Huang diffuse scattering (HDS), which decreases as $1/g^2$ in the vicinity of the Bragg peaks, as discussed further on, much as with thermal diffuse scattering [16–18], where $\mathbf{Q} = \mathbf{G}_{hk\lambda} \pm \mathbf{g}$ and $\mathbf{G}_{hk\lambda}$ is a reciprocal lattice vector.

When the static Debye-Waller factor, $e^{-2M_{st}}$, is due to the presence of large anharmonic displacements, it may differ substantially from its thermal counterpart. Krivoglaz [16] derived the proper expression for such cases, shown below in Equation 7.1:

$$M_{st} \rightarrow c \sum_{m} (1 - \cos \mathbf{Q} \cdot \delta_m), c << 1 . \tag{7.1}$$

Here, c is the concentration of defects, and the sum is over all m neighbors of the defect where displacement off the lattice sites is δ_m. This may give rise to the unusual effect of an oscillatory Debye-Waller factor in which the peaks can decrease and then increase as ($\mathbf{Q} \cdot \delta$) increases, and this effect may also obtain at the surface of crystals whose atoms show large displacements.

When I was at MIT a few years ago, the late Morris Cohen presented me with a very interesting problem, namely the determination of the nearest-neighbor dipole distortion about the octahedrally sited C atom in Fe-C martensite. This tetragonal phase, quenched from the fcc austenite phase, in which the C atom is in tetrahedral sites, is responsible for the strength of steel through the distortion dipole. Applying the conventional theory for a random tetragonal polycrystalline structure, the *c* axis is larger, and the (002) reflection comes at a lower angle than the (200, 020) pair arising from equal *a* and *b* axes, and their intensity ratio should thus be 2:1. We knew that the sample from U.S. Steel was excellent, that is, fine-grained and random. But the intensity ratio (002)/(200,020) was not 2:1, but rather ~4:1, and this was not due to texture effects [1]. The (112)/(211,121) also was much larger than expected. Comparing these doublets was very useful because we did not have to be concerned with any other diffraction effects. These large ratios were due to the very large nearest-neighbor dipole distortion, and Krivoglaz's expression had to be used.

All of the other displacements (save δ_1) were determined from the calculations of Keating and Goland [20] using a Johnson potential. But the dipole distortion needed to be determined, and from the data analysis, using Equation 7.1 for the first five neighbors, we obtained a value for the extension of the Fe-C distance along the *c* axis of δ_1 = 0.5 Å, which was interesting because the Fe_3C phase also has that expanded Fe-C distance, but with a bonding expected to be strongly ionic. More practically, this dipole distortion is responsible for the hardness of steel. (Using conventional theory for the nearest-neighbor dipole distortion yielded a value of δ_1 ~0.9 Å, which was patently incorrect.)

Another quite early work, done at MIT in collaboration with R. E. Newnham [2], was the determination of the Cr position within the oxygen octahedral cage of Al_2O_3, where the Cr substitutes for Al to produce the beautiful pink-red color of a ruby crystal. However, Cr_2O_3 is a green compound, while pure corrundum (Al_2O_3) is clear. The question of the exact Cr position then arose out of an extensive Weisenberg photographic study of a blood-red 4 at.% Cr crystal, grown by A. Linz in the von Hippel Laboratory for Insulation Research using a flame fusion technique. We had an excellent but distorted Al_2O_3 crystal and two heavily distorted (extinction-free) 4% Cr crystals. The question facing us was where the Cr^{+3} substitutional ion resided along the octahedral *z* axis (the *c* axis of the crystal). Because the structure factor for the 00l reflections contains a phase factor of exp $-(2\pi i 6lz)$, it was extremely sensitive to tiny shifts in the *z* axis position of the Al or Cr (neither of which was centered in the octahedron). Using Al_2O_3 crystals to obtain an accurate diffractometer (ω-scan) refinement of the Al^{+3} position based on the 00 · 6, 00 · 12, 00 · 18, 00 · 24, and 00 · 30 reflections, we could even refine the value of the actual ionicity of "Al^{+3}" and obtain an Al Debye-Waller factor. These numbers were then used for the Al atoms in the Cr-doped crystal, and in Figure 7.1, from Moss and Newnham [2], is the resulting reliability-factor (R-factor) versus Cr position. Tables 7.2 and 7.3 from Moss and Newnham [2] show the actual data, and it is clear that an excellent fit would be obtained if the larger

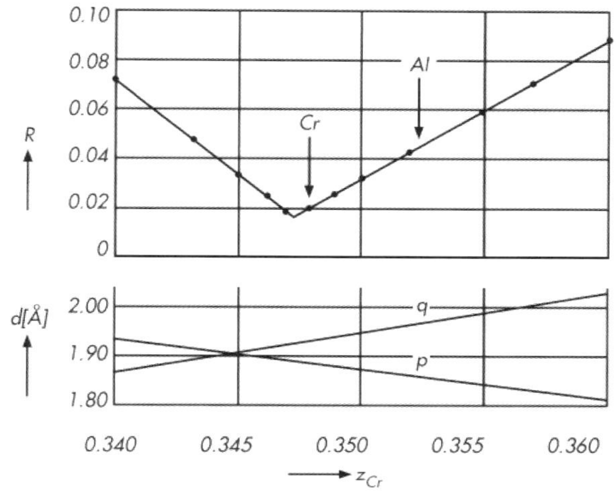

Figure 7.1. *The R-factor and chromium-oxygen distances for 4% ruby plotted as a function of z_{Cr}, the Cr atomic coordinate; q is the distance to the three oxygens shared with the closest Al neighbor, and p represents the distance to the other oxygen triad. Arrows indicate the Al and Cr coordinates derived from the X-ray intensity data.*

Cr ion moved closer to the center of the octahedron. As we noted at the time, this gives rise to a crystal field shift of the Cr levels. Were this not so, rubies would be pale green!

3 Bragg and Diffuse Scattering

Another example in which scattering data was used not only to infer atom size but also to restrain models for its diffusion is the very dilute solution of Au in Pb, whose rapid diffusion (similar to the interstitial diffusion in liquids) was pioneered by David Turnbull and his students (see Warburton and Moss [3] [and references therein] and Bolze et al. [4,5]) and was studied as well by Decker et al. [21],

TABLE 7.2

Summary of $|F_o|^2$ values, corrected for polarization and absorption

		4% RUBY			
$h\,k \cdot l$	AL_2O_3	CRYSTAL A	CRYSTAL B	AVERAGE	RATIO RUBY/AL_2O_3
0 0 · 6	2205	1217	1259	1238	0.56 ± .01
0 0 · 12	18524	21652	20910	21281	1.15 ± .02
0 0 · 18	16968	18032	16550	17291	1.02 ± .04
0 0 · 24	279	399	322	361	1.29 ± .13
0 0 · 30	7543	8592	8544	8568	1.13 ± .01

TABLE 7.3

Comparison of observed and calculated structure factors of corundum and ruby

$h\ k\ l$	0 0 · 6	0 0 · 12	0 0 · 18	0 0 · 24	0 0 · 30
Al_2O_3					
$\|F_o\|$	47.0	136.1	130.3	16.7	86.9
F_o (z_{Al} = .3523)	–41.6	136.6	–134.7	–16.7	–87.4
4% ruby					
$\|F_o\|$	35.2	145.9	131.5	19.0	92.6
F_o (z_{Cr} = .347)	–31.5	142.6	–132.1	–17.3	–90.9
F_o (z_{Cr} = .352)	–33.5	138.1	–137.4	–21.2	–91.1
F_o (z_{Cr} = .360)	–38.2	130.3	–141.9	–19.0	–83.8

Vanfleet [22], and Stracke and Herzig [23]. Our first publication [3] was solely on the variation of lattice parameters along the then accepted solubility line, where the Au solubility limit at the eutectic line is of the order of ~1000 ppm, or 0.1%.

The question facing us was the site occupancy of Au (interstitial or substitutional) and the displacements exerted by the dissolved Au atom on the Pb host lattice (either positive or negative, i.e., either expanding or contracting the lattice). The first study by Warburton and Moss [3] on a 900-ppm crystal showed clearly that, once our measured thermal expansion of the Pb was corrected for, the solid solution *decreased* in volume with increasing Au (increasing *T* along the phase boundary). This certainly ruled out an interstitial siting, but some questions remained about the bulk behavior, even though the experiments were done in vacuo.

To this end, two companion studies were done by the group of J. S. Peisl in Munich [4,5], in which both lattice parameter values were measured with neutrons using a very accurate backscattering technique and HDS was performed, first with X-rays and then in bulk with neutrons. The general trend of Warburton and Moss [3] was recovered as regards the lattice parameter versus temperature, but no increase was seen on entering the single-phase field. (A new phase boundary was used from Dworschak et al. [24].) The first paper concentrated on the quenched state in which the Au atoms seem not to precipitate into Pb_3Au despite their rapid diffusion at higher *T*. Both fully annealed (at higher *T*) and quenched crystals containing 900 ppm of the Au were studied using X-ray HDS. In such studies, the HDS in the neighborhood of a Bragg reflection is given by

$$I_{HDS} \sim c(F / V_c)^2 \,|G \cdot \tilde{\mu}(g)|^2 , \tag{7.2}$$

where c is the concentration, F is the structure factor including a thermal Debye-Waller factor, V_c is the unit cell volume, and G is, again, the scattering wave vector to the Gth reciprocal lattice (Bragg) point [4]. $\tilde{\mu}$ (g) is the Fourier transform of the real-space displacement field $\mu(\mathbf{r})$ ($\mathbf{g} = \mathbf{Q} - \mathbf{G}$, where $\mathbf{Q}$ is the scattering wave vector) and the scattering peaks at the Bragg points falling off, similar to one-phonon thermal diffuse scattering (TDS) as g^{-2}. $\tilde{\mu}$ ($\mathbf{g}$) contains the information about the strength and symmetry of the asymptotic displacement field about an isolated defect (Au in this case). It can be written explicitly for a cubic crystal in terms of the trace of the double-force tensor (Kanzaki forces) and the elastic constants of the crystal. The vector **g** is associated with a wave vector in reciprocal space and characterizes the transform of the displacement field where the HDS is valid for small **g** near the Bragg point.

Because of the cubic symmetry of the defect, the HDS can be written using the $(\mathrm{TrPij})^2$ for **g** and **G** in the <110> direction. Using the value of the lattice parameter change with composition (Δa)/a, one may also recover the TrPij and thus retrieve both the sign and strength of Pij.

Essential to this study was the combination of the two measurements

$$I_{HDS} = k_1 c_z \left(\mathrm{TrPij}\right)^2 \tag{7.3}$$

and

$$(\Delta a) / a = k_2 c_z \mathrm{TrPij}, \tag{7.4}$$

where k_1 and k_2 are known constants, and c_z is the concentration of defects of number z (1 for singles, 2 for dimers, etc.). From the combined measurement of I_{HDS} and (Δa)/a, the quantities c_z and TrPij were determined.

The crystal was supplied by G. Gruzalski of the Turnbull group (noted in References 4 and 5) and also contained 900 ppm Au. By comparing the quenched and fully annealed state, the distortion-induced diffuse scattering could be evaluated. (It did, indeed, have a slope of g^{-2} as in Figure 7.2 [4].) The results of lattice parameter (Δa)/a above yielded a value of TrPij = −10.5 ± 1eV, while the HDS yielded |TrPij| = 11.9 ± 0.5 eV, strongly supporting the negative value for TrPij and agreeing quite well. *z* was assumed to be 1.0; that is, only *single* substitutional sites were filled!

The measurements of HDS with neutrons are more difficult because of resolution effects (Pb is, however, an excellent neutron scatterer), but they too yielded |TrPij| = 9.8 ± 2.0 eV, in nice agreement with the X-ray work and confirming the fact that the surface sensitive X-ray work was characteristic of the bulk.

It was therefore necessary to rule out higher order clusters, such as Au_4 (see Warburton and Moss [3]). However, it is certainly possible that the quenched state could consist of single substitutionals and dimers of less than 20%, as predicted in general by Krivoglaz [19], where the thermodynamic and elastic conditions for the formation of $n + 1$ atoms on n sites are discussed in detail. Such a result would also be in agreement with the available diffusion data of Vanfleet [22].

The final question regarding this system is the temperature dependence of the lattice parameters. For this measurement, the neutron back-scattering spectrometer at the former TU-Garching reactor was used. Once again, the falloff of lattice parameters with increasing Au in solution was observed (following the Dworschak et al. phase boundary [24]). The final result combined all of the HDS and (Δa)/a data; Figure 7.3 could then be constructed, in which, for the quenched and fully annealed state, only Pb_3Au could be seen. However, with increased annealing temperature, the equilibrium solute concentration of Au in Pb increases with increasing temperature. (Again, without annealing, the quenched state retains essentially all of the Au in solution.) There is an acceptance range for singlets and dimers

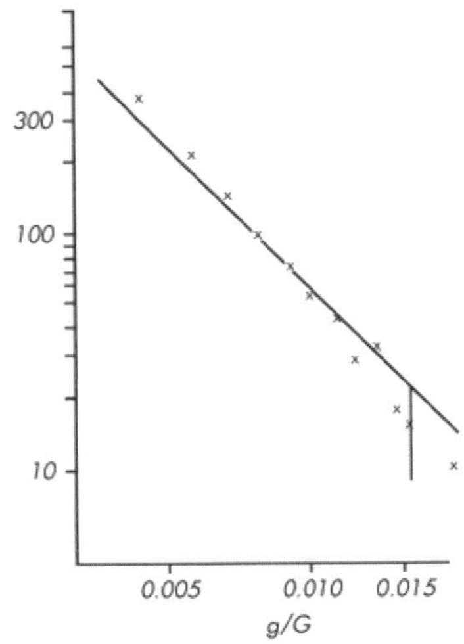

Figure 7.2. *Distortion-induced diffuse scattering close to (660) (difference between the quenched and fully precipitated states) against g/G on a double logarithmic scale. The straight line has a slope of –2.*

that evolves with increasing T. The only break observed in the data was in the concentration of singlets, which increase, and dimers, which appear to decrease while the concentration of Pb_3Au continually decreases up to the eutectic, as expected from the phase diagram.

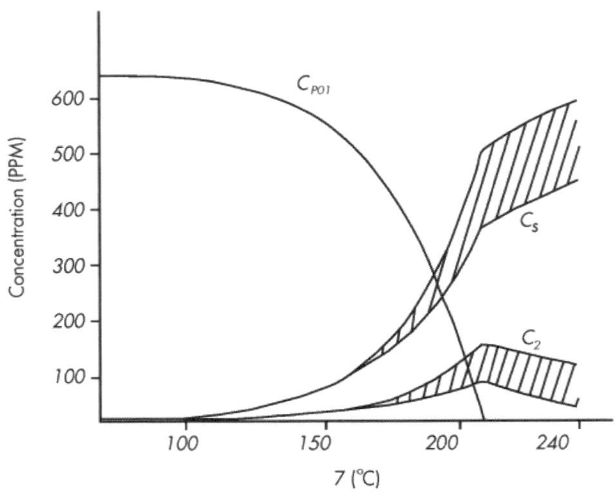

Figure 7.3. *Temperature dependence of the concentration of dimers c_2 and single-substitutional defects c_s.*

The mechanism of rapid diffusion would seem to involve a "kick-out" and capture mechanism [25] with interstitial paths containing Au for essentially unmeasurably brief times. The stability of the quenched single substitutional atoms is still remarkable.

The next example is the determination by Reinhard et al. [6] of the local atomic order or short-range order (SRO) (clustering in this case) and species-dependent displacements in an Fe-Cr crystal quenched from above the sigma (σ) phase. Were it not for the intervening σ phase, this system would constitute a classic phase-separating system. Because Fe and Cr are so close in the periodic table with very similar lattice parameters [a_0(Cr) > a_0(Fe)], one may perform a 3 – λ experiment in which anomalous X-ray scattering (AXS) is used to shift the scattering factors

$$f = f_0 + f' + if'', \tag{7.5}$$

and it is f', the real part of the dispersion correction, that can be diminished by several (five to six) electrons at ~20 eV below the K absorption edge. As a result, $f_{Fe} >> f_{Cr}$ below the Cr K-edge, $f_{Cr} > f_{Fe}$ below the Fe K-edge, and $f_{Fe} \sim f_{Cr}$ when one measures well below all of the K-edges. This gives us the separate experiments from which we can, using a least-squares method, extract the SRO and the species-dependent size effect (SE) while avoiding the region of strong HDS or TDS. Actually, the quenched state had been previously shown (see Reinhard et al. [6]) to preserve the atomic arrangements at the equilibrium temperature, and these were of the clustering variety, as noted above.

This presented a problem in measurement because, in such cases, the intensity peaks at the Bragg points. The experiment utilized an elegant focusing analyzer crystal [26], which spread the various scattered energies, at a fixed 2θ, along the wire of a position-sensitive detector, and these energy scans were automatically collected and analyzed separately for each data point to eliminate both Compton and resonant Raman scattering.

Finally, after several procedural corrections and checks, a reliable set of data was obtained for the three energies. The SRO can simply be written as

$$I(\mathbf{q}) = \sum_n \alpha_n e^{i\mathbf{q}\cdot\mathbf{r}_n}, \tag{7.6}$$

where $I(\mathbf{q})$ ($\mathbf{q} \equiv \mathbf{g}$ in Equations 7.3 and 7.4) is in conventional Laue units $Nc_A c_B (f_B - f_A)^2$ and α_n is the Warren-Cowley SRO parameter or pair correlation function. This contrast factor $(f_B - f_A)^2$ is what enters into our calculations at all energies.

In addition, there is the SE intensity, which contains only $(f_B - f_A)$, and its sign can thereby be tuned by working at the three energies. As Dietrich and Fenzl [27] note, $I_{SRO} \propto <\sigma_i\sigma_j>$, or the concentration-concentration operators (correlation functions), and $I_{SE} \propto <\sigma_i\mu_j>$, or the concentration-displacement correlations. Neglecting the I_{HDS}, we can see how the three data sets can be combined to yield unambiguous values for the SRO and SE terms. I will not discuss in detail either the analysis or the host of results, but while the diffuse scattering due to SRO is symmetric in reciprocal space, the SE scattering can severely distort this symmetry. In fact, the SE terms produce peaks or minima at ~2/3<111>, and equivalent points, that are very similar to ω-phase peaks as observed in Nb-Zr alloys [28–30]. Figure 7.4 [6] shows the SE scattering peaking at ~1/3<111> and 2/3<111> positions but of opposite sign at the two K-edges (Fe and Cr), and this is an excellent demonstration of the utility of the AXS·3 – λ method. This is not surprising; the σ phase resembles in some detail the simpler, more familiar ω phase. In addition, inelastic neutron scattering showed *elastic* peaks in the quenched Fe-Cr crystal at ~2/3<111>; that is, the atomic size disparity, however small it is, was sufficient to produce pseudoembryos of an ordered ω-like phase. This is an excellent example of the enormous importance of atom size in disordered alloys. Table 7.4 [6] shows the *x*-axis displacements for Fe-Fe, Cr-Cr, and Fe-Cr (given by a sum rule for each shell; the *y*- and *z*-axis displacements are the same for cubic crystals). What is remarkable here is the unavoidable fact that, while a_0(Cr) > a_0(Fe) in their pure bcc phases, the nearest-neighbor Cr-Cr displacement, $\delta_{Cr\text{-}Cr}$, off the average lattice separation is negative and considerably less than its pure phase distance, and the magnitude is greater than either δFe-Fe or δFe-Cr. This is certainly a very nonintuitive result and requires theory to provide more insight.

In a more recent paper [31], my colleagues and I indicated that, in the ordered σ phase, there are several Cr-Cr "nearest-neighbor" distances that are considerably smaller than the Fe-Fe distance in pure Fe, and the shortest of them is the Cr-Cr distance of 2.265 Å. What, then, do we make of atom size in this Fe-Cr alloy, where, in the ordered σ phase, the Cr-Cr nearest-neighbor distance is the smallest of all when in the pure bcc elements a_0(Cr) > a_0(Fe)? Clearly, even in the disordered (clustered) phase, there are hints of the ordered σ phase.

The final brief example of an unusual atom size effect occurs in very dilute Al-TM (Fe, Ti) liquids. Concentrating solely on the Al-Ti liquid, which contains a very deep eutectic near the Al melting point and a rapidly rising liquidus to the melting point of Al_3Ti at ~1000°C. Using the thermodynamic data of Gebhardt et al. [32], Turnbull [7] showed that the TM partial molar volumes were large, negative, and well below the values of the pure elements. From this he argued that the observations could best be understood through the clustering, in the liquid phase, of a large number at Al atoms about the TM atom. While recent synchrotron data on levitated liquid alloys of Al-Ti and Al-Fe showed no particularly anomalous behavior [33], it is in the dynamics that we should search for the strong dehancement

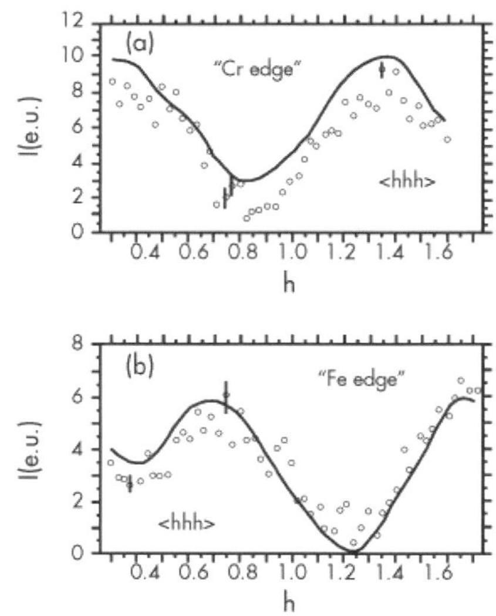

Figure 7.4. *Measured and reconstructed intensities in the <hhh> direction, where the HDS is experimentally avoided. (a) Cr edge. (b) Fe edge.*

of self-diffusion expected from the formation of these clusters in liquids, where the viscosity can be shown to increase enormously over pure aluminum. In fact, using the Einstein relation,

$$\eta / \eta_0 = 1 + 2.5x_v , \qquad (7.7)$$

TABLE 7.4

Displacement parameters ($x_{\ell mn}^{A,B}$), where A, B is Fe, Cr and ℓmn represents a neighbor shell in the bcc crystal; cubic symmetry dictates that the *y* and *z* components are the same [6].

lmn	$< x_{\ell mn}^{FeFe} >$	$< x_{\ell mn}^{CrCr} >$	$< x_{\ell mn}^{FeCr} >$
111	−0.00070(4)	−0.00190(5)	0.00172(4)
200	−0.00029(9)	0.00268(12)	−0.00120(8)
220	−0.00022(4)	−0.00050(6)	0.00038(4)
311	0.00018(5)	0.00007(6)	−0.00013(4)
131	−0.00022(3)	0.00011(4)	0.00008(2)
222	−0.00053(5)	0.00039(6)	0.00013(4)
400	0.00009(9)	0.00063(13)	−0.00034(8)
331	−0.00013(4)	0.00016(4)	0.00000(5)
133	−0.00005(4)	−0.00005(6)	0.00005(4)

where η_0 is the viscosity of pure Al, η is for the dilute alloy, and x_v is the volume fraction of the "spherical" clusters of Al about, say, Ti, at $\chi_{Ti} = 0.0028$, Gebhardt et al. [32] calculate that $\eta/\eta_0 = 1.21$ at 800°. This represents a cluster of ~80 atoms about a Ti atom {!} and must surely be an overestimate.

Our experiment, therefore, is the measurement of the *q*-dependent, quasi-elastic broadening of the neutron scattering at low *q*, which gives directly a measure of Al self-diffusion. Small-angle neutron scattering will also be done to attempt an estimate of the proposed cluster size (eighty does seem rather large). The alloys will be contained in a cylinder of ^{11}BN, whose isotopic substitution for ^{10}B will reduce the absorption of neutrons to a negligible value.

4 Conclusion

I have tried, through selected examples, to highlight the importance of atom size in alloys and other compounds. The origin of the "atom size" must surely lie in the electronic structure, and the size in turn can have a profound effect on phase stability and properties of a host of metals, semiconductors, and oxides.

Acknowledgments

This work has been supported over the years through a number of funding agencies, prominent among which are the U.S. Department of Energy, Basic Energy Sciences; the National Science Foundation, Division of Materials Research; ARPA and Defense Advanced Research Projects Agency (DARPA); the State of Texas; and (with special thanks) the University of Houston, where I have happily resided for 34 years with their gracious and valued support. Supported collaborations with the groups of Professor J. S. Peisl of Munich (approximately 20 years beginning in 1979 with a Senior Scientist Award) and Professor H. Dosch of Stuttgart (through the support of the Alexander von Humboldt Stiftung) are also acknowledged. In Stuttgart, the Max-Planck Institute für Metallforschung provided support for visits over several years. The University of Munich also supported trips over the 20-year period noted above.

References

1. S. C. Moss, "Static atomic displacements in iron-carbon martensite," *Acta Metall.* 15 (1967): 1815–1826.
2. S. C. Moss and R. E. Newnham, "The chromium position in ruby," *Zeitschrift für Krist.* 120 (1964): 359–363.
3. W. K. Warburton and S. C. Moss, "Lattice parameter changes due to Au dissolution in Pb," *Phys. Rev. Lett.* 42 (1979): 1694–1697.
4. T. Bolze, H. Metzger, J. Peisl, and S. C. Moss, "X-ray and neutron diffraction studies of the structure of dilute solutions of Au in Pb, I: The quenched state," *J. Phys. F: Met. Phys.* 14 (1984): 1073–1084.
5. T. Bolze, H. Metzger, J. Peisl, and S. C. Moss, "X-ray and neutron diffraction studies of the structure of dilute solutions of Au in Pb, II: Neutron measurements of bulk equilibrium lattice parameters," *J. Phys. F: Met. Phys.* 14 (1984): 1084–1094.
6. L. Reinhard, J. L. Robertson, S. C. Moss, G. E. Ice, P. Zschack, and C. J. Sparks, "Anomalous X-ray scattering study of local order in bcc Fe0.53Cr0.47," *Phys. Rev. B* 45 (1992): 2662–2676.
7. D. Turnbull, "The gram-atomic volumes of alloys of transition metals with Al and Si," *Acta Metall.* 38 (1990), 243–249.
8. N. F. Mott, "The structure of metallic solid solutions: Introductory lecture," in *Metallic Solid Solutions*, ed. J. Friedel and A. Guinier (New York: W. A. Benjamin, 1963), 1–6.
9. P. A. Wolff, "Localized moments in metals," *Phys. Rev.* 124 (1961): 1030–1035.
10. S. C. Moss and B. L. Averbach, "Local atomic configurations in a Au-Ni alloy," in *Proceedings of the Conference on the Small Angle Scattering of X-rays*, Syracuse, NY (June 24–26, 1965), ed. H. Brumberger (New York: Gordon and Breach Science Publishers, 1967).
11. T. B. Wu and J. B. Cohen, "Clustering of a Au-Ni alloy above the miscibility gap," *Acta Metall.* 31 (1983): 1929–1935.
12. H. Reichert, A. Schöps, I. B. Ramsteiner, V. N. Bugaev, O. Shchyglo, A. Udyansky, H. Dosch, M. Asta, R. Drautz, and V. Honkimäki, "Competition between order and phase separation in Au-Ni," *Phys. Rev. Lett.* 95 (2005): 235703.
13. J. A. Rodriguez, S. C. Moss, J. L. Robertson, J. R. D. Copley, D. A. Neumann, and J. Major, "Neutron scattering studies of short-range order, atomic displacements, and effective pair interactions in a null-matrix $^{62}Ni_{0.52}Pt_{0.48}$ crystal," *Phys. Rev. B* 74 (2006): 104115.
14. W. Hume-Rothery, *Electrons, Atoms, Metals and Alloys* (New York: Dover, 1963).
15. S. C. Moss, "Scattering studies of real materials," *2001 Von Hippel Award Presentation, Materials Research Society Bulletin*, 27 (2002): 539–549.
16. M. A. Krivoglaz, *Theory of X-ray and Thermal Neutron Scattering from Real Crystals*, English translation (New York: Plenum Press, 1969); an extensive updating has also appeared in English translation in two volumes: *X-ray and Neutron Diffraction in Nonideal Crystals* and *Diffuse Scattering of X-rays and Neutrons by Fluctuations* (Berlin, Heidelberg: Springer-Verlag, 1996).
17. B. E. Warren, *X-ray Diffraction* (New York: Dover, 1990).
18. A. Guinier, *X-ray Diffraction in Crystals, Imperfect Crystals and Amorphous Bodies* (New York: Dover, 1994).

19. M. A. Krivoglaz, "Defects in solid solutions stabilized upon cooling," *Sov. Phys.-Solid State* 12 (1971): 1955.
20. D. T. Keating and A. N. Goland, "Atomic displacements in iron martensite," *Acta Metall.* 15 (1967): 1805–1814.
21. D. L. Decker, J. G. Melville, and H. B. Vanfleet, "De-enhancement of gold diffusion in lead by impurities in the lead," *Phys. Rev. B* 20 (1979): 3036–3044.
22. H. B. Vanfleet, "Interstitial substitutonal model for the anomalous diffusion in Pb and Pb alloys," *Phys. Rev. B* 21 (1980), 4340–4348.
23. E. Stracke and C. Herzig, "Formation of slowly diffusing gold defects in thermomigration experiments of gold in lead," *Phys. Stat. Sol.* 66 (1981): 189–197.
24. F. Dworschak, C. Herzig, and J. N. Mundy, "The solid solubility of Au in Pb," *J. Phys. F: Met. Phys.* 10 (1980): 367–375.
25. F. Spaepen, private communication with author.
26. G. E. Ice and C. J. Sparks Jr., "Mosaic crystal x-ray spectrometer to resolve inelastic backgound from anomalous scattering experiments," *Nuclear Instruments and Methods in Physics Res. Sec. A* 291 (1990): 110–116.
27. S. Dietrich and W. Fenzl, "Correlations is disordered crystals and diffuse scattering of X-rays or neutrons," *Phys. Rev. B* 39 (1989): 8873–8899; S. Dietrich and W. Fenzl, "Diffuse scattering of X-rays or neutrons from binary alloys and null matrices," *Phys. Rev. B* 39 (1989): 8900–8906.
28. S. C. Moss, D. T. Keating, and J. D. Axe, "Neutron study of the beta-omega instability in Zr80Nb20," in *Phase Trans.*, ed. L. E. Cross (New York: Pergamon Press, 1973), 179–188.
29. J. D. Axe, D. T. Keating, and S. C. Moss, "Anomalous neutron scattering from Zr–8% Nb," *Phys. Rev. Lett.* 35 (1975): 530–533.
30. J. D. Axe, A. Heidemann, W. S. Howells, S. C. Moss, and R. Pynn, "Diffuse neutron scattering from an Ω-Phase ZrNb alloy," *Solid State Commun.* 24 (1977): 743–745.
31. S. C. Moss, J. L. Robertson, D. A. Neumann, and L. Reinhard, "Anomalous static displacements and their relation to lattice dynamics in Fe0.53Cr0.47," NSF/CNRS Workshop on Alloy Theory, *Comput. Mater. Sci.* 8 (1997): 33–38.
32. E. Gebhardt, M. Becker, and S. Dorner, *Zeitschrift Metallk* 44 (1953): 510; 44 (1953): 573.
33. I. Pozdnyakova, L. Hennet, G. Mathiak, J. Brillo, D. Zanghi, J. F. Brun, S. Brassamin, A. Bytchkov, V. Cristiglio, E. Véron, et al., "Structural properties of molten dilute aluminium-transition metal alloys," *J. Phys. C* (in press).

8

Elastic Diffuse Scattering of Alloys: Status and Perspectives

Bernd Schönfeld[1]

Gernot Kostorz[2]

1. *Laboratory of Metal Physics and Technology, Department of Materials, ETH Zurich, 8093 Zurich, Switzerland*
2. *Department of Physics, ETH Zurich, 8093 Zurich, Switzerland*

Abstract

Two areas of current research in the investigation of the microstructure of bulk alloys by elastic diffuse X-ray or neutron scattering will be addressed. First, alloys with a large atomic-size difference that still show signs of local order in the solid solution despite the absence of any intermetallic phase in the phase diagram have attracted attention. Results for Ni-Au and Ni-Re will serve to illustrate this case. For both alloys, small-angle scattering is found to be decisive in the identification of alloy parameters in addition to wide-angle scattering. Second, the determination of the species-dependent static atomic displacements exploiting measurements at various X-ray wavelengths will be summarized and illustrated with recent results for Cu-Mn.

1 Introduction

Diffuse scattering from substitutional solid solutions has become quite an elaborate research tool over the years, both for X-rays and for neutrons. On the way toward analyzing local order in terms of pair correlation functions (Warren-Cowley short-range order parameters and species-dependent static atomic displacements), much success has been achieved, as reported in several reviews [1–5].

For sufficiently detailed studies, single crystals are mandatory. If, as is frequently the case, modulations between the Bragg reflections are small, means have to be exploited to optimize the scattering contrast, that is, the difference in the atomic scattering factors and the elastic coherent scattering lengths of the constituents. With thermal neutrons, this is achieved by employing isotopic replacement, with X-rays by choosing a wavelength close to an absorption edge. Quantitative evaluation may still suffer from the lack of data in adequate ranges of reciprocal space, additional inelastic scattering contributions, errors in referring scattering intensities to an absolute scale, the problem of splitting up the elastic diffuse scattering into its contributions due to atomic order and static displacements, and limited sample quality and purity. Thus it comes as no surprise that contradictory results have been reported in the literature. Information on the static atomic displacements is repeatedly missing because the species dependence is hard to evaluate owing to intercorrelations. It has to be kept in mind that displacements are only treated to a certain degree of approximation under the assumption of small displacements. An indicator for the quality of atomic order is available (since there is only one atom at a lattice site); for displacements, an analogous simple criterion is not established.

The difficulty of investigating alloys with a large difference in atomic size by diffuse scattering was revealed in the investigations of Guinier-Preston (GP) zones in Al-rich Al-Cu (with less than about 2.5 at.% Cu and aging below 373 K). The system is known to exhibit a series of intermediate states, two coherent preprecipitates called GPI and GPII zones, the intermediate state Θ', and the equilibrium and incoherent precipitate Θ ($CuAl_2$). The question was how to separate strong displacement scattering from short-range order scattering. For the case of GPI zones, the prevailing uncertainties in this procedure led to the postulation of two models. In textbooks, GPI zones are described as single-layered $\{100\}$ planes exclusively consisting of Cu atoms, as originally proposed by Gerold [6] more than fifty years ago. A long-ranging displacement field is required to reproduce the X-ray diffuse scattering along the $\langle 100 \rangle$ streaks. Somewhat later, Matsubara and Cohen [7], using synchrotron radiation, obtained a larger fraction of multilayered Cu planes within their planar stacking model. In an attempt to resolve these ambiguities, Müller et al. [8] showed by diffuse neutron scattering and by employing samples with the Cu-65 and Cu-63 isotopes that the range of scattering vectors used in the evaluation is decisive for the result obtained. Certainly data closest to the origin of reciprocal space are more reliable because the approximate way to treat displacement scattering is progressively less justified if one measures at larger scattering vectors. A section from a model crystal shown in Figure 8.1 (see Schönfeld [4] and references therein) demonstrates that the model of single-layered Cu planes most suitably describes the actual situation of GPI zones.

Quite generally, any diffuse scattering study will benefit from the combination of wide-angle and small-angle scattering. The combination is mandatory for diffuse scattering from alloys that show strong displacement scattering. Investigations of two such systems, Ni-Au and Ni-Re, will be addressed in this chapter. Certainly such a combination in scattering is required even more if the modulation in the short-range order term—that is, any deviation from the scattering of a statistically uncorrelated arrangement—is small. The robustness of separating displacement scattering into the corresponding species-dependent displacements is still under discussion inasmuch as accompanying calculations are mostly missing. The progress with systems that show small lattice distortions will be addressed and illustrated by recent results from Cu-Mn in which the tunability in X-ray energy at storage rings was exploited.

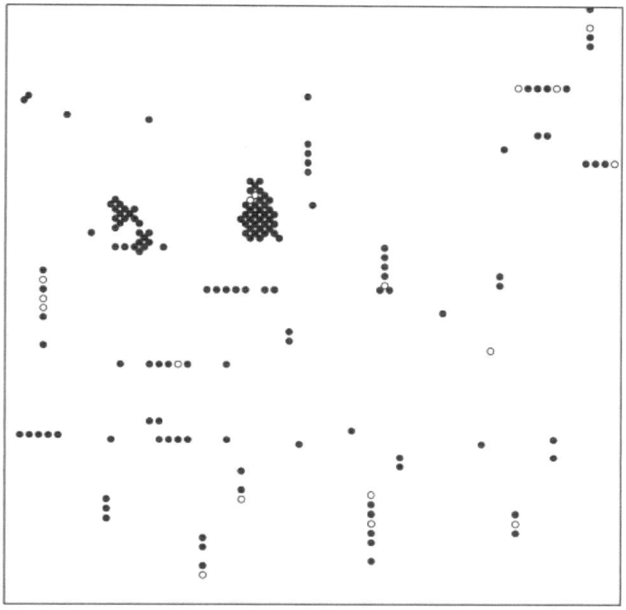

Figure 8.1. *An h_1h_20 plane of an Al-1.75 at.% Cu crystal, modeled with the short-range order parameters given by Malik [9]. Only GPI zones are shown. Filled symbols: Cu atoms, open symbols: Al atoms. After Schönfeld [4].*

2 Theory

For the analysis of elastic diffuse scattering, two approaches may be used; a real-space approach and a reciprocal-space approach. Both have their merits and deficiencies. Subsequently, the presentation will be restricted to binary cubic alloys for which nearly exclusively investigations were performed.

In the real-space approach, the static atomic displacements $\delta_{lmn} = \mathbf{r}_{lmn} - \mathbf{r}^{o}_{lmn}$ in neighboring shell lmn (the indices lmn are given with respect to the usual cubic axes, $\mathbf{r}_{lmn}$ describes the actual position, and $\mathbf{r}^{o}_{lmn}$ is the nearby position of the average lattice) cannot be written in a close expression and have to be treated in an approximate way. Assuming that the product of static displacements $\delta^{\mu\mu}_{lmn}$, μ = A,B of the A-B alloy and scattering vector $\mathbf{k}$ ($k = 4\pi\sin\theta/\lambda$, where 2θ is the scattering angle and λ the wavelength) is small, a series expansion in this small quantity is done within the phase factor between any two scattering atoms, yielding in second order for the elastic diffuse scattering $I_{\mathrm{diff}}(\mathbf{h})$:

$$I_{\mathrm{diff}}(\mathbf{h}) = I_{\mathrm{SRO}}(\mathbf{h}) + I_{\mathrm{SE}}(\mathbf{h}) + I_{\mathrm{H}}(\mathbf{h})\,. \tag{8.1}$$

Here, the scattering vector is already written dimensionless in reciprocal lattice units (RLUs), $\mathbf{h}$ = $(h_1,h_2,h_3) = \mathbf{k}/(2\pi/a)$ (a = lattice parameter). Scattering intensities are usually presented in Laue units

(LUs), where 1 LU is the scattering from a statistically uncorrelated arrangement. For binary cubic alloys, this expansion yields one Fourier series in short-range order scattering $I_{\mathrm{SRO}}(\mathbf{h})$, six in size-effect scattering $I_{\mathrm{SE}}(\mathbf{h})$, and eighteen in Huang scattering $I_{\mathrm{H}}(\mathbf{h})$. The validity of this series expansion has to be considered carefully when alloys with a large atomic size difference are studied. If one is only interested in $I_{\mathrm{SRO}}(\mathbf{h})$, neutrons may be preferred over X-rays, as the number of unknown Fourier series is then reduced from twenty-five to ten (since the species dependence of the static atomic displacements is not accessible in a single neutron-scattering experiment). This situation was addressed by Borie and Sparks [10]. If quadratic displacement scattering $I_{\mathrm{H}}(\mathbf{h})$ may not be omitted because of large displacements, it is advantageous to measure data for positions within a so-called separation volume (placed as close as possible to the origin of reciprocal space) and to use a separation algorithm. Such an algorithm was introduced by Gragg and Cohen [11]; it exploits the different symmetries of the Fourier series in reciprocal space and their different weighting with the scattering vector **h**. A least-squares fitting approach is less suited because $I_{\mathrm{H}}(\mathbf{h})$ and $I_{\mathrm{SRO}}(\mathbf{h})$ show the same cosine modulation in reciprocal space and are thus highly correlated.

To determine the species-dependent static atomic displacements $\delta_{lmn}^{\mu\mu}$, X-rays are preferred. Here, the species dependence of the displacements is accessible through the slightly different variation of the atomic scattering factors with scattering vectors. An extension of the Borie-Sparks procedure was introduced and implemented by Georgopoulos and Cohen [12]. In this procedure, a simple separation algorithm for intensities within a separation volume is not required, but highly overdetermined systems of equations for symmetry-equivalent positions have to be solved. This has to be done for any position on a given grid within the repeat volume of the Fourier series $I_{\mathrm{SRO}}(\mathbf{h})$. Thus five to ten times more positions have to be investigated than within the Borie-Sparks procedure. Mathematical stabilization techniques are required to avoid a mutual balancing of large static atomic displacements [4,12].

If the components of an alloy do not differ much in the number of electrons, the large energy dependence of the atomic scattering factors close to absorption edges is used to vary the magnitude and even the sign of the scattering contrast by tuning the X-ray energy close to values of the K absorption edges of both constituents. In addition, a third measurement is done under a "null-Laue" condition to determine experimentally all scattering contributions that do not depend on the scattering contrast. This is especially valuable for separating $I_{\mathrm{H}}(\mathbf{h})$, with its large number of Fourier series. This so-called 3λ method was introduced and implemented by Ice and Sparks [13].

In the reciprocal-space approach, a crystal with "gray" atoms is taken as a reference crystal. Deviations in site occupation from the average composition and in the displacements from the average lattice positions are considered. Elastic diffuse scattering is then expressed by the Fourier transforms of both deviations, $\sigma_{\mathbf{k}}$ and $\delta_{\mathbf{k}}$, as

$$I_{\mathrm{diff}}(\mathbf{h}) = N^2 \,|\, (c_A f_A + c_B f_B)\mathbf{k}\cdot\delta_{\mathbf{k}} - (f_A - f_B)\sigma_{\mathbf{k}} \,|^2 , \tag{8.2}$$

where N is the number of atoms and c_μ, $\mu = A,B$, are the atomic fractions [14]. The square of the second term is short-range order scattering, the square of the first term is Huang scattering, and the mixed term is size-effect scattering. In this approach, the long-ranging displacement field is again considered without any limitation in range. However, no species dependence in the displacement field is provided. This problem is only overcome when a dilute alloy is present. Then the displacements within a crystal may be considered as a linear superposition of the displacement fields of single defects within the matrix of majority atoms.

3 Local Order and Local Decomposition in Ni-Au

Following one of the Hume-Rothery rules, the binary system Ni-Au has been considered as a prototype of a phase-separating alloy because the atomic sizes of the constituents differ considerably (by about 15%). The Ni-Au system is characterized by a broad miscibility gap, and the solid-solution range above it stretches contiguously from pure Ni to pure Au (Figure 8.2a). Up to now, a series of diffuse scattering experiments were performed and electronic structure calculations were done yielding partly conflicting results about whether local order or local decomposition is present in the solid solution. A recent X-ray investigation will be presented and compared with previous results from single crystals. The presentation is restricted to the case of rapidly quenched-in states and to solid solutions at elevated temperatures. The cases of Ni-rich Ni-Au and Ni-60 at.% Au are treated separately.

3.1 Ni-rich Ni-Au

On the Ni-rich side, diffuse neutron scattering from Ni-8.4 at.% Au was done at 1083 K, about 100 K above the miscibility gap. The Ni-58 isotope was preferred over natural nickel for two reasons: (1) to increase the scattering contrast by a factor of 6.4 and (2) to reduce the elastic incoherent scattering otherwise comprising 98% of the elastic diffuse scattering. Figure 8.3 shows elastic diffuse scattering along $h00$; intensities were always calibrated using the incoherent scattering of vanadium. Data from ^{58}Ni-8.4 at.% Au were taken with neutrons of three incoming energies [14.68(50) meV, 4.76(15) meV, and 1.28(26) meV] and at two stations (the triple-axis spectrometer DrüchaL used in the elastic mode and the SANS-I facility with 20% energy resolution, both at PSI, Villigen). Consistently, a large intensity increase toward the direct beam is seen, much larger than the scattering from a random solid solution of the same composition of 1 LU. Without any further evaluation, this indicates the presence of local decomposition. This conclusion is supported by a detailed three-dimensional scattering investigation [16,17]. Figure 8.4 shows that short-range order scattering strongly peaks at the Γ position. Also, the Warren-Cowley short-range order parameters of the first five shells are positive, with a leading value of the nearest-neighbor shell (Table 8.1). A configurational analysis of modeled crystals showed the presence of small Au agglomerates (comprising approximately five atoms) in the Ni-rich matrix.

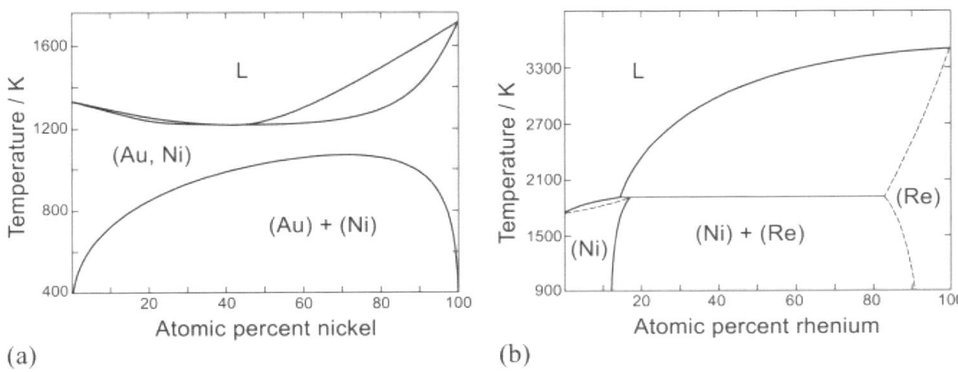

Figure 8.2. *Phase diagrams of (a) Ni-Au and (b) Ni-Re according to Massalski [15]. (Redrawn with permission of ASM International.)*

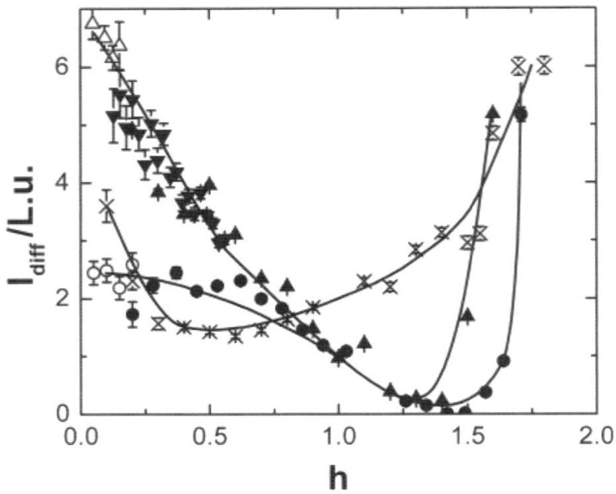

Figure 8.3. *Elastic diffuse neutron scattering along h00 in Laue units of ^{58}Ni-8.4 at.% Au at 1083 K (triangles), ^{60}Ni-7.0 at.% Au at 1083 K (crosses), and ^{58}Ni-60 at.% Au at 1013 K (circles). Small-angle scattering (open symbols), wide-angle scattering (filled up-triangles, filled circles, and crosses for E = 14.68 meV, down-triangles for E = 4.76 meV) [17,18]. Lines are drawn to guide the eye. (Figure 8.3 of Portmann et al. [17] redrawn with permission of the American Physical Society.)*

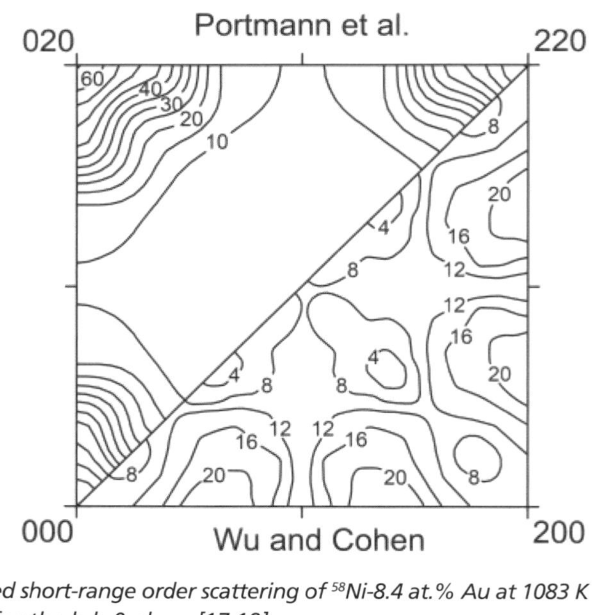

Figure 8.4. *Recalculated short-range order scattering of ^{58}Ni-8.4 at.% Au at 1083 K and of Ni-60 at.% Au at 1023 K, in 0.1 LU for the $h_1h_2 0$ plane [17,19].*

TABLE 8.1

Warren-Cowley short-range order parameters α_{lmn} of ^{58}Ni-8.4 at.% Au at 1083 K [17]. For Ni-60 at.% Au, results are provided for a state quenched from 1123 K using two evaluation methods—Georgopoulos-Cohen (GC) and Borie-Sparks (BS)—for a state at 1023 K from Wu and Cohen [19] and calculated parameters (1600 K) from Wolverton et al. [29]. In comparison, values for the three superstructures L1$_2$, L1$_0$, and Z1 are given.

lmn	α_{lmn}						
	Ni-8.4 AT.% Au	Ni-60 AT.% Au				L1$_2$, L1$_0$	Z1
	[17]	GC	BS	[19]	[29]		
000	1.153(47)	0.855(13)	0.907(7)	1.047(92)	1.0000	1	1
110	0.220(16)	−0.018(6)	0.050(4)	0.039(45)	−0.0235	−1/3	1/9
200	0.135(18)	0.161(5)	0.160(3)	0.148(39)	0.1208	1	5/9
211	0.016(9)	−0.000(3)	0.001(2)	−0.081(27)	−0.0134	−13	−1/3
220	0.001(6)	0.004(3)	−0.007(2)	−0.057(27)	−0.0021	1	1/9
310	0.019(6)	0.033(3)	0.025(2)	0.020(24)	0.0164	−1/3	1/9
222	−0.013(6)	−0.025(3)	−0.040(2)	−0.030(26)	−0.0195	1	−1/3
321	−0.002(4)	0.019(2)	0.004(1)	0.039(17)	0.0056	−1/3	−1/3
400	0.022(7)	0.035(4)	0.039(2)	−0.018(35)	0.0334	1	1
330	−0.001(5)	0.007(3)	0.000(2)	−0.084(25)	−0.0032	−1/3	1/9
411	0.017(4)	0.002(3)	0.003(1)	−0.022(20)	0.0121	−1/3	1/9
420	0.010(3)	−0.004(2)	−0.004(1)	0.027(18)	−0.0018	1	5/9
233	−0.006(4)	0.002(2)	−0.007(1)	−0.003(17)	−0.0062	−1/3	−1/3

Figure 8.3 also shows elastic diffuse neutron scattering along *h*00 from a ^{60}Ni-7.0 at.% Au single crystal taken at the triple-axis spectrometer D10 (ILL, Grenoble) in the elastic mode using incoming neutrons of 14.68 meV in energy [18]. Again, data were taken at 1083 K and calibrated using the incoherent scattering of vanadium. An intensity increase toward the Γ position is observed. This time, the dip in the elastic diffuse scattering along *h*00 is shifted from a value with $h > 1$ toward $h < 1$; as the sign of $b_{Ni} - b_{Au}$ changes from +6.8 fm with the Ni-58 isotope to −4.8 fm with the Ni-60 isotope, the sign of the linear displacement scattering term is reversed.

In conclusion, there is broad experimental evidence that local decomposition is present in Ni-rich Ni-Au solid solutions.

3.2 Ni-60 at.% Au

On the Au-rich side, diffuse scattering experiments were focused on alloys with compositions near the minimum freezing point of the liquid, that is, Ni-60 at.% Au. Wu and Cohen [19] performed the first three-dimensional diffuse X-ray measurement (with Cu K_α radiation) at 1023 K, about 30 K above the miscibility gap. Data were calibrated by measuring an Al powder standard. Scattering within 0.45 RLU around Bragg reflections had to be discarded because of contaminations from the tails of the Bragg reflections. After a smooth extrapolation of the diffuse scattering toward the Γ position and separation of the scattering contributions according to the Georgopoulos-Cohen method, the authors obtained broad diffuse maxima in the short-range order scattering term at ~0.6 0 0 positions (Figure 8.4). The resulting nearest-neighbor Warren-Cowley short-range order parameter α_{110} = 0.039(45) has a slightly positive value, while α_{200} = 0.148(39) is clearly positive (Table 8.1). Thus the alloy may be considered to be either locally ordered or locally decomposed if one takes the location of the diffuse maximum or the sign of α_{110} as criterion. From their modeled crystals, Wu and Cohen [19] concluded that platelike and rodlike Ni clusters are present. Ni-rich plates (typically single-layered, comprising typically ~10 Ni atoms) are located on {100} planes about 6–7 Å apart, consistent with the position of the maximum in short-range order scattering.

To get better experimental access to short-range order scattering near the Bragg reflections, small-angle neutron scattering (SANS) investigations at SANS-I (PSI, Villigen) were performed by Portmann et al. [17]. Scattering intensities at 1023 K from polycrystalline material smoothly fit to the data taken at the triple-axis spectrometer RITA-II (PSI, Villigen) in the elastic mode from a ^{58}Ni-60 at.% Au single crystal (Figure 8.3). There are no indications of local order, but a smooth increase toward the Γ position is found. This increase, however, is much less pronounced than noted for ^{58}Ni-8.4 at.% Au, thus indicating a lower tendency toward local decomposition with an increasing Au fraction. With ~2.5 LU, a slightly larger intensity is obtained at ~0.6 0 0, where Wu and Cohen [19] observed the diffuse maximum (of ~2.0 LU) in $I_{SRO}(\mathbf{h})$. As no separation of $I_{SRO}(\mathbf{h})$ is possible on the basis of such a restricted linear scan—and linear displacement scattering adds to the total scattering in this range—the experimental data of Wu and Cohen [19] and Portmann et al. [17] are considered compatible along $h00$.

Repeatedly, measurements were performed for Ni-60 at.% Au alloys that were quenched from the solid solution. It is evident that different quench rates will introduce uncertainties in any comparison. The goodness of a quench may only be judged on the basis of detailed investigations on the ordering kinetics, for example, by measurements of the electrical resistivity. To the authors' knowledge, this information, however, is missing for the Ni-Au system.

A wide-angle diffuse X-ray scattering experiment for an h_1h_20 plane was done by Moss [20,21] using Cu K_α radiation. A Ni-60 at.% Au sample was investigated that was quenched from 1163 K within about 2 seconds. No diffuse maximum between Bragg reflections was noted, only an irregularly shaped plateau slightly displaced from the 310 position. A shoulder at the low-angle side of fundamental $h00$ reflections was also found about 0.5 RLU off the Bragg position (note that with X-rays the asymmetry in $h00$ scans is as with the Ni-60 isotope in Figure 8.3). No evaluation of the calibrated scattering was performed because data were only collected within a single reciprocal-lattice plane. Also, small-angle X-ray scattering (SAXS) raw data from a thin foil 23 μm in thickness yielded weak humps at about h = 0.5 smeared along $h00$. Scattering was not further evaluated because nonmonochromatic X-rays were employed. It was concluded that modulations in gold-rich and nickel-rich regions seem to be present in quenched alloys, with a modulation wavelength of about two unit cells.

Diffuse scattering patterns of the h_1h_20 plane of Ni-60 at.% Au quenched to room temperature were presented by Reichert et al. [22] using X-rays of 60 to 90 keV in energy. Outside the range of the beamstop ($h > 0.23$ RLU) they observed a diffuse scattering pattern with increased intensity at ~0.6 0 0 and around 110 positions, the latter attributed to $2k_F$ maxima ($2k_F$ spans the flat parts of the Fermi surface across ⟨110⟩). Temperature-dependent scans along $h00$ revealed that a maximum at ~0.6 0 0 is barely visible in raw data for states within the solid solution. However, the maximum will become visible with aging. Thus its presence after a quench indicates the quality of the quench. The tendency toward order was related to the tendency to form the Z1 structure [23] that shows a periodicity of two fcc unit cells along its z axis. No increase in intensity toward the direct beam was reported for $h > 0.23$ RLU. Still, phase separation was claimed to be seen in the complex scattering pattern close to the Γ position. Because intensities still include thermal diffuse scattering beside elastic diffuse scattering and data are given in arbitrary units, no quantitative comparison with the previous results is possible.

3.3 Renewed X-ray Diffuse Scattering from Ni-60 at.% Au

A three-dimensional wide-angle scattering experiment was recently done by Schönfeld. A single crystal of nominally Ni-57.5 at.% Au was grown by the Bridgman technique in an Al_2O_3 crucible. A slice was homogenized at 1123 K for fourteen days and rapidly quenched in brine. Subsequently, the surface was prepared by mechanical polishing with 1 μm Al_2O_3 solution and electropolishing in methanol with 10% HCl. In a final step, the sample temperature was rapidly raised to 1123 K and rapidly quenched by a He gas stream to avoid mechanical damage of the surface. The sample concentration as determined by X-ray fluorescence using standards then amounted to Ni-60(1) at.% Au.

The measurements were done with Mo K_α radiation. About 8000 positions were investigated with scattering vectors between 1.5 and 5.5 RLU. Raw data within an h_1h_20 plane are given in Figure 8.5, exhibiting the well-known asymmetry along $h00$ across fundamental reflections. No diffuse maximum between Bragg positions is resolved. A slight increase displaced off the 130 position resembles the plateau noted by Moss [20]. Scattering was calibrated using polystyrene and the calculated inelastic scattering contributions were subtracted. The elastic diffuse scattering was analyzed employing the Georgopoulos-Cohen and the Borie-Sparks techniques. Based on the fitted short-range order parameters (Table 8.1; a set of twenty-two neighboring shells had to be considered), short-range order scattering was calculated (Figure 8.6). Several points should be noted:

1. It is not possible to attribute any superiority to one of the two separation techniques. The weak modulation in diffuse scattering indicates that a low number of Fourier series might be sufficient and thus favors a Borie-Sparks evaluation. On the other hand, the large displacement scattering favors a Georgopoulos-Cohen evaluation, and the agreement between the fitted components of the Fourier series and the intensities for any set consisting of symmetry-equivalent positions was better than with a Borie-Sparks evaluation. For any decision, however, a distinct increase in the number of positions measured would first be required.
2. The global maximum of short-range order scattering is located at the Γ position. This conclusion was reached by including the SANS data of Portmann et al. [17] in the evaluation because X-ray data within 0.25 RLU around any Bragg reflection could not be used. Without SANS data, both evaluation techniques show discrepancies in their behavior close to the Γ position, supporting either a predominance of local order or local decomposition.

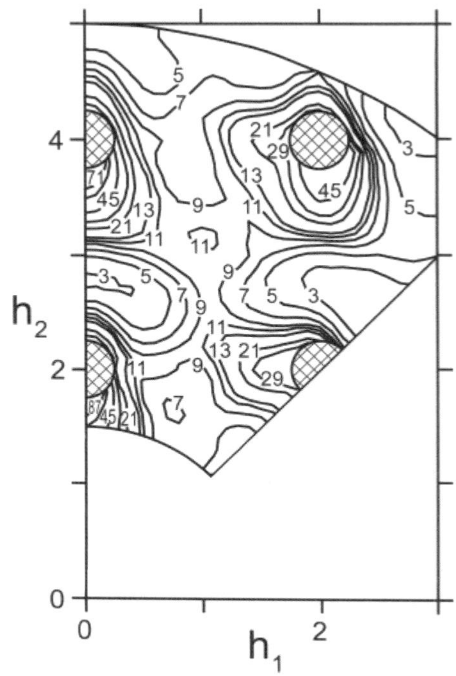

Figure 8.5. *Raw data of diffuse X-ray scattering of Ni-60 at.% Au quenched from 1123 K to room temperature in units of 10^3 counts.*

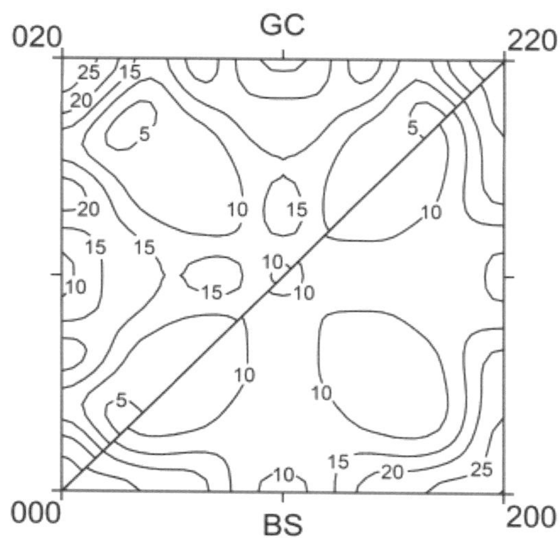

Figure 8.6. *Recalculated short-range order scattering of Ni-60 at.% Au quenched from 1123 K to room temperature, in 0.1 LU for the h_1h_20 plane. Diffuse X-ray scattering was evaluated by the methods of Georgopoulos-Cohen (GC) and Borie-Sparks (BS).*

3. Increased intensity or a broad shoulder of $I_{SRO}(\mathbf{h})$ = 2.0 to 2.3 LU is found at ~0.6 0 0 depending on the evaluation technique employed. This value is close to ~2.0 LU given by Wu and Cohen [19] for this region.
4. There are also weak and broad $2k_F$ maxima of ~1.5 LU at 1 ~0.65 0 positions corresponding to $2k_F \approx 1.934$ Å^{-1}. This experimental value for Ni-60 at.% Au is nicely obtained starting from the value in pure gold ($2k_F \approx 2.289$ Å^{-1} along ⟨110⟩ by Coleridge and Templeton [24]); within a free-electron approach, the number of valence electrons per atom is reduced to e/a = 0.6 (a valence of 0 is usually attributed to Ni when applying Hume-Rothery rules [25]), yielding $2k_F \approx$ 1.931 Å^{-1}. A minimum, but no maximum intensity as reported by Reichert et al. [22], is noted at 110 positions, even when the "cloverleaf" around 110 is of comparable intensity.

Transient long-range ordered states have been reported within the miscibility gap; that is, the free energy of the ordered states gets lowered by ordering before reaching the equilibrium two-phase state. By selected-area diffraction in transmission electron microscopy (TEM) investigations, Zhao and Notis [26] showed the presence of the $L1_0$ structure for Ni-50 and Ni-60 at.% Au for two cases—first, when reannealing a sample with a two-phase microstructure at 763 K, and second (in some cases), during aging of a single-phase solid solution within the miscibility gap—but its formation then seemed to be restricted to very thin sections of TEM foils. As ordered-phase particles were only seen close to surfaces or phase interfaces, it was concluded that strain relaxation is required. For Ni-75 and Ni-60 at.% Au, Reichert et al. [22] noted the presence of the $L1_2$ and Z1 structures on the basis of the location of diffuse maxima in X-ray diffraction and accompanying first-principle calculations.

The Warren-Cowley short-range order parameters of Table 8.1 indicate local order as well as local decomposition, with the sign of α_{110} changing between the Borie-Sparks and the Georgopoulos-Cohen evaluation. This feature confirms that in Ni-60 at.% Au, both tendencies are competitive. Otherwise, the signs of the leading α_{lmn} do not differ significantly between both evaluations and, as already known from the parameters of Wu and Cohen [19], α_{200} is dominant. One also notes that the features of the α_{lmn} of the locally decomposed state, investigated by Portmann et al. [17], are clearly different (see Figure 8.4).

Ordered structures expected for the scattering patterns are $L1_2$ (Au_3Ni) based on the $2k_F$ maxima and Z1 (Au_3Ni) based on the maximum at ~0.6 0 0. A direct comparison between the signs of the experimental α_{lmn} and those of any superstructure is hampered by the concurrent presence of local decomposition and the generally low values of α_{lmn}. The closest agreement in sign may be claimed for Z1, considering the shells lmn = 310 and 222 (Table 8.1). An abundance analysis for the nearest-neighbor configurations (Clapp configurations [27]) is better suited for identifying local features of superstructures (Table 8.2). The largest enhancement factors in configurations with those of a statistically uncorrelated arrangement are found for $L1_2$, $L1_0$, and Z1. Agglomerates indicating local decomposition are also seen. Note that the Clapp configurations C1 (for Au atoms around Au) and C129 (for Ni atoms around Ni) are not attributed to Ni_3Au with $L1_2$ structure, as an enhancement of this type of order was not found by Portmann et al. [16] for Ni-rich Ni-Au. Only NiAu with $L1_0$ order is compatible with the data.

3.4 Electronic-Structure Calculations

Electronic-structure calculations were started by Wolverton and Zunger [30] and Asta and Foiles [28] to better rationalize the unexpected features observed in Ni-Au, the short-range order tendency

TABLE 8.2

Abundance analysis for selected Clapp configurations with largest enhancement factors (with respect to a statistically uncorrelated arrangement). Results are provided for the two evaluation schemes: Georgopoulos-Cohen (GC) and Borie-Sparks (BS). Possible structures are also indicated. $L1_2$ and Z1 are attributed to Au_3Ni.

CLAPP CONFIGURATION	ENHANCEMENT FACTOR	ABUNDANCE IN %	STATE
Ni around Ni (BS)			
C144	9.7	0.06	decomposition
C143	6.3	0.41	~decomposition
C142	5.3	0.42	~decomposition
C1	4.9	0.59	$L1_2$
C129	4.3	0.21	~$L1_0$
Au around Au (BS)			
C1	23	0.03	~$L1_0$
C144	8.4	1.2	decomposition, Z1
C143	4.4	4.8	~decomposition, ~Z1
C129	4.2	0.48	$L1_2$
Ni around Ni (GC)			
C1	5.6	0.68	$L1_2$
C16	4.9	0.66	$L1_0$, Z1
C129	3.6	0.17	~$L1_0$
Au around Au (GC)			
C1	11	0.01	~$L1_0$
C129	6.0	0.69	$L1_2$
C16	5.0	0.17	$L1_0$

at elevated temperatures as revealed by the investigations of Wu and Cohen [19], and the presence of phase decomposition at low temperature. This situation was attributed to the dominant volume deformation energy (because of the large lattice mismatch between the constituents), which is present in addition to the usually dominant constant-volume spin-flip energy [29,30]. As the sum of both energies determines the ground state, decomposition occurs in Ni-Au. The constant-volume spin-flip energy on the other side is responsible for the ordering tendency seen in short-range order scattering. The same conclusions as in this first-principle study were reached by Asta and Foiles [28] within their embedded-atom method, which also ascribed the unexpected order-decomposition features in Ni-Au to the competition between chemical and relaxation-energy contributions.

In Table 8.1, short-range order parameters calculated from first principles by Wolverton et al. [29] are given for an equilibrium state of Ni-60 at.% Au within the solid-solution regime. The agreement with the present experimental parameters (though not representing a state of thermal equilibrium) is slightly better than with those of Wu and Cohen [19]. The two differences mentioned by Wolverton et al. [29] concerning the sign of α_{110} and α_{400} are no longer present. Though differences in the data sets are small, the diffuse maxima are located at different positions. Also, there remains the problem that for Ni-rich Ni-Au, the electronic-structure calculations of Wolverton et al. [29] yield local order instead of local decomposition as observed by Portmann et al. [17]. A later inclusion of attenuation in the coherency strain by Wolverton et al. [31] led to a modification of the location for the calculated maximum for Ni-rich Ni-Au, now being located along the Γ-X line. The character of a prevailing local order remained unchanged.

X-rays seem less suited for deciding on local order or local decomposition than neutrons because linear displacement scattering shifts any intensity increase due to decomposition toward smaller scattering vectors (see Figure 8.3). (When the Ni-60 isotope is used instead of the Ni-58 isotope, the scattering contrast normalized to an average scattering factor for X-rays and neutrons is similar. Thus a similar weighting between $I_{SRO}(\mathbf{h})$ and $I_{SE}(\mathbf{h})$ is expected.) Also, SAXS from single crystals is difficult, as very thin foils are needed because of large absorption with a resulting sensitivity to the roughness of the foil surface and to double-Bragg scattering.

4 Local Order in Ni-Re

Ni-Re is a system with two solid solutions, an fcc Ni-rich and an hcp Re-rich phase. There is a peritectic reaction at 1893 K and 17.4 at.% Re. The atomic radii of the constituents differ by 10%, and thus a large two-phase regime is present (Figure 8.2b). Not much is known about the microstructure of the solid solutions, though the fraction of Re used in technical Ni-base superalloys has been steadily increased because it slows down the γ' coarsening kinetics and modifies the misfit between the γ matrix and the γ' precipitates [32]. Re is found in the γ' precipitates where it replaces Ni, but also at the interface, and preferentially within the γ matrix [33–35]. The local atomic arrangement in a Ni-rich Ni-Re solution is still unknown. Therefore, small-angle neutron scattering and diffuse X-ray wide-angle scattering experiments were performed.

Starting from an alloy with nominally Ni-10 at.% Re, a single crystal was grown by the Bridgman technique in an Al_2O_3 crucible. This turned out to be too difficult for a sufficiently large sample diameter of about 10 mm, and only large grains were obtained. The slice finally prepared for diffuse X-ray scattering measured 8 mm across. It was heated for seven days at 873 K and cooled in air. Its composition, as determined by X-ray fluorescence analysis with standards, amounted to Ni-9.4(4) at.% Re. The sample was mechanically polished down to 3 μm diamond grains. After this procedure, its mosaicity was 0.4° full width at half-maximum.

For the evaluation of the diffuse X-ray scattering, additional information is needed about the elastic constants. These were measured by the pulse-echo overlap method using a single crystal of Ni-5.4 at.% Re at room temperature. With the elastic constants of Pottebohm et al. [36] for pure Ni (c_{11} = 250 GPa, c_{12} = 151 GPa, c_{44} = 123 GPa), the values of c_{11} = 269(3) GPa, c_{12} = 184(3) GPa, c_{44} = 128(1) GPa are obtained by linear extrapolation for Ni-9.4 at.% Re. From these values, an overall (static and dynamic) Debye-Waller factor [37] $\exp[-2B(\sin(\theta/\lambda)^2]$ was determined with $B = 3.73 \times 10^{-3}$ nm^2.

For the neutron scattering, the Ni-58 isotope was employed. Because of the high melting temperature of Re, alloys were prepared using arc melting and subsequent floating zone melting to achieve good homogeneity and to minimize contamination by impurities. The sample composition was checked by neutron transmission measurements at the SANS facility (SANS-I at PSI), yielding a value of ^{58}Ni-13.0(6) at.% Re. Here good sensitivity is obtained by the large difference in the absorption cross sections of 89.7(1) barn for Re but only 4.6(3) barn for the Ni-58 isotope [38]. Scattering measurements were performed at room temperature with incoming neutrons of 8 Å in wavelength to avoid double-Bragg scattering, and a quenched-in state after homogenization at 1350 K was investigated. No indications of decomposition were found because the SANS scattering was essentially flat, with intensities of 0.7(1) LU within the range of 0.02 to 0.12 RLU in scattering vectors.

Diffuse X-ray scattering was performed with Mo K_α radiation. The diffuse scattering showed indications of fcc Debye-Scherrer rings. These high-intensity data points (about 200 out of 8500) were discarded in the evaluation. This was easily possible, as intensities were measured within a contiguous volume of $h_1 > h_2 > h_3$ with scattering vectors being 1.5 to 5 RLU in magnitude. The various contributions to the elastic diffuse scattering were obtained by a Borie-Sparks and a Georgopoulos-Cohen evaluation. Short-range order scattering was fitted by six α_{lmn} (Table 8.3); the recalculated intensity is shown in Figure 8.7.

There are a few points to be noted:

1. The modulation in short-range order scattering is small, indicating a small deviation in the microstructure from a statistically uncorrelated arrangement. The type of local order is not well established because the maximum is low (reaching at most 1.5 LU) and forms a shell around 110 that comprises the $1\frac{1}{2}0$ position. Such a value is at the limit of resolution, as previously noted in the investigation of Pt-Rh alloys [40]. In the latter case, as with the present investigation, data at small scattering angles have to be measured to decide on the basic question of

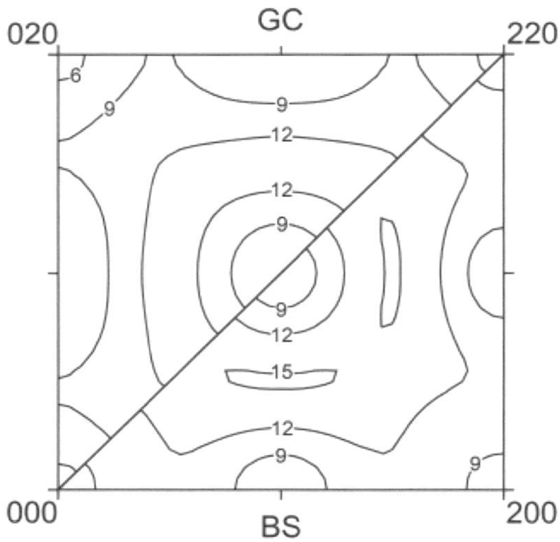

Figure 8.7. *Recalculated short-range order scattering of Ni-9.4 at.% Re air cooled from 873 K to room temperature in 0.1 LU for the h_1h_20 plane. Diffuse X-ray scattering was evaluated by the methods of Georgopoulos-Cohen (GC) and Borie-Sparks (BS).*

TABLE 8.3

Warren-Cowley short-range order parameters α_{lmn} of Ni-9.4 at.% Re air cooled from 873 K. The two evaluation methods—Georgopoulos-Cohen (GC) and Borie-Sparks (BS)—were employed.

lmn	α_{lmn}	
	GC	BS
000	1.062(11)	1.218(5)
110	−0.018(4)	−0.024(2)
200	−0.028(4)	−0.008(2)
211	0.011(2)	0.012(1)
220	−0.025(3)	−0.036(1)
310	−0.004(2)	0.003(1)

whether an alloy is locally ordered or decomposed. Electronic-structure calculations would be highly welcome to demonstrate whether Ni-Re exemplifies the situation attributed to Ni-Au by Wolverton et al. [31].

2. There is also clear evidence for the presence of local order from the negative sign of α_{110}. Its value amounts to about 20% of the possible minimum value of −0.104. This is one-half to one-third of the values previously noted in Ni-rich solid solutions of Ni-Ti or Ni-Al with comparable Ni fractions, where maxima in $I_{SRO}(\mathbf{h})$ reached values up to 2.5 LU [41]. The lower degree of local order may simply be the result of the slightly higher aging temperature that was chosen, as diffusion is claimed to be slow for Re in Ni [43]. Slow Re diffusion was also claimed to be responsible for the pileup of Re at the interface between the γ matrix and the γ' precipitates when the precipitates grow [34].
3. The presence of local order in Ni-rich Ni-Re is unexpected. Previous atom-probe ion microscopy consistently revealed the presence of small Re clusters in single-crystalline superalloys like CSMX-2 and PWA 1480 with 1.3 at.% Re (substituting W) or CSMX-4 (with 1 at.% Re). The γ matrix then contained ~3 at.% Re and showed Re agglomerates of a few atoms.

5 Species-Dependent Static Atomic Displacements in 3λ Investigations

Before the 3λ method was introduced, the Georgopoulos-Cohen technique had been the main method for determining static atomic displacements in concentrated alloys. Concerning systems with strong distortions, Al-rich Al-Cu (12.5%), Cu-rich Cu-Be (10.7%), and Al-rich Al-Zn (~7%) were investigated by the group of J. B. Cohen [7,44,45] (the differences in atomic sizes are given in brackets). These decomposing systems all form GP zones (platelets on {100} planes for the first two alloys and spheres/ellipsoids for the latter alloy). Static atomic displacements up to several tenths of the lattice parameter a were obtained. Note that with $c_A/c_B + \alpha_{lmn} \approx 0$, the way to determine the static atomic displacements is ill-conditioned, as the product of both terms forms the Fourier coefficients in the underlying Fourier series of linear displacement scattering, a situation repeatedly met in Al-Cu. Also, for interpreting displacements in the case of platelets, a model is required because the cubic symmetry is locally lost.

The 3λ investigation, on the other hand, yields static atomic displacements for weakly distorted systems because the components of the systems to be investigated have to be close in atomic number Z to fully exploit this procedure. Alloys with ΔZ of 2 (Ni-Fe and Fe-Cr) [46–48] as well as 4 (Ni-Cr and Cu-Mn) [49,50] were investigated and typical displacements were in the range of 10^{-2} to $10^{-4}a$. To judge the quality in determining displacement parameters, one has to remember that the published standard deviations of the static displacements are usually only based on counting statistics. Thus performing a comparative study with the data from a 3λ measurement was indicated. This was done with the raw data for Ni-Cr and Cu-Mn and with partly treated data for Fe-Cr. The conclusions are as follows:

1. The larger variation in the ratio of the atomic scattering factors of the constituents contributes much to the stability in data evaluation. Still, one has to note that the number of shells considered relevant for the displacement parameters is quite limited (5 to 10 shells) despite the long-ranging character of any displacement field. A reason may be that the systems investigated are all weakly distorted.
2. The relative calibration of the measurements at three wavelengths presents a larger problem. With an estimated ±5%, the uncertainty in calibration at any wavelength is not negligible. For a relative calibration, the strong thermal diffuse scattering close to Bragg reflections [49,50], which is common to all samples, was employed. Also, an assumed value of short-range order scattering close to fundamental reflections [46] may serve as a guide.
3. The experimental separation of thermal diffuse scattering and Compton scattering versus any calculation within the Georgopoulos-Cohen separation scheme faces the difficulty of how to properly analyze Compton scattering in an energy-resolved spectrum at small scattering angles. A general difficulty is the separation of resonant Raman scattering when measuring close to the absorption edges. A conclusion from Schönfeld et al. [50] was that distances in the range 10 to 20 eV from the absorption edge are needed to avoid additional intensity uncertainties.

As an example for the closeness by which displacement scattering was determined in the two evaluation strategies, the case of Cu-17.2 at.% Mn is considered [50]. Here diffuse X-ray scattering was registered at three energies (6526 eV, 8969 eV, and 8909 eV) suited for an evaluation by the 3λ technique. The measurements were done at room temperature for a state quenched from 483 K. Nearest-neighbor displacements were found to be dominant, being largest among Cu-Mn pairs. The sign and magnitude of the displacement parameters closely agree with those found by the Georgopoulos-Cohen method for those shells that are characterized by a single displacement component (≠ 0). To avoid the energy dependence of any X-ray displacement scattering, Figure 8.8a shows the situation expected for the case in which experiments would have been performed with neutrons (this implies a simple rescaling). Differences between the two evaluations up to ~0.6 LU are seen within this scattering region, revealing the differences in the displacement parameters. For comparison, the case of Ni-20 at.% Cr that was also investigated at three energies (5975 eV, 8327 eV, and 8097 eV) is shown in Figure 8.8b [49]. The scattering patterns from both evaluation strategies, again rescaled for the case of neutron scattering of an alloy with natural Ni, show close agreement. Differences found in the sign of nearest-neighbor displacements are probably due to the large resonant Raman scattering $K\text{-}M_{II}M_{III}$ at the Ni-K absorption edge.

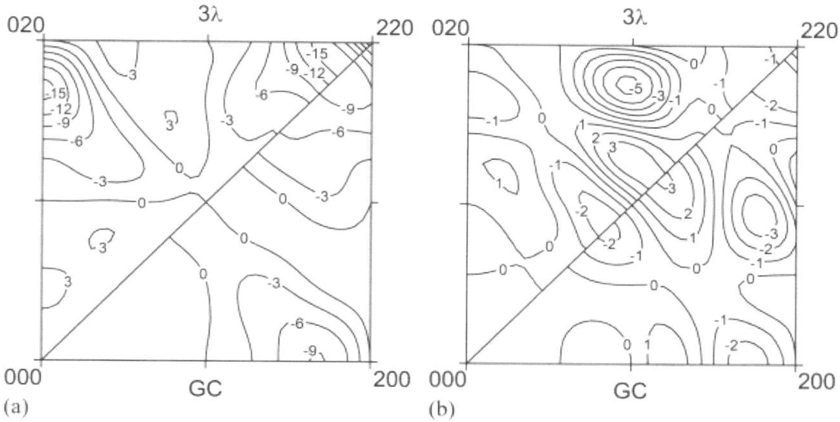

Figure 8.8. *Linear displacement scattering for (a) Cu-17.2 at.% Mn aged at 483 K [50] and (b) Ni-20 at.% Cr aged at 744 K [49] calculated for the case of diffuse neutron scattering. Atomic displacement parameters from a Georgopoulos-Cohen (GC) and a 3λ evaluation were employed.*

Unfortunately, additional information beyond data for close neighbors from extended X-ray absorption fine structure investigations is still missing but urgently required.

6 Conclusions

Local atomic arrangements in the two strongly distorted alloys Ni-Au and Ni-Re show signs of local order and local decomposition. For a Ni-rich Ni-Re solid solution, it was established that local order is present. For solid solutions of Ni-Au in thermal equilibrium, the situation is more complex; the weight of local order in comparison with local decomposition varies (increases) with increasing Au fraction. While only local decomposition is observed for Ni-~8 at.% Au, until now it remained unclear whether local order or local decomposition prevails for a state of thermal equilibrium of Ni-60 at.% Au. There are obviously limitations to extract individual alloy parameters with high precision for alloys with larger atomic displacements.

The static atomic displacements in weakly distorted alloys are determined with increasing confidence from a 3λ-type X-ray scattering experiment. A combination of measurements at several wavelengths will add to a robust determination of the species dependence of displacements even when a "null-Laue" measurement might not be feasible.

Acknowledgments

The authors are grateful to E. Fischer for his support in growing the single crystals. They thank Professor J. Löffler for his support and encouragement. They also thank the Institut Laue-Langerin, Grenoble, and the Paul Scherrer Institute, Villigen, for beam time. This work was partially supported by the "Schweizerischer Nationalfonds zur Förderung der wissenschaftlichen Forschung".

References

1. G. Kostorz, "X-Ray and Neutron Scattering", in *Physical Metallurgy*, ed. R. W. Cahn and P. Haasen (Amsterdam: North-Holland, 1996), 1115.
2. W. Schweika, *Disordered Alloys: Diffuse Scattering and Monte Carlo Simulations* (Berlin: Springer, 1998).
3. G. E. Ice and C. J. Sparks, *Annu. Rev. Mater. Sci.* 29 (1999): 25.
4. B. Schönfeld, *Progr. Mater. Sci.* 44 (1999): 435.
5. V. M. Nield and D. A. Keen, *Diffuse Neutron Scattering from Crystalline Materials* (Oxford: Clarendon Press, 2001).
6. V. Gerold, *Z. Metallkde.* 45 (1954): 599.
7. E. Matsubara and J. B. Cohen, *Acta Metall.* 33 (1985): 1945.
8. P. P. Müller, B. Schönfeld, G. Kostorz, and W. Bührer, *Acta Metall.* 37 (1989): 2125.
9. A. Malik, "Decomposition in Aluminium Alloys: Diffuse Scattering and Crystal Modelling" (PhD diss., ETH Zürich, 1995).
10. B. Borie and C. J. Sparks, *Acta Crystallogr., Sect. A: Cryst. Phys., Diffr., Theor. Gen. Crystallogr.* 27 (1971): 198.
11. J. E. Gragg and J. B. Cohen, *Acta Metall.* 19 (1971): 507.
12. P. Georgopoulos and J. B. Cohen, *J. Physique Colloq.* 38 (1977): C7–191.
13. G. E. Ice, C. J. Sparks, A. Habenschuss, and L. B. Shaffer, *Phys. Rev. Lett.* 68 (1992): 863.
14. M. A. Krivoglaz, *Diffuse Scattering of X-Rays and Neutrons by Fluctuations* (Berlin: Springer, 1996).
15. T. B. Massalski, *Binary Alloy Phase Diagrams* (Materials Park, OH: ASM International, 1990).
16. M. J. Portmann, B. Schönfeld, G. Kostorz, and F. Altorfer, *Phys. Rev. B* 65 (2001): 024110.
17. M. J. Portmann, B. Schönfeld, G. Kostorz, F. Altorfer, and J. Kohlbrecher, *Phys. Rev. B* 68 (2003): 012103.
18. B. Schönfeld, G. Kostorz, and G. McIntyre, unpublished results.
19. T. B. Wu and J. B. Cohen, *Acta Metall.* 31 (1983): 1929.
20. S. C. Moss, "Local Order in Solids", in *Local Atomic Arrangements Studied by X-Ray Diffraction*, ed. J.B. Cohen and J. E. Hilliard (New York: Gordon and Breach, 1966), 95.
21. S. C. Moss and B. L. Averbach, "Local Atomic Configurations in a Gold-Nickel Alloy", in *Small-Angle X-Ray Scattering*, ed. H. Brumberger (New York: Gordon and Breach, 1967), 335.
22. H. Reichert, A. Schöps, I. B. Ramsteiner, V. N. Bugaev, O. Shchyglo, A. Udyansky, H. Dosch, M. Asta, R. Drautz, and V. Honkimäki, *Phys. Rev. Lett.* 95 (2005): 235703.
23. C. Dressler, G. Abadias, P. Bayle-Guillemaud, A. Marty, I. Schuster, J. Thibault, and B. Gilles, *Appl. Phys. Lett.* 72 (1998): 2241.
24. P. T. Coleridge and I. M. Templeton, *Phys. Rev. B* 25 (1982): 7818.
25. P. Haasen, *Physical Metallurgy* (Cambridge: Cambridge University Press, 1986).
26. J.-C. Zhao and M. R. Notis, *Metall. Mater. Trans. A* 30 (1999): 707.
27. P. C. Clapp, *Phys. Rev. B* 4 (1971): 255.
28. M. Asta and S. M. Foiles, *Phys. Rev. B* 53 (1996): 2389.
29. C. Wolverton, V. Ozolins, and A. Zunger, *Phys. Rev. B* 57 (1998): 4332.
30. C. Wolverton and A. Zunger, *Comput. Mater. Sci.* 8 (1997): 107.
31. C. Wolverton, V. Ozolins, and A. Zunger, *J. Phys.: Condens. Matter* 12 (2000): 2749.
32. A. F. Giamei and D. L. Anton, *Metall. Trans. A* 16 (1985): 1997.
33. J. Deges, A. Schneider, R. Fischer, and G. Frommeyer, *Mater. Sci. Eng. A* 353 (2003): 80.
34. P. J. Warren, A. Cerezo, and G. D. W. Smith, *Mater. Sci. Eng. A* 250 (1998): 88.
35. N. Wanderka and U. Glatzel, *Mater. Sci. Eng. A* 203 (1995): 69.
36. H. Pottebohm, G. Neite, and E. Nembach, *Mater. Sci. Eng.* 60 (1983): 189.
37. M. A. Krivoglaz, *X-Ray and Neutron Diffraction in Nonideal Crystals* (Berlin: Springer, 1996).
38. V. F. Sears, *Neutron News* 3 (1992): 26.
39. D. Blavette, P. Caron, and T. Khan, *Scripta Metall.* 20 (1986): 1395.
40. Ch. Steiner, B. Schönfeld, M. J. Portmann, M. Kompatscher, G. Kostorz, A. Mazuelas, T. Metzger, and J. Kohlbrecher, *Phys. Rev. B* 71 (2005): 104204.
41. R. Bucher, B. Schönfeld, G. Kostorz, and M. Zolliker, *Phys. Stat. Sol. A* 175 (1999): 527.
42. B. Schönfeld, L. Reinhard, G. Kostorz, and W. Bührer, *Acta Mater.* 45 (1997): 5187.

43. M. S. A. Karanuratne, P. Carter, and R. C. Reed, *Mater. Sci. Eng. A* 281 (2000): 229.
44. Y. M. Koo and J. B. Cohen, *Acta Metall.* 37 (1989): 1295.
45. D. R. Haeffner and J. B. Cohen, *Acta Metall.* 37 (1989): 2185.
46. X. Jiang, G. E. Ice, C. J. Sparks, L. Robertson, and P. Zschack, *Phys. Rev. B* 54 (1996): 3211.
47. J. L. Robertson, G. E. Ice, C. J. Sparks, X. Jiang, P. Zschack, F. Bley, S. Lefebvre, and M. Bessiere, *Phys. Rev. Lett.* 82 (1999): 2911.
48. L. Reinhard, J. L. Robertson, S. C. Moss, G. E. Ice, P. Zschack, and C. J. Sparks, *Phys. Rev. B* 45 (1992): 2662.
49. B. Schönfeld, G. E. Ice, C. J. Sparks, H.-G. Haubold, W. Schweika, and L. B. Shaffer, *Phys. Stat. Sol. B* 183 (1994): 79.
50. B. Schönfeld, H. Roelofs, G. Kostorz, J. L. Robertson, P. Zschack, and G. E. Ice, *Phys. Rev. B 77 (2008): 144202.*

9

X-RAY DIFFUSE SCATTERING NEAR BRAGG REFLECTIONS FOR THE STUDY OF CLUSTERED DEFECTS IN CRYSTALLINE MATERIALS

Bennett C. Larson

Materials Science and Technology Division, Oak Ridge National Laboratory, Oak Ridge, TN 37831, USA

Abstract

The fundamental aspects of diffuse scattering from lattice defects within the coherent wave theory are reviewed, and the use of Huang and asymptotic diffuse scattering for the determination of the size, type, and internal structure of clustered defects is discussed. The scattering amplitude for diffuse scattering from lattice defects is presented within the single defect approximation, and the physical interpretation of the asymptotic diffuse scattering from dislocation loops and precipitates is discussed in terms of local Bragg scattering. The importance of using detailed and accurate displacement fields for calculating diffuse scattering cross sections for dislocation loops and coherent precipitates is emphasized, and examples of the measurement of size distributions for vacancy and interstitial loops and for coherent precipitates in Cu are revisited. In addition, the use of submicrometer-diameter X-ray microbeams to perform depth-resolved diffuse scattering measurements of vacancy and interstitial loop distributions in 10 MeV self-ion-implanted silicon single crystals is discussed.

1 Introduction

The analysis of diffuse scattering near Bragg reflections represents a powerful method for the study of defect clusters that induce lattice distortions in crystalline materials. Krivoglaz, Dederichs, and Trinkaus [1–6] were instrumental in analyzing the form and intensity of the Huang diffuse scattering resulting from the long-range elastic distortion fields surrounding point defects and defect clusters.

Their work further considered the scattering in the so-called asymptotic scattering region (beyond the Huang scattering region) that is characterized by local Bragg scattering from the relatively large distortions near defect clusters [5,7,8]. Although the asymptotic diffuse scattering associated with the large lattice displacements near defect clusters cannot be fully specified in closed form, the physical interpretation of asymptotic diffuse scattering in terms of particle size- and strain gradient-broadened Bragg-like scattering from locally rotated or dilatated regions near defect clusters is clear. Fortunately, accurate diffuse scattering cross sections can be obtained for this region through numerical integration of lattice sums and the use of numerically calculated elastic displacement fields for dislocation loops and precipitates [8–12].

The ability to calculate accurate diffuse scattering cross sections for defect clusters has been key for using diffuse scattering as a quantitative tool to investigate defect clusters in crystals [8–11]. X-ray diffuse scattering has been used in a large number of studies of irradiation-induced and intrinsic defect clusters in crystals, as discussed in original papers and reviews [13–27]. Thermodynamic precipitation of coherent second-phase particles is of both fundamental and technological importance for advanced materials. Alloy precipitation and both electron and neutron irradiations produce defects distributed uniformly throughout bulk samples. However, ion irradiations typically involve penetration depths of only a few microns below surfaces, and because ion damage is strongly energy dependent, they produce defect distributions that are strongly depth dependent. Understanding the generation and annealing characteristics of ion irradiation-induced defects is important for simulating high-dose neutron irradiation damage, and ion damage has become increasingly important in relation to ion implantation processing steps incorporated into electronic device manufacture.

In this chapter, we discuss the fundamental concepts of diffuse scattering, the information available on defect clusters from diffuse scattering near Bragg reflections, and the application of submicron spatial resolution X-ray microbeams for depth-resolved studies of ion implantation-induced defects in silicon. The results of depth-dependent differential diffuse scattering measurements, made in cross sections on self-ion-irradiated silicon, are presented in terms of size distributions and point defect concentrations for vacancy and interstitial loops, and the relative merits of cross section and plan-view diffuse scattering techniques are discussed.

2 Diffuse Scattering and Coherent Wave-Scattering Theory

X-ray scattering from lattice defects is generally quite weak; therefore, the cross section for scattering of an incoming plane wave $e^{i\mathbf{k}_o \cdot \mathbf{r}}$ with wave vector $\mathbf{k}_0$ into an outgoing wave $e^{i\mathbf{k} \cdot \mathbf{r}}$ with wave vector $\mathbf{k}$ is usually written in terms of the kinematic or first Born approximation with no provision for multiple scattering. In the first Born approximation, the differential cross section for an ensemble of atoms is given by

$$\frac{d\sigma(\mathbf{K})}{d\Omega} = \left|V(\mathbf{K})\right|^2, \tag{9.1}$$

where $\mathbf{K}=\mathbf{k}-\mathbf{k}_0$ and $V(\mathbf{K})$ is the Fourier transform of the scattering potential $V(\mathbf{r}) = \sum_i r_e f_i(\mathbf{K})\delta(\mathbf{r}-\mathbf{r}_i)$.

The atomic scattering length for X-rays, $r_e f_i(\mathbf{K})$, is given by the product of the classical electron radius, r_e, and the atomic form factor $f_i(\mathbf{K})$. The lattice sites are specified by $\mathbf{r}_i = \mathbf{r}_i^o + \mathbf{s}_i$, in which $\mathbf{r}_i^o$ represents the undistorted lattice site and $\mathbf{s}_i \equiv \mathbf{s}(\mathbf{r}_i)$ specifies the displacement from the ith periodic lattice site. The scattering amplitude is then defined by

$$\frac{d\sigma(\mathbf{K})}{d\Omega} = \left| r_e f(\mathbf{K}) \right|^2 \left| A(\mathbf{K}) \right|^2 = \left| \sum_i r_e f_i(\mathbf{K}) e^{i\mathbf{K}\cdot\mathbf{r}_i} \right|^2 . \quad (9.2)$$

This is the total elastic scattering cross section for the sample and so contains Bragg reflections for average periodic $\mathbf{r}_i$ expanded by image forces and diffuse scattering components generated by deviations $\mathbf{s}(\mathbf{r}_i)$ from periodic lattice sites (e.g., by phonons, point defects, clusters of point defects, second-phase precipitates, or dislocations) [1,3,17]. As pointed out by Krivoglaz [1], line defects such as dislocations are a special case in that they are not localized defects and their slowly decaying strain field falls off only as $1/r$. Therefore, the static Debye-Waller factor diverges for crystals with dislocations as a signature of the destruction of long-range periodicity. The remaining strain-broadened and particle-size-broadened scattering may be strongly peaked, but according to the strict definition of Bragg scattering from long-range periodic structures, it represents diffuse scattering that is often referred to as local Bragg-like or Stokes-Wilson scattering within the deformation microstructure [28]. Although dislocation loops will be addressed here in terms of point-defect clusters in otherwise long-range periodic crystals, diffuse scattering from dislocation lines and tangles is fundamentally different in origin from that of point defects, dislocation loops, and clusters and will not be considered here.

The distinction between sharply peaked (coherent) Bragg scattering and the more smoothly varying (incoherent) diffuse scattering is not easily defined within the kinematical first Born approximation theory discussed above because Huang scattering from both thermal vibrations and lattice defects diverges at Bragg reflections. On the other hand, the scattering processes are, in fact, well defined quantum-mechanically in terms of changes of any quantum number in the state of the sample, and, accordingly, the separation is well defined within the formal theory of scattering, which inherently includes multiple scattering. As discussed by Lax [29] and in more detail by Dederichs [5,17], the separation of the Bragg (coherent) scattering from the diffuse (incoherent) scattering for crystalline samples containing localized defects is well defined in terms of the so-called coherent wave theory, which is based on the formally exact Lippman-Schwinger equation [17]. Although a rigorous and complete presentation of the coherent wave analysis is outside the scope of this chapter, it is instructive to recall the elegant arguments of the coherent wave theory that led intuitively to the so-called single-defect approximation, which forms the underlying basis of diffuse scattering investigations of statistically distributed lattice defects.

In outlining the coherent wave approach, we follow Lax [29] and Dederichs [2,17] and consider a periodic assembly of atoms with a statistical distribution of point defects or point-defect clusters that generate frozen-in displacements by their elastic distortion fields. The separation between Bragg and diffuse scattering within the coherent wave formalism is identified by performing a formal (theoretical) interference experiment to determine what part of the scattered wave remains quantum-mechanically coherent within the scattering process and hence is able to interfere with the incoming wave. Defining ψ to be the solution to the scattering of an incoming wave, ϕ, by a particular statistical distribution of defects within the sample, the intensity (density) of scattered radiation is given by $\psi^* \psi = |\psi|^2$. The experimentally observed scattering from a statistically random distribution of defects is given by averaging $|\psi|^2$ over all possible lattice defect configurations. Denoting this statistical average by $\langle\ \rangle$, the measured intensity is then given by $\langle|\psi|^2\rangle$.

The formal interference experiment to determine what part of ψ remains coherent with the incident beam and which part represents diffuse scattering is defined by

$$\langle|\phi+\psi|^2\rangle = \langle\phi^*\phi+\phi^*\psi+\psi^*\phi+\psi^*\psi\rangle, \tag{9.3}$$

which, after performing the implied statistical averages and adding and subtracting $\langle\psi^*\rangle\langle\psi\rangle$, leads to the important result that

$$\langle|\phi+\psi|^2\rangle = |\phi+\langle\psi\rangle|^2 + \langle|\psi|^2\rangle - |\langle\psi\rangle|^2 . \tag{9.4}$$

Equation 9.4 indicates that only $\langle\psi\rangle$ remains absolutely coherent with ϕ and that the remaining fraction $\langle|\psi|^2\rangle - |\langle\psi\rangle|^2$ is incoherent or, according to Lax [29], only relatively (i.e., not absolutely) coherent with the incoming beam ϕ. We note that the symbolic form of the incoherent term $\langle|\psi|^2\rangle - |\langle\psi\rangle|^2$ is just the form that is associated with fluctuation scattering (or departures from mean field behavior). It is natural then to associate the coherent term with the Bragg scattering term, where $\langle\psi\rangle$ indicates that the Bragg scattering that remains coherent is reduced by a Debye-Waller factor (as will be seen below) that is introduced by the averaging over statistically distributed defect distributions. Equation 9.4 then leads to the following intuitive separation of the diffuse scattering (or fluctuation scattering) cross section from the total scattering cross section by subtracting the Bragg scattering:

$$\left[\frac{d\sigma(\mathbf{K})}{d\Omega}\right]_{Diffuse} = \left[\frac{d\sigma(\mathbf{K})}{d\Omega}\right]_{Total} - \left[\frac{d\sigma(\mathbf{K})}{d\Omega}\right]_{Bragg} \tag{9.5}$$

Equation 9.4 assures this to be rigorously true in a quantum-mechanical sense and not simply by an empirically derived result. As will be mentioned later, however, separating coherent scattering, inelastic (Compton) scattering, and thermal diffuse scattering from defect diffuse scattering is not at all a simple or straightforward task experimentally, even though it is well defined theoretically.

Now, if one associates ψ with $\sum_i r_e f_i e^{i\mathbf{K}\cdot\mathbf{r}_i}$ and understands $f_i = f_i(\mathbf{K})$ for shorthand, the following expressions are obtained for the diffuse scattering components:

$$\langle|\psi|^2\rangle = \left\langle\left|\sum_i r_e f_i e^{i\mathbf{K}\cdot\mathbf{r}_i}\right|^2\right\rangle = \left\langle\sum_{i,j} r_e^2 f_i f_j e^{i\mathbf{K}\cdot(\mathbf{r}_i-\mathbf{r}_j)}\right\rangle, \tag{9.6}$$

$$\left|\langle\psi\rangle\right|^2 = \left|\left\langle \sum_i r_e f_i e^{i\mathbf{K}\cdot\mathbf{r}_i} \right\rangle\right|^2 = \left\langle \sum_i r_e f_i e^{i\mathbf{K}\cdot\mathbf{r}_i} \right\rangle \left\langle \sum_j r_e f_j e^{-i\mathbf{K}\cdot\mathbf{r}_j} \right\rangle . \quad (9.7)$$

If we further write $\mathbf{K} = \mathbf{H} + \mathbf{q}$, where $\mathbf{H}$ is a reciprocal lattice vector and $\mathbf{q}$ is a vector within the Brillouin zone, and recall that $\mathbf{r}_i = \mathbf{r}_i^o + \mathbf{s}_i$, we obtain with $\mathbf{H}\cdot\mathbf{r}_i^o = 2\pi n$,

$$\left[\frac{d\sigma(\mathbf{K})}{d\Omega}\right]_{Diffuse} = \sum_{i,j} r_e^2 f_i f_j e^{i\mathbf{q}\cdot(\mathbf{r}_i^o - \mathbf{r}_j^o)} \left[\left\langle e^{i\mathbf{K}\cdot(\mathbf{s}_i - \mathbf{s}_j)}\right\rangle - \left\langle e^{i\mathbf{K}\cdot\mathbf{s}_i}\right\rangle\left\langle e^{-i\mathbf{K}\cdot\mathbf{s}_j}\right\rangle\right] . \quad (9.8)$$

Following Dederichs [2,3,17], after detailed consideration of overlapping defect displacement fields for statistically random defect sites and the implications of image forces, which implies that $\mathbf{r}_i$ now refers to the average expanded (or contracted) lattice positions, Equation 9.8 yields

$$\left[\frac{d\sigma(\mathbf{K})}{d\Omega}\right]_{Diffuse} = \left|\sum_i r_e f_i e^{-L(\mathbf{K})} e^{i\mathbf{q}\cdot\mathbf{r}_i}\left[e^{i\mathbf{K}\cdot\mathbf{s}(\mathbf{r}_i)} - 1\right]\right|^2 . \quad (9.9)$$

Including the scattering from the defect atoms located at position $\mathbf{r}_j^d$ (with respect to average lattice positions) generating the displacements $\mathbf{s}(\mathbf{r}_i)$, Equation 9.9 becomes

$$\left[\frac{d\sigma(\mathbf{K})}{d\Omega}\right]_{Diffuse} = \left|\sum_j^{n^d} r_e f_j^d e^{i\mathbf{H}\cdot\mathbf{r}_j^d} e^{i\mathbf{q}\cdot\mathbf{r}_j^d} + \sum_i r_e f_i e^{-L(\mathbf{K})} e^{i\mathbf{q}\cdot\mathbf{r}_i}\left[e^{i\mathbf{K}\cdot\mathbf{s}(\mathbf{r}_i)} - 1\right]\right|^2 , \quad (9.10)$$

where n^d is the number of atoms in the defect/cluster and L($\mathbf{K}$) is the static Debye-Waller factor given by $L(\mathbf{K}) = c\sum_i [1 - \cos(\mathbf{K}\cdot\mathbf{s}(\mathbf{r}_i))]$, with c representing the fractional concentration of defects (i.e., defects per atom) within the lattice.

Equation 9.10 corresponds to the so-called single-defect approximation for randomly distributed defects, where the first term represents the direct or Laue scattering from the atoms in the defect cluster (i.e., dislocation loop, coherent precipitate, or loose aggregate) and the second term represents the scattering from the surrounding atoms that are displaced $\mathbf{s}(\mathbf{r}_i)$ from their regular lattice positions by the elastic distortion field surrounding the defect clusters.

As discussed above, the single-defect approximation refers to the specific case of a statistically large number of defects that are randomly distributed within an otherwise undistorted periodic crystal. The simplicity of the formula and the ability to consider only the distortion field for a single defect in the calculation of the cross section are due to the fact that the cross-terms in the scattering between individual defects cancel identically for randomly distributed defect sites. It should be emphasized also that the scattering formalism specifically assumes that the displacement fields of all defects overlap completely since their displacements fall off as $1/r^2$ with no cutoff. Hence each defect impacts the positions of atoms in every part of the sample, and the forces at the edges of the sample lead to the expansion/contraction of the average periodic lattice. Implicit in the process is the assumption that the density and strength of the defects are low enough that they do not distort the crystal lattice so much as to preclude the notion of an average periodic lattice, which implies that within a dynamical diffraction extinction distance for the periodic average lattice there is a statistically random distribution of defects.

3 Evaluation of Diffuse Scattering Cross Sections for Small *q* and Large *q*

In order to understand and interpret diffuse scattering from lattice defects, methods to perform the lattice sums involved in the cross sections in Equation 9.10 have been the focus of considerable attention for more than half a century, both in closed form and numerically [1–8,10–12,20]. The direct (or Laue) scattering term in the amplitude is a rather straightforward Fourier transform of the atoms comprising the defects themselves, and so presents little difficulty [5,11]. The scattering amplitude associated with the distortion of the surrounding crystal (the second term in Equation 9.10) is more complex, however, and requires a combination of analytic and numerical tools for evaluation. For small $\mathbf{q}$, the exponential inside the bracket in the distortion scattering function

$$\sum_i e^{i\mathbf{q}\cdot\mathbf{r}_i}\left[e^{i\mathbf{K}\cdot\mathbf{s}(\mathbf{r}_i)}-1\right] \tag{9.11}$$

can be expanded to obtain the so-called Huang scattering amplitude,

$$A_H(\mathbf{K}) = i\mathbf{K}\cdot\mathbf{s}(\mathbf{q}), \tag{9.12}$$

which can be evaluated analytically even for elastically anisotropic materials [3,5,11] and produces the well-known $1/q^2$ dependence of the Huang scattering intensity close to Bragg reflections. The ability to evaluate this leading term analytically is critical for both large and small q because the Huang amplitude is sensitive to the long-range elastic distortions and so converges very slowly in r. By removing the leading term from the sum and evaluating it analytically, the remaining portion of Equation 9.11 is given by

$$\sum_i e^{i\mathbf{q}\cdot\mathbf{r}_i}\left[e^{i\mathbf{K}\cdot\mathbf{s}(\mathbf{r}_i)}-1\right] \equiv i\mathbf{K}\cdot\mathbf{s}(\mathbf{q}) + \sum_i e^{i\mathbf{q}\cdot\mathbf{r}_i}\left[e^{i\mathbf{K}\cdot\mathbf{s}(\mathbf{r}_i)}-1-i\mathbf{K}\cdot\mathbf{s}(\mathbf{r}_i)\right], \tag{9.13}$$

which in sine and cosine notation is given by

$$\sum_i e^{i\mathbf{q}\cdot\mathbf{r}_i}\left[e^{i\mathbf{K}\cdot\mathbf{s}(\mathbf{r}_i)}-1\right] \equiv i\mathbf{K}\cdot\mathbf{s}(\mathbf{q}) + \sum_i e^{i\mathbf{q}\cdot\mathbf{r}_i}\left[\cos\left(\mathbf{K}\cdot\mathbf{s}(\mathbf{r}_i)\right)-1+i\sin\left(\mathbf{K}\cdot\mathbf{s}(\mathbf{r}_i)\right)-i\mathbf{K}\cdot\mathbf{s}(\mathbf{r}_i)\right]. \tag{9.14}$$

This form indicates immediately that only terms of order $\left(\mathbf{K}\cdot\mathbf{s}(\mathbf{r}_i)\right)^2$ and higher remain in the summation. Therefore, after subtracting the Huang amplitude from the summation and evaluating it analytically utilizing double-force tensors and Fourier-transformed elastic Greens functions [5], the second term falls off as $1/r^4$ or faster ($\mathbf{s}(\mathbf{r})\sim 1/r^2$) so that volume integrals converge as $1/r^2$ and so can be integrated without difficulty using numerically calculated displacement fields $\mathbf{s}(\mathbf{r})$ as discussed below.

The Huang scattering amplitude, given in closed form by Equation 9.12, is of course very strong for small q due to the fact that the amplitude diverges as $1/q$ [2,6,11]. The high intensity and the ability to specify accurately the angular dependence of the Huang scattering for both point defects and point defect clusters (loops, precipitates) makes the Huang scattering component attractive for investigating the local symmetry of lattice defects. However, Huang scattering information is, in practice, most powerful for point defects or very small clusters [2–6,16,30], where direct imaging techniques such as transmission electron microscopy (TEM) are not applicable. This is because the Huang scattering amplitude does not provide direct information on cluster sizes except through the often-used (but actually rather crude [1,11,18]) approximation of $R \sim 1/q_c$ (or $R \sim 1/\left(\sqrt{Hbq_c}\right)$), where b is the Burgers vector for dislocation loops and q_c corresponds to the crossover from the Huang scattering $1/q^2$ dependence to the asymptotic scattering $1/q^4$ falloff for larger q. Even then the determination is heavily weighted toward larger sizes in distributions—the fourth power of the radius for loops and the sixth power of the radius for precipitates. Moreover, because of the square of the amplitude, it does not provide direct information on the vacancy versus interstitial nature of loops when both types are present.

For clustered defects such as dislocation loops or precipitates, Huang scattering measurements provide symmetry information [3,18], but if combinations of loops and precipitates or other types of structural deformation are present, symmetry information on individual types of defect clusters is not available. For sizes above $R \sim 10$ Å, TEM measurements are, of course, able to provide detailed information on defect cluster symmetry, morphology, and size for deformed and polycrystalline samples as well as for samples containing a mixture of defect types. On the other hand, when both vacancy and interstitial loops are present (such as in neutron-, electron-, or ion-irradiated materials), obtaining size distributions for each type of loop separately from TEM measurements becomes a very tedious process [14,15]. Moreover, when defect clusters are too small to be imaged definitively by TEM or for investigations such as cluster nucleation and sequential thermal growth studies, where nondestructive measurements are desired, X-ray diffuse scattering in the so-called asymptotic (large q) region beyond the Huang scattering region provides critical information not available by TEM [7,14].

Although scattering cross sections cannot be calculated in closed form for the large q asymptotic scattering regime, accurate numerical calculations can be made using Equation 9.13 [2,7,8,10,11,14,32]. As discussed originally by Stokes and Wilson in connection with straight dislocation distributions [28] and as shown for clustered defects by Trinkaus [5,7,8] within the infinitesimal distortion center approximation (i.e., assuming lattice displacements increase as $\sim 1/r^2$ down to zero r), diffuse scattering from the strongly deformed volumes close to defect clusters can be characterized as Bragg-like scattering from locally dilatated and rotated regions. Detailed cross section calculations using Equation 9.13 for both dislocation loops and coherent precipitates have verified the local Bragg scattering interpretation [8,10,11]. However, the numerical calculations further showed that it is critically important to use the actual deformation fields near dislocation loops and inside precipitates rather than to assume a $\sim 1/r^2$ dependence for all r. This is because not only do the actual distortion fields depart markedly from $1/r^2$ for r less than the radius R but also the form of the actual displacement fields yield constant or nearly constant lattice parameters for $r < R$, which in turn introduces strong distortions in the $1/q^4$ dependence predicted by the infinitesimal distortion center approximation [5,7–11].

Accordingly, it is imperative to use accurate (numerically calculated) scattering cross sections based on actual defect cluster displacement fields for quantitative diffuse scattering investigations of defect cluster concentrations, sizes, and size distributions. As discussed in detail in the next section, the strong intensity distortions in the $1/q^4$ dependence are very informative; in fact, they provide direct size information for both dislocation loops and precipitate particles, and they also make it possible to extract separate size distributions for interstitial and vacancy loops. Moreover, the information on cluster size obtained from scattering derived from lattice positions $r < R$ is far more detailed and far more accurate than information on sizes determined using the above discussed transition from the Huang scattering ($1/q^2$) to the asymptotic scattering ($1/q^4$ or faster) regime.

As mentioned following Equation 9.5, the separation of Bragg scattering, electronic inelastic (i.e., Compton) scattering, and quasielastic thermal diffuse scattering from defect diffuse scattering experimentally is in reality a rather subjective process and, if not carried out properly, can have profound effects on the smallest q and largest q forms of the measured defect diffuse scattering patterns. At small q (close to the Bragg reflection) the defect diffuse scattering is overwhelmed by thermal diffuse scattering and by the Bragg tails, and at large q the defect cluster diffuse scattering is again dominated by thermal diffuse scattering and Compton inelastic scattering as the defect diffuse scattering falls off as $1/q^4$ or faster. For irradiation-induced defect clusters, it is possible to perform detailed measurements before the irradiation in most cases, and for precipitation from alloys, it is possible to perform such measurements before and after the precipitation process. However, for samples with intrinsic defect clusters in which the sample does not exist without the clusters, the background subtraction must be done with care, and in general, measurements at very small q and very large q should not be allowed to dominate the analyzed results in terms of very large clusters or very small defect clusters, respectively.

4 Asymptotic Diffuse Scattering

The distortion fields surrounding coherent precipitates and prismatic dislocation loops are depicted schematically in Figure 9.1. The illustrative lattice models in the Figure depict the distorted regions surrounding and within a spherical coherent precipitate for atoms that are undersized compared to the host and the distorted region near a planar interstitial loop. The precipitate model illustrates specifically that the lattice distortions associated with an undersized, coherent, second-phase particle (such as Co in Cu [31]) correspond to a smaller (and uniform) lattice parameter inside the particle. The precipitate model shows further that the coherency strain related to the smaller precipitate volume pulls the surrounding lattice inward toward the particle and thus produces a positive strain in the surrounding host lattice.

The interstitial loop model emphasizes that interstitial loops have compressed regions (negative strain) above and below the loop plane and only small expanded regions near the dislocation line that defines the edge of the loop. Of course, vacancy loops are just the opposite; they generate an expanded region normal to the loop plane as the surrounding lattice planes relax into the vacant loop area and only a small compressed region just inside the dislocation line at the circumference of the loop. The fact that interstitial loops are extrinsic defects in which the defect atoms occupy nonlattice positions and vacancy loops are intrinsic defects in which the missing atoms are on lattice sites leads to an asymmetry between vacancy and interstitial loops [8,11]. In addition, the compressed region at the circumference is inside the dislocation line for vacancy loops and the expanded region is outside the dislocation line for interstitial loops, which leads to a slight asymmetry in the cross sections for vacancy and interstitial loops as well.

Considering first the case of dislocation loops, the local Bragg scattering interpretation [7,8,14,32] of scattering from the asymptotic region contains direct information on the vacancy/interstitial nature of loops through sensitivity to the sign of the lattice strain in the direction normal to the loops. Numerical calculations of the lattice distortions around loops in elastically anisotropic cubic crystals have shown that the loop normal displacements depend nearly linearly on r within a volume of radius R [33] and that the strain (given by the gradient of the displacements) in regions normal to the plane of edge-type loops is given by $\Delta d / d \approx -b / 4R$ for metals [14,33], where b is the magnitude of the Burgers vector of the loop, taken to be positive for interstitial and negative for vacancy loops.

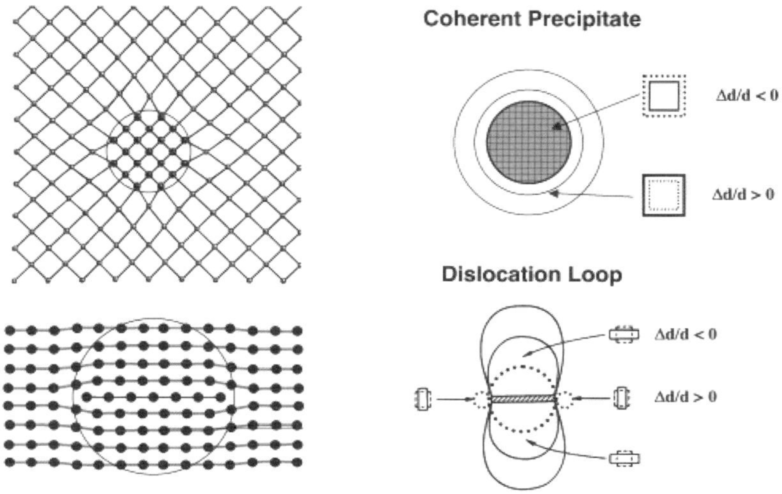

Figure 9.1. *Schematic view of the lattice distortions around coherent precipitates and interstitial loops. The precipitate lattice parameter is smaller than the host, yielding displacements that expand the surrounding lattice. The dislocation loop introduces a compressive strain normal to the loop plane in the surrounding host lattice. The circles enclose regions of nearly constant lattice parameter.*

Therefore, for loops oriented normal to H, there is a positive modulation in the $\sim 1/q^4$ falloff of the asymptotic scattering at the position $q = bH / 4R$, which is followed by a falloff more rapid than $1/q^4$ in the intensity for larger q values. Since it is the $1/r^2$ dependence of $\mathbf{s}(\mathbf{r})$ closer and closer to defect cluster centers that gives rise to the $1/q^4$ intensity for larger and larger q values in the asymptotic scattering region, the local Bragg scattering pileup of the intensity is due to a nearly uniform lattice parameter inside the circle containing the loop in Figure 9.1, which is caused by the displacements reaching a limiting value of $s(\mathbf{r}) = b / 2$ as r goes to zero (i.e., at the loop plane) [33].

Figure 9.2 shows numerical calculations of the diffuse scattering cross section for both vacancy and interstitial loops in Cu using Equation 9.13 [11,14,33], with lattice displacements calculated using anisotropic elasticity [33]. Since the cross sections are scaled by q^4/R^2, the lack of an extended flat region indicates the lack of an extended $1/q^4$ falloff region, and the nearly equal peak heights independent of size indicate that the scattering is proportional to the number of point defects contained in loops (i.e., loop volume = $b\pi R^2$). In this form, it is possible to see directly the departure of the scattering from an extended $1/q^4$ dependence [14] to a local Bragg scattering peak at a loop radius-dependent position of $q = bH / 4R$, as discussed above. The fact that vacancy loops scatter predominantly at negative q and interstitial loops scatter predominantly at positive q is also a direct result of the local Bragg scattering nature of the scattering, for which the compressive strain near interstitial loops puts the scattering at q higher than that of the average lattice and the expansive strain near vacancy loops puts the scattering at lower q than the average lattice Bragg position. This result is very significant for the asymptotic scattering region because the sign of b (i.e., vacancy or interstitial type) can be determined by the direction

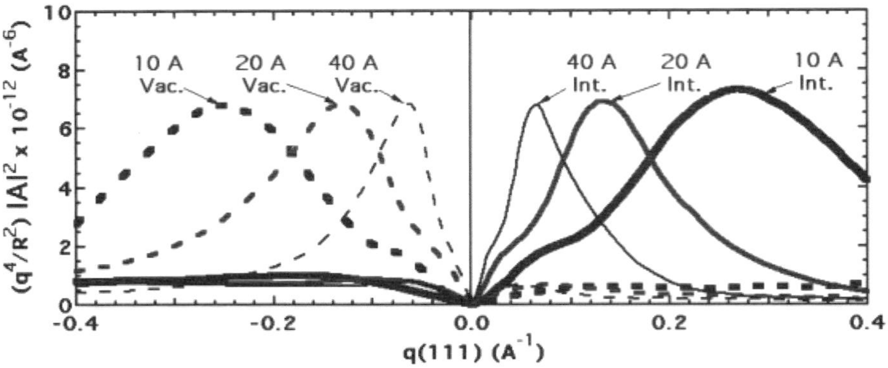

Figure 9.2. *Diffuse scattering cross sections calculated for vacancy and interstitial loops scaled by q^4/R^2 as a function of the loop radius for q values radial along the [111] direction at the (222) reciprocal lattice point for Cu.*

of q relative to H, and the radius R can be inferred from the magnitude of q. Because there are no large regions with expanded lattice parameters for interstitial loops, there is no significant scattering for negative q values associated with interstitial loops, and vice versa for vacancy loops [14].

Turning now to asymptotic diffuse scattering for coherent precipitates, one finds that different but entirely analogous local Bragg scattering effects occur. As depicted schematically in the model of a precipitate lattice at the top of Figure 9.1, the $\sim 1/r^2$ form of the distortion field terminates abruptly at the radius of the precipitate; inside the precipitate, the displacements fall linearly to zero as a direct result of the fact that coherent precipitate lattices have a (constant) lattice parameter different from the host. Of course, regions with a constant lattice parameter generate local Bragg scattering at a radial position $q = -\varepsilon H$ relative to the host Bragg position, where ε is the fractional difference between the lattice parameter of the precipitate and the host [10,12]. In contrast to the case for dislocation loops, however, the departure from $1/r^2$ for precipitates does not simply redistribute the intensity within the $1/q^4$ region. The strain of the precipitate is of the opposite sign to that generated in the host lattice, which puts the local Bragg scattering from the precipitate particle on the opposite side of the host Bragg position from the local Bragg scattering from the highly distorted region of the surrounding host lattice. These effects have long been appreciated in general [1,23,24], but they have been analyzed in more detail and discussed quantitatively in connection with numerical diffuse scattering calculations for precipitates in an isotropic elastic material by Iida et al. [10]. Figure 9.3 shows diffuse scattering cross section calculations for the (400) reciprocal lattice point for 25, 50, and 75 Å precipitates with internal strain $\varepsilon = -1.5\%$. Scaled by q^4/R^3, the form of the cross sections appears rather similar to those in Figure 9.2 for interstitial loops. In the sense that the q^4-scaled asymptotic scattering has a single peak on the positive q side for both precipitates and interstitial loops, it is a direct indication that they are both characterized by regions with lattice parameters compressed compared to the host lattice. However, in the case of interstitial loops, the compressed region is a result of distortions in the host lattice generated by the insertion of a single extra plane of atoms and represents a local Bragg scattering pileup of the $1/q^4$ scattering due to the loss of the $1/r^2$ displacement field near the loop. In the precipitate case, the compressed lattice is the intrinsic lattice structure of the second-phase precipitate, and the $1/q^4$ scattering from the (expanded) distorted host lattice is on the opposite side of the reciprocal lattice

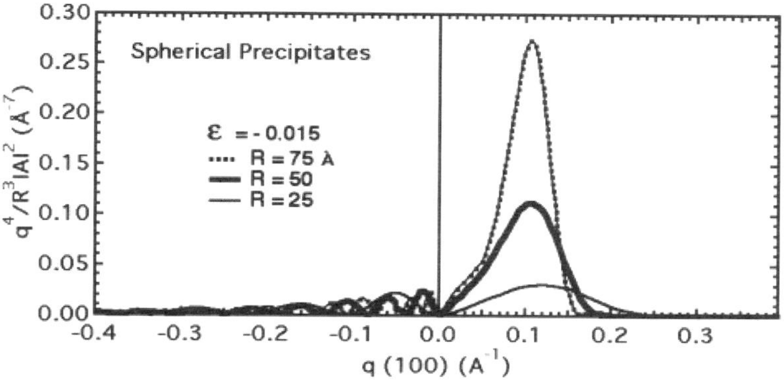

Figure 9.3. *Diffuse scattering cross section calculations scaled by q^4/R^3 for spherical precipitates with three different radii and a strain of $\varepsilon = -0.015$; the calculations were made for q along the [100] direction at the (400) reciprocal lattice point for isotropic elasticity [9].*

point. We comment that in the infinitesimal defect approximation for asymptotic diffuse scattering [5,7,10], there is no direct scattering from the defect cluster itself, and only a smooth $1/q^{10/3}$ falloff is predicted. Although this is a major shortcoming of the infinitesimal defect approximation in terms of analyzing experimental measurements, the physical insight provided by the model is very useful.

We note that in both the precipitate and dislocation loop cases, diffuse scattering radial along H contains local Bragg scattering with particle size-broadened shapes in the sense that larger clusters have smaller widths in q. The correspondence between particle size and the peak width can be only approximate, of course, because the diffuse scattering distribution is not solely Bragg scattering. In the second-phase particle case, the position of the peak, the width of the peak, the strength of the peak, and the frequency of the oscillations on the opposite side of the Bragg position all provide important information on the size and the lattice parameter of the particle, and hence on the strength of the distortions. For dislocations, the position and the width of the q^4-scaled asymptotic scattering peak provide whatever information is available on the loop size and the vacancy/interstitial nature of the loops, since the scattering on the opposite side of the periodic lattice Bragg position is relatively weak and unstructured due to the lack of significant volumes with strains of the opposite sign for loops.

5 Asymptotic Diffuse Scattering Measurements

The thrust of this chapter is to present an overview of the underlying formalism and the measurement and analysis techniques associated with X-ray diffuse scattering investigations of clustered defects. Accordingly, the remainder of the presentation focuses on selected examples of asymptotic diffuse scattering measurements and analyses for the cases of coherent precipitates and dislocation loops. A review of asymptotic (or Stokes-Wilson) diffuse scattering applications is not attempted; rather, illustrative examples of measurements performed radially along H will be selected.

Figure 9.4 shows diffuse scattering measurements made near the (400) reflection using 1.39 Å X-rays at the Cu absorption edge (to minimize Cu scattering compared to Co) on Cu(1%)Co single

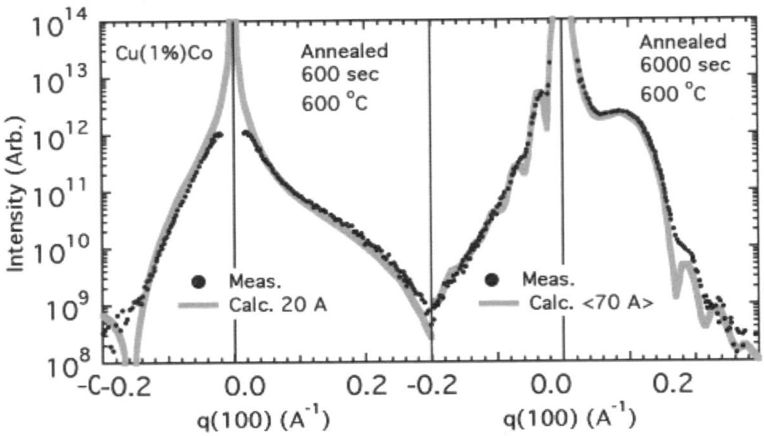

Figure 9.4. *Diffuse scattering measurements (dots) made for the wavelength of the Cu K-edge at the [400] reflection of Cu(1%)Co crystals after aging for 600 and 6000 seconds at 600°C to grow spherical coherent Co precipitates. The thick gray lines represent diffuse scattering calculations made using* $\varepsilon = -0.014$ *for 20 Å (left) and a combination of 65, 70, and 75 Å radius precipitates (weighted to contain equal volumes for each radius) in the right panel.*

crystals after thermal aging at 600°C for 600 seconds and 6000 seconds. Cu-Co is a prototype system known to form coherent spherical precipitates that increase in size by a ripening process that has been the subject of extensive investigation [35], including both X-ray diffuse scattering and small-angle neutron diffuse scattering [12,34] on a similar sample annealed for 61200 seconds at 570°C. Figure 9.6 shows diffuse scattering profiles for samples annealed for 600 and 6000 seconds compared with anisotropic elasticity calculations [12] based on face-centered cubic Co precipitates with $\varepsilon = -1.4\%$ for 20 Å radius precipitates and for an average $\langle R \rangle$ of 70 Å (i.e., weighted for equal volumes of 65, 70, and 75 Å precipitates) for the short and long anneals, respectively. Although the cross section calculations in Figure 9.4 do not represent fits to the measured intensities, they show that by using accurate, numerically determined scattering cross sections, remarkable agreement with measured diffuse scattering profiles can be achieved.

Figure 9.5 shows a plot of these Cu(1%)Co measurements scaled by q^4 to emphasize experimentally the fact established by the cross section calculations in Figure 9.3 that the local Bragg scattering from the precipitate particles themselves dominates the scattering in the asymptotic diffuse scattering region. The plots in Figures 9.3 and 9.5 show that scaling diffuse scattering measurements from clustered defects by q^4 provides a natural weighting for performing least-squares fitting analyses to extract defect cluster size distributions from the scattering. With the q^4 scaling, the parabolic form of the Huang scattering (which is often distorted either by imperfect/mosaic host crystals or by the lack of sufficient detector resolution both perpendicular to and in the scattering plane) that does not contain direct size information is present but not heavily weighted, while the Laue scattering containing direct size and strain information is enhanced. Results of small-angle scattering, as well as asymptotic diffuse scattering, and also references to other types of investigations of size distributions for this system are discussed elsewhere [12,31,34,35] and will not be discussed further here except to emphasize that the

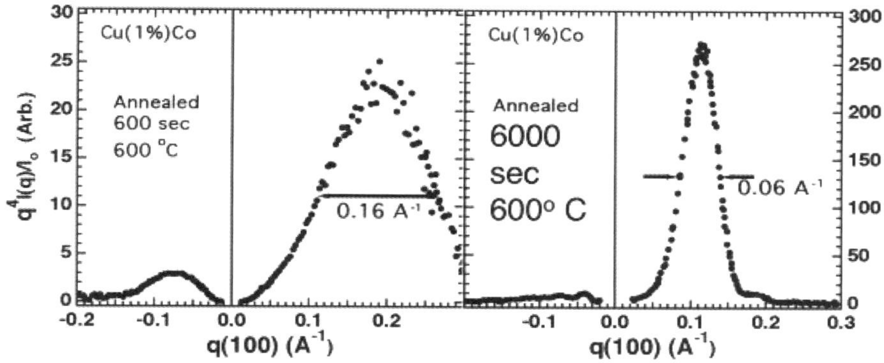

Figure 9.5. *Plots of the diffuse scattering measurements in Figure 9.4 scaled by q^4. The arrows indicate the full width at half maximum of the positive-q peaks, which are dominated by the direct (or Laue) scattering from the precipitates.*

size-dependent scaling of the intensities makes it important to obtain high-precision data and to use accurate cross sections when detailed information is desired for both large and small sizes.

The above measurement and analysis considerations hold in general for dislocation loops and other types of clustered defects as well as for coherent precipitates. Accordingly, Figure 9.6 shows the results of asymptotic X-ray diffuse scattering investigations of high-energy ion irradiation-induced dislocation loops in a Cu single crystal [14,32]. Diffuse scattering intensities measured in symmetric Bragg geometry along the [111] direction near the (222) Bragg reflection are shown scaled by q^4 out to q valu`es of 0.3 Å^{-1}. In contrast with the form of the scattering from coherent precipitates (Figure 9.5 above), we note that the q^4 scaled diffuse scattering intensity is of comparable magnitude for negative and positive q. This is, of course, due to the fact that displacement damage produces both vacancy and interstitial defects, and, as is well known, distributions of both vacancy and interstitial loops are formed in room-temperature irradiations. As discussed in connection with the loop cross section calculations in Figure 9.2, it is possible to determine qualitatively just from the positions of the peaks in the scattering that the interstitial loops have larger radii because they peak at a smaller q value. It is further possible to surmise that the total number of vacancies and interstitials condensed into loops in their respective size distributions is approximately equal, judging by the nearly equal intensities for positive and negative q, again as discussed in connection with Figure 9.2 above. While we only referenced detailed size distribution analyses for Co precipitates in Cu(1%)Co [12,34], it is of interest with respect to the nonuniform depth distribution for ion irradiated materials to discuss methods for determining absolute cluster concentrations as well as their sizes.

As is well known [14], diffuse scattering measurements made in the symmetric Bragg geometry in samples with uniform distributions of defects are related quantitatively to the scattering cross sections for a distribution of defect cluster sizes through

$$I(\mathbf{K}) = \frac{I_o}{2\mu_o} \sum_i c_i(R_i) \frac{d\sigma_i(\mathbf{K}, R_i)}{d\Omega} (\Delta\Omega) \;, \tag{9.15}$$

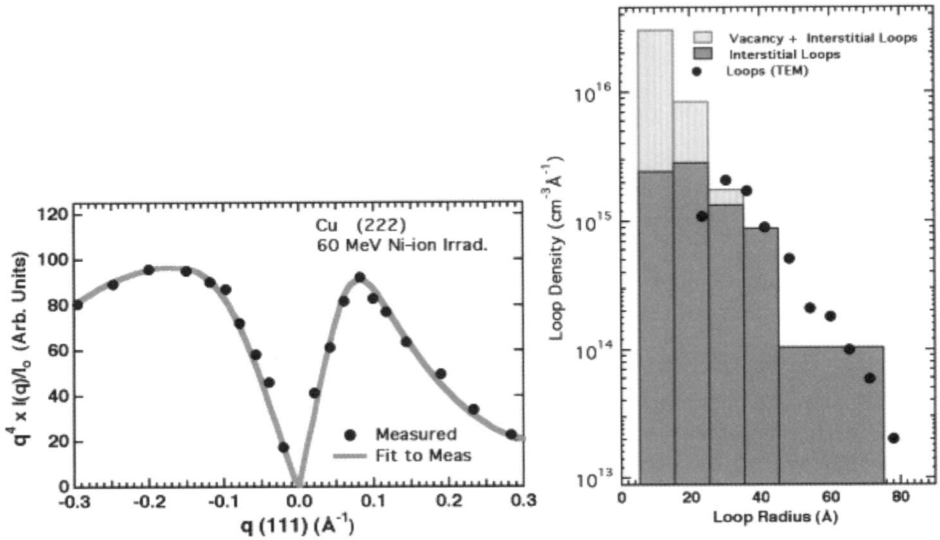

Figure 9.6. *Diffuse scattering measurements (scaled by q^4) made at the (222) reciprocal lattice point along the [111] direction for Cu irradiated with 1.3×10^{13} Ni-ion per cm^2 (left). The solid line represents a fit to the measurements with the concentrations of 10, 20, 30, 40, and 60 Å radius loops as parameters. The histograms in the right panel show the concentrations of vacancy plus interstitial loops (light shade) and the concentrations of interstitials only (dark shade). In addition, TEM measurements made on these samples show that nearly all of the vacancy loops were missed by the TEM measurements.*

where I_o is the incident beam power, μ_o is the linear absorption coefficient, $c_i(R_i)$ is the volume concentration of defect clusters of radius R_i, and $\Delta\Omega$ is the solid angle subtended by the detector. For the case of ion irradiations, the defect distribution is uniform in the plane of the surface but is highly nonuniform in depth, with penetrations generally on the order of a few microns. Ion irradiation damage is strongly energy dependent and is dominated initially by electronic excitations that produce relatively small amounts of atomic displacements, but as the ion energy decreases, the displacement damage increases dramatically near the end of the range of the ions in the sample. Without the benefit of spatial depth resolution as discussed in the next section, it is possible to use damage energy calculations to model the profile of the displacement damage, from which the diffuse scattering intensity can be related to damage at particular depths under the assumption that the nature and type of damage is not a function of depth. Equation 9.15 then yields [14]

$$I(\mathbf{K}) = \frac{I_o}{\sin(\theta)} \sum_i c_i(t_o, R_i) \frac{d\sigma_i(\mathbf{K}, R_i)}{d\Omega} (\Delta\Omega) \int_0^\infty \frac{c_i(t)}{c_i(t_o)} e^{-2\mu_o t / sin(\theta)} dt \qquad (9.16)$$

where t is the depth in the crystal, $c_i(t, R_i)$ is the depth dependence of the concentration of loops of radius R_i, and $c_i(t_o, R_i)$ is the concentration at the reference depth t_o of interest. Although this assumption has been shown to be reasonable for ion damage in metals (as discussed in the next section on spatially resolved diffuse scattering using microbeams), this is not the case for self-ion irradiation of perfect crystal semiconductors.

Making use of Equation 9.16 for Cu irradiated to a dose of 1.2×10^{13} Ni/cm^2 with 60 Mev Ni ions and fitting the measurements with scattering cross sections for prismatic vacancy and interstitial loop radii of 10, 20, 30, 40, and 60 Å, the dislocation loop size distributions (for $t_o = 5.6$ μm, the depth of the damage peak near the end of range of the Ni ions) shown at the right-hand side of Figure 9.6 were obtained. Since the units on the concentrations are expressed as loops/cm^2/Å, the area under the distributions provides the total number of loops. As discussed in more detail elsewhere [14], the average size of the vacancy loops is 12 Å, the average size of the interstitials is 22 Å, and the total number of defects in the form of loops is 6×10^{19} cm^{-3}, in which the number of vacancies and interstitials were found to be equal within ~6% (interstitials/vacancies = 47/53). Considering the logarithmic scale for the loop concentrations in Figure 9.6, we note that there are more interstitial loops than vacancy loops for the 30, 40, and 60 Å size bins, but there are approximately twice as many 20 Å vacancy loops as interstitial loops and ~15 times more 10 Å vacancy loops than interstitial loops. This trend is observed for irradiation-induced defects in metals in general [14] and is apparently a result of greater mobility for interstitial loops providing a mechanism for coalescence.

Finally, it is of interest to compare the TEM results in Figure 9.6 with the X-ray diffuse scattering results. Although the TEM results reported in the study by Narayan and Larson [22] are in good agreement with the X-ray results for loops 30 Å or more in radius, the TEM study reported no loops below a 20 Å radius. Although it is known that electron microscopy techniques are capable of imaging dislocation loops down to ~5 Å radius, it seems clear that the TEM size distribution is not complete, even at 20 Å in this case. It is possible that this is due to nonideal sample imaging conditions in the preparation of cross section samples, vacancies not fully collapsed into loops (i.e., stacking-fault tetrahedral), or a preferential loss of small loops to the surface of the thin foil sample. As discussed previously, in cases where TEM measurements separated the vacancy and interstitial loop components, it is not unusual to find a sizeable imbalance in the total vacancy and interstitial in dislocation loops [15]. The X-ray measurements in Figure 9.6 do not prove that the vacancy component is in dislocation loop form, but they do demonstrate that the vacancy component is present in the form of collapsed clusters with net distortion fields similar to vacancy loops. As mentioned above, separating the size distributions of vacancy and interstitial loops is tedious using TEM, and this is one area in particular in which asymptotic diffuse scattering represents a very attractive tool. This will be an important attribute in the spatially resolved diffuse scattering measurements of Si-ion-implanted Si, to be discussed below.

6 Spatially Resolved Diffuse Scattering Measurements on High-Energy, Self-Ion-Implanted Silicon

In the diffuse scattering techniques and examples discussed above, the incident X-ray beams were of the order of millimeters in size, so spatial resolution was not a consideration. In this closing section, we discuss spatially resolved diffuse scattering measurements of vacancy- and interstitial-type defect clusters as a function of depth dependence in high-dose self-ion-implanted Si single crystals. This system is of interest technologically in connection with the so-called vacancy implanter effect observed in high-energy ion implantation processing of electronic devices [36]. The underlying science is integrally connected with a fundamental understanding of energetic particle displacement of atoms in crystalline materials and the subsequent evolution and interactions of the displaced atoms.

Figure 9.7 shows a schematic of the scattering geometry for depth-resolved diffuse scattering measurements using X-rays focused to 0.7 μm by a Fresnel zone plate [37] on the XOR/UNI ID33 beamline at the Advanced Photon Source. This cross section geometry makes possible depth-dependent measurements of defect clusters in Si irradiated with 9×10^{16} 10 MeV Si ions at a temperature of 300°C. By cleaving the implanted Si normal to the implanted surface, diffuse scattering measurements were made as a function of depth at the (220) reflection with a monochromatic X-ray microbeam [27] by translating the crystal along the [001] surface normal direction. In this cross section geometry, used recently for depth-resolved measurements of strain in MoN coatings on Si [38], not only is it possible to make submicron-resolution depth-resolved measurements on the defect distributions, but also the scattering intensity comes from sample volumes that are ~10–100 μm along the beam, which enhances the defect scattering intensity significantly (~10–100 times) depending on the photoelectric absorption length. In plan-view geometry, where the incident beam is directed into the surface, submicron-resolution measurements would collect diffuse scattering from only submicron lengths along the beam.

Figure 9.8 shows a crystal translation scan monitoring the total diffuse scattering at $q = 0.04$ Å^{-1} in the [220] direction as a function of depth for the implanted Si crystal. A weak plateau of diffuse scattering is present in the first two to three microns below the surface. The plateau is followed by an order of magnitude higher peak in the diffuse scattering centered at a depth of 4.25 μm, which corresponds to the end of range of the implanted silicon ions. TEM measurements [39] readily identify a dense band (0.3–0.4 μm thickness) of interstitial clusters at the implanted ion end of range, but they do not see the vacancy clusters that, from indirect measurements [31], are believed to be present near the surface. It is understood that the implanted ions constitute an excess supply of interstitial atoms that do not have corresponding vacancies with which to annihilate. In addition, there is strong evidence that the forward momentum of the interstitials (as they are created by the impinging implanted ions), combined with an increasing damage density with depth, leads to a local imbalance in the number of vacancies and interstitials in favor of vacancies near the surface. Because vacancy-rich regions give rise to strongly enhanced thermal diffusion, these issues are critical for device manufacturing considerations; unfortunately, they have been difficult to understand quantitatively because of the difficulty of identifying directly the presence of vacancy clusters.

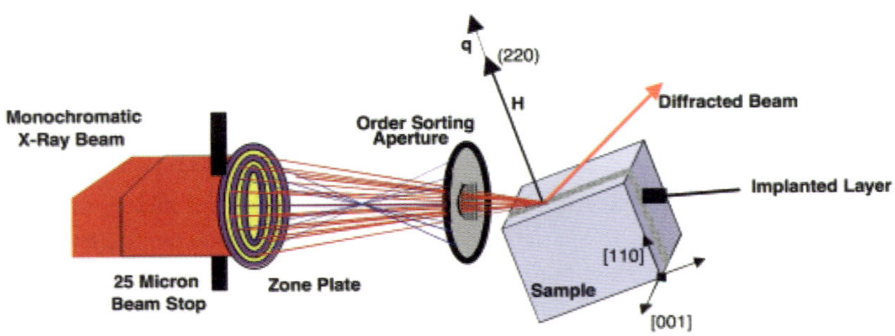

Figure 9.7. *Microbeam diffuse scattering geometry showing the Fresnel zone plate focusing element, a wire central beamstop, an aperture to limit third-order and higher order focused X-rays, and the implanted sample orientation and scattering geometry.*

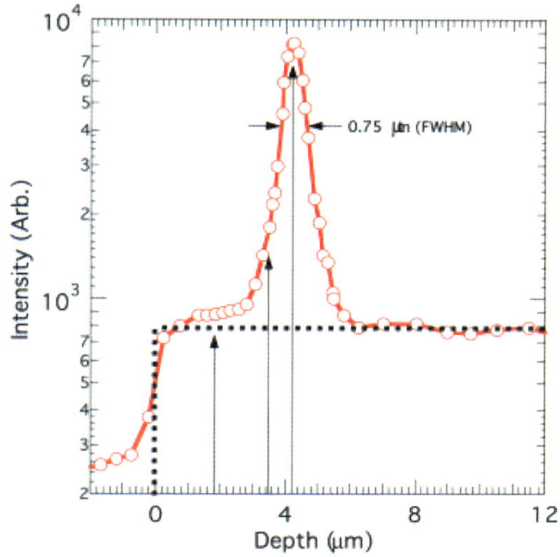

Figure 9.8. *Depth translation scan of the diffuse scattering at q = 0.04 Å^{-1} for 10 MeV self-ion-implanted silicon. The dashed lines indicate the sample edge at the defect-free silicon scattering level.*

The three vertical arrows in Figure 9.8 indicate depth positions at which the diffuse scattering measurements plotted in Figure 9.9 were performed. The scattering intensity located at d = 4.25 μm strongly dominates the scattering for the range of q values measured. It can be inferred qualitatively from the diffuse scattering curves in Figure 9.9 that the defect clusters located at 4.25 μm (the end of range of the implanted Si) are predominantly of the interstitial type because the much stronger intensity on the positive q side corresponds to interstitial loops. On the other hand, the relatively weaker diffuse scattering intensities at d = 3.45 and 1.80 μm have more nearly equal intensities for positive and negative q, indicating the presence of an equal or larger fraction of vacancy-type loops for these depths. Analyzing the data quantitatively, the solid lines in Figure 9.9 represent least-squares fits to the measured data for vacancy and interstitial loops with radii 3.8, 11.5, 19.2, 26.9, 34.6, 42.2, and 49.9 Å; these radii correspond to integer numbers of silicon unit cells for {111} loops. To limit the number of free parameters, the concentrations in Equation 9.16 for R = 26.9 and 34.6 Å were constrained to have the same number of point defects and similarly for R = 42.2 and 49.9 Å.

The size distributions corresponding to the solid line fits in Figure 9.9 are plotted in Figure 9.10. The results are presented in the form of histograms of the relative concentrations of vacancy and interstitial loops centered at the radii of the cross section calculations. Also included in Figure 9.10 are values for the relative number of point defects, N'_{PD}, contained in the loop distributions. The quantitative fits to the diffuse scattering are consistent with the (above) qualitative interpretation of the data in the sense that the number of interstitial loops in the distribution (and the number of point defects in the interstitial loops) is overwhelmingly larger than the number for vacancies at d = 4.25 μm. This is consistent with the expected oversupply of interstitials in this region due to the end of range of the implanted Si ions. At d = 1.80 μm, on the other hand, the number of vacancy loops and the total number of vacancies stored in loops is larger than the corresponding number for interstitials.

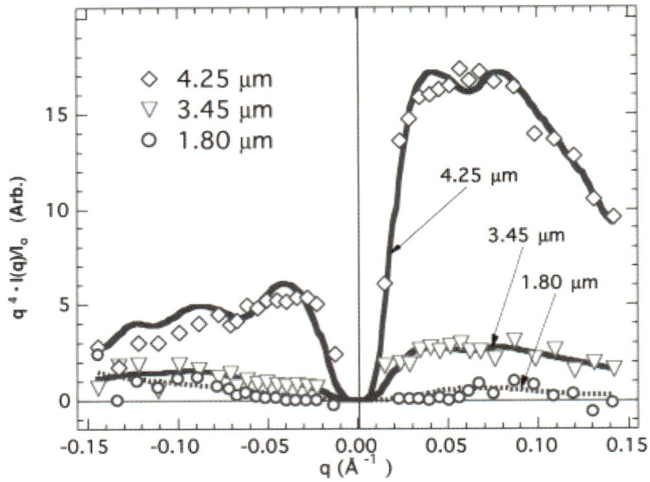

Figure 9.9. *Depth-resolved diffuse scattering weighted by q^4. Measurements as a function of q were performed at three depths identified by arrows in Figure 9.8. The solid lines represent least-squares fitting of the measured data to determine size distributions of vacancy and interstitial loops.*

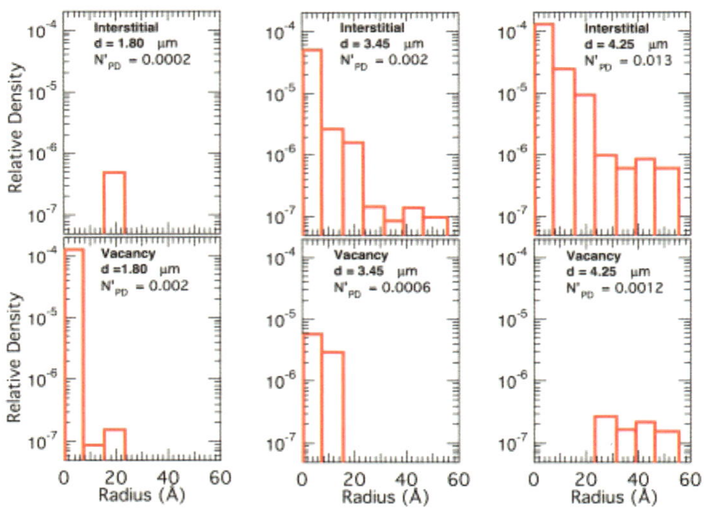

Figure 9.10. *Vacancy and interstitial loop size distributions corresponding to the three measurements in Figure 9.9. Relative concentrations of point defects N'_{PD} contained within the loop distributions are provided.*

Of course, the results for the 1.80 μm position are less reliable than for the other two positions measured because of the low intensities, indicating the need for additional measurements. Nevertheless, within the measurement uncertainty in the 1.80 μm measurements, we note that the vacancy components for both d = 3.45 and 1.80 μm trend toward a higher ratio of vacancies to interstitials and that the number of interstitials is lowest near the surface, both of which are consistent with expectations for this region. Considering the extremely high defect density in the narrow (~0.3–0.4 μm) defect band observed by electron microscopy [39], it is likely that the interstitial component at d = 3.45 μm contains a nonnegligible contribution from the tail of the spatial resolution of the microbeam; however, the measurements for d = 1.80 μm should not be significantly impacted by the tails of the spatial resolution even though the scattering intensities are quite low.

Although these depth-resolved diffuse scattering results are in need of improved statistical precision in the measurements as well as higher spatial resolution, they suggest new directions for diffuse scattering investigations as microbeam capabilities develop, and the results demonstrate the potential for new information from depth-resolved studies of ion implantation-induced defects in particular. Just as with spatially integrated measurements, spatially resolved X-ray diffuse scattering measurements are complementary to electron microscopy measurements in their sensitivity to small defect clusters, and they provide a method for direct separation of the vacancy and interstitial components that has been difficult for TEM at small sizes.

7 Summary and Concluding Remarks

The underlying formalism that has been developed for diffuse scattering generated by lattice defects in crystalline materials provides both a physical understanding of the origin of diffuse scattering from clustered defects and a sound basis for quantitative analyses of experimental measurements in terms of the type, size, and concentration of defect clusters. The Huang scattering close to Bragg reflections provides direct information on the local defect symmetry through the sensitivity of the Huang scattering to the form of the long-range elastic distortions. On the other hand, the asymptotic diffuse scattering at somewhat larger distances q from reciprocal lattice points provides a local Bragg-scattering mapping of the highly deformed lattice in the immediate neighborhood of defect clusters.

When analyzed using accurate, numerically calculated scattering cross sections based on actual deformation fields for both large and small distances from defect clusters, it is possible to extract detailed information on the size distributions and concentrations of defect clusters. The sensitivity of local Bragg-like scattering to the sign of the elastic strain (i.e., $\pm$ strain $\rightarrow$ $\mp q$) provides intuitive insight into the presence of vacancy or interstitial loops, and similarly, into the presence of coherent precipitates with undersized/oversized lattices with respect to the host lattice. Moreover, the magnitude of the distorted lattice strains and particle-size broadening combine to provide remarkable sensitivity to cluster sizes. The R^2 and R^3 weighting of asymptotic diffuse scattering cross sections for dislocation loops and coherent precipitates, respectively, conveniently renders the overall scattering sensitive to the total number of point defects contained in the size distribution. However, this weighting of the asymptotic scattering and the fact that the Huang scattering is weighted by the fourth and sixth powers of R necessitates the use of both highly accurate cross sections and accurate diffuse scattering measurements over the entire q range in order to extract reliable size distributions. This is particularly important for broad size distributions with widely varying sizes and for bimodal size distributions with disparate sizes.

Accurate characterization of defect cluster distributions should, in principle, be made using diffuse scattering distributions from measurements around a number of different types and orders of reciprocal lattice positions. Although comprehensive measurements of full three-dimensional volumes in q space are now feasible with modern synchrotron X-ray sources and high-resolution X-ray area detectors, such measurements and detailed analyses of the resulting measurements for clustered defects do not seem to be made in present practice. In lieu of comprehensive measurements to constrain fits and analyses, combinations of TEM measurements and X-ray diffuse scattering measurements can be exploited to reduce ambiguities and increase the accuracy of diffuse scattering results. In particular, when more than one type of defect cluster are either present or potentially present, TEM measurements can provide valuable information within their resolution limits. Conversely, for small cluster sizes and defect distributions containing both vacancy and interstitial loops, diffuse scattering measurements can often provide information on the sizes below the resolution limit for TEM investigations of clustered defects.

As discussed in the previous section, the availability of submicrometer-diameter X-ray beams at synchrotron sources now provides the possibility of spatially resolved diffuse scattering measurements. The spatial resolution of X-ray microbeams or nanobeams will remain complementary to electron microscopy in view of the sub-Ångstrom spatial resolution capability of new high-resolution electron microscopy systems; however, TEM sample sectioning techniques are inherently destructive of the sample environment. Moreover, there are entire classes of materials issues associated with the so-called mesoscale regime (approximately tenths of microns to hundreds of microns) for which spatial resolution of a few Ångstroms is not required, but nondestructive investigations with submicron spatial resolution over, say, tens of microns are needed; grain sizes of structural materials are often in the range of a few microns to tens of microns, for instance.

The example discussed above of 10 MeV self-ion-implanted Si is the type of case in which the sample-sectioning process for TEM may have led to the dissolution or surface escape of small vacancy clusters; nondestructive microbeam techniques preserve the bulk nature of the sample associated with the implantation. There are, of course, issues to deal with in terms of microbeam or nanobeam X-ray investigations that perhaps have not been adequately addressed at this point. For instance, the assumption of a statistically large number of defects/clusters within the diffracting volumes may not be fulfilled for precipitates of large size and sparse density, and the impact of edge effects associated with diffraction from beams with submicron to a few nanometers beams have not been carefully addressed up to now. In general, detailed experimental and theoretical investigations definitively testing the limits of even nonspatially resolved diffuse scattering techniques would be useful; most investigations appearing in the literature understandably focus on the perceived successes of diffuse scattering measurements, and convincing demonstrations of the limits of applicability are difficult to find in the literature.

Establishing definitive limits for diffuse scattering from statistical distributions does not necessarily define limitations of X-ray scattering investigations; rather, they tend to point to entirely new opportunities, such as the transition to the regime of imaging individual lattice defects rather than entire ensembles of defect clusters for larger defect aggregates that often occur in low densities, for instance. Moreover, forward-direction [40] and high-angle [41] coherent diffraction direct Fourier inversion X-ray techniques are being developed that offer the possibility of three-dimensional imaging with resolution from a few nanometers down to atomic resolution. To be sure, direct inversion imaging (i.e., so-called lensless imaging) is at present not close to atomic resolution, nor is it even imminent for X-rays because significant increases in beam brilliance are needed. However, the brilliance of new and planned synchrotrons continues to increase, and there are no fundamental limitations precluding such resolutions [42] using large numerical aperture X-ray geometries associated with high-angle diffraction. After all, crystallography techniques involving (large-angle) Bragg reflections have provided atomic resolution information on periodic structures for nearly a century.

Acknowledgments

The author acknowledges the vital contributions of collaborators J. Barhorst, J. Chung, P. Ehrhart, M. Galloway, T. Haynes, G. Ice, S. Iida, R. Kalayanaraman, J. Lewis, S. Ohr (deceased), W. Schmatz (deceased), S. Spooner, J. Tischler, H. Trinkaus, M. Yoon, F. Young Jr. (deceased), W. Yun, and P. Zschack in the diffuse scattering investigations discussed in this chapter, and A. Meldrum and V. Venezia are thanked for access to TEM measurements. Oak Ridge National Laboratory (ORNL) research was sponsored by the U.S. Department of Energy (DOE), Basic Energy Sciences Division of Materials Sciences and Engineering. The Advanced Photon Source and the National Synchrotron Light Source are supported by the DOE Office of Science.

References

1. M. A. Krivoglaz, *Theory of X-ray and Neutron Scattering by Real Crystals* (New York: Plenum Press, 1969).
2. P. H. Dederichs, *Phys. Rev. B* 1 (1970): 1306.
3. P. H. Dederichs, *Phys. Rev. B* 4 (1971): 1041.
4. H. Trinkaus, *Phys. Stat. Sol.* 54 (1972): 209.
5. P. H. Dederichs, *J. Phys. F: Met. Phys.* 3 (1973): 471.
6. H. Trinkaus, *Phys. Stat. Sol. (b)* 51 (1972): 307.
7. H. Trinkaus, *Z. Angew. Phys.* 31 (1971): 229.
8. P. Ehrhart, H. Trinkaus, and B. C. Larson, *Phys. Rev. B* 25 (1982): 834.
9. S. M. Ohr, *J. Appl. Phys.* 43 (1972): 1361.
10. S. Iida and B. C. Larson, *J. Mater. Res.* 3 (1988): 267.
11. B. C. Larson and W. Schmatz, *Phys. Stat. Sol. (b)* 99 (1980): 267.
12. B. C. Larson, S. Iida, J. Z. Tischler, J. D. Lewis, G. E. Ice, and A. Habenschuss, *Characterization of Defects in Materials*, ed. R. W. Siegel, R. Sinclair, and J. R. Weertman, vol. 82 (Pittsburgh: Materials Research Society, 1987), 73.
13. P. Ehrhart, and R. S. Averback, *Phil. Mag. A* 60 (1989): 283.
14. B. C. Larson and F. W. Young, *Phys. Stat. Sol. (a)* 104 (1987): 273.
15. B. C. Larson and J. F. Barhorst, *Defects in Semiconductors*, ed. J. Narayan and T. Y. Tan (Amsterdam: North Holland, 1981), 151.
16. P. Ehrhart, *J. Nuc. Mat.* 216 (1994): 170.
17. P. H. Dederichs, *Solid State Physics* 27 (1972): 135.
18. B. C. Larson, *J. Appl. Cryst.* 8 (1975): 150.
19. H. Peisl, *J. Appl. Cryst.* 8 (1975): 143.
20. K. Nordlund, P. Partyka, R. S. Averback, and I. K. Robinson, *J. Appl. Phys.* 88 (2000): 2278.
21. Y. Zhong, Y. Bailat, R. S. Averback, S. K. Ghose, and I. K. Robinson, *J. Appl. Phys.* 96 (2004): 1328.
22. J. Narayan, and B. C. Larson, *J. Appl. Phys.* 48 (1977): 4536.
23. M. A. Krivoglaz, *Phys. Met. Metall.* 9 (1960): 641.
24. M. A. Krivoglaz and K. P. Ryaboshapka, *Phys. Met. Metall.* 16 (1963): 641.
25. A. P. Shpak, V. B. Molodkin, S. I. Olikhovskii, Y. M. Kyslovskyy, O. V. Reshetnyk, T. P. Vladimirova, E. G. Len, A. I. Nizkova, V. M. Venger, and S. V. Dmitriev, *Phys. Stat. Sol. (a)* 204 (2007): 2651.
26. R. I. Barabash and M. A. Krivoglaz, *Phys. Met. Metall.* 45 (1978): 1.

27. M. Yoon, , B. C. Larson, . J. Z. Tischler, T. E. Haynes, J.-S. Chung, G. E. Ice, and P. Zschack, *Appl. Phys. Lett.* 75 (1999): 2791.
28. A. R. Stokes and A. J. C. Wilson, *Proc. Phys. Soc.* 56 (1944): 174.
29. M. Lax, *Rev. Mod. Phys.* 23 (1951): 287.
30. P. Ehrhart and W. Schilling, *Phys. Rev. B* 8 (1973): 2604.
31. B. C. Larson and W. Schmatz, *Phys. Rev. B* 10 (1974): 2307.
32. B. C. Larson and F. W. Young Jr., *Point Defects and Defect Interactions in Metals*, ed. J. Takamura, M. Doyama, and M. Kiritani (Tokyo: University of Tokyo Press, 1982), 679.
33. S. M. Ohr, *Phys. Stat. Sol. (b)* 64 (1974): 317.
34. S. Spooner, S. Iida, and B. C. Larson, *Characterization of Defects in Materials*, ed. R. W. Siegel, R. Sinclair, and J. R. Weertman, vol. 82 (Pittsburgh: Materials Research Society, 1987), 79.
35. W. Wagner, J. Piller, H.-P. Degischer, and H. Wollenberger, *Zeitschrift für Metallkunde* 76 (1985): 693.
36. R. Kalayanaraman, T. E. Haynes, O. W. Holland, H.-J. L. Gossmann, C. S. Rafferty, and G. H. Gilmer, *Appl. Phys. Lett.* 79 (2001): 1983.
37. B. Lai, W.-B. Yun, D. Legnini, Y. Xiao, J. Chrzas, P. J. Viccaro, V. White, S. Bajikar, D. Denton, F. Cerrina, E. Di Fabrizio, M. Gentili, L. Grella, and M. Baciocchi, *Appl. Phys. Lett.* 61 (1992): 1877.
38. G. Chen, D. Singh, O. Eryilmaz, J. Routbort, B. C. Larson, and W. Liu, *Appl. Phys. Lett.* 89 (2006): 172104.
39. V. C. Venezia and A. Meldrum, private communication (2000).
40. J. Miao, T. Ishikawa, E. H. Anderson, and K. O. Hodgson, *Phys. Rev. B* 67 (2003): 174104.
41. I. A. Vartanyants, I. K. Robinson, I. McNulty, C. David, P. Wochner, and T. Tschentscher, *J. Synchrotron Rad.* 14, 453. (2007).
42. B. C. Larson and B. Lengeler, *MRS Bulletin* 29 (2004): 152.

10

Probing Phonons and Phase Transitions in Solids with X-Ray Thermal Diffuse Scattering

Ruqing Xu[1,2]

Hawoong Hong[2]

T.-C. Chiang[1,2]

1. *Department of Physics, University of Illinois at Urbana-Champaign, 1110 West Green Street, Urbana, IL 61801, USA*
2. *Frederick Seitz Materials Research Laboratory, University of Illinois at Urbana-Champaign, 104 South Goodwin Avenue, Urbana, IL 61801, USA*

Abstract

X-ray scattering by thermally populated phonons in crystals gives rise to thermal diffuse scattering (TDS). Measurements of TDS intensities as functions of momentum transfer and sample temperature can provide valuable information about the bonding of the crystal lattice, including phonon dispersion relations and interatomic forces. This method has a long history, and after being largely neglected for decades, it has received renewed interest in recent years. With the advent of intense synchrotron X-ray sources, the method has become a powerful tool for materials research. In this chapter we review the basics of TDS together with recent experimental work on various systems. Examples are presented to illustrate the power of TDS as a measurement tool for phonon dispersion relations and as a probe of phase transitions that involve phonons. The method is well suited for detailed studies of critical behavior and the properties of soft modes, and the information gained from such studies is useful for a basic understanding of the competing interatomic interactions, instabilities, and fluctuation effects associated with material transformations.

1 Introduction

When scattered by a perfect, infinite crystal lattice, an incident monochromatic X-ray beam gives rise to discrete Bragg peaks in reciprocal space. If the scattering lattice is imperfect, however, diffuse scattering can occur. In real crystals, lattice imperfection may occur for a variety of reasons, including stress, impurities, vacancies, dislocations, stacking faults, grain boundaries, and so on, but the most universal cause is the thermal vibration of the constituent atoms. Thermal vibration is intrinsic to any system; its amplitude remains finite at absolute zero temperature due to zero-point vibration. The diffuse scattering caused by atomic thermal vibration is referred to as *thermal diffuse scattering* or *temperature diffuse scattering*, both abbreviated as TDS. Often regarded as an undesired background in X-ray crystallography work, TDS actually has a rich structure in reciprocal space and contains valuable information about the lattice dynamics, or phonons.

Historically, X-ray TDS was the first method used to map phonon dispersion relations of a crystal. In the following sections, we will begin with a brief account of the historical development of this method. This is followed by a brief derivation and a collection of key formulae for TDS intensity calculations, organized for easy reference. Next, a discussion of recent experimental work is presented to illustrate the methodology and selected research topics of prevailing interest. Ongoing work on high-order scattering and the development of direct-inversion algorithms will also be discussed.

2 History

Studies of the interaction of X-rays with a thermally vibrating lattice date back to nearly a century ago. Shortly after von Laue's milestone experiment of crystallographic diffraction of X-rays, Debye [1] theorized in 1913 that thermal motion of a crystal lattice should give rise to diffuse scattering accompanied by a decrease of the Bragg peak intensities. Debye's initial work, however, was incomplete and contained errors [2]. Following notable contributions from Schrödinger [3], Debye [4], and Brillouin [5] on this subject matter, Faxén [6] and Waller [7] presented in 1923 detailed derivations of the effects based on Debye's idea. The work was later extended by Waller to include scattering calculations for general types of lattices [8]. The theory by Faxén and Waller was semiclassic; namely, the atomic motion was treated within the framework of classical mechanics, while the vibration amplitudes were computed from quantum theories. Nonetheless, their results were later confirmed to be correct through rigorous quantum mechanical derivations by Waller [9] and Ott [10]. For a comprehensive review of the early theoretical development, readers are referred to the report by Born published in 1943 [2].

Experimentally, TDS was first discovered in the 1920s as unexpected "extra reflections" in X-ray Laue pictures. The phenomenon attracted great interest and attention, and numerous experiments followed [11]. A number of explanations were proposed [2,11]; among them, the Faxén-Waller theory was eventually confirmed to be correct by the comprehensive study of Laval in 1938 and 1939 [12] and by numerous experiments of others. Many of the details can be found in the review articles of Lonsdale published in 1942 [11] and 1943 [13]. According to Lonsdale [13], the terms "diffuse scattering" and "temperature diffuse scattering" were probably first coined by Zachariasen [14].

The idea of making use of X-ray TDS as a probe of lattice dynamics was proposed by Lonsdale in 1942 [11]. But it was not until 1948 that the method was first applied by Olmer [15] to a study of single crystal aluminum. In the decade following this pioneering work, numerous systems were examined by TDS, including, among others, α-Fe by Curien [16], β-brass by Cole and Warren [17], AgCl by Cole [18], Zn by Joynson [19], Cu by Jacobsen [20], Al by Walker [21], and KCl by Boccara [22]. Application of this method was limited, however, because of its low efficiency and low accuracy. The rate of data acquisition was rather slow due to the small cross section of TDS, and those early measurements were typically limited to high-symmetry directions. Quantitative analysis was difficult with the limited computational power at that time, and modeling typically involved force constants up to the second- or third-nearest neighbors only, resulting in limited accuracy. With the advent of neutron scattering as a direct and more convenient method for phonon measurements, the TDS method was soon largely abandoned. Nonetheless, a few studies were carried out in the ensuing years, including the work on NaCl by Buyers and Smith [23], V by Colella and Batterman [24], InAs by Orlova [25], VSi by Sirota and Orlova [26], and Ni by Ice [27].

With recent advances in X-ray sources and related instrumentation, TDS has seen a revival in interest. High-brightness X-ray beams are now available at third-generation synchrotron-radiation sources, and two-dimensional X-ray detectors allow parallel measurements with excellent quantum efficiency over a large solid angle. These enhanced capabilities make it possible to carry out highly precise and efficient TDS measurements. Also, modern computational capabilities have enabled efficient quantitative analysis of experimental data based on sophisticated modeling. In 1999, Wu et al. [28] reported a study of TDS patterns obtained from single crystals of Si using intense X-rays derived from a third-generation light source. The main features in the observed TDS patterns were well explained by the phonon dispersion relations of Si. Similar patterns were also observed by Hastings [29] at about the same time. Later, Holt et al. [30,31] successfully extracted the phonon dispersion relations of Si from measured TDS images using a model that included atomic force constants up to the sixth neighbors. Similar studies were later performed on Nb [32] and δ-Pu [33]. In another experiment, Holt et al. [34] made detailed studies of a soft phonon mode associated with a charge-density wave transition in $TiSe_2$, which has long been a difficult problem for neutron scattering because of interference from an intense central peak. The same idea was applied to studies of an antiferrodistortive phase transition in $SrTiO_3$ by Holt et al. [35] and a martensitic phase transformation in δ-Pu by Xu et al. [36]. With continued improvements in instrumentation and analysis algorithms, TDS is becoming an increasingly useful tool, and it has great potential for applications in materials characterization at the microscale and nanoscale under extreme conditions.

3 Theory

We present here a brief derivation of key TDS formulae for crystals with a multiatom basis. For further details, one is referred to Xu et al. [37] and Warren et al. [38]. Consider a crystal lattice with N unit cells and n atoms per unit cell. Some basic notations are defined as follows:

$\mathbf{R}_m$ = position vector of the mth unit cell

τ_s = basis vector of the sth atom within a unit cell

$\mathbf{R}_{m,m'} = \mathbf{R}_{m'} - \mathbf{R}_m$

$\tau_{s,s'} = \tau_{s'} - \tau_s$

$\mathbf{u}_{m,s}$ = displacement of the sth atom in the mth unit cell

μ_s = mass of the sth atom

$\omega_{\mathbf{k},j}$ = angular frequency of the jth phonon mode with wave vector $\mathbf{k}$

$\mathbf{e}_{\mathbf{k},j,s}$ = eigenvector associated with the sth atom for phonon mode $(\mathbf{k}, j)$

The potential energy of the crystal about its ideal periodic configuration is expanded in terms of the atomic displacements according to

$$U = U_0 + \frac{1}{2} \sum_{m,m'=1}^{N} \sum_{s,s'=1}^{n} \sum_{\alpha,\beta} \left[-\Phi_{s,s'}^{\alpha,\beta}\left(\mathbf{R}_{m,m'}\right) \right] u_{m,s}^{\alpha} u_{m',s'}^{\beta} + \ldots, \tag{10.1}$$

where α and β denote Cartesian indices x, y, and z, and $\Phi_{s,s'}^{\alpha,\beta}\left(\mathbf{R}_{m,m'}\right)$ are the Born-von Kármán (BvK) force constants. The equation of motion of the atoms is given by

$$\mu_s \frac{d^2}{dt^2} \mathbf{u}_{m,s} = \sum_{m',s'} \Phi_{s,s'}\left(\mathbf{R}_{m,m'}\right) \cdot \mathbf{u}_{m',s'} . \tag{10.2}$$

One seeks wavelike solutions with wave vector $\mathbf{k}$. This leads to an eigenvalue equation:

$$\mathbf{D}\left(\mathbf{k}\right) \cdot \varepsilon_{\mathbf{k},j} = \omega_{\mathbf{k},j}^2 \varepsilon_{\mathbf{k},j} . \tag{10.3}$$

The eigenvector $\varepsilon_{\mathbf{k},j}$ has $3n$ components

$$\varepsilon_{\mathbf{k},j} \equiv \begin{pmatrix} \mathbf{e}_{\mathbf{k},j,s=1} \\ \mathbf{e}_{\mathbf{k},j,s=2} \\ \cdots \\ \mathbf{e}_{\mathbf{k},j,s=n} \end{pmatrix}, \tag{10.4}$$

which satisfy the orthonormality relation

$$\varepsilon_{\mathbf{k},j} \cdot \varepsilon_{\mathbf{k}',j'}^{*} = \sum_{s} \mathbf{e}_{\mathbf{k},j,s} \cdot \mathbf{e}_{\mathbf{k}',j',s}^{*} = \delta_{\mathbf{k},\mathbf{k}'} \delta_{j,j'} . \tag{10.5}$$

For crystals with a multiatom basis, two common definitions exist for the dynamical matrix $\mathbf{D}(\mathbf{k})$ [37,39], which could be a source of confusion [31,40,41]. Since the final expression of TDS intensities depends on the choice of $\mathbf{D}(\mathbf{k})$, care should be taken to make a clear distinction between the two choices. In one choice, the dynamic matrix is periodic (p) in reciprocal space, and in the other, it is not (np). They are given in terms of the force constants by

$$D_{s,s'}^{\alpha,\beta}\left(\mathbf{k}\right)^{(p)} \equiv -\sum_{m'} \frac{\Phi_{s,s'}^{\alpha,\beta}\left(\mathbf{R}_{m,m'}\right)}{\sqrt{\mu_s \mu_{s'}}} \exp\left(i\mathbf{k} \cdot \mathbf{R}_{m,m'}\right) \tag{10.6}$$

and

$$D_{s,s'}^{\alpha,\beta}\left(\mathbf{k}\right)^{(np)} \equiv -\sum_{m'} \frac{\Phi_{s,s'}^{\alpha,\beta}\left(\mathbf{R}_{m,m'}\right)}{\sqrt{\mu_s \mu_{s'}}} \exp\left[i\mathbf{k} \cdot \left(\mathbf{R}_{m,m'} + \tau_{s,s'}\right)\right]. \tag{10.7}$$

The corresponding lattice waves are given by

$$\mathbf{u}_{m,s}^{(p)}(t) = \mathrm{Re}\left\{\sum_{\mathbf{k},j} \frac{1}{\sqrt{\mu_s}} a_{\mathbf{k},j} \mathbf{e}_{\mathbf{k},j,s}^{(p)} \exp\left[i\left(\mathbf{k}\cdot\mathbf{R}_m - \omega_{\mathbf{k},j}t + \phi_{\mathbf{k},j}\right)\right]\right\} \tag{10.8}$$

and

$$\mathbf{u}_{m,s}^{(np)}(t) = \mathrm{Re}\left\{\sum_{\mathbf{k},j} \frac{1}{\sqrt{\mu_s}} a_{\mathbf{k},j} \mathbf{e}_{\mathbf{k},j,s}^{(np)} \exp\left[i\left(\mathbf{k}\cdot(\mathbf{R}_m + \tau_s) - \omega_{\mathbf{k},j}t + \phi_{\mathbf{k},j}\right)\right]\right\}, \tag{10.9}$$

where $a_{\mathbf{k},j}$, normalized by $\sqrt{\mu_s}$, is the amplitude of the lattice vibration mode ($\mathbf{k}$,j), and $\phi_{\mathbf{k},j}$ is an arbitrary phase factor. Although both definitions of the dynamical matrix yield the same set of eigenvalues, $\omega_{\mathbf{k},j}^2$, there is a phase difference between their eigenvectors,

$$\mathbf{e}_{\mathbf{k},j,s}^{(p)} = \mathbf{e}_{\mathbf{k},j,s}^{(np)} \exp\left(i\mathbf{k}\cdot\tau_s\right), \tag{10.10}$$

which must be included in the TDS intensity calculation. In this chapter we use the periodic definition only and will drop the superscript p hereafter. The choice is completely arbitrary. For TDS formulae derived with the nonperiodic definition of the dynamical matrix, one may refer to Xu et al. [37].

Intensity distribution of a linearly polarized X-ray beam scattered by a single electron is given by the Thomson scattering formula [38],

$$I_e = I_{incident} \frac{e^4}{m_e^2 c^4 d^2}\left[\sin^2\phi + \cos^2\phi\cos^2(2\theta)\right], \tag{10.11}$$

where 2θ is the scattering angle, ϕ is the angle between the scattering plane and the plane defined by the polarization and the direction of the incident beam, and d is the distance between the scattering center and the detector.

The intensity of the X-ray scattered by a crystal at scattering vector $\mathbf{q}$ is derived from a sum of the scattering amplitudes from all electrons in the system

$$I(\mathbf{q}) = I_e\left\langle\left|\sum_{m=1}^{N}\sum_{s=1}^{n} f_s \exp\left[i\mathbf{q}\cdot\left(\mathbf{R}_m + \tau_s + \mathbf{u}_{m,s}\right)\right]\right|^2\right\rangle, \tag{10.12}$$

which can be simplified to yield

$$I(\mathbf{q}) = I_e \sum_{\substack{m,m'\\ s,s'}} f_s f_{s'} e^{-i\mathbf{q}\cdot\left(\mathbf{R}_{m,m'} + \tau_{s,s'}\right)} e^{-\frac{1}{2}\left\langle\left[\mathbf{q}\cdot\left(\mathbf{u}_{m,s} - \mathbf{u}_{m',s'}\right)\right]^2\right\rangle}, \tag{10.13}$$

where f_s is the usual atomic scattering factor of the sth atom in a unit cell, and $\langle\ldots\rangle$ denotes averaging over time. In going from Equation 10.12 to 10.13, the following relationship was used:

$$\left\langle \exp\left(i\mathbf{q}\cdot\mathbf{u}\right)\right\rangle = \exp\left[-\frac{1}{2}\left\langle\left(\mathbf{q}\cdot\mathbf{u}\right)^2\right\rangle\right]. \tag{10.14}$$

This holds for a harmonic lattice. Proofs of this relationship can be found in the work of Ott [10], Born and Sarginson [2,42], and Mermin [43].

Using Equation 10.14 and assuming that each mode has an independent random phase, the time average in the last exponent in Equation 10.13 can be evaluated. The result is

$$\frac{1}{2}\left\langle\left[\mathbf{q}\cdot\left(\mathbf{u}_{m,s}-\mathbf{u}_{m',s'}\right)\right]^2\right\rangle = M_s + M_{s'} - \sum_{\mathbf{k},j}\frac{\left|a_{\mathbf{k},j}\right|^2}{2\sqrt{\mu_s\mu_{s'}}}\left(\mathbf{q}\cdot\mathbf{e}_{\mathbf{k},j,s}\right)^*\left(\mathbf{q}\cdot\mathbf{e}_{\mathbf{k},j,s'}\right)e^{i\mathbf{k}\cdot\mathbf{R}_{m,m'}}, \tag{10.15}$$

where M_s or sometimes its exponential, $\exp(-M_s)$, is known as the Debye-Waller factor:

$$M_s = \frac{1}{4\mu_s}\sum_{\mathbf{k},j}\left|a_{\mathbf{k},j}\right|^2\left|\mathbf{q}\cdot\mathbf{e}_{\mathbf{k},j,s}\right|^2. \tag{10.16}$$

For systems with cubic symmetry, M_s depends on the magnitude of $\mathbf{q}$ only:

$$M_s = \frac{q^2}{12\mu_s}\sum_{\mathbf{k},j}\left|a_{\mathbf{k},j}\right|^2\left|\mathbf{e}_{\mathbf{k},j,s}\right|^2. \tag{10.17}$$

Note that, except for monatomic crystals, $\mathbf{e}_{\mathbf{k},j,s}$ is not normalized in general for each individual atom s (see Equation 10.5).

The amplitude of each lattice mode is determined by the Bose-Einstein distribution of the corresponding phonons and is given by

$$\left|a_{\mathbf{k},j}\right|^2 = \frac{2\hbar}{N\omega_{\mathbf{k},j}}\left(\frac{1}{e^{\hbar\omega_{\mathbf{k},j}/k_BT}-1}+\frac{1}{2}\right) = \frac{\hbar}{N\omega_{\mathbf{k},j}}\coth\left(\frac{\hbar\omega_{\mathbf{k},j}}{2k_BT}\right). \tag{10.18}$$

Substituting Equations 10.18 and 10.15 into Equation 10.13, we arrive at the following results for the scattered X-ray intensity:

$$I(\mathbf{q}) = NI_e\sum_{m=1}^{N}\sum_{s,s'=1}^{n} f_s f_{s'} e^{-M_s-M_{s'}} e^{-i\mathbf{q}\cdot\left(\mathbf{R}_m+\tau_{s,s'}\right)}\exp\left(G_{m,\mathbf{q}}^{s,s'}\right) \tag{10.19}$$

and

$$G_{m,\mathbf{q}}^{s,s'} \equiv \frac{\hbar}{2N\sqrt{\mu_s\mu_{s'}}}\sum_{\mathbf{k},j}\frac{1}{\omega_{\mathbf{k},j}}\coth\left(\frac{\hbar\omega_{\mathbf{k},j}}{2k_BT}\right)\left(\mathbf{q}\cdot\mathbf{e}_{\mathbf{k},j,s}\right)^*\left(\mathbf{q}\cdot\mathbf{e}_{\mathbf{k},j,s'}\right)e^{i\mathbf{k}\cdot\mathbf{R}_m}. \tag{10.20}$$

Here we have replaced the double summation $\sum_{m,m'=1}^{N}$ by a single summation $N\sum_{m=1}^{N}$ in view of the translational symmetry. The quantity G defined by Equation 10.20, sometimes referred to as the phonon structure factor, involves a sum over all phonon modes within the first Brillouin zone.

With $\omega_{\mathbf{q},j}$ and $\mathbf{e}_{\mathbf{q},j}$ determined for all ($\mathbf{k}$, j) in a crystal by diagonalizing the dynamic matrix, the scattered X-ray intensity as a function of scattering vector $\mathbf{q}$ can be computed from Equations 10.19 and 10.20. Evaluation of these equations can be a numerically demanding task because it involves two nested three-dimensional sums—one in real space and the other in reciprocal space. If one hundred points are included for each one-dimensional summation, a straightforward calculation would include 10^{12} terms at just one point in reciprocal space. A standard simplifying approach is to expand the exponential $\exp\left(G_{m,\mathbf{q}}^{s,s'}\right)$ into a power series. This leads to an expansion of the scattered X-ray intensity into a series:

$$I = I_0 + I_1 + I_2 + \ldots. \tag{10.21}$$

For a large crystal, $N \to \infty$, and

$$\sum_{m=1}^{N} e^{i\mathbf{k}\cdot\mathbf{R}_m} = N\sum_{l}\delta\left(\mathbf{k}-\mathbf{K}_l\right), \tag{10.22}$$

where $\mathbf{K}_l$ is a reciprocal lattice vector. Note that this expression is not applicable to clusters or nanocrystals, an issue to be discussed further later. The zeroth-, first-, and second-order terms of the scattered intensity become

$$I_0\left(\mathbf{q}\right) = N^2 I_e \sum_{s,s'} f_s f_{s'} e^{-M_s - M_{s'}} e^{-i\mathbf{q}\cdot\boldsymbol{\tau}_{s,s'}} \sum_{l}\delta\left(\mathbf{q}-\mathbf{K}_l\right), \tag{10.23}$$

$$I_1\left(\mathbf{q}\right) = \frac{\hbar N I_e}{2}\sum_{j}\frac{1}{\omega_{\mathbf{q},j}}\coth\left(\frac{\hbar\omega_{\mathbf{q},j}}{2k_B T}\right)\left|\sum_{s}\frac{f_s}{\sqrt{\mu_s}} e^{-M_s}\left(\mathbf{q}\cdot\mathbf{e}_{\mathbf{q},j,s}\right)e^{-i\mathbf{q}\cdot\boldsymbol{\tau}_s}\right|^2, \tag{10.24}$$

and

$$I_2\left(\mathbf{q}\right) = \frac{\hbar^2 N v I_e}{8}\int\frac{d^3\mathbf{k}}{\left(2\pi\right)^3}\sum_{j,j'} W_{\mathbf{k},j} W_{\mathbf{q}-\mathbf{k},j'}\left|\sum_{s}\frac{f_s}{\mu_s} e^{-M_s} e^{-i\mathbf{q}\cdot\boldsymbol{\tau}_s}\left(\mathbf{q}\cdot\mathbf{e}_{\mathbf{k},j,s}\right)\left(\mathbf{q}\cdot\mathbf{e}_{\mathbf{q}-\mathbf{k},j',s}\right)\right|^2, \tag{10.25}$$

where v is the volume of a real-space unit cell and $W_{\mathbf{k},j}$ is given by

$$W_{\mathbf{k},j} \equiv \frac{1}{\omega_{\mathbf{k},j}}\coth\left(\frac{\hbar\omega_{\mathbf{k},j}}{2k_B T}\right). \tag{10.26}$$

The integral in Equation 10.25 is over the first Brillouin zone. The hyperbolic cotangent factors in the above equations depend on temperature and are related to the phonon population of each mode.

The zeroth-order term, Equation 10.23, corresponds to Bragg diffraction at $\mathbf{q} = \mathbf{K}_l$. The intensity of each Bragg peak is reduced by temperature-dependent Debye-Waller factors. The intensities that are "taken away" are redistributed in reciprocal space as thermal diffuse scattering. The higher order terms in the expansion I_1, I_2, and so on are called *first-order TDS*, *second-order TDS*, and so on and are generally nonzero for any scattering vector **q**. First-order TDS, given by Equation 10.24, is often the dominant contribution to the total TDS intensity under typical experimental conditions. It does not involve any three-dimensional summation and is therefore straightforward to evaluate. The reason for this simplicity is that each first-order scattering event at **q** involves just one phonon of any of the $3n$ branches at the same wave vector **q** because of momentum conservation. The phonon can be either emitted or absorbed, and the process is the same as the usual one-phonon Raman scattering. At absolute zero temperature, phonon absorption is quenched, but phonon emission remains allowed. As the lattice temperature rises, phonon population increases and I_1 increases, but eventually the Debye-Waller factor dominates and the first-order intensity diminishes at sufficiently high temperatures (assuming that the crystal does not melt). The intensity is transferred to higher order scattering processes. The higher order terms have similar temperature dependencies, but they peak at different temperatures.

Experimental TDS images are often recorded with a transmission Laue geometry using a two-dimensional detector. As an illustration of what one might expect from an experimental observation, Figures 10.1a and 10.1d show theoretical two-dimensional images in linear and logarithmic scales, respectively, of first-order TDS from a Cu single crystal at 300 K oriented with its (100) direction parallel to an incident X-ray beam of energy 42 keV. The detector covers an angular span of ±21° in both x and y directions. Each pixel in the images corresponds to a unique momentum transfer (or scattering vector **q**) on the Ewald sphere in reciprocal space. The intensity of each pixel is mostly determined by the thermal population factor $\coth(\hbar\omega/k_B T)$ together with the additional factor of $1/\omega$;

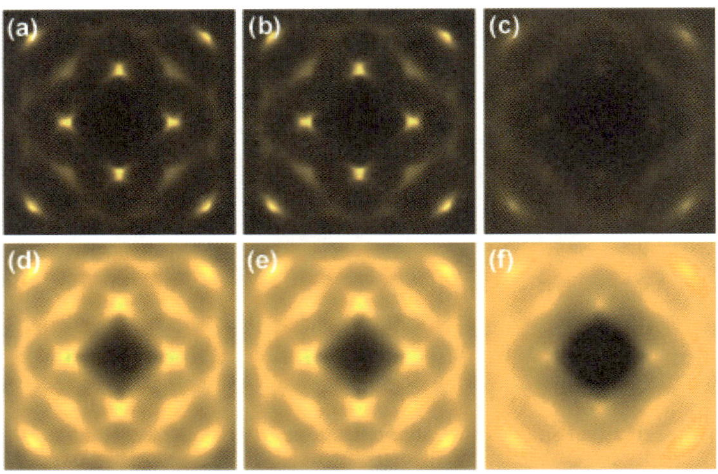

Figure 10.1. *Theoretical TDS images of Cu(100) at 300 K for an incident X-ray energy of 42 KeV. The detector covers an angular span of ±21° in both x and y directions. A linear intensity scale is used for (a) first-order calculation, (b) all-order calculation, and (c) high-order contribution amplified by a factor of 10. (d)–(f) The same images presented using a logarithmic intensity scale.*

both favor low-frequency phonons. Thus the bright areas generally correspond to regions in reciprocal space where phonon frequencies are relatively low, typically near reciprocal lattice points. The contrast in scattering intensity can be quite large, and linearly scaled images (such as Figure 10.1a) can be dominated by just a few bright spots. It is often preferable to present the images using a logarithmic intensity scale in order to bring out the rich details of TDS patterns (Figure 10.1d).

The second- and higher order terms, or *high-order TDS* collectively, correspond to multiphonon scattering processes. Since the multiple phonons can share the momentum transfer in any infinite number of ways, a calculation of the scattering intensity must involve all phonons. Specifically, the second-order TDS intensity, given by Equation 10.25, involves a summation over the first Brillouin zone. The expression is roughly a weighted self-convolution of first-order scattering with the two involved phonon modes constrained by momentum conservation. The convolution tends to smear out details. For this reason, the diffuse scattering intensity becomes more blurred as the scattering order increases. The summation over the Brillouin zone makes numerical evaluation of high-order contributions much more time consuming. Fortunately, first-order scattering often dominates.

Because of the computational complexity, high-order TDS is often neglected or treated approximately. In Olmer's method [15,38], the crystal was assumed to be isotropic and the lattice dynamics were approximated by a Debye model. In other words, linear dispersion relations with the same speed of sound for all phonon branches were assumed. Walker [21] improved Olmer's method by adopting dispersion curves of a linear chain and assuming different speeds of sound for transverse and longitudinal waves while still keeping the isotropic assumption. Such approximations provide reasonable estimates for the magnitude of high-order TDS intensities but cannot reproduce the details.

With modern computing capabilities, it has become possible to routinely perform TDS calculations to all orders without resorting to an expansion. Figures 10.1b and 10.1e show the results of such an "all-order" calculation derived from a sum of a first-order calculation based on Equation 10.24 and a high-order calculation based on

$$I_{\mathrm{HO}}(\mathbf{q}) = NI_e \sum_{m=1}^{N} \sum_{s,s'=1}^{n} f_s f_{s'} e^{-M_s - M_{s'}} e^{-i\mathbf{q}\cdot(\mathbf{R}_m + \tau_{s,s'})} \left[\exp\left(G_{m,\mathbf{q}}^{s,s'}\right) - G_{m,\mathbf{q}}^{s,s'} - 1 \right]. \quad (10.27)$$

The above equation is obtained from Equation 10.19 with the zeroth and first-order terms subtracted away. The real-space summation needed in this high-order calculation was performed for a cluster of Cu atoms contained in a box of size 40 × 40 × 40 in units of $a/2$, where a is the lattice parameter. The size was chosen such that the second-order intensity, evaluated with the exact expression in Equation 10.25, agreed with that computed for this finite cluster to within an accuracy of 1%. Since high-order patterns are fairly diffuse, good accuracy can be achieved with a modest cluster size. Using a larger cluster would increase the accuracy, but the computation time goes up as the cube of the size of the cluster. The first-order contribution can be evaluated directly without this cluster approximation, and it is therefore calculated separately. The zeroth-order term for the cluster would lead to a set of Bragg peaks with a finite width related to the cluster size. The tails of the Bragg peaks could leak into the calculated TDS patterns, causing confusion. For this reason, the Bragg term is also subtracted away in Equation 10.27.

As seen in Figure 10.1, the computed all-order and first-order images are essentially identical to the eye. This suggests that high-order scattering is much weaker under the assumed conditions. Figures 10.1c and 10.1f are images derived from the difference between the all-order and first-order calculations, amplified by a factor of ten. This image contains contributions from all high-order terms but is dominated by second-order scattering. Its appearance is roughly a blurred version of the first-order image, which is consistent with the discussion above that the second-order image is roughly a self-convolution of the first-order image.

How important is high-order scattering under typical experimental conditions? It would be very desirable if a first-order analysis suffices. This depends on the phonon structure factor G as defined by Equation 10.20, which is the quantity involved in the expansion. It scales as q^2 and increases linearly as a function of T in the high temperature limit (relative to the Debye temperature). Thus high-order effects are expected to become more important at larger q and higher T. As an example, Figure 10.2a plots the calculated first-order, high-order, and all-order (total) TDS intensities of Cu at 300 K along an arbitrarily chosen direction of (1, 0.2, 0.02) in reciprocal space. Figure 10.2b presents the computed percentage of high-order intensity relative to the total intensity at two different temperatures, 100 and 300 K. Clearly, high-order effects are more important at larger q and higher T. Assuming that a 3% accuracy is desired, the results show that one should limit q to within $\sim 4\pi/a$ at room temperature, or $\sim 6\pi/a$ at 100 K. Note that T enters Equation 10.20 through the ratio $\hbar\omega/k_BT$, meaning that the temperature limit for a given accuracy depends on the overall scale of the phonon energies of the crystal. A rule of thumb is that high-order effects are less important for crystals with a higher Debye temperature. To illustrate this point, we show in Figure 10.2c the computed percentage of high-order TDS intensity relative to the total intensity for Si at the same two temperatures and for the same arbitrarily chosen direction in reciprocal space. There is much less high-order scattering in Si. The Debye temperature of Si is 625 K, which is about twice the value for Cu, or 344 K.

The analysis above shows how to compute TDS intensities from phonon dispersion relations and polarization vectors. The idea of using TDS as a tool for determining phonon dispersion relations requires a reverse analysis. In other words, a procedure is needed to deduce the phonon dispersion relations from the measured TDS intensities. This is discussed below and illustrated by selected examples.

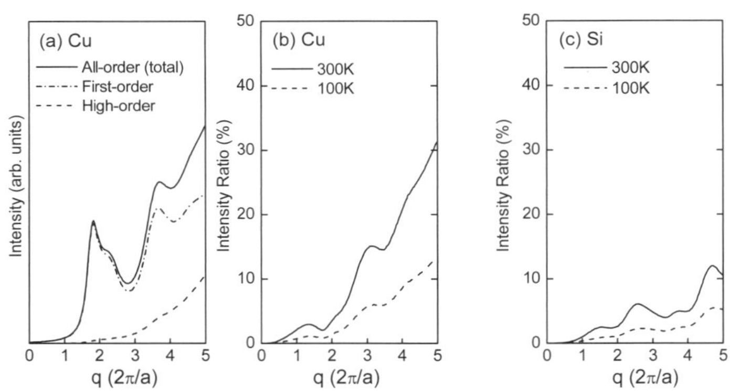

Figure 10.2. *(a) Theoretical all-order (total), first-order, and high-order TDS intensities of Cu at 300 K for* **q** *along (1.00, 0.20, 0.02). (b) The ratio of high-order TDS to total TDS intensities at 300 and 100 K for Cu along the same direction. (c) The same ratio for Si along the same direction at 300 and 100 K.*

4 Experiments

Early TDS experiments [15–27] were performed with X-ray tubes; the scattered X-rays were detected by Geiger counters [19–21] or later by scintillation detectors [23,24]. For the purpose of intensity normalization, an additional detector often was used to measure the intensity of scattering of the incident beam from an amorphous material placed between the X-ray source and the sample. Because of the small cross section of TDS, data collection was a slow process; typical times for data collection per reciprocal-space point were at least an hour and could be much longer for higher frequency phonons [21,23]. The measurements were typically limited to selected high-symmetry directions, and the resolution in reciprocal space was modest. The TDS method fell out of favor soon after neutron scattering became readily available in the 1960s. Revival of the TDS method was made possible by the development of intense synchrotron-radiation sources and two-dimensional detectors (such as charge-coupled devices and image plates). The tremendous increases in data acquisition rate and the capability of modern computers for data analysis make TDS a powerful and highly competitive tool for phonon studies.

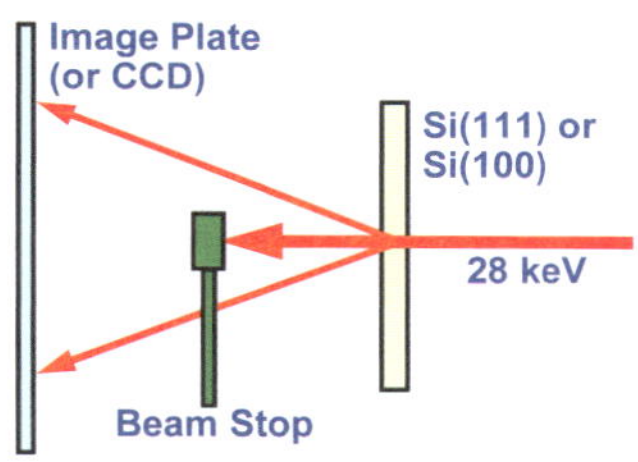

Figure 10.3. *A schematic setup for taking TDS images.*

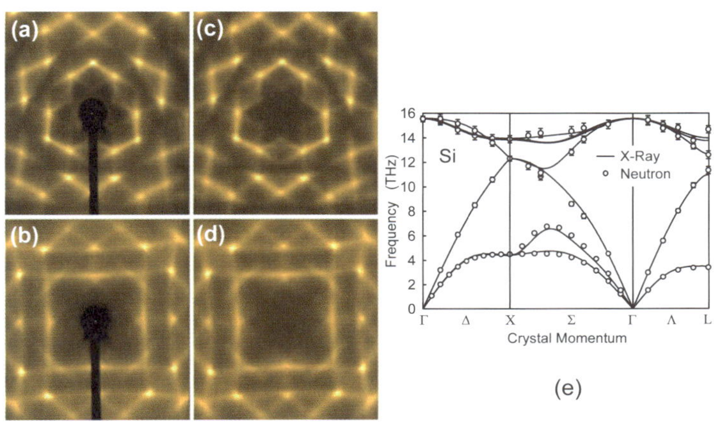

Figure 10.4. *(a) and (b) Experimental TDS images from Si(111) and Si(100), respectively. (c) and (d) Theoretical first-order TDS fits. (e) Si phonon dispersion curves; the circles are neutron scattering data, and the solid curves are results derived from TDS.*

As an example, Figure 10.3 shows a very simple transmission Laue setup employed by Holt et al. [30,31] for taking TDS images from Si. The experiment was performed at the Advanced Photon Source using an undulator beamline. The samples were at room temperature, and the signal was recorded with an image plate. For a better sampling of the reciprocal space, two different crystal orientations, (111) and (100), were used in the experiment, with each crystal oriented perpendicular to the incident X-ray beam. The recorded TDS images from the two crystals, presented in Figures 10.4a and 10.4b using a logarithmic intensity scale, show rich patterns with the expected threefold and fourfold symmetries, respectively. Each picture was taken with a 10-second exposure time; thus the data acquisition rate can be very high in this type of measurement.

Thermal diffuse scattering image analysis procedures in prior work, including the experiment on Si, were generally as follows. A Born-von Kármán model with force constants for a number of nearest neighbors was employed to calculate the phonon eigenvalues and eigenvectors. Theoretical TDS patterns were then computed within the first-order approximation in accordance with Equation 10.24. The computed results were compared with the experimental images; a least-squares procedure was employed to optimize the force constants until a best fit was achieved. The calculation of the theoretical TDS intensities included additional factors [30–32], as indicated by

$$I_{theory}(\mathbf{q}) = \cos(2\theta)T(\theta,\phi)A\left[I_1(\mathbf{q}) + BI_e(\theta,\phi) + C(\theta)\right] + D, \tag{10.28}$$

where I_1 is the first-order intensity from Equation 10.24. T accounts for intensity attenuation by the sample itself. A is an intensity normalization factor. The terms containing B and C together are included to approximately describe background intensity arising from high-order, defect, and Compton scattering. C is modeled by a slowly varying function of the angle, while I_e includes the angular dependence associated with scattering by a polarized beam (Equation 10.11). D is an overall constant background from detector noise. The cos(2θ) factor accounts for solid angle conversion associated with planar projection of the scattering pattern onto a flat detector. Other than the force constants, quantities that were unknown or could not be precisely measured were also set as fitting parameters. These included the three Euler angles for the crystal orientation, parameters for intensity normalization and the background functions, and so on.

In the case of the Si work, force constants up to the sixth neighbors were employed in the fit. Theoretical TDS images that gave the best fit to the experimental ones are shown in Figures 10.4c and 10.4d. Apart from the shadows of the beamstop, the experimental and calculated images are essentially identical to the eye. Phonon dispersion relations of Si derived from the final force constants from the fits are shown in Figure 10.4e. These are in excellent agreement with prior neutron scattering results, demonstrating that the TDS method works well on data that can be acquired rapidly.

Holt et al. [32] performed a similar experiment on Nb. TDS images were recorded in three different crystal orientations. The deduced phonon dispersion relations, presented in Figure 10.5, show a reasonable agreement with neutron scattering data. Some discrepancies are evident, which were partly attributed to the limitation of the lattice dynamics model used in the analysis. This problem was already known from the prior neutron scattering studies. The dispersion relations of Nb have a number of pronounced Kohn anomalies that cannot be well reproduced by a force-constant model unless a very large number of atomic shells are included.

A very important area of application for TDS is phase transitions. The efficiency of the method makes it feasible to take data at closely spaced intervals of temperature, pressure, or applied field. Holt

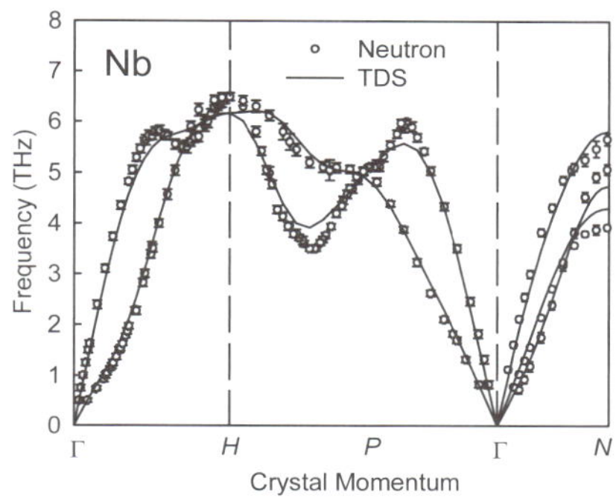

Figure 10.5. *Phonon dispersion relations of Nb. Circles are neutron scattering data; solid curves are derived from TDS.*

et al. performed such studies on $TiSe_2$ [34] and $SrTiO_3$ [35]. $TiSe_2$ has a charge-density wave transition at ~200 K, which results in a commensurate (2 × 2 × 2) superlattice at lower temperatures. The transition involves a softening of the L_1^- phonon mode at the zone boundary. Neutron scattering experiments were attempted in an effort to determine the detailed temperature dependence of this soft mode, but the results were largely inconclusive because of issues of a poor signal-to-noise ratio and an intense central peak. Figures 10.6a and 10.6b show linearly scaled TDS transmission images of this material at 291 and 193 K, respectively, with their corresponding logarithmically scaled images shown in Figures 10.6c and 10.6d. The area indicated by the circle in each picture is close to an *L* point, and clearly the TDS intensity is much higher near the transition temperature, suggesting the softening of a phonon mode. For a detailed analysis around and at the *L* point, Holt et al. took line scans in reciprocal space along the *A-L-A* direction with a point detector. An increase of the TDS intensities around the *L* point was observed as the sample temperature was lowered from room temperature toward the transition temperature. A Bragg peak emerged at the transition temperature and became more intense as the temperature was reduced further. Quantitative analysis was performed by fitting the observed

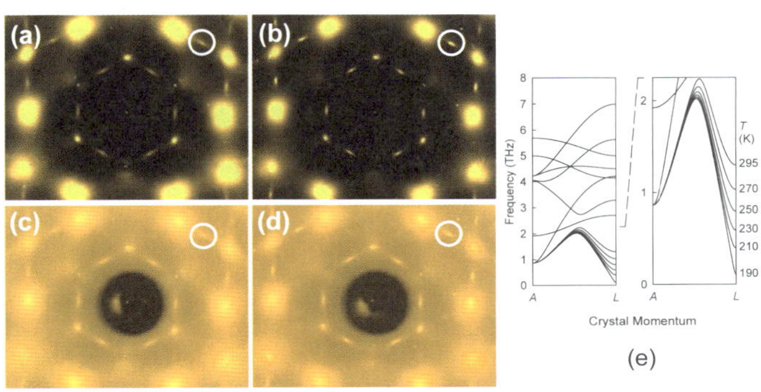

Figure 10.6. *(a–d) X-ray TDS images of $TiSe_2$. (a) and (b) Linearly scaled images at 291K and 193 K, respectively; (c) and (d) corresponding logarithmically scaled images. The circles highlight an area near an L point where phonon softening causes the TDS intensity to increase at lower temperatures. (e) Phonon dispersion curves of $TiSe_2$ as a function of temperature deduced from TDS.*

TDS line shapes using a force-constant model; one of the force constants, associated with the softening of the L_1^- mode, was allowed to vary as a function of temperature. The resulting dispersion curves along the *A-L* direction at various temperatures are presented in Figure 10.6e. The L_1^- mode softens with a power law in agreement with the prediction of a Landau mean-field theory, which is expected for this system.

Similar to $TiSe_2$, $SrTiO_3$ has a second-order structural phase transition. It is an antiferrodistortive transition involving the softening of a zone-boundary phonon mode, R_{25}, with a transition temperature of ~105 K. In the $SrTiO_3$ experiment [35], TDS line scans along the *M-R-M* direction were carried out at various temperatures across the phase transition. Unlike $TiSe_2$, fluctuation effects are important in this system, and the detailed behavior of the soft mode cannot be described by a Landau mean-field theory. From a dense data set near the transition temperature, the critical exponent γ associated with phonon softening was determined to be 1.38 ± 0.05, which is in excellent agreement with the predicted value of 1.386 ± 0.004 from a renormalization group calculation.

A TDS study of fcc δ-Pu (alloyed with 0.6 wt% of Ga for stabilization) was performed by Xu et al. [36] following an earlier inelastic X-ray scattering observation [44] at room temperature of a soft transverse phonon mode at the *L* point on the Brillouin zone boundary. The phonon dispersion curves determined from the inelastic X-ray scattering experiment are shown in Figure 10.7a. It was suggested, since this soft mode at *L* was related to a lattice-shearing mechanism, that this mode might be responsible for the martensitic transition of the material to the α′ phase at ~170 K. Further softening of this mode was suspected for decreasing sample temperatures toward the transition temperature. The TDS experiment was performed with a polycrystalline sample because single crystals were not available. TDS images were recorded from two crystallites of (111) and (110) orientations at various temperatures between room temperature and 200 K. Figures 10.7b and 10.7c show the recorded TDS images from the (111) grain at 307 and 200 K, respectively. The images show powder rings arising from oxide inclusions in the crystal. These oxide inclusions are hard to avoid because Pu is toxic and highly

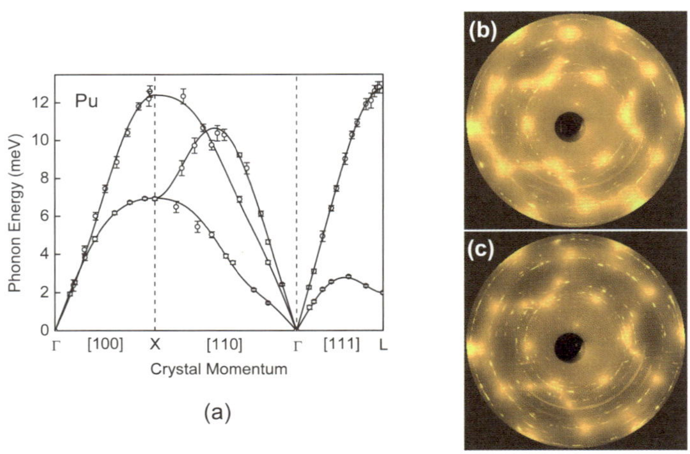

Figure 10.7. *(a) Phonon dispersion curves of δ-Pu (0.6 wt% Ga) measured by inelastic X-ray scattering. (b) and (c) TDS images for a (111) grain at 307 and 200 K, respectively.*

sensitive to air exposure. The rings were masked out in the data analysis using a software algorithm. A force-constant analysis of the data led to the conclusion that no further softening of this transverse (111) phonon branch occurred in this temperature range, indicating that kinetic phonon softening is not responsible for this first-order phase transition.

5 Discussion

The above summary of some recent experimental results serves to illustrate the utility of TDS. Compared with neutron scattering and inelastic X-ray scattering, presently the two other major techniques of phonon measurements, TDS has the distinct advantages of high data acquisition rates and experimental simplicity. Neutron scattering generally requires a large crystal, and the data acquisition time is typically much longer. Inelastic X-ray scattering involves much more complicated instrumentation, and measurements are generally very slow, even with state-of-the-art detector arrays. For example, the phonon dispersion curves presented in Figure 10.7a took a full thirty days of beam time for data collection at the European Sychrotron Radiation Facility (ESRF).

Modern materials of interest are often not available in the form of large single crystals. Furthermore, they may have naturally occurring domain structures that are important to their functions. In such cases the methods based on X-rays are much better choices than neutron scattering since synchrotron X-rays can be easily focused down to micron sizes or even smaller. Typical data acquisition times for TDS are 1 ~ 100 seconds per image exposure, or several minutes for a reciprocal-space line scan with a point detector. Such high data rates can be especially advantageous for cases requiring a large data set, such as detailed temperature, pressure, or applied field dependencies near a phase transition. This is likely a promising area for future applications of TDS. Once the basic phonon structure is known (from TDS or other methods), it would be fairly straightforward to carry out TDS intensity analysis to extract changes in the phonon spectrum associated with phase transition. The modeling might involve just one or a few variable force constants relevant to the problem, and the numerical analysis is generally a much simpler problem than directly inverting TDS patterns to yield the entire set of force constants. Most phase transitions involve substantial participation of the lattice; even those driven nominally by electronic properties are often accompanied by a significant lattice distortion. TDS is thus a useful tool for a broad spectrum of materials issues related to electronic and lattice instabilities.

There are a number of issues with TDS image analysis that require further work. One issue is the large amount of computation often needed for extracting phonon dispersion curves. A TDS image typically contains 10^5 ~ 10^6 pixels, and two or more images along different directions are usually needed for sufficient sampling in reciprocal space. While theoretical calculations of high-order or all-order TDS can be done within a reasonable time for a given scattering vector, the time required can become substantial for analyzing an entire set of TDS images. Furthermore, least-squares image fitting requires iterative optimization and a large number of recalculations of the images, and the total time for a complete analysis can be well beyond the typically available computational capacity. Fortunately, in many cases a first-order analysis is sufficient, and this can be done on a personal computer. The high-order contribution, if weak, can be approximately accounted for by a background function, resulting in an acceptable or negligible error. A rule of thumb was mentioned earlier, namely, one should keep the temperature significantly below the Debye temperature and limit the range of q to the first few zones. Compton scattering is another contribution to the scattering that needs to be accounted for. Again, this can be treated as a background function, or it can be computed and included in the theoretical model. More complicated and under less control is background scattering derived from materials nearby (including the environmental gas, if any, containment walls, windows, and so on), but these can all be modeled by a slowly varying background function.

A second issue is that numerical fitting of images can involve a very large number of parameters. The number of force constants, depending on the structure of the crystal and the accuracy required, can be several tens, even for a crystal with just one or two atoms per unit cell. The fitting can get trapped in local minima in the multidimensional configuration space. It is usually necessary to try different initial guess values of the parameters to minimize such risks.

A more fundamental issue of this technique is the limitation of the model of the lattice dynamics. In Nb, for example, it is known that a force-constant model with up to the eighth neighbors cannot reproduce all of the essential features in its phonon dispersion curves. Without such prior knowledge, TDS analysis based on simpler models with fewer neighbors could lead to results with large errors. In general, one should perform such analysis for increasing numbers of neighbors until a satisfactory and consistent fit is obtained. A large number of fitting parameters, however, tends to increase the chances of spurious solutions and trapping at local minima.

It would be highly desirable, therefore, if some general and simple approach could be developed to determine phonon dispersion relations directly from TDS data without relying on any specific force-constant models. Ideas have been proposed [31,45], but they generally require absolute measurements of the TDS intensities (and assume negligible high-order scattering). It is difficult in practice to make absolute intensity measurements, and it is often difficult to separate out background scattering.

A good solution is to measure the temperature dependence of the TDS intensity. Assuming that the background is independent of temperature, it can be eliminated by taking intensity differences. The first-order formula, Equation 10.24, depends on T, $\omega_{\mathbf{q},j}$, and $\mathbf{e}_{\mathbf{q},j}$. By selecting high-symmetry directions and specific scattering geometries, $\mathbf{q} \cdot \mathbf{e}_{\mathbf{q},j}$ can be predetermined based on symmetry considerations, and contributions from different phonon branches can be disentangled, making it

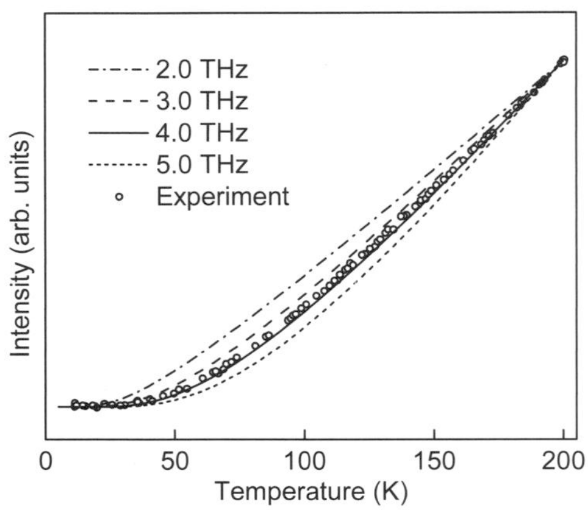

Figure 10.8. *Temperature dependence of Cu TDS intensity (circles) at* $\mathbf{q} = (2\pi/a)(1.15, 1.15, 1.15)$*, where the longitudinal phonon frequency is 3.6 THz. The curves are theoretical results assuming a phonon frequency of 2, 3, 4, and 5 THz as indicated. These curves are rescaled and shifted individually to emphasize the shape differences.*

possible to determine the frequencies of the individual phonon branches directly. As an illustration, Figure 10.8 plots the experimentally measured TDS intensity of Cu as a function of temperature at $\mathbf{q} = 2\pi/a(1.15,1.15,1.15)$, where the longitudinal phonon, with a frequency of 3.6 THz, is the only contributing mode. Also plotted are theoretical results assuming a phonon frequency of 2.0, 3.0, 4.0, and 5.0 THz. These theoretical curves are normalized and shifted to coincide with the experimental results at the two ends of the measurement temperature range. The differences among the theoretical curves are significant, and it is evident that the experimental results do correspond to 3.6 THz by interpolation. This analysis shows that, indeed, phonon frequencies can be directly determined from TDS in this manner. Effectively, one is comparing background-subtracted TDS intensities at different temperatures; absolute-intensity measurements thus become unnecessary. Note that the temperature range of the experiment was limited to 200 K (~60% of the Debye temperature), which was chosen to limit high-order scattering to ~2% (see Figure 10.2 and related discussion). Further details of this method as demonstrated by a study of copper can be found in Xu et al. [46].

6 Summary

This chapter reviews some key theoretical and experimental aspects of X-ray TDS as a probe of phonons and phase transitions. A set of essential formulae needed for performing intensity calculations is presented for easy reference. TDS images can be analyzed by fitting with a force-constant model to yield phonon dispersion relations, provided high-order scattering is negligible. This approach is illustrated by examples. Rules of thumb are developed as a guide to estimate the relative importance of high-order scattering. Likely the most promising area of applications of TDS is the study of phase transitions. TDS measurements are useful for addressing questions regarding interatomic forces, instabilities, and fluctuation effects. While the broad utility of TDS as a materials characterization tool is evident, the method has its limitations. Further research is needed to extend and expand its capabilities. For example, direct inversion of TDS data for phonon frequency determination is an area of great importance and interest. Other areas for further development may include advanced analysis methods and efficient numerical algorithms to handle high-order scattering, systems with large unit cells and complex structures, alloys, surfaces and interfaces, and nanoscale systems.

Acknowledgments

This work is supported by the U.S. Department of Energy (grant DE-FG02-07ER46383). We also acknowledge partial support from the U.S. National Science Foundation (grant DMR-05-03323) and the Petroleum Research Fund, administered by the American Chemical Society, for personnel, equipment, and beamline operations. The Advanced Photon Source, where some of the experimental work was performed, is supported by the U.S. Department of Energy (grant W-31-109-ENG-38)

References

1. P. Debye, *Verh. Dtsch. Phys. Ges.* 15, 678, 738 (1913): 857.
2. M. Born, *Rep. Prog. Phys.* 9 (1943): 294.
3. E. Schrödinger, *Phys. Z.* 15 (1914): 79; ibid., 15 (1914): 497.
4. P. Debye, *Ann. Phys.* 43 (vol. 348 in whole series) (1914): 49.

5. L. Brillouin, *Ann. Phys. Paris* 17 (1922): 88.
6. H. Faxén, *Z. Phys.* 17 (1923): 266.
7. I. Waller, *Z. Phys.* 17 (1923): 398.
8. I. Waller (PhD diss., Uppsala University, 1925).
9. I. Waller, *Z. Phys.* 51 (1928): 213.
10. H. Ott, *Ann. Phys.* 23 (1935): 169.
11. K. Longsdale, *Proc. Phys. Soc.* 54 (1942): 314.
12. J. Laval, *C. R. Acad. Sci.* 207 (1938): 169; ibid., 208 (1939): 1512; *Bull. Soc. Franç. Minér.* 64 (1941): 1.
13. K. Longsdale, *Rep. Prog. Phys.* 9 (1943): 256.
14. W. H. Zachariasen, *Phys. Rev.* 57 (1940): 597–602; ibid., 59 (1941): 766.
15. P. Olmer, *Acta Crystallogr.* 1, 57; 1948, *Bull. Soc. Franç. Minér.* 71 (1948): 145.
16. H. Curien, *Acta Crystallogr.* 5 (1952): 393.
17. H. Cole and B. E. Warren, *J. Appl. Phys.* 23 (1952): 335.
18. H. Cole, *J. Appl. Phys.* 24 (1953): 482.
19. R. E. Joynson, *Phys. Rev.* 94 (1954): 851.
20. E. H. Jacobsen, *Phys. Rev.* 97 (1955): 654.
21. C. B. Walker, *Phys. Rev.* 103 (1956): 547.
22. N. Boccara, *C. R. Acad. Sci.* 250 (1960): 1025.
23. W. J. L. Buyers and T. Smith, *Phys. Rev.* 150 (1966): 758.
24. R. Colella and B. W. Batterman, *Phys. Rev. B* 1 (1970): 3913.
25. N. S. Orlova, *Phys. Stat. Sol. (b)* 93 (1979) 503; ibid., 103 (1981) 115; ibid., 119 (1983): 541.
26. N. N. Sirota and N. S. Orlova, *Doklady Akademii Nauk Belarusi* 33 (1989): 110.
27. G. Ice, private communication with T.-C. Chiang.
28. Z. Wu, H. Hong, R. Aburano, P. Zschack, P. Jemian, J. Tischler, H. Chen, D.-A. Luh, and T.-C. Chiang, *Phys. Rev. B* 59 (1999): 3283.
29. J. Hastings, unpublished. For a related study of KMnF3, see A. Gibaud, D. Harlow, J. B. Hastings, J. P. Hill, and D. Chapman, *J. Appl. Cryst.* 30 (1997): 16.
30. M. Holt, Z. Wu, H. Hong, P. Zschack, P. Jemian, J. Tischler, H. Chen, and T.-C. Chiang, *Phys. Rev. Lett.* 83 (1999): 3317.
31. M. Y. Chou and M. Choi, *Phys. Rev. Lett.* 84 (2000): 3733; M. Holt and T.-C. Chiang, ibid., 84 (2000): 3734.
32. M. Holt, P. Czoschke, H. Hong, P. Zschack, H. K. Birnbaum, and T.-C. Chiang, *Phys. Rev. B* 66 (2002): 064303.
33. J. Wong, M. Wall, A. J. Schwartz, R. Xu, M. Holt, H. Hong, P. Zschack, and T.-C. Chiang, *Appl. Phys. Lett.* 84 (2004): 3747.
34. M. Holt, P. Zschack, H. Hong, M. Y. Chou, and T.-C. Chiang, *Phys. Rev. Lett.* 86 (2001): 3799.
35. M. Holt, M. Sutton, P. Zschack, H. Hong, and T.-C. Chiang, *Phys. Rev. Lett.* 98 (2007): 065501.
36. R. Xu, J. Wong, P. Zschack, H. Hong, and T.-C. Chiang, *Europhys. Lett.* 82 (2008): 26001.
37. R. Xu and T.-C. Chiang, *Z. Krist.* 220 (2005): 1009.
38. B. E. Warren, *X-Ray Diffraction* (Reading, MA: Addison-Wesley, 1966).
39. H. M. J. Smith, *Phil. Trans. Roy. Soc. A* 241 (1948): 105.
40. F. Herman, *J. Phys. Chem. Solids* 8 (1959): 405.
41. C. Patel, W. F. Sherman, and G. R. Wilkinson, *J. Phys. C* 17 (1984): 6063.
42. M. Born and K. Sarginson, *Proc. Roy. Soc. A* 179 (1941): 69.
43. N. D. Mermin, *J. Math. Phys.* 7 (1966): 1038.
44. J. Wong, M. Krisch, D. L. Farber, F. Occelli, A. J. Schwartz, T.-C. Chiang, M. Wall, C. Boro, and R. Xu, *Science* 301, 1078; G. Lander, *Science* 301 (2003): 1057.
45. N. W. Ashcroft and N. D. Mermin, *Solid State Physics* (Philadelphia, PA: Sauders College Publishing, 1976), 795.
46. R. Xu, H. Hong, P. Zschack, and T.-C. Chiang, *Phys. Rev. Lett.* 101 (2008): 085504.

Part III

Scattering from Complex and Disordered Materials: Experiments and Theory

11

Diffuse Scattering from Molecular Crystals

D. J. Goossens[1,2]

T. R. Welberry[1]

1. Research School of Chemistry, Australian National University, Canberra, 0200, Australia

2. Department of Physics, Australian National University, Canberra, 0200, Australia

Abstract

Diffuse scattering can reveal information about the two-body correlations in crystal structures. As such, it acts as a probe of the local ordering, whereas Bragg peaks are descriptive of the average long-range ordering. This average is made up of numerous local configurations whose population cannot be determined from analysis of the Bragg peaks. Diffuse scattering can be used to examine this population and so offers a way to look at crystal structures in great detail. This is particularly the case when making use of the three-dimensional distribution of diffuse scattering. However, diffuse scattering is very weak and broad and is often of similar intensity to the experimental background, so data collection is demanding. Further, disorder can take on many forms, and local configurations are not constrained by the average crystallographic symmetry. This means that many of the tools used in Bragg scattering analysis are not applicable to the study of diffuse scattering. These twin difficulties have until now limited the study of diffuse scattering to a specialist field within structural science.

Recent years have seen the development of an approach to the modeling of diffuse scattering from molecular crystals that pays particular attention to systems in which the molecules show internal flexibility. This chapter outlines the approach through discussion of a number of examples of diffuse scattering studies using X-rays and neutrons. Systems discussed include benzil ($C_{14}H_{10}O_2$), racemic, and optically pure ibuprofen ($C_{13}H_{18}O_2$); pentachloronitrobenzene ($C_6Cl_5NO_2$); and the adduct of hexamethylenetetramine and azelaic acid ($C_6H_{12}N_4{\cdot}C_9H_{16}O_4$).

1 Introduction

The structure of a material is one of the most fundamental pieces of information about it. A key question is: How are the atoms arranged? In crystalline materials, whether a single crystal or a powder, this question is most commonly answered by a diffraction experiment and analysis of the Bragg peaks. Analysis of intensity, width, and line shape of Bragg peaks reveals a wealth of information pertaining to single-site averages. For example, Bragg peak analysis can reveal the average occupancy of a site and what the average occupancy of a neighboring site might be, but it cannot indicate how the occupancy of one site affects the other; yet plainly it makes crystal chemical sense that occupancies (and displacements) on neighboring sites will not be independent. Substituting an atom inevitably changes the energetic state of the surrounding atoms. This may favor (for example) a clustering of substituents, a contraction or expansion of the lattice, or any combination of effects that makes the local environment different from the average environment.

When "crystal structure" is referred to, it most often means "crystal structure as determined by an analysis of the Bragg scattering." Here, this will be referred to as the "Bragg structure." Often the Bragg structure is quite adequate for advancing the understanding of the material's properties, and in many cases there is little structural detail present beyond that supplied by the Bragg analysis.

The weak diffracted intensity that lies underneath and between the Bragg peaks is referred to as "diffuse scattering" because it is generally (though not always) weak and broad—diffused through reciprocal space. Its distribution is sensitive to two-body correlations in the structure [1] and hence can, returning to the example above, give an indication of how the occupancy of one site influences that of its neighbor. Diffuse scattering offers a way of looking at the population of local environments out of which arises the overall long-range average obtained from Bragg scattering. Diffuse scattering offers a way of testing models of the population of local configurations. Since it is these local configurations that are reflective of the crystal chemistry, this is valuable information.

This chapter will discuss diffuse scattering from single crystals. Diffuse scattering has been of interest for many years [2,3]. However, its study has long been problematic. The reasons for this are twofold. First, because its distribution depends on the short-range order (SRO) in the structure, determination of the unit cell is no longer the end point in "solving" the structure. When considering SRO, unit cells can no longer be considered as identical. Thus solving for the atomic positions directly is difficult or impossible because there are too many unknown parameters.

Second, until recently it has been difficult to collect enough high-quality data for detailed analysis. Because the forms taken by disorder are not constrained by the space group symmetry of the unit cell (the average is so constrained, but the local deviations from the average are not), a large region of reciprocal space must be mapped if a model is to be adequately evaluated. Mapping a range of sections extending into all three reciprocal spatial dimensions is desirable, as is measuring out to relatively large scattering vectors. Point counters are not suitable for reciprocal space mapping, but area and line detectors have the problem that the data show a large dynamic range, and measuring diffuse scattering in the presence of large Bragg peaks or backgrounds can be a problem [4,5].

The difficulties inherent in data collection have largely been overcome through the use of synchrotron sources, and for large crystals it has been shown that useful data can be collected at both spallation and constant wavelength neutron sources [6–10]. Powerful software now exists for extracting and reconstructing reciprocal space sections [11–13].

However, analysis is still far from routine. Many studies of the diffuse scattering from inorganic solids such as metals and oxides have been undertaken [1,10], though many problems are still very challenging. Various authors have made use of the limited case of studying a handful of particular modulations in the structure through examination of limited regions or directions in reciprocal space [14,15]. Others have tackled the elucidation of magnetic SRO by studying neutron magnetic diffuse scattering, a valuable but flux-limited technique [16,17].

This chapter will focus on the analysis of diffuse scattering data from molecular crystals, including crystals made up of molecules with internal degrees of freedom—flexible molecules. These have been shown to be particularly intractable in structure prediction studies [18]. Such crystals can show many forms of disorder—substitutional, orientational, thermal, and combinations of these.

2 The Modeling Procedure

Diffuse scattering gives information on the average SRO in the material. Hence, while the expected intensity of Bragg peaks can be calculated from a single unit cell, the distribution of the diffuse scattering can be calculated from a model that is large enough to contain a statistically valid sample of the population of SRO configurations. One approach to this is to construct a "model crystal," typically of the order of $32 \times 32 \times 32$ unit cells or preferably larger (limited only by time and computing power), and then to allow a population of local configurations to be imprinted on the model via some algorithm that allows the atoms or molecules to interact and therefore become correlated. Once the model has reached equilibrium (or not, depending on the study of interest), its diffuse diffraction pattern can be calculated and compared with the observed.

The calculation of diffuse scattering is well established [19,20]. Creating the model crystal is a more open-ended question given the vast range of forms disorder can take. The studies presented here center on molecular crystals. In terms of algorithm implementation, this allows a number of assumptions. First, unlike an inorganic solid made up of, for example, corner- or edge-connected polyhedra, the molecular crystal can be considered as a set of discrete structural units—the molecules. As a corollary to this, interactions between molecules can be considered as weak compared to those within molecules, meaning that to a first approximation molecules or parts of molecules can be held rigid. Third, this approximation can be refined by allowing a limited number of motions within molecules. By limiting these to a few key degrees of freedom, the number of variables used to describe the system can be held in check. The form of molecular description chosen is the *z*-matrix [21], in which the molecule is described in terms of an internal coordinate system. If the position of each atom is forced to depend on that of the previously defined atoms, then segments of the molecule can be made to bend or rotate simply by allowing the coordinate(s) of one key atom to alter. The molecule as a whole is placed into the crystal frame of reference by "external" variables, the position of the origin atom (a 3-vector, $\mathbf{x}$), and the orientation in space of the first defined bond within the molecule, given by a rotation matrix or a quaternion (a normalized 4-vector, $\mathbf{q}$). Hence, if the molecule has n internal degrees of freedom (rotations around single C-C bonds, for example), which can be written as an n vector, $\mathbf{i}$, then the whole molecule can be described by its *z*-matrix plus $3 + 3 + n$ variables (not $3 + 4 + n$ due to normalization of $\mathbf{q}$). Figure 11.1 shows an example of a molecule, in this case benzil ($C_{14}H_{10}O_2$) [22]. Table 11.1 shows its *z*-matrix.

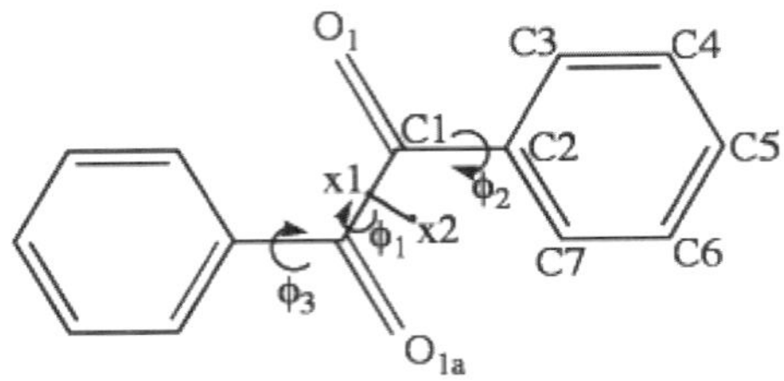

Figure 11.1. *Schematic diagram of a benzil molecule with the three internal degrees of freedom noted as ϕ_1 to ϕ_3. H atoms omitted.*

Molecules are allowed to interact via interatomic potentials of the form

$$E_{\mathbf{v}} = F_{\mathbf{v}}(d_{\mathbf{v}} - d_{\mathbf{v}0})^2, \tag{11.1}$$

which can be interpreted as the energy of a contact vector, **v**, connecting (atoms on) neighboring molecules via harmonic springs. The "spring constants," the $F_{\mathbf{v}}$, parameterize the interactions (only "effective" interactions, because Equation 11.1 is a crude approximation and not all atom-atom interactions are used, just a subset large enough to ensure propagation of molecular displacements). $d_{\mathbf{v}}$ is the instantaneous length of an interatomic contact vector and $d_{\mathbf{v}0}$ is its "ideal" length, generally determined from a Bragg analysis. The displacive component of the energy of a single molecule can be thought of as the sum over all such interactions connecting to it, while the displacive component of the energy of the whole model crystal is the sum over all the contact vectors in the crystal. By connecting molecules with such springs, displacements (and conformational alterations) can propagate through the model crystal, inducing SRO and giving rise to structured diffuse scattering. This propagation is done by equilibrating the model crystal at finite temperature using a Monte Carlo (MC) algorithm. In this algorithm, a molecule is selected at random and its energy is calculated. The variables that describe the molecule (**x**, **q**, and **i**) are given small random shifts, and the energy is redetermined. The shifted configuration is then kept or rejected based on a Metropolis-like [23] MC algorithm. Such an MC simulation may be referred to as a "displacive" simulation. However, such a mechanism cannot in general model occupational disorder, in which different molecules, or different orientations of the same molecule, are ordered over short range. In such cases an initial modeling of these occupancies is performed in which typically an occupancy structure is constructed via an additional MC program. This may be based on an Ising-type energy term

$$E = JS_iS_j \tag{11.2}$$

or some other appropriate potential. Plainly, if more than two states are possible, then the Ising Hamiltonian will not suffice. In Equation 11.2, for a two-state system, for example, $S_i = \pm 1$, depending on the state of molecule *i*. The sign of *J* will determine whether the associated two molecules are correlated positively or negatively, and the magnitude of *J* relative to the simulation temperature, *T*, controls the extent of the correlations subject to any frustration. The occupancy variables need not be two-state

TABLE 11.1

The *z*-matrix used for benzil

LABEL	*l*	DISTANCE FROM *l* (Å)	*m*	ANGLE TO *lm* (°)	*n*	ANGLE TO PLANE *lmn* (°)	
x1							
x2	1	0.124					
C1	1	0.761	2	85.420			
O1	3	1.216	1	115.786	2	35.750	
C2	3	1.450	4	123.495	1	177.242	
C3	5	1.406	3	120.117	4	−3.976	*
C4	6	1.364	5	120.755	3	178.777	
C5	7	1.380	6	119.625	5	−0.005	
C6	8	1.386	7	121.150	6	−1.139	
C7	9	1.396	8	119.261	7	2.041	
C1a	1	0.761	2	94.578	3	−179.993	
O1a	11	1.216	1	115.789	2	73.100	*
C2a	11	1.450	12	123.494	1	177.234	
C3a	13	1.406	11	120.121	12	−3.972	*
C4a	14	1.364	13	120.758	11	78.775	
C5a	15	1.380	14	119.625	13	−0.007	
C6a	16	1.386	15	121.150	14	−1.143	
C7a	17	1.396	16	119.259	15	2.047	
D1a	18	1.08	13	−120.0	14	0.0	
D2a	17	1.08	18	−120.0	13	0.0	
D3a	16	1.08	17	−120.0	18	0.0	
D4a	15	1.08	16	−120.0	17	0.0	
D5a	14	1.08	15	−120.0	16	0.0	
D1	6	1.08	7	−120.0	8	0.0	
D2	7	1.08	8	−120.0	9	0.0	
D3	8	1.08	9	−120.0	10	0.0	
D4	9	1.08	10	−120.0	5	0.0	
D5	10	1.08	5	−120.0	6	0.0	

Note: THE STARRED ENTRIES ARE THE TORSIONAL ANGLES ALLOWED TO VARY TO IMPLEMENT THE INTERNAL DEGREES OF FREEDOM, ϕ_1 TO ϕ_3. DEUTERIUM ATOM POSITIONS ARE IDEALIZED. EACH ATOM IS DEFINED RELATIVE TO THOSE DEFINED PREVIOUSLY *l*, *m*, AND *n*. FOR EXAMPLE, C4 IS 1.364 Å FROM ATOM 6 (C3) AND THE C4-C3 BOND MAKES AN ANGLE OF 120.755° TO THE BOND FROM ATOM 6 TO ATOM 5 (C3-C2). THE PLANE DEFINED BY C4, C3, AND C2 (ATOMS 7, 6, AND 5) MAKES AN ANGLE OF 178.357° TO THE PLANE DEFINED BY ATOMS 6, 5, AND 3 (C3, C2, AND C1). X1 AND X2 ARE DUMMY ATOMS ADDED TO ALLOW THE MOLECULE TO ORIGINATE ON ITS CENTER OF SYMMETRY; THIS IS NOT STRICTLY NECESSARY BUT CAN HELP IN TROUBLESHOOTING THE MODEL.

variables—one study examined the distribution of orientations of a molecule that showed six different orientations [24].

This occupancy structure is then fed into the displacive simulation to construct the final model. This is quite appropriate for systems in which the occupancy disorder is "frozen in" during crystallization or quenching. In systems in which the orientations of the molecules can interconvert at the temperature of the scattering measurement, it would be preferable to use a simulation into which occupancy and displacement enter simultaneously. This is at present an unwieldy computational task in most, but not all, instances.

In a system that shows occupancy ordering, the interatomic distances determined from Bragg scattering (the d_0) are in truth averages over possible occupancy configurations. This must be allowed for when performing the displacive simulation. Considering a system in which a molecule can take on one of two orientations—labeled, for example, A and B—and a given intermolecular vector, **v**, has an average length determined from Bragg scattering as $d_{\mathbf{v}0}$, then

$$d_{\mathbf{v}0} = c_{\mathbf{v}\mathrm{AA}} d_{\mathbf{v}\mathrm{AA}} + c_{\mathbf{v}\mathrm{AB}} d_{\mathbf{v}\mathrm{AB}} + c_{\mathbf{v}\mathrm{BA}} d_{\mathbf{v}\mathrm{BA}} + c_{\mathbf{v}\mathrm{BB}} d_{\mathbf{v}\mathrm{BB}}, \tag{11.3}$$

where $c_{\mathbf{v}\mathrm{AA}}$ is the concentration of vectors from orientation A to A, $c_{\mathbf{v}\mathrm{AB}}$ is the concentration that goes from A to B, and so on. Hence there are, in a sense, four contact vectors in a set associated with **v**. Thus

$$c_{\mathbf{v}\mathrm{AA}} + c_{\mathbf{v}\mathrm{AB}} + c_{\mathbf{v}\mathrm{BA}} + c_{\mathbf{v}\mathrm{BB}} = 1, \tag{11.4}$$

so Equation 11.3 can be rewritten using size-effect parameters $\varepsilon_{\mathbf{v}ij}$, where $d_{\mathbf{v}ij} = d_{\mathbf{v}0}(1 + \varepsilon_{\mathbf{v}ij})$ to give

$$0 = c_{\mathbf{v}\mathrm{AA}} \varepsilon_{\mathbf{v}\mathrm{AA}} + c_{\mathbf{v}\mathrm{AB}} \varepsilon_{\mathbf{v}\mathrm{AB}} + c_{\mathbf{v}\mathrm{BA}} \varepsilon_{\mathbf{v}\mathrm{BA}} + c_{\mathbf{v}\mathrm{BB}} \varepsilon_{\mathbf{v}\mathrm{BB}}, \tag{11.5}$$

leading to an energy contribution of the form

$$E_{\mathbf{v}ij} = F_{\mathbf{v}} (d_{\mathbf{v}} - d_{\mathbf{v}0}(1 + \varepsilon_{\mathbf{v}ij}))^2, \tag{11.6}$$

so that the (displacive) energy of the whole crystal is

$$E = \sum_{\mathbf{v}} F_{\mathbf{v}} \sum_{ij} \left(d_{\mathbf{v}} - d_{\mathbf{v}0}(1 + \varepsilon_{\mathbf{v}ij}) \right)^2, \tag{11.6a}$$

where the sum is over all instances of all types of contact vectors present in the model. For a molecule's energy, the sum is restricted to the vectors associated with a single molecule.

Equation 11.5 means that there is at least one constraint linking the size-effect parameters of the four "subtypes" that combine to give the vector in question. The size-effect parameters give a means of modeling the component of the distribution of intensity in a diffuse diffraction pattern due to the fact that the desirable molecular separation depends on the mutual orientation or type of the two molecules. Orientation is distinguished from conformation. Conformational change occurs when a fragment of a molecule twists, bends, or stretches; orientation occurs when the whole molecule flips over or rotates, something which cannot be modeled by modifying one of the internal degrees of freedom.

This shows one of the advantages of diffuse scattering analysis: the d_0 from the Bragg study is an average over configurations (the average itself may be a completely unphysical value), but modeling the diffuse data and incorporating the ε_{ij} allows some ability to extract details of that population of configurations and to relate it back to physically and chemically reasonable atomic separations.

When a molecule shows internal flexibility, whether a torsional rotation, a bond angle change, or a bond stretching (the latter is generally a higher energy motion), there will be an energy penalty as the internal flexing stretches or compresses intermolecular vectors. In addition, such a motion will alter the *intra*molecular energy, and this is modeled in three ways. First, contact vectors can be placed within a molecule, which gives another energy contribution of the form of Equation 11.1. Second, for example, dihedral angles can have a torsional-pendulum type of energy term of the form

$$E_i = G_i(\Delta\phi_i)^2, \tag{11.7}$$

where G is the force constant and $\Delta\phi$ is the deviation away from the "average" position. Third, intramolecular degrees of freedom can interact on a pairwise basis

$$E_{ij} = G_{ij}(\Delta\phi_i\Delta\phi_j), \tag{11.8}$$

which allows that a twist in one place in the molecule might make a twist elsewhere more or less energetically favorable. Bond angles and lengths can be treated similarly if desired.

If a model crystal whose diffuse scattering reproduces the observed can be found, then it may be supposed that (1) the occupancy structure is appropriate (if applicable), (2) the $F_{\mathbf{v}}$ (and associated ε_{ij}) have appropriate values (and the set of contact vectors chosen is reasonable), and (3) the G_i and G_{ij} have appropriate values (and the molecules are allowed appropriate flexibility). From this, conclusions can be drawn about the dynamics of the molecules in the crystal state and about the local-scale structures that give rise to the global average. The structure of the model crystal can be examined to look for correlations between and within molecules to build up a picture of the population of local configurations, something beyond Bragg analysis.

While it may be desirable in some respects to use more "realistic" potentials than those shown here, this introduces the problem of increasing the number of free parameters in the modeling. At present, typically 20 to 30 F_i are required, plus the G_i and G_{ij}, plus any ε_{ij}, meaning it is quite possible for a model to have more than 40 parameters. If Equation 11.1 is replaced with a form requiring more parameters, then the number of free parameters in the model increases dramatically.

Once a useful model has been found interactively, it may be automatically refined [22,25–27] using a least-squares approach.

3 Example Studies

3.1 Benzil

Benzil is an example of a molecular system of relatively high symmetry in which there is no occupational disorder but the molecules show internal flexibility. Benzil, $C_{14}H_{10}O_2$ (Figure 11.1), crystalizes in the space group $P3_121$, a = 8.402 Å, c = 13.655 Å (hexagonal setting) at 300 K [28]. It is isostructural with

α quartz. On cooling, benzil undergoes a phase transition to a ferroelectric ferroelastic phase at 83.5 K [29–33]. Figure 11.1 shows a molecule of benzil with its three internal degrees of freedom noted.

Benzil has long been known to exhibit strong, highly structured diffuse scattering [34]. However, the average structure shows no evidence of static disorder, suggesting that the structured scattering is essentially the result of highly and anisotropically correlated thermal motions. Figure 11.2 shows diffuse scattering data from benzil. Figures 11.2a and 11.2b show X-ray data collected on a home laboratory system [35], while 11.2c shows neutron scattering data collected using the SXD instrument at the ISIS spallation neutron source [36]. The persistence of the lines of scattering from layer to layer indicates that they are in fact planes, indicating one-dimensional correlations of displacements ("chains of molecules") in directions perpendicular to these planes (one such plane is indicated by an arrow in Figure 11.2a). The large rectangular regions of strong scattered intensity (circled in Figure 11.2a) indicate that the chains can shear relative to each other. Figure 11.3b shows the *ab* plane of benzil with the directions of these correlated chains superimposed. It can be seen that they coincide with the long axes of the phenyl rings when projected onto the *ab* plane. One layer of these molecules has been extracted in Figure 11.3b to show this more clearly.

In modeling the data from benzil, it was assumed that the force constants on the internal degrees of freedom ϕ_2 and ϕ_3 would be the same given that the two halves of the molecule are related by symmetry in the average structure. Further, as there is no occupancy SRO structure present, there is no need for size-effect parameters. When the network of contact vectors was constructed, this resulted in eleven types of contact vectors, which are described in Table 11.2 (derived from Table 2 of Welberry et al. [27] and Table 1 of Goossens et al. [26]). This is a relatively small number of vector types, largely due to the (relatively) high symmetry of the structure. Allowing for one torsional spring constant on ϕ_1 (F_{12}), a second on ϕ_2 and ϕ_3 (F_{13}), a third to allow interactions between pairs ϕ_1 & ϕ_3 and ϕ_1 & ϕ_2 (F_{15}),

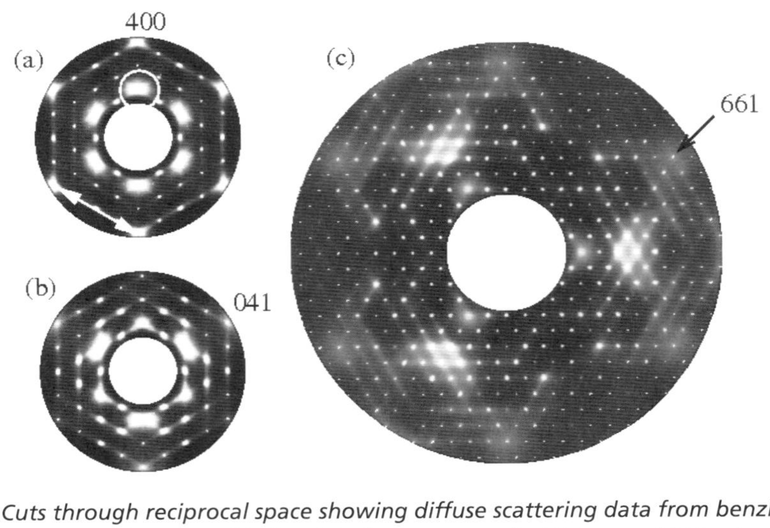

Figure 11.2. *Cuts through reciprocal space showing diffuse scattering data from benzil. (a) The hk0 plane measured with X-rays. (b) The hk1 plane measured with X-rays. (c) The low scattering vector region of the hk1 plane measured with neutrons (large scattering vector when compared to (a) and (b)). Black is lowest intensity, white is maximum. Some Bragg reflections are noted to indicate scale and orientation.*

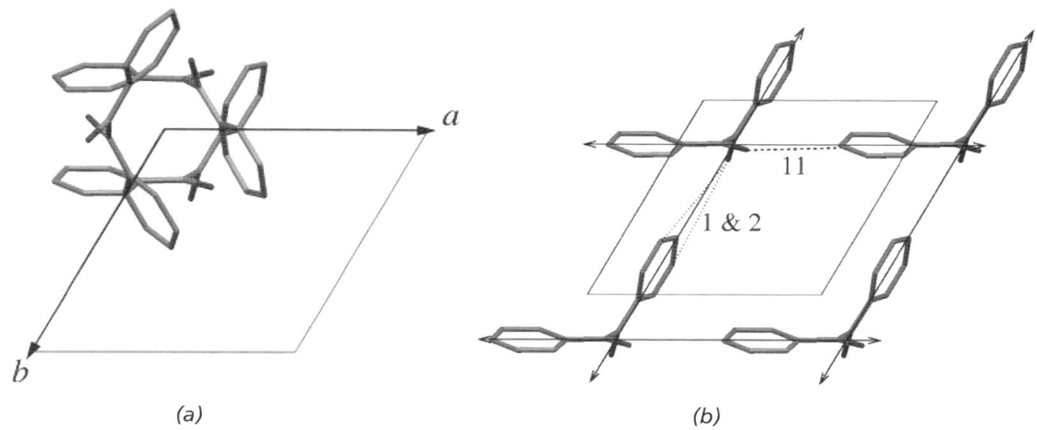

Figure 11.3. *Structure of benzil viewed down the c axis (H atoms omitted). (a) Unit cell contents. (b) A single layer of molecules showing, with arrows, (some of) the directions perpendicular to the planes of scattering apparent in the data in Figure 11.2. Also shown are examples of contact vector types 1, 2, and 11. They are all present on all phenyl rings but have been separated here for clarity.*

and a fourth for interactions between ϕ_2 & ϕ_3 (F_{14}), there are fifteen parameters in the model. Table 11.1 shows the *z*-matrix used.

The model was initially improved interactively until it approximated the data, then an automatic refinement process was implemented based on comparing the calculated diffraction patterns with those measured using X-rays [27]. This proved successful in giving an excellent fit—except for one artificiality. It included an isotropic increase to the atomic displacement parameters (ADPs) that was implemented *after* the model had been run. Essentially, the ADPs from the disorder model were too small given the Bragg scattering data, but the limited reciprocal space extent of the X-ray data (Figures 11.2a and 11.2b) meant that the magnitudes of the ADPs were poorly constrained. This is reasonable when it is recalled that conventional crystallography largely relies on the dependence of Bragg peak height on scattering vector magnitude (and direction) to determine the ADPs, and in the X-ray data the scattering extends out only to 400-type reflections. This was corrected through refinement of an SRO model against neutron *and* X-ray diffuse scattering data simultaneously [26].

This proved a successful approach because (1) the neutron scattering data extend much further out into reciprocal space (data out to 12 0 0-type reflections were used, although due to the high energies of some neutrons in the spallation pulse, even higher order reflections are measurable) [6] and (2) the neutron is sensitive to the atoms at the molecular extremities (H, or, in the neutron case, D), which undergo the greatest librations and are therefore more sensitive to the molecular motions.

Using comparison with the neutron data as a constraint on the model removed the need for the artificially increased ADPs and had the added bonus of giving new insight into the internal degrees of freedom due to the second point noted above. In the modeling and fitting, the model crystal was brought into thermal equilibrium using the defined interactions, and from this equilibrated model

TABLE 11.2

Contact vector definitions and values of spring constants (from simultaneous refinement against X-ray and neutron data) for modeling diffuse scattering from benzil. One example of each vector type is given; fuller details are available in Tables 2 and 3 of Welberry et al. [27].

CONTACTS		CELL TRANSLATION			SYMMETRY EQUIVALENT			
From atom	To atom	*x*	*y*	*z*	*s*	Length (Å)	Spring constant label	Spring constant (k_BT Å^{-2})
O3	C7	0	1	0	1	4.50	F_1	3.90
O3	C8	0	1	0	1	4.01	F_2	1.20
C8	C10	0	−1	0	1	4.34	F_3	0.25
O3	C9	1	1	0	2	3.35	F_4	4.4
O3	C8	0	0	0	2	3.64	F_5	0.67
O3	O4	0	0	0	2	4.04	F_6	0.82
C7	C8	0	−1	0	2	3.99	F_7	2.11
C7	C9	0	0	0	2	4.23	F_8	2.11
C7	C10	0	0	0	2	3.73	F_9	4.1
C7	C9	0	0	0	3	4.46	F_{10}	2.9
O3	C14	0	1	0	1	3.42	F_{11}	1.4
								Spring constant (k_BT deg.$^{-2}$)
Torsional force constant on ϕ_1							F_{12}	0.000
Torsional force constant on ϕ_2 and ϕ_3							F_{13}	0.140
Force constant on cross term for ϕ_2 and ϕ_3							F_{14}	0.000
Force constant on cross terms for ϕ_2 & ϕ_3 and ϕ_1 & ϕ_3							F_{15}	0.0067

crystal, the X-ray and neutron diffuse scattering were calculated. The resulting images were compared pixel by pixel with observations to determine a "goodness of fit" which was then subject to refinement via a least-squares process [22].

Figure 11.4 shows the final calculated data that can be compared with Figure 11.2. The fit is good, as can be seen.

Once the model has been refined, it can be analyzed to reveal the structural details of interest. Figure 11.5 shows one possibility: correlations in displacements as a function of neighbor distance. It can be seen that displacements in the direction of the vector connecting neighbors tend to be more strongly correlated than transverse displacements, that $\Delta\phi_1$ is negatively correlated for molecules adjacent along the *c* axis, and it is uncorrelated for neighbors in the *a* direction (and in the *hk* plane

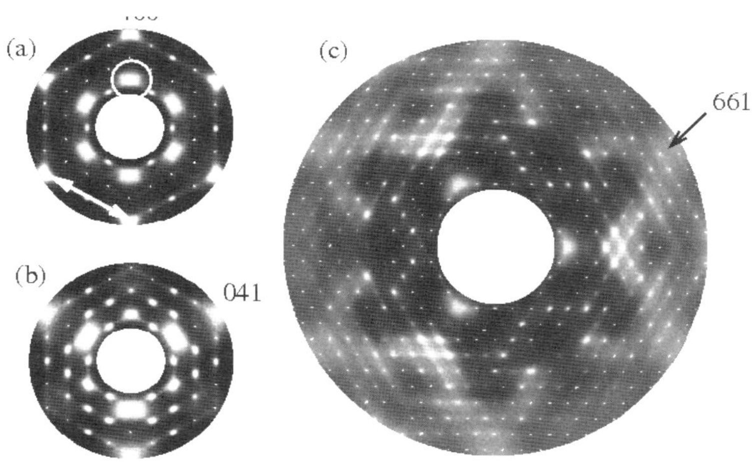

Figure 11.4. *Calculated diffuse scattering from benzil. (a) The hk0 plane as measured with X-rays. (b) The hk1 plane as measured with X-rays. (c) The hk1 plane as measured with neutrons.*

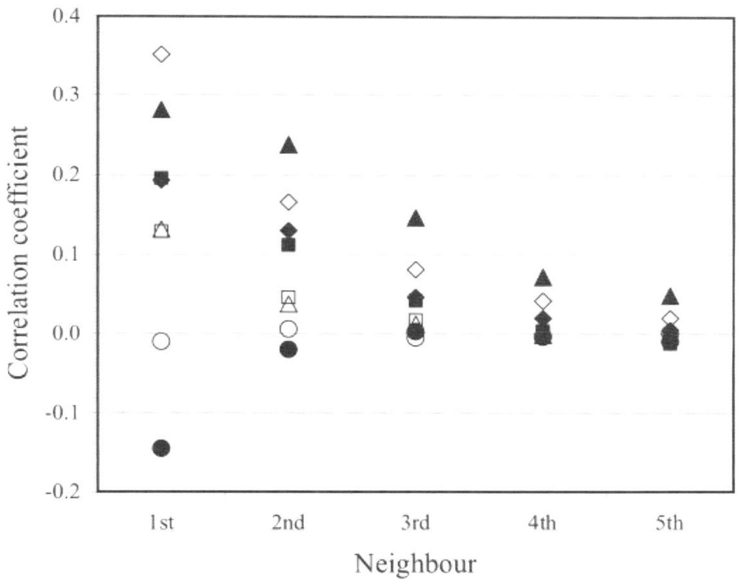

Figure 11.5. *Some short-range correlations in benzil as revealed by diffuse scattering. Hollow symbols apply to neighbors in the a direction, while filled symbols apply to neighbors along the c axis. Diamonds show correlations between a-direction displacements of neighbors, triangles give correlations between c-direction displacements, and boxes give y-direction displacements, where y completes a Cartesian set of coordinates. Circles give correlations in the values of ϕ_1 along a and c.*

generally). Various correlations and cross-correlations can be determined to gain similar insight into the SRO structure of the crystal.

Atomic displacement parameters can be calculated from the model by direct summation over the atomic population, allowing comparison with ADPs found by "conventional" crystal structure refinement. Further, an estimate of the expected frequency of librational vibration of the phenyl rings can be obtained from the refined values of the torsional force constants [26,27]. This estimate, 38.0 cm^{-1} [27] to 35.3 cm^{-1} [26], proves remarkably close to both an *ab initio*-derived value (38–40 cm^{-1}) [27] and a value experimentally observed using spectroscopic techniques such as Raman scattering (39 cm^{-1}) [37]. Further, a key interaction was found to be that acting from the *para* H atoms (D3 and D3a) to the oxygen atoms (examples noted by dashed lines in Figure 11.3b); this is contact vector 11, and while its spring constant is not particularly large, the fitting showed that it was crucial that this vector be included. Figure 11.3b shows that vector types 1, 2, and 11 all contribute to propagating the displacement correlations that most heavily influence the observed diffuse scattering.

The force constants in Table 11.2 show that neighboring internal degrees of freedom interact significantly ($F_{15} > 0$). Therefore the energy penalty exacted by the molecule scissoring around the central C-C bond is through (1) the correlation of that bond's dihedral angle (ϕ_1) with ϕ_2 and ϕ_3 and (2) the changes in contact vector lengths that result from the motion. Adding in the neutron data allowed differentiation of the values of F_{12} and F_{13} (which could be left identical when fitting only the X-ray data).

3.2 3,3′-dimethoxybenzil

3,3′-dimethoxybenzil (33DMOB, $C_{16}H_{14}O_4$, Figure 11.6) is a molecule related to benzil but showing two additional degrees of freedom and a very different (and lower symmetry) crystal structure. It also does not show occupational ordering. It has been examined, along with 4,4′-dimethoxybenzil [38] and 2,2′-dimethoxybenzil, as part of the ongoing program to explore the uses of analysis of diffuse scattering in shedding light on the behavior of solids made up of flexible molecules. In this case, the diffuse scattering data were collected using a synchrotron light source (the 1-ID-C beamline at the Advanced Photon Source [APS] in Chicago, IL), and three planes of scattering are reproduced in Figure

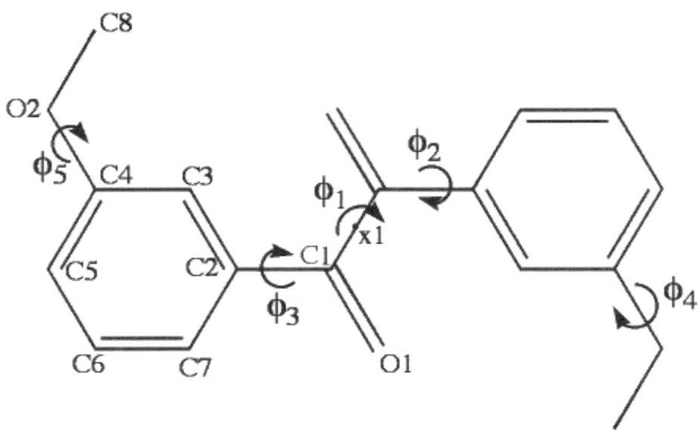

Figure 11.6. *A schematic diagram of a molecule of 3-3′-dimethoxy benzil showing the five internal degrees of freedom (ϕ_1 to ϕ_5), all of which are rotations around C-C single bonds. H atoms are omitted. The diagram also shows the atom labeling.*

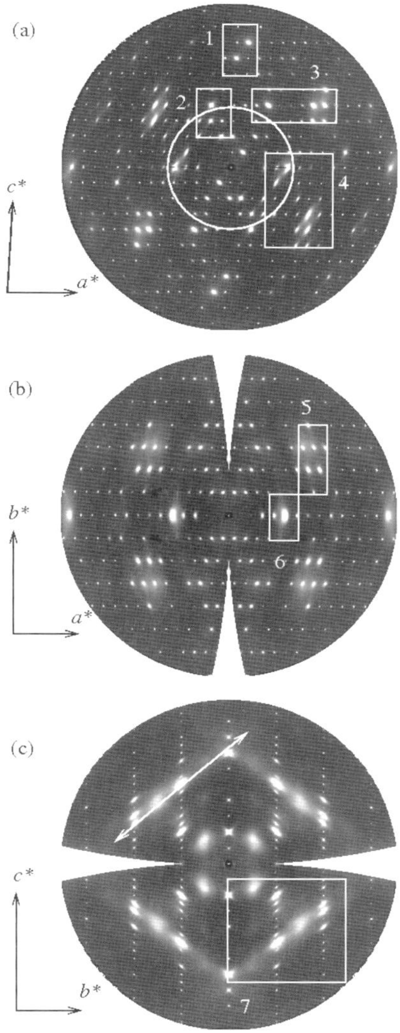

Figure 11.7. *Diffuse scattering data from 3,3′-dimethoxybenzil. (a) The h0l plane. (b) The hk0 plane. (c) The 0kl plane. Boxes indicate regions of data used in automated refinement [22] and the circle in (a) indicates the extent of data collected in the home laboratory using the PSD system [35].*

11.7: (a) *h*0*l*, (b) *hk*0, and (c) 0*kl*. The crystal structure is monoclinic, with space group *C*2/*c* and lattice parameters a = 20.0790 Å, b = 4.8086 Å, and c = 13.7747 Å and β = 92.3620° [22].

Based on symmetry, three torsional spring constants are required for five internal degrees of freedom, whereas the five torsional angles give rise to the possibility of ten pairwise cross-terms among these degrees of freedom, although symmetry reduces this to six different types. This reduces to four or even two when considering adjacent degrees of freedom only.

The synchrotron data have one of the advantages of neutron diffraction data—they extend considerably further out into reciprocal space than the home laboratory X-ray data, in this case by a factor of approximately 2.5 (the extent of the home laboratory data are noted on Figure 11.7a by a white circle). Also (admittedly, after considerable image processing [5]), the signal-to-noise ratio is better than that in either laboratory X-ray or neutron data.

The data in this case do not show the pronounced highly anisotropic rod- or planelike features seen in benzil. Instead they show elongated thermally induced blobs around the Bragg positions but no large linear or planar features extending across many reciprocal cells. There are quite highly structured features in some regions of reciprocal space (boxed in Figures 11.7a–c) that suggest that interactions are stronger along certain directions than others (in particular along the direction perpendicular to the white arrow in Figure 11.7c) but perhaps not as pronounced as in benzil.

In this case, rather than simultaneously refining against laboratory and synchrotron X-ray data, automatic refinements using solely the synchrotron data were used to produce a final model of the SRO in 3,3′-dimethoxybenzil. It was concluded that the material can be considered as a stack of sheets of molecules. The sheets lie parallel with a plane defined by the *b* and *a-c* (*a* minus *c*) directions and as such lie along a face diagonal of the unit cell. This was determined because the interaction constants within these sheets are strong, whereas those between sheets are relatively weaker. The strongest interactions propagate along the *a-c* direction and relate to the methoxy groups on the periphery of the molecule. The strongest *correlations* between molecules run along the *b* direction, *not a-c*; this is not unreasonable. While *a-c* shows the strongest interactions, *b* shows the highest *density* of significant interactions. This is reasonable because the molecule's longer axis is parallel to *a-c* when projected onto the (*a-c*)*b* plane so that they present a greater cross section to each other along *b*, but the chemically active methoxy groups connect along *a-c*. The synchrotron data allowed reliable values for the torsional spring constants to be refined, which was not the case when relying on home laboratory X-ray data. These torsional constants showed that the phenyl group torsional angles (ϕ_2 and ϕ_3) are the stiffest and are negatively correlated with their most closely associated methoxy groups; this makes sense because contact vectors to the methoxy groups are relatively strong, and so when the phenyl ring is allowed to rotate, the methoxy groups are (to a degree) "pinned" to their positions by the contact vectors, resulting in what amounts to a methoxy rotation that negates the rotation of the phenyl ring, at least to some extent given that the two rotation axes are not parallel.

Hence analysis of diffuse scattering reveals information about the behavior of the molecules in the crystalline state. As with benzil, various correlations can be abstracted from the model crystal. Figure 11.8 shows the calculated diffuse scattering for the same three planes as in Figure 11.7.

As an example of the sort of two-body correlation information available from diffuse scattering analysis, Figure 11.9 shows correlations between displacements on neighboring molecules as a function of the direction of the displacement. It shows polar plots in which the angle, θ, gives the angle of the displacements being considered on the two molecules in question relative to the horizontal axis (*c* in Figure 11.9a and *b* in Figure 11.9b) in the plane shown, and the length of the vector gives the magnitude of the correlation between displacements in the direction defined by θ. This is for a pair of molecules connected by a given contact vector (in other words, a pair related by a particular symmetry element). So, for example, Figure 11.9a shows that within the *ac* plane, for neighbors stacked along the *a-c* direction, the *a-c* displacements are the most highly correlated, and Figure 11.9b applies to neighbors stacked along *c* and shows that, for displacements in the *bc* plane, those roughly pointing along $b + c$ are most highly correlated.

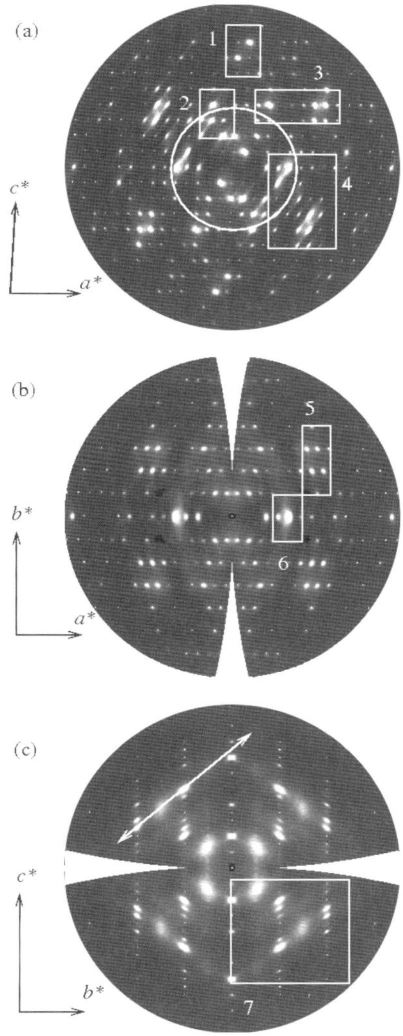

Figure 11.8. *Diffuse scattering data from 3,3′-dimethoxybenzil calculated from the final refined model. (a) The h0l plane. (b) The hk0 plane. (c) The 0kl plane.*

3.3 Ibuprofen

Ibuprofen ($C_{13}H_{18}O_2$) is a third example of a flexible molecular system, again without occupancy structure but in this case also without large rotating molecular groups like phenyl rings; it is rather like a single phenyl ring with flexible "arms" hanging off of it. It does, however, show enantiomerism with both racemic "S-(±)" and optically pure "S-(+)" forms. Both have been studied using diffuse scattering

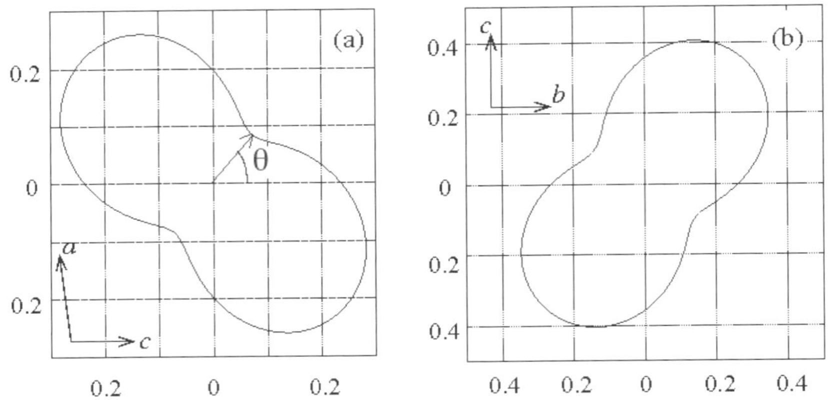

Figure 11.9. *A representation of the correlations between molecular displacements on adjacent molecules of 33DMOB. See text for details.*

coupled to MC modeling [39,40]. Figure 11.10 shows a simple schematic diagram of an ibuprofen molecule indicating the labeling of the atoms and the internal degrees of freedom used.

The crystal structures of S-(+) and S-(±) ibuprofen are summed up in Table 11.3. The racemate is slightly more tightly packed (~306 Å^3 molecule^{-1} to 312 Å^3 molecule^{-1}).

TABLE 11.3

Structural information for racemic S-(±) and optically pure S-(+) ibuprofen

TYPE	*a* (Å)	*b* (Å)	*c* (Å)	β (°)	SPACE GROUP	Z
Racemic [41]	14.667	7.886	10.730	99.362	$P2_1/c$	4
Optically pure [42]	12.456	8.0362	13.533	112.86	$P2_1$	4

In both structures the ibuprofen molecules dimerize. In S-(±), the dimerized pairs project most strongly onto the *ab* plane, whereas in S-(+) it is the plane perpendicular to –101 (Figure 11.11). Figure 11.12 shows the diffuse scattering from reciprocal space sections analogous to these two planes. Figure 11.12a shows the *hk*0 plane of S-(±) measured using synchrotron X-rays, and Figure 11.12b shows the *hkh* plane of S-(+) measured using Mo K_α radiation and a mar345 detector in the home laboratory (and, as a result, the two cuts differ in signal-to-noise and maximum |Q|). Both planes show considerable structure, though occupancy ordering is apparent in neither. This is not surprising in S-(+) because there is only one "type" of molecule present. It can be explained for S-(±) by the dimerization—while there are two different scattering objects (S-(+) and S-(–)), they strongly form into dimers that *are* all identical on average.

In both cases the diffuse scattering shows considerable structure, with the diffuse regions elongated. The narrow axes of the blobs coincide with the long axes of the dimers, and box-shaped regions of scattering on the reciprocal axes indicate the intersection of streaks. In neither case, however, does the scattering extend strongly from one reciprocal lattice point to the next—the features are elongated blobs two or three times longer than they are wide rather than fully fledged streaks as seen in benzil.

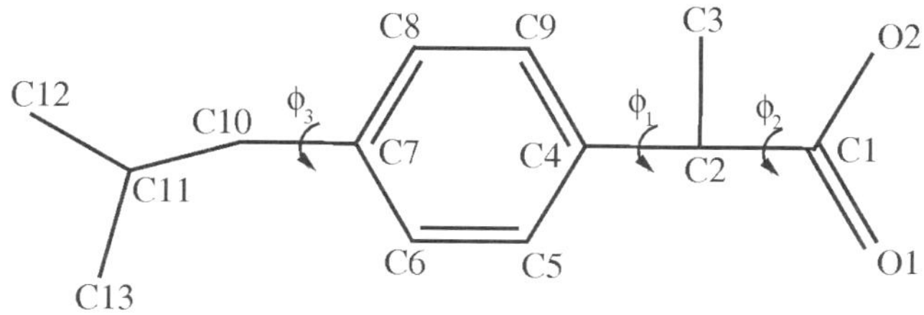

Figure 11.10. *Schematic diagram of an ibuprofen molecule indicating the labeling of the atoms and the internal degrees of freedom used. H atoms omitted.*

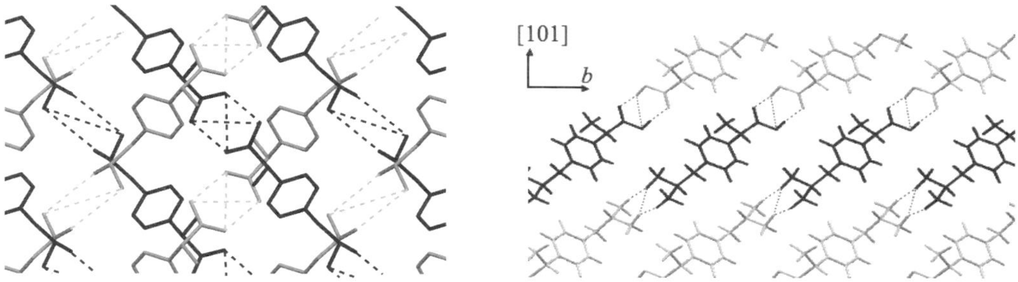

(a) (b)

Figure 11.11. *(a) Two planes of ibuprofen dimers (gray and black) projected onto the ab plane of S-(±) (H atoms omitted). (b) A single plane of dimers projected onto the plane of S-(+) perpendicular to [–101]. In both cases, dashed lines indicate the contact vectors that give rise to the chains of molecules.*

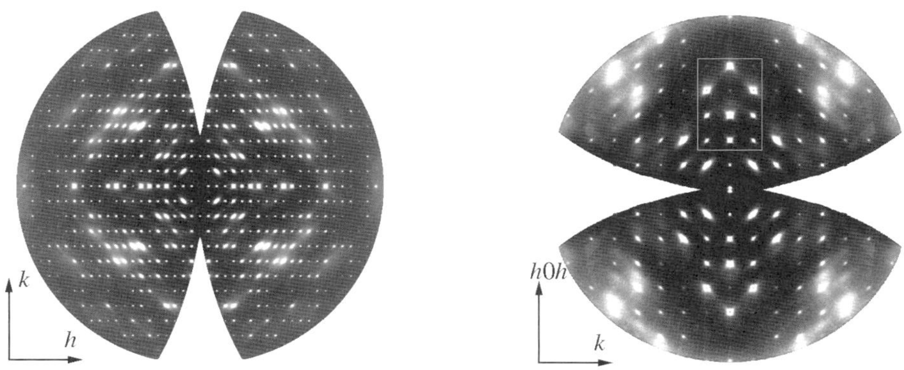

(a) (b)

Figure 11.12. *(a) The hk0 plane of diffuse scattering from S-(±) ibuprofen measured using synchrotron X-rays at the Advanced Photon Source. (b) The hkh plane of S-(+) measured in the home laboratory (a strongly structured area of scattering is noted in (b)).*

This suggests that the chains of molecules are not particularly long—little longer than the dimers themselves—and that interactions between chains are significant.

Figure 11.13 shows calculated diffuse scattering in the two planes shown in Figure 11.12. In both models the strongest interactions in the crystal are those that cause dimerization, and other strong

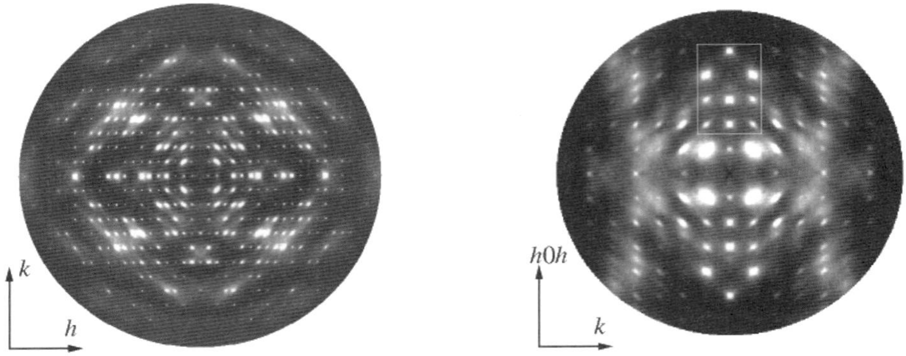

(a) (b)

Figure 11.13. *(a) Calculated scattering from the hk0 plane of diffuse scattering from S-(±) ibuprofen and (b) for the hkh plane of S-(+). In both models the dimerizing interaction is strongest.*

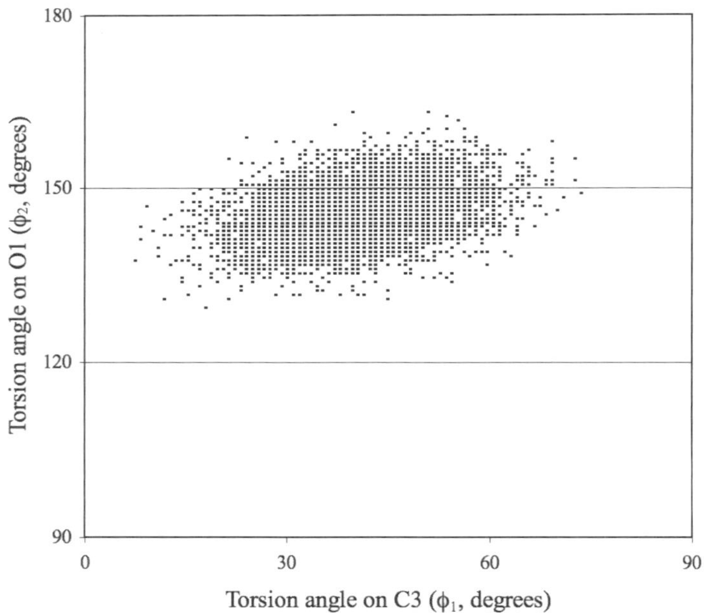

Figure 11.14. *A scatter plot of the distribution of the values of the dihedral angles on C3 (ϕ_1) and O1 (ϕ_2) for S-(±) that give minimum molecular energy. Dashed lines show the equilibrium values.*

interactions are those at the opposite ends of the molecules to the -COOH groups, which cause the dimers to form into chains. The other aspect of the modeling was that the molecule should be allowed to flex. In particular, the torsional degrees of freedom on C3, O1, and C11 were important (see Figure 11.10). Interestingly, the rotations on C3 and C11 were found to correlate negatively despite being separated by the phenyl ring. This is because both affect the orientation of the phenyl ring via their interactions with the environment, and the phenyl ring has to find a position that compromises between these influences (and best accommodates its own interactions with its surroundings). This has an interesting implication. Crystal structure prediction and molecular conformation prediction for a material may proceed by exploring typical conformations of various chemical groups as found in the literature (some aspects are discussed in Shankland et al. [43]); but if the fragment of interest cannot be decoupled from the rest of the molecule, such tools become less powerful.

Figure 11.14 shows a scatter plot of the dihedral angles on C3 (ϕ_1) and O1 (ϕ_2) of S-(±) that minimize the molecular energies; that is, once the final model crystal was established and equilibrated, each molecule was visited and its energy was minimized with respect to ϕ_1 and ϕ_2. In essence, this is a simple way to plot an aspect of the energy landscape inhabited by the molecules. The figure shows a positive correlation between the two angles and shows that the energy of the molecule is much more sensitive to the angle on C3 (torsional rotation around the C1-C2 bond) than to the angle on O1 (torsional rotation around C2-C4). This can be compared with Figure 2 of Shankland et al. [43], which shows a similar result derived from database mining.

3.4 *Pentachloronitrobenzene (PCNB, $C_6Cl_5NO_2$)*

Pentachloronitrobenzene (PCNB, $C_6Cl_5NO_2$) is a relatively simple molecule (Figure 11.15). Diffuse scattering from a crystal of PCNB was modeled without allowing for molecular flexibility, although in principle the NO_2 group could be allowed to rotate [24]. It does, however, show an occupancy structure with associated size-effects. Hence it offers somewhat different challenges from the examples discussed so far. In the average structure of PCNB at room temperature, each molecular site shows sixfold symmetry [44]. This means that the molecule can be, in a sense, written as $6(Cl_{5/6}(NO_2)_{1/6})$ with, in the average structure, the Cl and NO_2 groups superimposed.

Diffuse scattering data were collected at a number of locations around the world, but the diffraction patterns to be discussed here were recorded at the Advanced Photon Source on the 1-ID-C beamline out to $|Q| = 10.7$ Å^{-1}. Two cuts of data are shown in Figure 11.16: the $0kl$ and $hk0$ sections.

The $0kl$ data show a column of strong, asymmetric diffuse regions centered on the 0 0 $3n$ Bragg peaks (n = integer). The inner blobs show "wings" that become less pronounced for higher order reflections. Also very distinctive are the diffuse oblongs distributed between Bragg peaks (arrowed) and the diamond-shaped distributions of intensity that appear around many of the Bragg positions when $k \neq 0$. The $hk0$ cut shows an overall intensity modulation reflective of the sixfold symmetry of the molecule and shows a much less anisotropic distribution of diffuse intensity than the $0kl$ layer—blobs of scattering are somewhat elongated in directions perpendicular to the radial, but the effect is not strong.

The lack of a superlattice shows that the position of the NO_2 group on the molecule is not ordered from molecule to molecule in a long-range ordered fashion; this does not mean its distribution is random. Hence many schemes were explored in which the positions of the NO_2 groups on adjacent molecules were correlated—for example, it is plausible that they might try to avoid each other since $-NO_2$

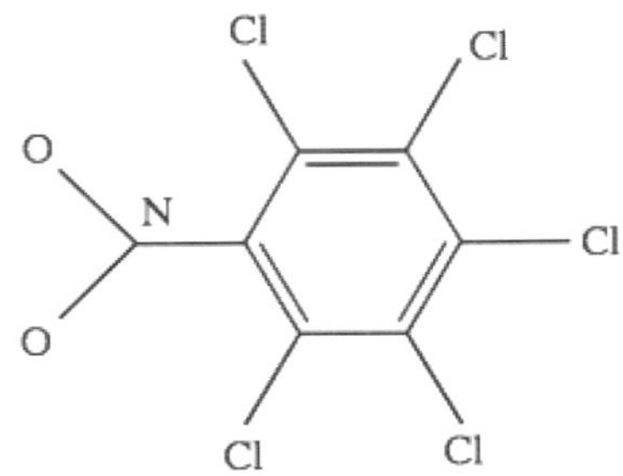

Figure 11.15. *A schematic diagram of a molecule of PCNB.*

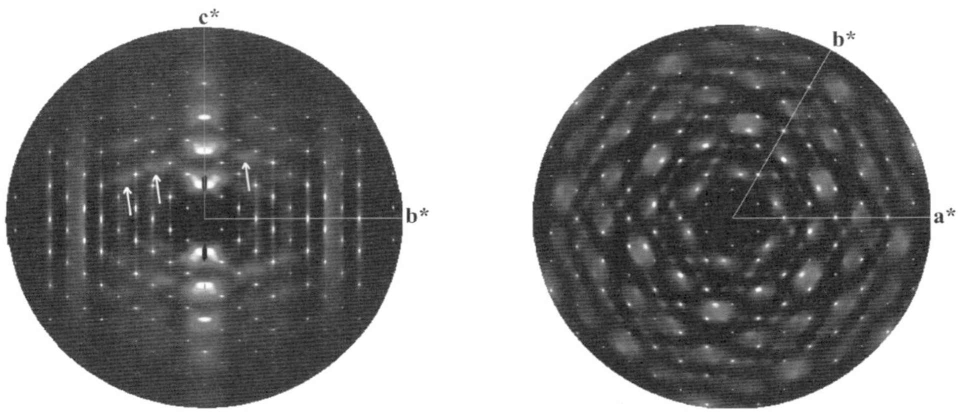

(a) (b)

Figure 11.16. *Observed diffuse scattering from PCNB. (a) The h0l plane. (b) The hk0 plane.*

is physically a substantially larger group than –Cl. Indeed, in the average structure O-O distances of 2.39 Å, 2.525 Å, and 2.763 Å are seen, which are substantially shorter than the combined van der Waals radii (~3.04 Å). On the other hand, Cl-Cl distances in the average structure are at least 3.578 Å, whereas the sum of the van der Waals radii gives 3.5 Å. Much effort was put into trying to determine an occupancy structure that could be fed into the displacive MC simulation. Oddly, given that the difference between 2.39 Å and 3.04 Å is substantial, the model that eventually best matched the data suggested that the occupational ordering *was* random and that the mismatch in average separation and plausible chemical separation of atoms was met entirely through size-effect parameters, ε_{ij}.

These size-effects were applied to contact vectors that have a component along *c*—the ones whose average lengths were too short relative to the van der Waals radii. The vectors to which the size-effect was applied are illustrated in Figure 11.17. Since the NO_2 groups were not allowed to rotate, the vectors were connected to the nitrogen atoms rather than the oxygen atoms. Vectors within the *ab* plane did not need to be size-effected, suggesting that the main phenomenon that needed to be induced was a tipping of the molecules out of the *ab* plane. So, in the final model, the position of the NO_2 on each PCNB ring was set at random and fixed, and the crystal was equilibrated displacively

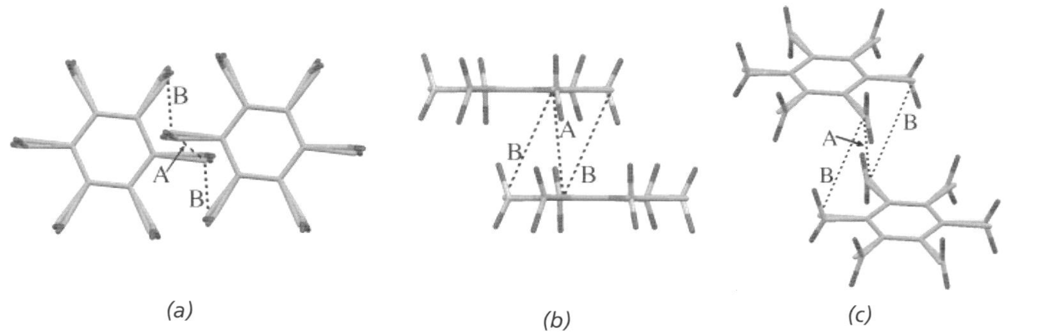

Figure 11.17. *The two contact vectors to which size-effect was applied, viewed (a) down the c axis, (b) down the a axis, and (c) obliquely.*

under the condition that NO_2 groups connected by the vectors shown in Figure 11.17 try to push each other apart, as do NO_2 and Cl groups, but contacts between Cl and Cl try to be shorter than the average structure value in order to maintain the overall average separation. The actual size-effect values are shown in Table 11.4. Such size-effects seem extreme, but in truth the average actual induced shifts away from the atomic separations suggested by the average structure are (for vector A) 0.587 Å for N-N (separation is 3.817 Å in the average structure), –0.084 Å for Cl-Cl (separation is 4.035 Å in the average structure), and 0.282 Å for Cl-N (separation is 3.951 Å in the average structure); they are far from the full values of the size-effect parameters due to the competition between interactions. Figure 11.18 shows the distribution of the deviations in interatomic separation for vector A; the half widths of the distributions are about 0.6 Å.

TABLE 11.4

Size-effects on the two types of contact vectors responsible for tipping the PCNB molecules out of the *ab* plane

VECTOR TYPE (SEE FIGURE 11.17)	ε_{NN}	ε_{NCL}	ε_{CLN}	ε_{CLCL}
A	+1.5	+0.75	+0.75	–0.36
B	+2.0	+1.0	+1.0	–0.48

Note: ALL THE ε_{ij} ARE IN Å.

Figure 11.19 shows the two planes of calculated scattering. Comparison with Figure 11.16 shows the high quality of the agreement. While some of the displacive spots are perhaps not quite as sharp as might be desired, there is very little intensity left unmatched, especially given the great simplicity of the model—it uses only nine parameters (five spring constants, F_i, and four independent size-effect terms, ε_{ij}). The relatively minor discrepancies between calculation and observation suggest that the model is indeed capturing the key structures in the system. Further parameters may be added—size-effects on other interactions, weak occupancy correlations between various types of neighbors, rotation on the

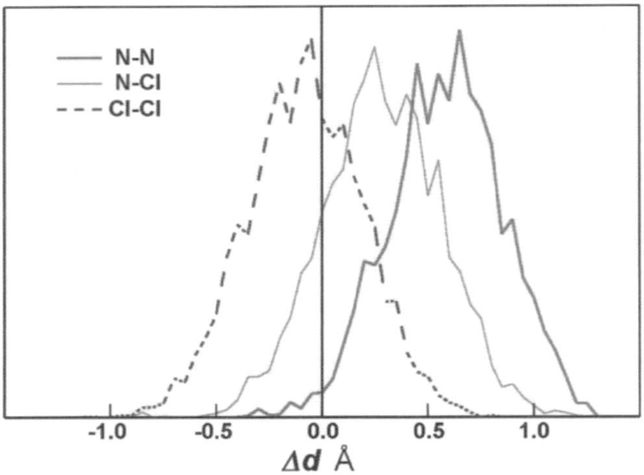

Figure 11.18. *The distributions of interatomic distances for contact vector type A; one histogram each for N-N and Cl-Cl. N-Cl and Cl-N have been combined into a single class.*

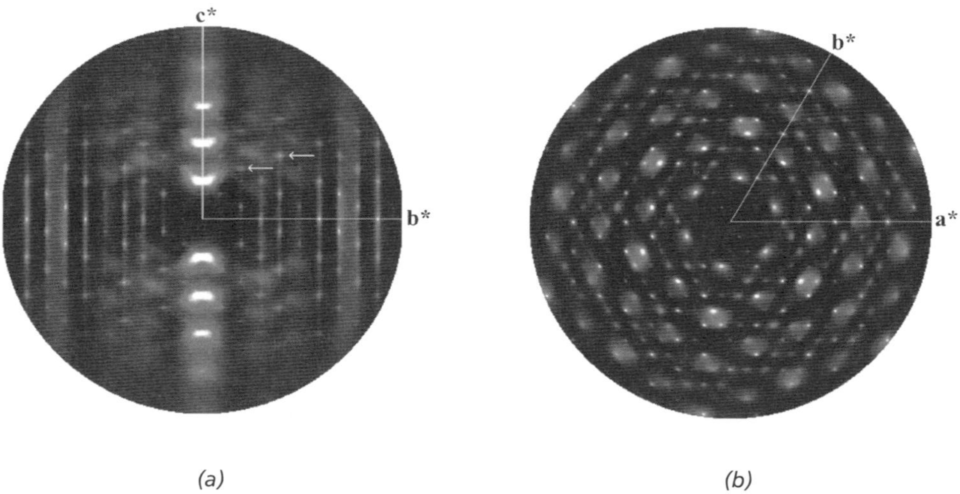

Figure 11.19. *The same planes of scattering as shown in Figure 11.16 but calculated from the final PCNB model. (a) The h0l plane. (b) The hk0 plane. Agreement is excellent.*

NO_2 group around the C-N bond (which may contribute to allowing the molecule to minimize its energy)—but the degree to which the model can be improved is so limited that there is little justification for adding further parameters at this point.

Hence diffuse scattering analysis on PCNB has shown that even though crystal chemistry suggests that correlations might be expected among the positions on the molecules of the NO_2 groups, there are in truth no significant occupancy correlations at all, and the packing is achieved through size-effect relaxation only. An interesting extension to this study would be to explore the occupancy structure in crystals that have been annealed in various ways. It is likely that the occupancy structure is established during crystallization and the crystal then accommodates that occupancy structure as best it can through size-effect relaxations. The crystal studied here was grown at room temperature from solution. It is possible that the crystal can reach a lower energy configuration if the NO_2 groups *are* able to avoid each other, whether by changing growth conditions (increased temperature, for example) or by holding a crystal at elevated temperature (melting point = 144°C) and annealing it. Such a difference, if not long-range ordered, would leave the Bragg scattering unchanged but would change the distribution of diffuse intensity.

3.5 HMTA, an Adduct of Hexamethylenetetramine and Azelaic Acid

HMTA, the adduct of hexamethylenetetramine and azelaic acid, is a very complicated system. It is an adduct of two sizable molecules, one of which is expected to show molecular flexibility, and also shows correlated occupancy structure *and* size-effect; hence it offers most of the complexities that might be encountered. While data on this material are limited to laboratory X-ray data, which was difficult to obtain because the crystals tended to degrade in the X-ray beam, it remains a useful example.

The two molecules are azelaic acid, HOOC-$(CH_2)_7$-COOH ("AZ"), and hexamethylenetetramine (urotropin, "HMT"), $(CH_2)_6N_4$. The average crystal structure is not well established due to its inherent complexity and the many phase transitions it goes through [45]. The best refinement suggests that the material at room temperature exhibits an orthorhombic space group *Bmmb* with a = 9.4157 Å, b = 26.124 Å, and c = 7.2034 Å [46]. There are four molecules of each type in the unit cell (Figure 11.20). The azelaic acid molecules appear to "flip" around their long axis. These flips are correlated, giving rise to an occupancy structure.

Figure 11.21 shows diffuse scattering data from HMTA, measured using the home laboratory-based position-sensitive detector (PSD) system. The data are not of the highest quality. Powder rings are in evidence due to the sample deterioration in the X-ray beam. However, there are a considerable number of interesting features. Some, such as the streaks in the 0*kl* plane, are the result of displacive correlations, while others, such as the boxed features in the 1*kl* plane, come from the occupancy correlations. The fact that they occur at half-index positions indicates some type of supercell, and this in turn suggests negative correlation in the molecular flips. The strong diagonal streaks suggest correlations in the displacements running parallel to the length of the azelaic acid molecules.

The cell doubling must occur at least in the c direction, and it has been suggested [46,47] that the cell is doubled along both c and b, which is interesting because the azelaic acid molecule neighbors are separated by HMT molecules in the b direction. This suggests that the HMT molecules are transmitting the occupancy of one AZ site to another—acting as microscopic mechanical switches.

The simulation made use of three z matrices, one for the HMT molecules and one for each orientation of azelaic acid. Because the AZ molecule shows two possible states (orientations), the occupancies were modeled using Ising-like interactions (Equation 11.2) and the output was an array of correlated z-matrix labels. This array was then fed into a displacive simulation that applied displacements and small orientational deviations to each molecule and allowed these to interact through contact vectors. Size-effect was applied such that, for example, azelaic acid molecules of like orientation might be pushed further apart than the average structure suggests or pulled closer together.

The correlation structure of the azelaic acid orientations that was determined consisted of correlations of –0.5 among nearest neighbors along b and c and zero correlation among neighbors along

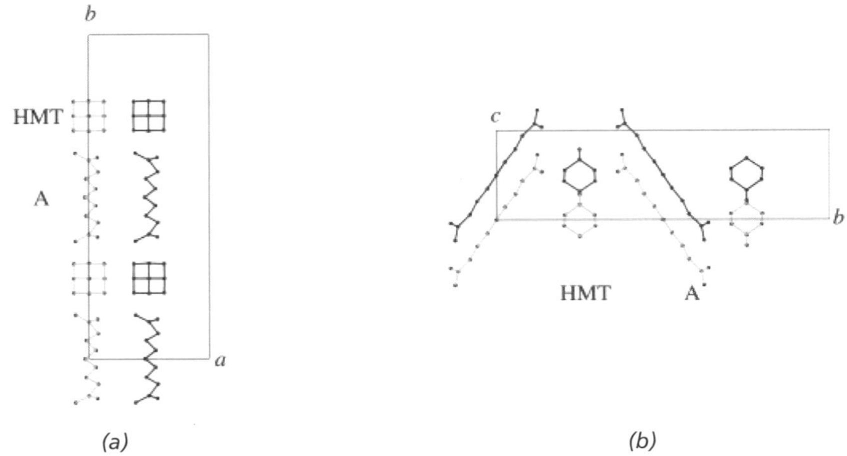

Figure 11.20. *Two views of the unit cell of HMTA showing one of the two possible orientations of azelaic acid. H atoms omitted. Grey molecules are half a unit cell above (or below) the black.*

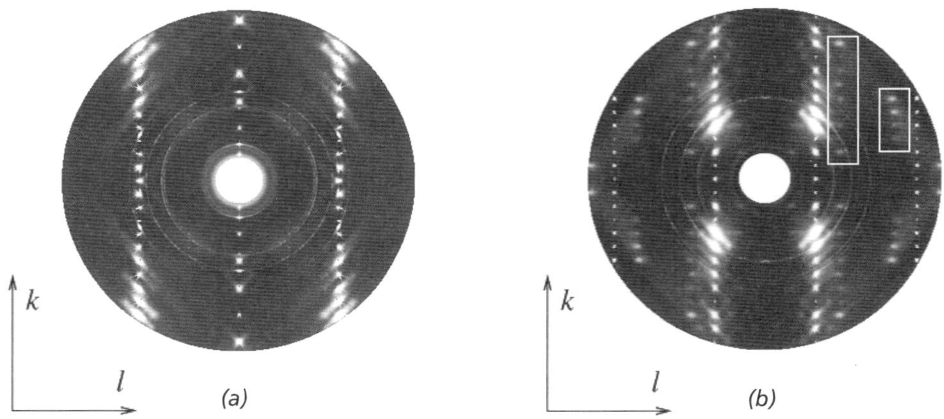

Figure 11.21. *Two cuts of observed diffuse scattering from HMTA: (a) 0kl and (b) 1kl.*

a. This is a development of an earlier model that suggested that there were positive occupancy correlations along *a* [46]. Figure 11.22 shows diffraction patterns calculated from a model in which the occupancy correlation structure noted above is implemented and the contact vectors that are more or less parallel to the azelaic acid backbones are a factor of 100 stronger than the others ("model 1"). Many of the key features of HMTA have been reproduced. The main point of disagreement here is the intensity of the spots boxed in Figure 11.22b; the relative intensity of spots on either side of a row of Bragg peaks can be adjusted through application of the size-effect.

Figure 11.23 shows the same two planes calculated from a model ("model 2") in which further springs along *a* have been added, and a size-effect on acid-to-acid contact vectors with a component

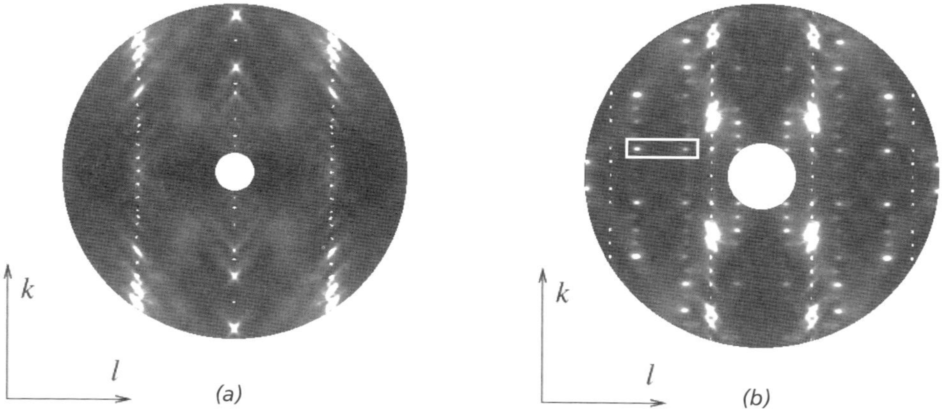

Figure 11.22. *Two cuts of calculated diffuse scattering from HMTA, model 1: (a) 0kl and (b) 1kl. Bragg peaks from the observed data have been ANDed with the calculated patterns.*

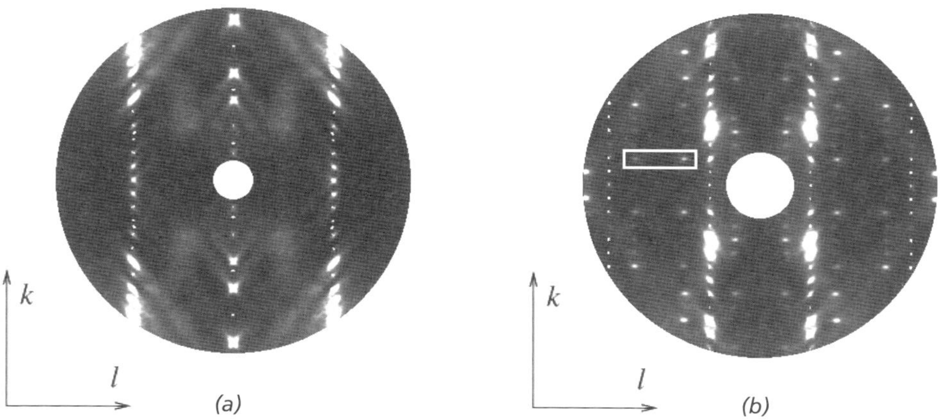

Figure 11.23. *Two cuts of calculated diffuse scattering from HMTA, model 2: (a) 0kl and (b) 1kl. Bragg peaks from the observed data have been ANDed with the calculated patterns.*

running along *c*, as noted in Figure 11.24, has also been added. This size-effect tries to pull molecules of like occupancy together while pushing molecules of unlike occupancy apart. Examination of the resulting interatomic distances from this MC simulation (Figure 11.25) shows that the size-effect has shifted the centers of the distributions away from the equilibrium value by about 0.13 Å and reduced the intensity of the outer peak of the boxed pair relative to the inner peak. Although the ratio is not as extreme as in the data, the trend is now correct.

Examining structural correlations with the final model HMTA crystal shows a range of results. The AZ molecules in HMTA show a significant short-range orientational order in which molecules adjacent in *b* and *c* prefer to be of opposite orientation (correlations of –0.5 in both directions). When it is recalled that the *b* direction nearest neighbors are separated by HMT molecules, this indicates that

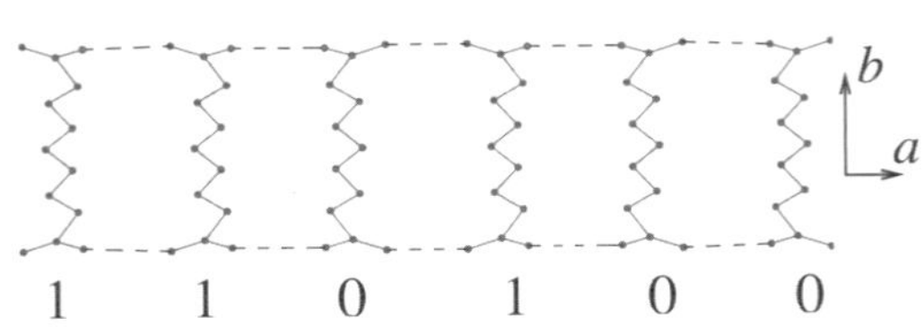

Figure 11.24. *Contacts (dashed lines) to which size-effect is applied to generate the model ("model 2") whose diffraction patterns are shown in Figure 11.23. The possible AZ orientations are denoted by 0 and 1, and 00, 01, 10, and 11 denote the four possible types of contacts.*

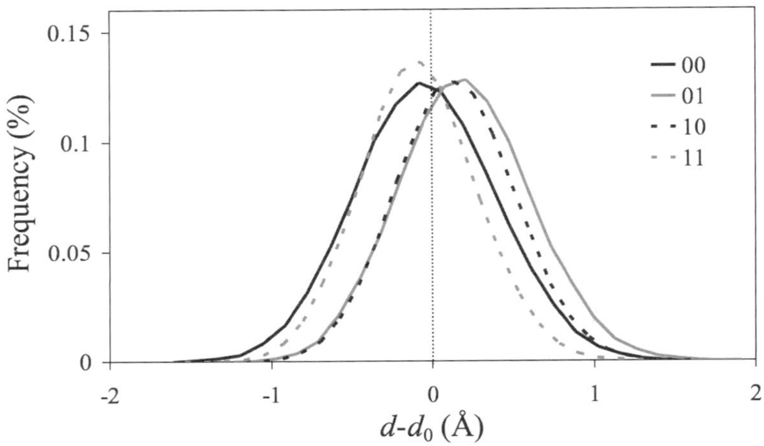

Figure 11.25. *Histograms of the lengths of the contact vectors to which the size-effect was applied. Centers of the distributions have shifted to ~±0.13 Å from the mean. Vectors are illustrated in Figure 11.24.*

the rigid HMT molecules are mediating the AZ-AZ interaction. The displacive results agree with this in that they show that there are significant correlations between AZ and HMT variables within the *bc* planes—for example, the correlation of the *x* coordinate of an HMT with that of the nearest *bc*-plane neighbor AZ is 0.17, whereas for the *y* coordinate this correlation is 0.38 and for *z* it is 0.51. These correlations coincide with the strongest displacive interactions, which were found to be in the diagonal directions within the *bc* plane along the lengths of the A molecules. Thus the crystal can be thought of as a series of weakly interacting planes stacked up along *a*.

This agrees with the occupancy structure because the best model was obtained with occupancies uncorrelated along *a*. It also agrees with the stacking-fault model of the structure of HMTA as put forward in Bonin et al.[46].

HMTA is a complex system. It contains two types of molecules showing occupational and displacive disorder. When this is combined with the possibility of molecular flexibility and size-effect, a system with a great many possible forms of disorder and SRO results. That it is possible to model the diffuse scattering at all, let alone with considerable agreement with the experiment, and then analyze the model meaningfully is encouraging.

4 Conclusions

Analysis of the diffuse scattering from molecular crystals is a complex problem, especially when the molecules show internal flexibility and occupancy structure as well as thermally activated motion. Through describing the molecule through the *z*-matrix and equilibrating a model crystal via an MC algorithm, key features of the diffuse scattering can be modeled and ascribed to the different degrees of freedom and their correlations. Although quantitative agreement between observed and calculated scattering can be difficult to obtain, valuable qualitative insights can nevertheless be found, and when good quantitative agreement *is* obtained, detailed analysis of the model can reveal details of intermolecular interactions, the energy landscape of the molecule in the crystalline environment, and even vibrational frequencies usually only available from spectroscopic techniques, such as infrared spectroscopy, Brillouin light scattering, or neutron inelastic scattering.

The MC technique is flexible in that it allows implementation and equilibration of very complicated models. Its drawback is that so many models are possible, and the many parameters that parameterize the interactions in the model must be determined, which is very far from trivial. Automated least-squares fitting [22,26,27,39] has been investigated but requires a model of good quality as a starting point, which in turn generally requires considerable user insight and a great deal of exploration of parameter space. Typically, the key interactions and degrees of freedom must be conjectured, and then the model must be explored sufficiently to determine whether it is capable of fitting the data. If it is, the model parameters may be interactively refined until good agreement with the data is reached, and then if the data are of sufficient quality, the process of automated refinement may begin. If the model is thought to be incapable of fitting the observations no matter what the parameter values may be set to, then it must be redesigned and the new model evaluated. This process can be lengthy, and strategies have been explored that it is hoped will aid in establishing an adequate model for further refinement. Genetic algorithms [48] offer one possible approach.

Another development is the use of three-dimensional data sets. At present, while the models are three-dimensional and the data collection is also three-dimensional, it is necessary to extract two-dimensional cuts of data, such as those shown here in various figures, for fitting. With the ever-increasing power of computers, it may soon be possible to directly calculate the full three-dimensional distribution of diffuse-scattered intensity in reciprocal space. Doing so, and fitting to the Bragg scattering as well as the diffuse scattering, would offer the most comprehensive test possible of any structural model and would allow a previously unattainable level of insight into the structures of materials.

The difficulties inherent in structure prediction of flexible molecular systems [18] show that the study of such systems is far from simple, and it is hoped that the analysis of diffuse scattering can begin to offer new insights into this important and diverse (and very common) class of molecular materials, including molecules of industrial and medical significance.

Acknowledgments

The authors greatly appreciate the financial support of the Australian Research Council, the Australian Synchrotron Research Program, the Access to Major Research Facilities Program, the Australian

Partnership for Advanced Computing, and the Australian Institute for Nuclear Science and Engineering. We thank A. P. Heerdegen for contributions to many aspects of this work and all coauthors of the papers that form the source material for this chapter.

References

1. T. R. Welberry, *Diffuse X-ray Scattering and Models of Disorder* (Oxford: Oxford University Press, 2004).
2. K. Lonsdale, *Acta Crystallogr.* 1 (1948): 12–20.
3. A. Guinier and R. Griffoul, *Acta Crystallogr.* 1 (1948): 188–193.
4. T. Weber, M. A. Estermann and H.-B.Bürgi, *Acta Crystallogr.* B57 (2001): 579–590.
5. T. R. Welberry, D. J.Goossens, A. P. Heerdegen, and P. L. Lee, *Z. Krist.* 220 (2005): 1052–1058.
6. T. R. Welberry, D. J. Goossens, W. I. F. David, M. J. Gutmann, M. J. Bull, and A. P. Heerdegen, *J. Appl. Cryst.* 36 (2003): 1440–1447.
7. D. A. Keen, M. J. Harris, and W. I. F. David, *Physica B* 241–243 (1998): 201.
8. D. Hohlwein, J.-U. Hoffmann, and R. Schneider, *Phys. Rev. B* 68 (2003): 140408.
9. W. Schweika and P. Boni, *Physica B* 297 (2001): 155–159.
10. W. Schweika, *Disordered Alloys: Diffuse Scattering and Monte Carlo Simulations* (New York: Springer Verlag, 1998).
11. M. A. Estermann and W. Steurer, *Phase Trans.* 67 (1998): 165–195.
12. S. Scheidegger, M. A. Estermann, and W. Steurer, *J. Appl. Cryst.* 33 (2000): 35–48.
13. D. A. Keen, M. J. Gutmann, and C. C. Wilson, *J. Appl. Cryst.* 39 (2006): 714–722.
14. S. K. Sinha, *Z. Kristallogr.* 219 (2004): 143–147.
15. L. Vasiliu-Doloc, S. Rosenkrantz, R. Osborn, S. K. Sinha, J. W. Lynn, J. Mesot, O. H. Seeck, G. Preosti, A. J. Fedro, and J. F. Mitchell, *Phys. Rev. Lett.* 83 (1999): 4393–4396.
16. T. J. Hicks, *Magnetism in Disorder* (Oxford: Oxford University Press,1995).
17. J. R. Stewart, K. H. Andersen, R. Cywinski, and A. P. Murani, *J. Appl. Phys.* 87 (2000): 5425–5430.
18. G. M. Day, W. D. S. Motherwell, H. L. Ammon, S. X. M. Boerrigter, R. G. D. Valle, E. Venuti, A. Dzyabchenko, J. D. Dunitz, B. Schweizer, B. P. van Eijck, et al., *Acta Crystallogr.* B61 (2005): 511–527.
19. B. D. Butler and T. R. Welberry, *J. Appl. Cryst.* 25 (1992): 391–399.
20. T. Proffen and R. B. Neder, *J. Appl. Cryst.* 30 (1997): 171–175.
21. W. J. Hehre, L. Radom, P. R. Schleyer, and J. A. Pople, *Ab Initio Molecular Orbital Theory* (New York: John Wiley, 1986).
22. D. J. Goossens, T. R. Welberry, and A. P. Heerdegen, *Z. Krist.* 220 (2005): 1035–1042.
23. N. Metropolis, A. W. Rosenbluth, M. N. Rosenbluth, A. H. Teller, and E. Teller, *J. Chem. Phys.* 21 (1953): 1087–1092.
24. L. H. Thomas, T. R. Welberry, D. J. Goossens, A. P. Heerdegen, M. J. Gutmann, S. J. Teat, C. C. Wilson, P. L. Lee, and J. M. Cole, *Acta Crystallogr.* B63 (2007): 663–673.
25. T. R. Welberry, T. Proffen, and M. Bown, *Acta Crystallogr.* A54 (1998): 661–674.
26. D. J. Goossens, T. R. Welberry, A. P. Heerdegen, and M. J. Gutmann, *Acta Crystallogr.* A63 (2007): 30–35.
27. T. R. Welberry, D. J. Goossens, A. J. Edwards, and W. I. F. David, *Acta Crystallogr.* A57 (2001): 101–109.
28. M. More, G. Odou, and J. Lefebvre, *Acta Crystallogr.* B43 (1987): 398.
29. J. C. Tolédano, *Phys. Rev. B* 20 (1979): 1147.
30. H. Terauchi, T. Kojima, K. Sakaue, and F. J. Tajiri, *Chem. Phys.* 76 (1982): 612.
31. P. G. Odou, M. More, and V. Warin, *Acta Crystallogr.* A34 (1978): 459.
32. D. J. Goossens, X. Wu, and M. Prior, *Solid State Commun.* 136 (2005): 543–545.
33. D. J. Goossens, T. R. Welberry, M. E. Hagen, and J. A. Fernandez-Baca, *Phys. Rev. B* 73 (2006): 134116.
34. K. Lonsdale and H. Smith, *Proceedings of the Royal Society of London, Series A. Mathematical and Physical Sciences* 179 (1942): 8–50.
35. J. C. Osborn and T. R. Welberry, *J. Appl. Cryst.* 23 (1990): 476–484.
36. D. J. Goossens, A. P. Heerdegen, T. R. Welberry, and M. J. Gutmann, *Physica B* 385–386 (2006): 1352–1354.

37. J. Hanuza, W. Sąsiadek, E. Kucharska, J. Michalski, M. Mączka, A. A. Kaminskii, A. A. Koronienko, E. B. Dunina, H. Klapper, J. Hulliger et al., *J. Raman Spectrosc.* 35 (2004): 224–235.
38. T. R. Welberry and A. P. Heerdegen, *Acta Crystallogr.* B59 (2003): 760–769.
39. D. J. Goossens, A. P. Heerdegen, T. R. Welberry, and A. G. Beasley, *Int. J. Pharm.* 343 (2007): 59–68.
40. P. A. Altin and D. J. Goossens, *Proceedings of the 31st Annual AIP Condensed Matter and Materials Meeting*, Feb. 2007, ISSN–1037–1214, http://www.aip.org.au/wagga2007/2007_2.pdf.
41. J. F. McConnell, *Cryst. Struct. Commun.* 3 (1974): 73.
42. L. K. Hansen, G. L. Perlovich, and A. Bauer-Brandl, *Acta Crystallogr.* E59 (2003): 1357.
43. N. Shankland, A. J. Florence, P. J. Cox, C. C. Wilson, and K. Shankland, *Int. J. Pharm.* 165 (1998): 107.
44. I. Tanaka, F. Iwasaki, and A. Aihara, *Acta Crystallogr.* B30 (1974): 1546–1549.
45. M. Hostettler, H. Birkedal, M. Gardon, G. Chapuis, D. Schwarzenbach, and M. Bonin, *Acta Crystallogr.* B55 (1999): 448–458.
46. M. Bonin, T. R. Welberry, M. Hostettler, M. Gardon, H. Birkedal, G. Chapuis, P. Möckli, C. A. Ogle, and K. J. Schenk, *Acta Crystallogr.* B59 (2003): 72–86.
47. D. J. Goossens and T. R. Welberry, *Comput. Phys. Commun.* 142 (2001): 387–390.
48. T. Weber and H.-B. Bürgi, *Acta Crystallogr.* A58 (2002): 526–540.

12

Diffuse Scattering and Phason Modes in Icosahedral Quasicrystals

Marc de Boissieu[1]

Sonia Francoual[2]

1. *Sciences de l'Ingéniérie, des Matériaux et des Procédés (SIMaP), Grenoble INP, CNRS, UJF, BP 75 38402 St Martin d'Hères Cedex, France*
2. *National High Magnetic Field Laboratory, Los Alamos National Laboratory, Los Alamos, NM 87545, USA*

Abstract

Quasicrystals (QCs) are long-range ordered materials but are not periodic. Their diffraction pattern displays sharp Bragg peaks as a signature of the long-range order but with rotational symmetries incompatible with lattice periodicity. Quasiperiodicity yields new, long-wavelength excitations called phasons, which are diffusive modes. In this chapter we review the experimental results providing evidence for the existence of diffusive phason modes in QCs with the icosahedral (i-) symmetry. After an introduction to the structural crystallographic models, the hydrodynamic and elasticity theories generalized to QCs are presented, which allow for a computation of the diffuse scattering caused by the phason fluctuations. The neutron and X-ray diffuse scattering measuring techniques are discussed in which we stress the assets of carrying out the measurements on an absolute scale. Experiments in the i-AlPdMn QC phase are reported where the phason elastic constants could be determined from an analysis of the phason diffuse scattering distribution and the diffusive dynamics of phason modes evidenced using the X-ray photocorrelation spectroscopy technique. Temperature studies in i-AlPdMn suggest an entropy stabilization of QCs. Discussions of other icosahedral phases bring out the fact that the phason diffuse scattering distribution and the phason spectrum are a unique feature of each of the investigated systems.

1 Introduction

Quasicrystals have been puzzling the scientific community since their discovery by Shechtman et al. more than twenty years ago [1]. Indeed, their diffraction pattern presents extremely sharp Bragg peaks, which are a signature of long-range order, but with rotational symmetries, such as the icosahedral one, incompatible with lattice periodicity. Although the first QCs were obtained by rapid quench from the melt, they can now be obtained as equilibrium phases in a few systems and grown as large, single grains using the Bridgman or the Czochralski technique. This is illustrated in Figure 12.1, which shows a single crystal in the ZnMgY system with clear dodecahedral facets and a fivefold X-ray Laue diffraction pattern obtained with a single grain of the i-AlPdMn phase. The first question to be answered is, of course, "Where are the atoms in a QC?" It is only recently that this question could get a satisfactory answer in the CdYb system, where a detailed understanding of the structure has been achieved.

As for any ordered material, the ideal description of the atomic structure of a QC gives rise to a diffraction pattern containing only Bragg peaks. Of course, a lot of phenomena may give rise to diffuse scattering in between the Bragg reflections: thermal vibrations, point defects, chemical disorder, dislocations, and so on. However, there are fluctuations specific to QCs called phasons, which also lead to diffuse scattering. As will be shown subsequently, phason modes are a key "ingredient" in the understanding of QCs, and they have been widely studied using diffuse scattering measurement techniques.

In this chapter we review studies of phason modes carried out in different icosahedral QCs. A detailed discussion of the diffuse scattering observed in decagonal phases can be found in Chapter 13 by T. Weber and W. Steurer and Chapter 14 by H. Abe. It is interesting to put these studies on phason modes in QCs in the broader perspective of phason modes in aperiodic crystals, where aperiodic crystals encompass incommensurately modulated phases, intergrowth or composites structures, and QCs. (A detailed introduction can be found in Janssen *et al.* [2] and a detailed review in de Boissieu *et al.* [3].) Indeed, although they are very different systems, phason modes originate from a same description of

Figure 12.1. *Left panel: Single crystal obtained by the flux method in the ZnMgY system. (Courtesy of P. Canfield [4].) Right panel: Fivefold Laue diffraction pattern obtained in the icosahedral AlPdMn phase. (Courtesy of W. Steurer.)*

aperiodic phases structures in a higher dimensional space. In section 2, we give a brief introduction to the high-dimensional description of aperiodic crystals using one-dimensional model systems and a few results on refined quasicrystalline atomic structures. Next, in section 3, we follow theoretical results on phasons modes in QCs: first, the hydrodynamic theory and its predictions on the diffuse scattering, and second, the atomic models. We also address in this section experimental techniques when measuring diffuse scattering—in particular the absolute scale normalization of the diffuse intensity. Section 4 is dedicated to results obtained in the i-AlPdMn phase: this is the only phase for which a complete study, including phason dynamics, has been carried out. Results obtained in other icosahedral phases are presented in section 5.

2 High-Dimensional Description of Quasicrystals

Quasicrystals belong to the broader category of aperiodic crystals, which encompasses incommensurately modulated (positional or chemical) phases, composites or intergrowth phases, and QCs. All aperiodic crystals can be described in a higher dimensional space where lattice periodicity is recovered [2]. This is illustrated in the simple case of a one-dimensional structure. Let us consider two model systems: (1) a one-dimensional incommensurately displacive modulated phase and (2) the one-dimensional Fibonacci chain, which is a one-dimensional QC toy model (although it does not include any "forbidden" rotational symmetry such as the fivefold symmetry). The modulated phase is described as a perturbation of a periodic structure by a sine wave. If the period of the modulation is incommensurate with the period of the lattice, the resulting structure is aperiodic. This can be represented in a two-dimensional periodic space, as illustrated in the left panel of Figure 12.2. The two-dimensional periodic lattice decomposes into the one-dimensional physical space, or parallel space $\mathbf{E}_{par}$, and a one-dimensional complementary perpendicular space $\mathbf{E}_{per}$. The two-dimensional lattice is decorated by a wavy line along the perpendicular direction, which holds for the modulation. The one-dimensional modulated phase is obtained as a cut by the parallel space of the decorated two-dimensional lattice.

The one-dimensional Fibonacci chain is characterized by two "tiles": a long one, L, and a short one, S, of length ratio L/S equal to τ, the golden mean, equal to 1.618.... One way of constructing

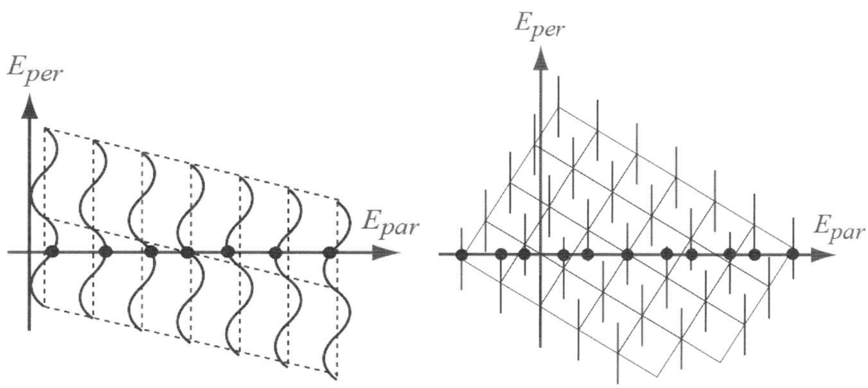

Figure 12.2. *Two-dimensional picture of a one-dimensional modulated phase (left panel) and the Fibonacci chain (right panel).*

the Fibonacci chain is by using the following substitution rules: $L \rightarrow LS$ and $S \rightarrow L$. When iterated at infinity, this leads to the chain *LSLLSLSLLSLLS...*, whose diffraction pattern consists of Bragg peaks. However, each Bragg peak can only be indexed using a linear combination of two reciprocal wave vectors whose lengths are incommensurate. This is the basis of the two-dimensional periodic picture of the one-dimensional QC shown Figure 12.2. The left panel illustrates the case of an incommensurately modulated phase, whereas the right one is for the Fibonacci QC chain. As for the modulated phase, the periodic square lattice is decomposed in two orthogonal, one-dimensional subspaces: the parallel space $\mathbf{E}_{par}$ or physical space and the perpendicular space $\mathbf{E}_{per}$. The slope of the parallel space is irrational with respect to the square lattice and is equal to $1/\tau$. Each node of the square lattice is decorated by a segment line lying along $\mathbf{E}_{per}$, of which the length is equal to the projection of the square lattice. The Fibonacci chain is obtained as the intersection of the decorated two-dimensional periodic lattice with the physical space.

This high-dimensional periodic description is very useful for determining the atomic structure and for understanding phason modes. This technique has been intensely developed for QCs with sophisticated refinement procedures, but it is only recently that a satisfactory atomic model has been obtained in the CdYb phase [5]. In the case of the icosahedral symmetry, the periodic space is six-dimensional because the diffraction pattern of the icosahedral QC can be indexed by a linear combination of six wave vectors pointing toward the vertices of an icosahedron. The parallel and perpendicular spaces are three-dimensional, and the analogs of the segment lines are three-dimensional objects called atomic surfaces.

The discovery of the first binary stable icosahedral QC $Cd_{5.7}Yb$ has been a breakthrough, making structural and chemical order analysis much simpler [6]. Moreover, for chemical compositions very close to the one of the icosahedral phase, periodic approximants can be synthesized. In the high-dimensional description, this corresponds to a tilt of the $\mathbf{E}_{par}$ space until it reaches a rational slope approximating the golden mean. It can be equivalently expressed in terms of a shear strain of the periodic lattice, with the shear strain along the perpendicular space direction: this is called a linear phason strain. The first rational approximations of $\tau = 1.618...$ are 1/1 and 2/1, and the Cd_6Yb 1/1 and the $Cd_{5.8}Yb$ 2/1 approximants have been successfully synthesized. These are cubic phases with a lattice parameter equal to 1.57 nm and 2.53 nm, respectively. Their structures have been solved using standard crystallographic techniques and can simply be described by a periodic packing of a large atomic cluster of diameter 1.57 nm containing 158 atoms [7,8]. The large clusters are connected along their twofold axes, where they share a face, and along their threefold axes, where they overlap. The different shells constituting the cluster are shown in Figure 12.3. Note that the shells are perfectly ordered chemically, with Yb atoms sitting on the vertices of an icosahedron. The structure of the icosahedral CdYb phase is built up with the same cluster connected along the same twofold and threefold axis but distributed quasiperiodically [5]. The distribution of the cluster centers in a fivefold plane is shown in Figure 12.3. As can be seen, packing is hierarchical with clusters grouped in a "cluster of clusters," forming a tenfold ring highlighted in light gray, which itself forms an inflated cluster.

3 Theory of Phason Modes

The name phasons was coined in 1971 by Overhauser [9] in the context of charge-density wave (CDW) systems. Studying the lattice excitations of an incommensurate CDW, he demonstrated that, because the free energy of the system is invariant under a phase shift of the modulation, there are new lattice

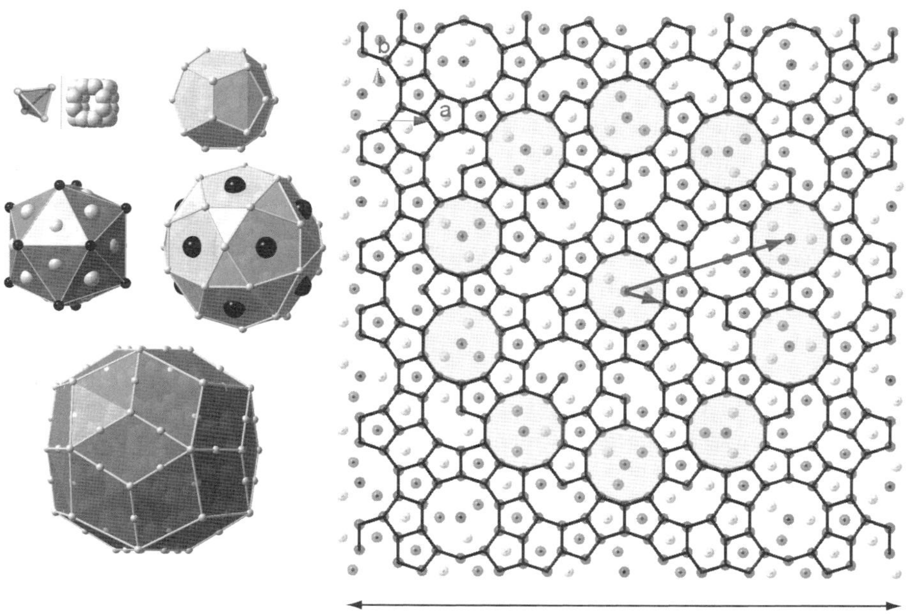

Figure 12.3. *Left panel: The successive shells of the large triacontahedral cluster containing 158 atoms in i-CdYb. Black spheres stand for Yb atoms. Right panel: Distribution of the cluster centers in a fivefold plane. The light gray disk emphasizes a cluster of clusters (tenfold ring), which forms an inflated cluster as shown by the two arrows. (From Takakura et al. [5].)*

excitations, which he named phasons, showing up around the satellite reflections and whose dispersion is acoustic-*like*. The nature and damping of those modes has been intensely studied at a microscopic level in a few incommensurately modulated phases [10].

Soon after the discovery of QCs, several groups pointed out the importance of phason modes [11–14]. Phason modes in QCs are a consequence of the hydrodynamic theory, which applies for all aperiodic phases. Their microscopic interpretation is, however, still a matter of debate.

3.1 Hydrodynamic Theory

The hydrodynamic theory is a continuum theory based on symmetry-breaking analysis that deals with long-wavelength excitations for which the frequency ω goes to zero as a function of some power of the excitation wavevector $\mathbf{q}$ in the limit $q \to 0$, that is, in the long-wavelength limit. The hydrodynamic theory of aperiodic crystals is best understood in the high-dimensional picture (see Figure 12.2). If the parallel space is translated along the perpendicular direction, a new aperiodic structure is generated. However, the new structure is indistinguishable from the initial one; in particular, its free energy is unchanged. This invariance of the free energy with respect to the perpendicular translation is referred to as the "phason degree of freedom" and is the exact counterpart of the translation invariance along the physical space (or $\mathbf{E}_{par}$) of the free energy of any solid (the translation being any vector). Looking at Figure 12.2, it can be realized that the translation along $\mathbf{E}_{per}$ leads to a change of the phase of the modulation in the case of the modulated phase. For the Fibonacci chain, the situation at the atomic scale is different and leads to local rearrangements such as $LS \rightarrow SL$, called "phason flips." This invariance of the free energy along perpendicular space is valid for all aperiodic phases and is a direct consequence of the parallel space having an irrational slope.

Using an expansion in density waves and continuity arguments, the hydrodynamic theory can then be developed [11,15–17]. From mode-counting arguments it can be shown that for *all aperiodic crystals*, there are phason modes that are not propagative modes but collective diffusive modes for which one can write

$$-i\omega = D_{Phason} q^2, \tag{12.1}$$

where D_{Phason} is the phason diffusion constant, which means that the phason mode amplitude decays exponentially with time with a characteristic time τ inversely proportional to q^2. We stress here that the diffusive nature of phason modes also holds for incommensurately modulated phases and is not specific to QCs. In fact, anharmonicity is crucial to reconcile the hydrodynamic and microscopic pictures of phason modes in modulated phases [3,18]. Indeed, in the case of incommensurately modulated phases, phason modes have a finite lifetime independent of q, leading to two regimes: a long-wavelength one for which the phason is diffusive, as predicted by hydrodynamics, and a shorter wavelength one for which the phason is a damped propagative mode.

The hydrodynamic theory relies on the hypothesis of a continuum approximation for which an infinitesimal displacement of the parallel space along the perpendicular direction does not cost energy. The validity of this hypothesis has been studied in detail in the Frankel-Kontorova-type model, which is a microscopic one-dimensional model for composite crystals where two sublattices interact and whose lattice constants are incommensurate. When the coupling between the two lattices is strong, this leads to the so-called analyticity breaking and the formation of a phason gap as q goes to zero [3,10]: in this case, the hydrodynamic theory is no longer valid. This problem has been dealt within a QC model system [19].

Within this continuum approximation, an elasticity theory of QCs can then be derived introducing the appropriate symmetry analysis. The elasticity relations of the icosahedral phase can be derived in the density-wave picture. The simplest atomic density requires six waves with wave vectors pointing toward the six vertices of an icosahedron with unfixed phases parameterized by $\mathbf{u}_{par}$ and $\mathbf{u}_{per}$ and the rigid translation along the parallel and perpendicular space. The theory relies on the two following hypotheses:

1. The free energy of the system is invariant under a uniform increment of $\mathbf{u}_{par}$ and $\mathbf{u}_{per}$.
2. An analytic expansion of the free energy is possible. Because we are at equilibrium, the first term in the expansion will be quadratic in the spatial gradient of $\mathbf{u}_{par}$ and $\mathbf{u}_{per}$, that is, $F = \alpha\,|\nabla \mathbf{u}_{par}|^2 + \beta\,|\nabla \mathbf{u}_{per}|^2$.

This means that there is a quadratic restoring force if the QC is brought out of equilibrium by applying a phason strain $\nabla \mathbf{u}_{per}$.

Using group theory arguments, it can be shown that the elastic free energy is the sum of three terms $F = F_{par} + F_{per} + F_{par,per}$. The first term is proportional to the squared "phonon" strain. It depends on the two Lamé coefficients λ and μ; because of the high symmetry of the icosahedral point group, this term is analogous to what is obtained for an isotropic solid. The second term is proportional to the squared "phason" strain and depends on two phason elastic constants, K1 and K2. Finally, the third term depends on both the phonon and phason strain and on a phonon-phason coupling constant, K3. Five elastic constants are thus required to account for the elasticity of icosahedral phases [11–13,17].

3.2 Diffuse Scattering and Phason Modes

Based on the elasticity theory of icosahedral QCs, the effect of equilibrium phonon and phason fluctuations on the diffraction pattern can be computed [20]. This follows the very same approach as the one used for computing the thermal diffuse scattering (TDS) due to phonons, which can be found in the textbook by Chaikin and Lubensky [21].

The scattered intensity at a point $\mathbf{Q}$ in reciprocal space is

$$S(\mathbf{Q}) = S_{Bragg}(\mathbf{Q}) + S_{Diff}(\mathbf{Q}), \tag{12.2}$$

where the first term corresponds to the Bragg scattering and the second one to the diffuse scattering. This expression holds only if the fluctuations leading to the diffuse scattering remain bounded as the system size R goes to infinity. More precisely, the system is described by a time and space average of the atomic density $<\rho(\mathbf{r})>$ and deviations with respect to this average, characterized by the two-body correlation function $<\rho(\mathbf{r}_i)\rho(\mathbf{r}_j)> - <\rho(\mathbf{r})>^2$. The two terms in Equation 12.2 correspond to the squared Fourier transform of the average density and to the Fourier transform of the two-body correlation, respectively.

The Bragg part of the scattered intensity is computed considering that the average structure is obtained by the convolution of Gaussian fluctuations with the "ideal" (i.e., without fluctuations) decorated lattice. In the case of QCs, the Bragg component is easily calculated in the high-dimensional periodic lattice decorated with atomic surfaces. Phonon fluctuations lead to a Gaussian broadening of the atomic surfaces along the parallel direction with a mean square amplitude $<u_{par}^2>$. Phason fluctuations lead to a Gaussian broadening in the perpendicular direction with a mean square amplitude $<u_{per}^2>$. Indeed, a phason fluctuation may be described equivalently either by an undulation of the cut space or by a fluctuation $\mathbf{u}_{per}(\mathbf{R})$ of atomic surfaces along the perpendicular direction. The Bragg peaks are thus delta functions located on the reciprocal lattice indexed with six-dimensional vectors $\mathbf{Q}_{6D} = (\mathbf{Q}_{par}, \mathbf{Q}_{per})$ and have an intensity of

$$S_{Bragg}(\mathbf{Q}_{par}) = S_{ideal}(\mathbf{Q}_{par}).\exp\left(-<u_{par}{}^2>Q_{par}{}^2\right).\exp\left(-<u_{per}{}^2>Q_{per}{}^2\right) \tag{12.3}$$

where $S_{ideal}(\mathbf{Q}_{par})$ is the Bragg intensity of the ideal QC without any fluctuations and where we have neglected the phonon-phason coupling term. The values of $<u_{par}{}^2>$ and $<u_{per}{}^2>$ can be computed from the hydrodynamic matrix $C(\mathbf{q})$ and correspond to the "phonon" and "phason" Debye-Waller terms. Phason mode fluctuations thus introduce a new Debye-Waller term. It can be shown that for three-dimensional QCs the phonon and phason Debye-Waller terms remain bounded, preserving the quasiperiodic long-range order.

The diffuse scattering measured at the point $\mathbf{Q}_{par} + \mathbf{q}$ can be computed within the hydrodynamic theory. It can be shown that it is related to the hydrodynamic matrix $C(\mathbf{q})$, by

$$S(\mathbf{Q}_{par}+\mathbf{q})=S_{Bragg}(\mathbf{Q}_{par}) < \mathbf{Q}_{6D} \,|\, C^{-1}(\mathbf{q})|\mathbf{Q}_{6D} >, \qquad (12.4)$$

In the case where the K3 phonon-phason coupling term can be neglected, this expression is written

$$S(\mathbf{Q}_{par}+\mathbf{q}) = S_{Bragg}(\mathbf{Q}_{par}).<\mathbf{Q}_{par}|C_{par,par}{}^{-1}(\mathbf{q})|\mathbf{Q}_{par}> + S_{Bragg}(\mathbf{Q}_{par}).<\mathbf{Q}_{per}|C_{per,per}{}^{-1}(\mathbf{q})|\mathbf{Q}_{per}>. \qquad (12.5)$$

The first term in this formula corresponds to the usual TDS, whereas the second term corresponds to the contribution of phason modes to the diffuse scattering and is sometimes referred to as phason diffuse scattering (PDS).

Although the TDS and PDS parts in Equation 12.5 have a very similar analytical form, they lead to very different intensity distributions.

The TDS intensity decays as $1/q^2$ along any given direction $\mathbf{q}$ from a Bragg peak positioned at $\mathbf{Q}$ having an intensity $I_{Bragg}(\mathbf{Q})$, and it scales as $I_{Bragg}(\mathbf{Q}).\, Q_{par}{}^2$. Because of the icosahedral symmetry featuring an isotropic medium, whatever $\mathbf{q}$, the three eigenvectors $\mathbf{e}_{par,i}(\mathbf{q})$ of the $C_{par,par}(\mathbf{Q})$ matrix in Equation 12.5, i = 1, 2, 3, have a simple angular relation to $\mathbf{q}$, being either collinear or orthogonal to it, and three single eigenvalues $\alpha_i(\mathbf{q}, \lambda, \mu)$ follow. This and the following scalar product $\Sigma_i\ (\mathbf{Q}_{par}.\mathbf{e}_{par,i})^2 / \alpha_i(\mathbf{q},\lambda,\mu)$ concealed beneath the first term in Equation 12.5 both yield a very recognizable TDS intensity distribution around Bragg peaks having, whatever $\mathbf{Q}$, the shape of a three-dimensional ellipsoid elongated in the two transverse directions to $\mathbf{Q}$.

The PDS intensity behaves differently, and three major features allow one to identify it unambiguously [20]:

1. The PDS intensity decays as $1/q^2$ along any given particular direction $\mathbf{q}$ from a Bragg peak positioned at $\mathbf{Q}$ and having an intensity $I_{Bragg}(\mathbf{Q})$.
2. It scales as $I_{Bragg}(\mathbf{Q}).Q_{per}{}^2$.
3. The PDS intensity distribution is highly anisotropic around a given Bragg peak, and it is different from one Bragg peak to another one as a consequence of the complex $\mathbf{q}$ dependence of the eigenvectors $\mathbf{e}_{per,i}(\mathbf{q})$ and eigenvalues $K_i(\mathbf{q}, K1, K2)$ of the $C_{per,per}(\mathbf{Q})$ matrix in Equation 12.5. It can be further stated that the anisotropy is mainly driven by the value of the ratio K2/K1 of the two phason elastic constants K1 and K2, which ranges between –0.60 and 0.75 in the icosahedral symmetry case. Whenever K2/K1 is negative and close to the lower limit value of –0.6, the PDS intensity is shown to be strengthened along directions parallel to the threefold axes around Bragg peaks, whereas whenever it is positive and close to the upper limit value of +0.75, the strengthening occurs along directions parallel to the fivefold axes around Bragg peaks.

We stress here that all three above conditions must be experimentally verified so that the measured diffuse scattering signal can be unambiguously attributed to phason modes.

The evolution of the diffuse scattering with temperature and its relationship with phase transitions has also been studied within the Landau theory [22–24] when a phason strain is applied and will be presented in section 4.2.1

3.3 Measuring Phason Diffuse Scattering

In any diffuse scattering measurement, minimizing the background and parasitic scattering is of great importance, so in general it is advisable to place the sample under vacuum to suppress the air scattering contribution. In the following, we insist on experimental details that are crucial for a proper characterization of the PDS.

The first complication in Equation 12.5 is that it contains two contributions that might be difficult to disentangle. The situation is thus much simplified if the measurement contains only the PDS contribution. This can be achieved using elastic neutron scattering and an energy analysis that eliminates the TDS contribution. This is best realized on a triple axis instrument, preferably using a cold source, which gives rise to a good Q and energy resolution. The measurement is then carried out at the elastic (zero-energy transfer) position of the triple axis, ensuring that only the PDS contribution is measured. This is the best experimental solution; however, it requires centimeter-size single-grain QC samples.

For measurements carried out at high temperatures (see section 4.2.2.), it is particularly important that the TDS does not contribute to the scattered intensity because the mean square atomic displacement $< u_{par}{}^2 >$ due to phonon modes grows as T in a first approximation and so does the TDS intensity.

When neutron scattering measurements are not feasible, an alternative is to carry out low-temperature X-ray measurements to reduce the TDS contribution. However, because of the zero-point motion of atoms, the TDS contribution is never zero. Diffuse X-ray scattering measurements carried out on an absolute scale are thus better suited. Absolute-scale measurements have been widely used for the study of TDS and phonon modes in metals [25–31] and in the field of order-disorder transitions in metallic alloys [30]. In such measurements, the collected scattered Bragg and diffuse scattering intensities are corrected so that the sample's structure factor and the sample's diffuse scattering cross section for one atom (or one sample volume element) are determined and expressed in electron units per atom (per $Å^3$). The sample's diffuse scattering cross section can then be compared directly to Equation 12.5 and the PDS contribution separated from the TDS contribution. Indeed, because the phonon elastic constants are known most of the time, the TDS is easily computed. Experimentally, absolute scale measurement of the diffuse scattering intensity requires an accurate knowledge of the values of the incident flux, of the irradiated sample volume, of the aperture of the detector slits, and of the distance between the sample and the detector during the measurement. Of importance is the fact that such measurements are the unique experimental road to a quantitative determination of the phason elastic constants in electron units/atom, thus allowing for a direct comparison with results from simulations on available atomic models.

When determining the phason elastic constants, not only must the $I_{Bragg}(\mathbf{Q}).Q_{per}{}^2 / q^2$ dependence be evidenced, but its **q** dependence must be tracked in a considered way [32]. At a different level, it is also important to get a reasonable estimate over which q range the hydrodynamic theory holds true, that is, on which range the $1/q^2$ decay of the intensity is observed. This imposes two requirements: on the low-q side, one must be sure that the contamination from the Bragg peak tail is not present, and on the high-**q** side, an accurate measurement of the weak signal and an estimate of the background must be carried out. The first criterion is best achieved with a high-Q resolution setup. On the other hand, the high-q region, where the signal becomes weak, is best measured with a coarse resolution, which will maximize the flux. As is usual in such a measurement, the best compromise is found out after successive tests consisting of changes in the instrumental resolution. For X-ray measurements carried out at the European Synchrotron Radiation Facility (ESRF), the measurable q range goes roughly from 0.03 Å^{-1} to 0.6 Å^{-1}. (*Note*: In the following, reciprocal space units are calculated as $Q = 4\pi\sin\theta/\lambda_X$, where 2θ is the scattering angle and λ_X the X-ray wavelength, and thus include the term 2π, so that the interlayer spacing is $d = 2\pi/Q$). For a fixed direction of **q**, Q-scans are best achieved with a point detector such as a scintillator and a set of appropriate attenuators, which can be introduced automatically during the scan. The entire intensity scale is measured this way, which in the case of X-ray synchrotron radiation might enlarge upon ten orders of magnitude, as shown in Figure 12.4.

As mentioned above, the shape anisotropy of the diffuse scattering is an important feature to evidence. The use of two-dimensional detectors might be thought as appropriate considering the experimental time issues. However, the large difference in magnitude between the Bragg intensities and the diffuse scattering intensities turns out to be a major difficulty in their use because charge-coupled

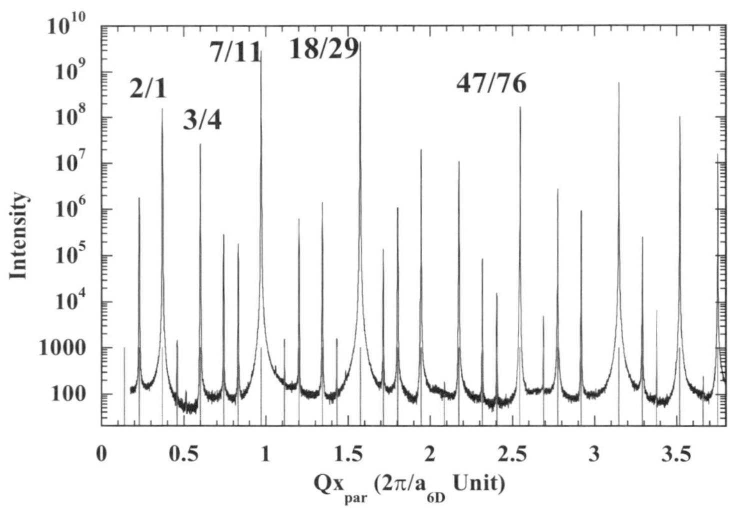

Figure 12.4. *Q-scan measured in the i-AlPdMn QC along a fivefold axis. The measurement has been carried out on the D2AM beamline at the ESRF. The intensity scale is per monitor counts and is roughly equivalent to the number of photons per second. The horizontal axis corresponds to the Q_x component of the reciprocal wave vector expressed in $2\pi/a_{6D}$, where a_{6D} is the six-dimensional lattice parameter equal to about 6.451 Å. A shorthand notation with two indices (N/M) is used to label the Bragg peaks. Note the non periodic distribution of the Bragg peaks and the large dynamical range between the strongest and weakest Bragg peaks. The diffuse scattering is well visible around the 18/29 and 7/11 peaks.*

devices (CCD), cameras, and imaging plates have a limited dynamical range of sixteen bits only. Also, the trajectory of the detector surface is not flat in the reciprocal space, making the **q** dependence analysis rather difficult. This is why the use of a point detector with a set of automated attenuators remains a pertinent solution. Note that the recent development of the so-called pixel detectors with high dynamical range and photon counting capability will certainly change the strategy for diffuse scattering measurements.

Finally, we stress that measuring the diffuse scattering on an absolute scale is the only way to have a quantitative comparison of the "amount" of diffuse scattering in different samples. This may turn out to be important when one is about to discuss the effects of heat treatment on a particular compound and when comparing different samples.

3.4 Atomic Models

As stated above, the hydrodynamic theory of QCs is a continuum theory, which does not give any clue on the precise atomistic realization of phason modes. However, from the high-dimensional description of QCs, phason modes are associated with fluctuations along the perpendicular space, leading to atomic jumps, which are somewhat correlated, although phason modes are not propagative.

The first studied atomic model was the three-dimensional version of the Penrose tiling. It is built up with two rhombuses: a "fat" and a "thin" one. The elementary "phason jump" is obtained inside a rhombic dodecahedron made out of four tiles (two of each kind) following a mirror symmetry (Figure 12.5 left panel). As a result, a vertice is moved in another position with the same local environment. The model studied by Tang [33] considers the ensemble of configurations reached by applying such a tile permutation: this has been called a maximally random tiling by Henley [34]. This should be considered, however, as a toy model because the Hamiltonian is given by the existence of the tiling only, regardless of the decoration and the interatomic forces. The interesting outcome of this simulation is that, in agreement with the hydrodynamic and elasticity theories, its diffraction pattern contains Bragg peaks and diffuse scattering. Since only phason modes are present, the diffuse scattering can be computed

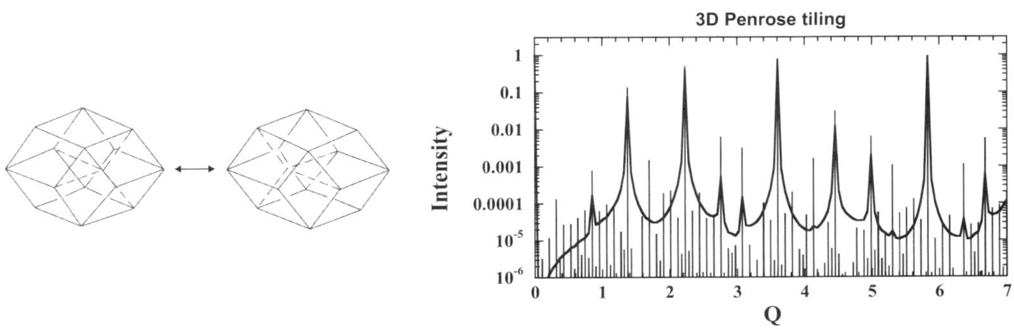

Figure 12.5. *Left panel: Phason flip inside a rhombic dodecahedron in the three-dimensional Penrose tiling. Right panel: Diffraction pattern of the ideal three-dimensional Penrose tiling (vertical bars) and of the randomized Penrose tiling (solid line). As can be seen, the diffraction pattern of the random tiling contains both a Bragg scattering part and a diffuse scattering part. The diffuse scattering part obeys the hydrodynamic theory. (From Henley et al. [37].)*

theoretically (Equation 12.5) and has been shown to be in agreement with the simulation once the proper phason elastic constants K1 and K2 are introduced. In this simple model, the "restoring" force in the free-energy expansion is the tile entropy. Indeed, the entropy is maximal in the random tiling QC and will diminish as soon as a phason strain is applied [34]. More sophisticated models have been studied since then [35], especially for decagonal QCs in which ab-initio-based calculations have been carried out. The entropy stabilization due to phason modes of a hypothetical B-Ru-Cu decagonal phase has even been predicted [36]. For icosahedral phases, models are still far from accounting for real systems.

To end this part, we illustrate, with a few examples, why phason modes are key "ingredients" in the understanding of the physics of QCs. The first subject is the growth of QCs. Indeed, how the long-range quasiperiodic order propagates is one of the most fascinating questions. The first question one may ask is whether it is possible to grow a QC with local rules by adding the atoms or clusters on the seed such that the resulting structure is a perfect QC. All model systems studied have shown that if local rules are used, the final structure always contains defects in terms of localized phason flips, which eventually can be annealed further [38]. In a completely different approach, the random tiling model considers that the entropy associated with phason fluctuations stabilizes the QC, this one being stable only at high temperatures, to transform into a periodic approximant at lower temperature [34]. These two extreme models are often quoted as the energy versus entropy stabilized model. The second field for which phason plays a role in dislocations. Indeed, the concept of dislocation can be generalized to QCs using the six-dimensional periodic description. A dislocation is thus associated with both a phonon and a phason distorsion field (see Janot [39] for an introduction). The phason field in that case is related to a distribution of phason strains (and not just phason modes) and leads to Bragg peaks broadening. Such a phason distorsion field has been evidenced experimentally [40].

4 i-AlPdMn Quasicrystal

We first present results obtained in the i-AlPdMn case for which a large set of experiments has been carried out. Indeed, this QC is a stable [41] one and can be grown as cm^3 size single grains using either the Bridgman [42] or the Czochralski technique [43,44]. A detailed study of the structural quality of this phase has been carried out showing extremely sharp Bragg reflections once the grown single grains are annealed [45]. In fact, the quality of this icosahedral phase is such that dynamical diffraction effects have been observed [46,47].

4.1 Phason Diffuse Scattering in i-AlPdMn: Room Temperature Studies

The first hint that PDS was present in the i-AlPdMn phase has been obtained by evidencing the presence of a phason Debye-Waller term in two ways: first in comparing low- and high-resolution measured integrated intensity, and second, by comparing collected data with a QC model. Both results showed a dependence with the $\mathbf{Q}_{per}$ component of the wave vector from which a perpendicular fluctuation of the atomic surfaces $<u_{per}{}^2>^{1/2}$ is found to be of the order of 1.2 Å [48,49].

To evidence the PDS, we used neutron scattering with an energy transfer analyzer and at the elastic position so that the TDS contribution was suppressed. Measurements have been carried out around a few Bragg reflections shown in the neutron diffraction pattern in Figure 12.6. In the case of elastic

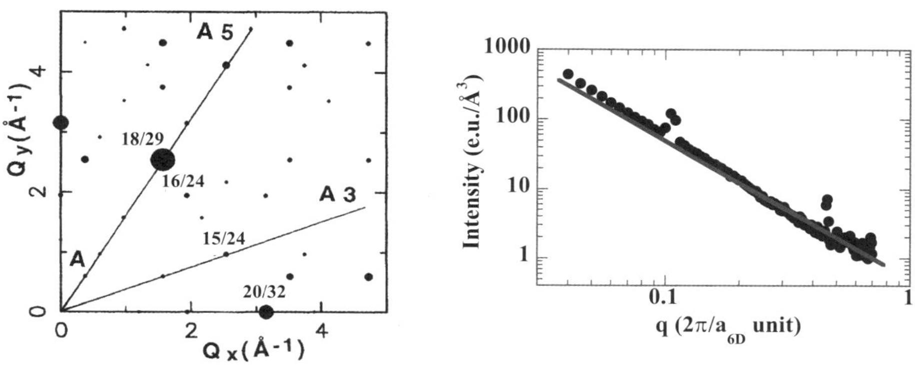

Figure 12.6. *Left: Neutron scattering intensity distribution in a twofold plane for the i-AlPdMn phase. The Bragg peaks are labeled by their N/M indices, which is a short hand notation for the full six-dimensional indexing scheme [50]. (These N and M indices only characterize the modulus of the wave vector of each reflection.) The 18/29, 15/24, and 20/32X reflections lie on a five-, three-, and twofold axis, respectively. Right: Log-log plot of the diffuse scattering intensity as a function of q, as measured on the D2AM beamline at the ESRF. The solid line is for the $1/q^2$ decay.*

static phonon distortion, the diffuse intensity is expected to display the same shape around all Bragg reflections because the icosahedral phase is elastically isotropic. This will not be the case if phason fluctuations are present: depending on the ratio of the phason elastic constants K2/K1, different shape anisotropies are expected around the different Bragg reflections. The PDS intensity being proportional to Q_{per}^2, we have studied reflections having different perpendicular components of their Bragg wave vector in order to evidence this property relation. The neutron Bragg intensity distribution in a twofold plane is shown Figure 12.6, together with the N/M indices of a few Bragg reflections [50].

Systematic diffuse intensity maps have been recorded around a few Bragg peaks and are shown Figure 12.7, where iso-intensity contours are plotted [51]. The same intensity scale has been selected for the four reflections so that the different panels can be directly compared.

As can be seen Figure 12.7, the intensity distribution displays strong anisotropies. Several observations point toward a phason component for this diffuse scattering. First, the 16/24 and 15/24 are two reflections having the same Bragg peak intensities but with Q_{per} components in a ratio 3 to 1. As can be observed in Figure 12.7b and c, there is a much larger amount of diffuse scattering intensity around the 16/24 reflection, as expected from Equation 12.5. Second, the 18/29 and 16/24 reflections are very close to each other in reciprocal space; any elastic (phonon) distortion would lead to a very similar shape of the diffuse scattering at the opposite of what is observed experimentally. This can be understood by computing the perpendicular component of these two reflections; they are almost orthogonal and thus lead to a very different scalar product "selection rule" in Equation 12.5. Finally, measuring two collinear reflections, it is possible to show that the intensity of the diffuse scattering decays as q^{-2} and scales with the product $I_{Bragg}.Q_{per}^2$.

These results are confirmed by a computation of the diffuse scattering using Equation 12.5. If only phason fluctuations are contributing to the diffuse scattering intensity and if the coupling constant K3 is neglected, then the shape anisotropy depends only on the ratio K2/K1. Using the measured Bragg

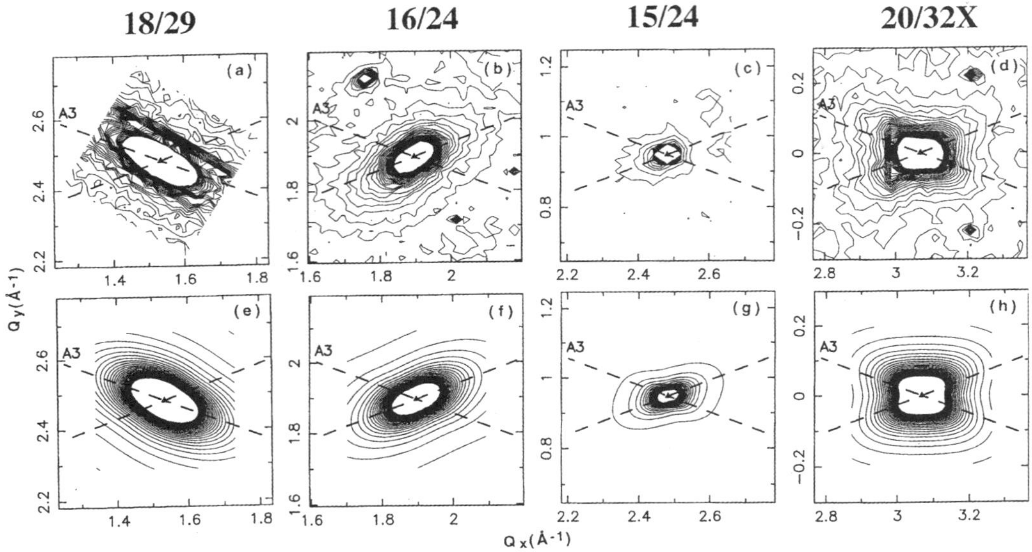

Figure 12.7. *Top panel (a–d): Isointensity contour plots measured by elastic neutron scattering around different Bragg peaks in the i-AlPdMn phase at room temperature. The indices are given in the N/M notation and refer to Figure 12.3. All contour lines are on the same scale. Bottom panel (e–h): Simulation of the diffuse scattering using as a single parameter the K2/K1 ratio equal to –0.52. The dashed lines indicate directions parallel to threefold axes.*

peak intensity, it is then possible to simulate the diffuse scattering as shown in Figure 12.7e–h, where a ratio $K2/K1 = -0.52$ has been used. As can be seen, there is a good agreement between the simulation and the experiment. A quantitative agreement was also achieved by comparing the intensity decay observed along different directions and various Bragg peaks.

The $1/q^2$ decay of the PDS signal has been evidenced both with neutron and X-ray scattering. Because of the higher resolution and higher flux available at X-ray synchrotron sources, the diffuse scattering decay could be measured on a larger range. A log-log plot of the diffuse scattering intensity measured around a twofold axis as a function of q is shown in the right panel of Figure 12.6. The solid line represents the expected $1/q^2$ decay of the intensity. From this measurement it could be deduced that this decay is valid roughly between 0.03 and 0.6 $Å^{-1}$, that is, a phason mode wavelength between 10 and 200 Å. Even larger phason mode wavelengths, up to 1000 Å, could be measured in a high-resolution setting. The $1/q^2$ decay of the PDS signal is also illustrated Figure 12.8, as seen around the twofold reflections 8/12 and 20/32.

The diffuse scattering was measured on an absolute scale using both neutron and X-ray scattering. In the case of X-ray scattering, in order to minimize parasitic scattering, the measurement is carried out on a large sample (1 cm in diameter) in reflection geometry with the sample under vacuum. To put the result on an absolute scale, we used either a calibrated Al powder or a combined precise measurement of

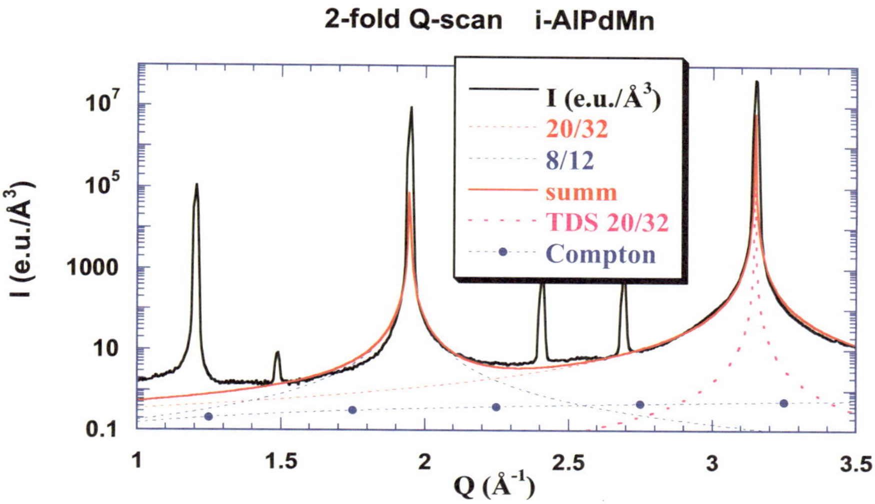

Figure 12.8. *Absolute scale intensity measurement of the diffuse scattering in the i-AlPdMn phase along a twofold axis. Because the measurement is on an absolute scale, the various contributions to the signal—TDS, Compton scattering, and PDS—can be separated. The PDS contribution is shown to dominate over all other contributions. Note the $1/q^2$ decay of the PDS signal around the 20/32 and 8/12 Bragg reflections.*

the direct beam intensity and of the aperture of the receiving slits [30]. Both normalization procedures lead to the same result. The absolute scale measurement is particularly important in the case of X-ray scattering, as illustrated in Figure 12.8. Indeed, it allows for the evaluation of the different contributions to the signal: TDS, Compton scattering, and PDS. As can be seen Figure 12.8, in the studied Q-range the signal is dominated by the PDS, which makes the interpretation of the data relatively simple. From this absolute scale measurement, the values of the two elastic constants K1 and K2 were evaluated. A value of $K1/k_BT$ equal to 0.1 and $K2/k_BT$ equal to –0.052 atom^{-1} has been found [52]. As already stated, this measurement procedure is completely analogous to the determination of phonon elastic constants or phonon dispersion relations using X-ray diffuse scattering [25,26].

The absolute scale measurements also allow for a comparison of different samples. In particular, it was important to check for the possible effect of the annealing treatment of the sample on the amount of diffuse scattering; it was found that the amount of diffuse scattering is independent on the sample heat treatment, which demonstrates that the PDS is an intrinsic property of the icosahedral phase [52].

Exploring a larger region of the reciprocal space showed that 95% of the observed diffuse scattering originates from PDS. In the left panel of Figure 12.9, the diffuse scattering intensity is measured in a plane perpendicular to a fivefold axis passing through the strong 18/29 fivefold reflection in a relatively large area of 4×4 Å^{-1}. As can be seen, the diffuse scattering intensity presents distinctive features, such as arcs of diffuse scattering. This can be accounted for only if the PDS originating from out of plane Bragg peaks is also considered, as shown on the simulation in the simulation in the right panel, which reproduces the experiment extremely well.

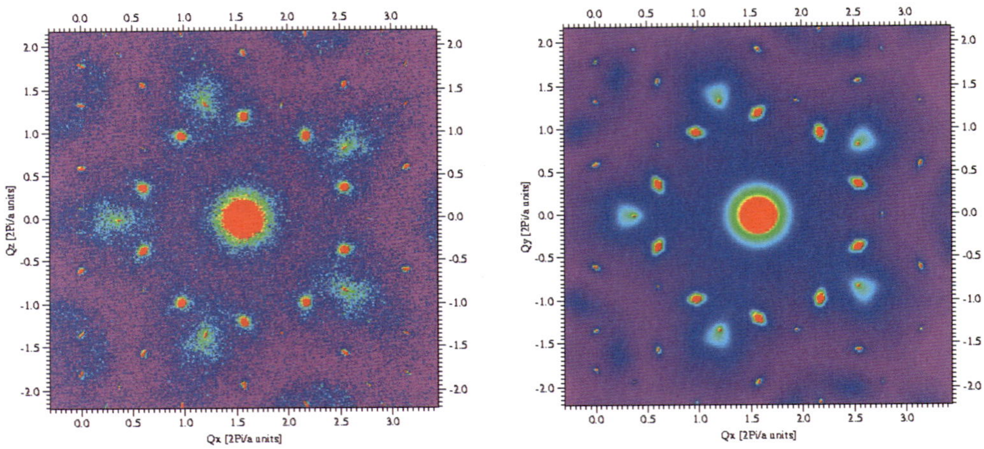

Figure 12.9. *Diffuse scattering in the i-AlPdMn phase. Left panel: Diffuse scattering measured around the strong fivefold 18/29 reflection in a plane perpendicular to the fivefold axis. Measurements were made on a rotating anode (Cu target) four-circle diffractometer using a point detector. Right panel: Simulation of the diffuse scattering taking into account the contribution from out-of-plane Bragg reflections. Both the shape and the intensity distribution of the diffuse scattering are well reproduced: note, for instance, the lobelike distribution seen on the outer part of the diffraction pattern.*

4.2 Temperature Dependence of the Diffuse Scattering

All measurements presented so far for the i-AlPdMn phase have been carried out at room temperature. Phason modes are probably frozen in at room temperature because the atomic motions over several interatomic distances are unlikely at room temperature from what is known of the atomic diffusion in Al-based alloys.

Understanding the origin of this PDS thus requires a high-temperature study. We first give a brief outline of some of the theoretical arguments because the temperature dependence of the diffuse scattering is related to the stabilization mechanism of QCs.

4.2.1 Phase Transition and Phason Modes

Two simple models have been proposed to explain the stability of QCs: the energy- or the entropy-stabilized model [19,34].

In the first model, the QC is considered as being the ground state at 0 K and is energetically stabilized. Whether this can be achieved using finite-range interatomic interactions only is still an open question. Simple Hamiltonians on toy models succeeded in having a quasicrystalline ground state at 0 K. When the temperature is increased, defects are introduced (e.g., vacancies, chemical disorder), but so are phason fluctuations. In this scheme, we should thus observe an increase of the diffuse scattering when the temperature increases. It is also believed that at low temperature the free energy is not analytical with respect to the phason strain. Instead of varying as $F = |\nabla\phi|^2$ (where ϕ is the phason strain), the free energy would go as $F \propto |\nabla\phi|$. In this case, the elasticity theory does not hold anymore, and Equation 12.5 can no longer be used. The free energy would then become analytical above a critical temperature T_c [19]. This "transition" only has consequences on the shape of the diffuse scattering because the quasiperiodic long-range order is preserved at all temperatures. Although the diffuse scattering intensity has not been computed below T_c, it would most likely look like a broad Lorentzian whose width would decrease as the temperature is raised. Above T_c we would recover the $1/q^2$ decay of the diffuse scattering. This transition between a nonanalytical and an analytical regime bears some similarities with the locked-unlocked transition studied by Aubry in the Frankel-Kontorova model [10].

The entropy stabilized model has been developed in the framework of the random tiling scenario [34]. It postulates that the QC is stabilized by the configurational entropy associated with phason fluctuations. In this scheme the QC is only stable at high temperatures and undergoes a phase transition toward a crystalline phase as the temperature is lowered [35]. This transition is accomplished via pretransitional soft phason modes. Using the Landau theory of phase transition, it is possible to predict the symmetry of the low-temperature phase and the phason strain distorsion necessary to accomplish the transition [22,23]. Similar to what occurs in displacive phase transition in which a soft phonon mode appears as a pretransitional fluctuation, a softening of the phason elastic constants might take place in the high-temperature regime when approaching the phase transition; the symmetry of the phason mode, which is then enhanced, is related to the symmetry of the low-temperature phase. Some information about the symmetry of the low-temperature phase can also be gained by looking at the behavior of the hydrodynamic matrix C (Equation 12.5) as a function of the ratio K2/K1 (if the coupling term K3 is neglected). There are two values of the ratio K2/K1 for which the matrix C^{-1} diverges, corresponding to the so-called hydrodynamic instabilities [24,53]. The upper limit K2/K1 = –0.75 is a threefold instability following a structural deformation along a threefold axis from the I_h icosahedral symmetry toward the D_{3d} symmetry, whereas the lower limit K2/K1 = +0.75 is a fivefold instability following a deformation along a fivefold axis toward the D_{5d} symmetry. Those hydrodynamic instabilities cannot be reached thermodynamically because a first-order transition occurs beforehand at –0.6 and +0.75. The phason mode softening still takes place depending, however, on how close K2/K1 is to the value at which the hydrodynamic instabilities occur. This softening induces an enhancement of the diffuse scattering along directions parallel to either a threefold or a fivefold axis.

To summarize, the two models can be distinguished by the relative weight they give to the energy and entropy terms stabilizing the QC. Both models predict a high-temperature region in which phason fluctuations are hydrodynamic and the diffuse scattering can be computed using Equation 12.5. However, the two models predict a different temperature dependence of the diffuse scattering [35]. In the case of the energetically stabilized scenario, the diffuse scattering should increase as the temperature is increased, and Equation 12.5 is only valid above a critical temperature T_c. In the random tiling scenario, there should be a low-temperature crystalline phase in which the QC is stable only in the high-T region. Starting from the high-temperature quasicrystalline phase, soft phason modes lead to an increase of the diffuse scattering as the temperature is lowered. Simulations were carried out in a model system by Mihalkovič and Henley [35].

4.2.2 Temperature Dependence of the Diffuse Scattering

The temperature dependence of the diffuse scattering has been studied by neutron scattering measurements between 25°C and 770°C using a triple-axis instrument to suppress the phonon contribution. As already said, this is particularly important because the mean square atomic displacement due to phonons is roughly proportional to T so that the TDS contribution might not be negligible in the high T region.

Figure 12.10 displays iso-intensity contours of the diffuse scattering measured around the 18/29 reflection at 200°C (Figure 12.10a) and at 770°C (Figure 12.10b) [54]. Two observations can be made: first, there is a clear decrease of the diffuse scattering intensity at high temperatures; second, the shape of the diffuse scattering is more "circular" at high temperatures. The decrease of the diffuse scattering intensity when T is increased has been checked by measuring systematic Q-scans around different reflections at several different temperatures: almost no intensity change is observed up to 500°C; from 500°C up to 770°C there is a continuous decrease of the diffuse scattering intensity with a factor 2 for some directions. In the meantime, the intensity of some weak reflections increases. Those results can be understood if we consider a change of the ratio of the phason elastic constant from K2/K1 = –0.52 at room temperature to K2/K1 = –0.4 at 750°C. Such an evolution reproduces the observed diffuse scattering quite well, in particular the more "circular" shape of the diffuse scattering in the high-T region, as shown in Figure 12.10c and d.

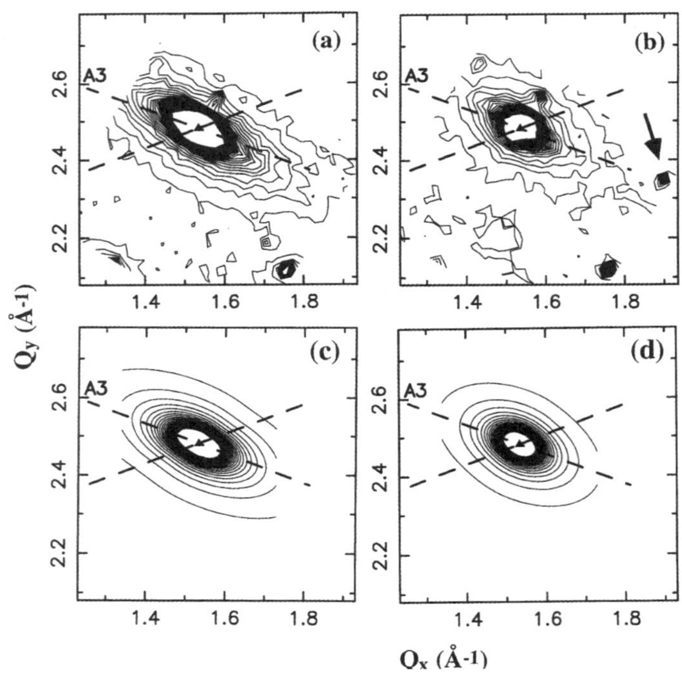

Figure 12.10. *Temperature dependence of the diffuse scattering measured around the 18/29 five-fold reflection at (a) 200°C and (b) 770°C. The change from 200°C to 770°C is well reproduced by a change of the phason elastic constants from (c) K2/K1 = –0.5 to (d) –0.4.*

A more detailed evolution of the phason elastic constants can be deduced because all data are measured on the same scale. Considering a simple linear T dependence of the phonon Debye-Waller factor, it is possible to compute the T dependence of the Bragg peak intensity. (A direct measurement of the intensity is not possible because of extinctions effects.) From this we deduce that K2 is T independent, and only K1 decreases as T is decreased. Thus we have a softening of the K1 phason elastic constant as the temperature is decreased.

These results are clearly in agreement with the random tiling scenario. However, in this scenario we should observe a transition toward a lower-symmetry phase at low temperature, which is not the case. The evolution of the phason elastic constants points toward a threefold-type instability. As already explained theoretically before, the Landau theory of phase transition predicts a low-temperature phase with D_{3d} symmetry. One possible explanation is that the transition is frozen in because of kinetics. The ground state of the system would be a crystalline phase with tetrahedral symmetry, but because atomic diffusion becomes too slow below 500°C, the system would be blocked in a metastable icosahedral state with frozen-in phason fluctuations.

Some hint that this is the case is given by a temperature study of the phase transition occurring for AlPdMn icosahedral phases with a different Pd/Mn content. When the Mn content goes from 9% to 8%, the icosahedral phase is stable only at high temperatures [55–57]. Below 650°C, there is a transition from the i-phase toward a phase of tetrahedral symmetry of extreme complexity, named $F2_M$. Although its symmetry is D_{3d}, the $F2_M$ phase is quasiperiodic. It can be described as a superstructure on the icosahedral phase with satellite reflections along directions parallel to the threefold axes. A detailed atomic description has not yet been achieved, but the supplementary superstructure reflections are clearly related to phason fluctuations because their intensity is correlated to the $\mathbf{Q}_{per}$ component of the corresponding main Bragg peaks. This $F2_M$ phase thus might be considered as the low-temperature state of lower symmetry.

Finally, we should keep in mind the complexity of the chemistry and of the stabilization mechanism of QCs. From experimental results, the entropy term has an important weight on the stability of QCs. A microscopic model of this mechanism is, however, still lacking, and chemical potentials are certainly important to consider, as indicated by the narrow composition range over which the stable QCs can be grown.

4.3 Dynamics of Phason Fluctuations

So far we have only considered the static behavior of phason modes in QCs. As explained in the introduction, the hydrodynamic theory predicts that phason modes are collective diffusive modes. This means that a phason fluctuation follows an exponential time decay with a characteristic time scale τ going as q^{-2}, where q is the phason mode wave vector. This is equivalent to a characteristic time scale going like λ^2, where λ is the phason mode wavelength. At the thermodynamic equilibrium, phason modes of all wavelengths are present in the QC following a Boltzman distribution. Since a distortion of the cut space results in atomic rearrangements, a phason mode with a wave vector $\mathbf{q}$ results in correlated atomic rearrangements. Suppressing this fluctuation requires atomic diffusion on long distances, which explains why the decay time should go as λ^2. This time dependence is supported by simulations on three-dimensional Penrose random tiling, which is a toy model for the study of quasicrystalline structures [33,58].

Coming to the i-AlPdMn phase, phason dynamics should be observed above 500°C, where we know that changes occur in the static diffuse scattering, and thus phason modes should be active. An obvious tool to study such phenomena is inelastic neutron scattering. Indeed, an exponential time decay of a fluctuation leads, by time and space Fourier transform, to a quasielastic signal of a Lorentzian shape in the S(Q,E) measured response function. The width of the Lorentzian is inversely proportional to the characteristic time scale.

Above 600°C, a quasielastic signal was observed by G. Coddens and coworkers in the neutron spectrum. However, this signal does not correspond to long-wavelength phason fluctuations because it is not correlated at all to the observed diffuse scattering. It was interpreted as the result of assisted atomic jumps involving one or two atoms in an elementary process [59].

Long-wavelength phason fluctuations are most likely too slow to be observable in the neutron time scale window. A study of their dynamics was thus carried out using the X-ray photon correlation spectroscopy technique [60]. This technique allows one to measure slow dynamics (10^{-3} to 1000 s) at the atomic scale. When a coherent X-ray beam is used to measure the diffuse scattering, the **q** dependence of the intensity does not show a smooth variation (resulting from an ensemble average) but displays strong intensity fluctuations called speckles. A coherent X-ray beam of reasonable flux can be obtained at third-generation synchrotron sources like ESRF using high-brilliance undulator sources. For instance, using the focusing optics of the ID20 beamline (ESRF), a flux of 5.10^8 photons per second is obtained through a 10-μm pinhole located just before the sample [61]. The diffuse scattering spectrum is recorded using a directly illuminated CCD camera located 1.85 m away from the sample and acting as a two-dimensional photon detector [62]. With such a setup, the partial coherence of the beam β was found equal to 0.08 and 0.03 for low- and high-angle reflections. A typical two-dimensional image of the diffuse scattering recorded close to a Bragg reflection is shown in Figure 12.11 [61]. Spikes and large-intensity fluctuations are clearly visible. The shape of the diffuse scattering is

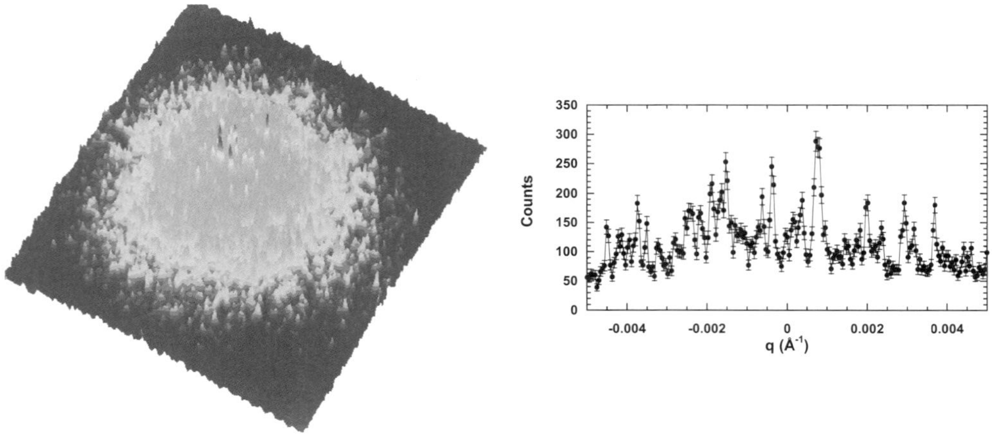

Figure 12.11. *Left: Two-dimensional CCD image of the diffuse scattering measured close to a two-fold Bragg reflection using a coherent X-ray beam. Right: Section through the two-dimensional image. The "spikes" and intensity fluctuations are much larger than the statistical ones shown by the error bar. The speckles are interference effects due to the coherent X-ray beam. [61]*

well reproduced considering K1 and K2 phason elastic constants only, which ensures that the observed signal is the one that is relevant, that is, the one due to hydrodynamic phasons.

Any time evolution of phason fluctuations will result in a time dependence of the speckle pattern. This is best evidenced by computing the intensity-intensity correlation function, defined as

$$F_{cor}(\mathbf{q},t)=\left\langle I(\mathbf{Q}_B+\mathbf{q},t).I(\mathbf{Q}_B+\mathbf{q},t+t')\right\rangle_{t'}/\left\langle I(\mathbf{Q}_B+\mathbf{q},t')\right\rangle_{t'}^{2}, \tag{12.6}$$

where $\mathbf{Q}_B$ is the Bragg peak position and $\mathbf{q}$ is the phason wave vector. We have the relation $F_{cor}(\mathbf{q},t)=[1+\beta g(\mathbf{q},t)]$, where $g(\mathbf{q},t)$ is the time dependence of a phason mode with a wave vector $\mathbf{q}$.

In situ measurements between room temperature and 650°C have been carried out in a furnace under secondary vacuum [63,64]. The measurements were performed in the i-AlPdMn phase near the 7/11 reflection, which has a Bragg angle 2θ of 24° and a component $\mathbf{Q}_{par}$ along the (1, τ, 0) fivefold axis. At each temperature, several q values were probed by rotating the θ incident angle with a 0.1° step, which corresponds to a q step of ~0.003 Å^{-1} along the (−τ, 1, 0) direction in reciprocal space. Below 500°C, the correlation function does not show any time dependence up to a time scale of the order of 10 to 15 minutes. Above 500°C, there is a decay of the correlation function. Left panel of Figure 12.12 displays the correlation function measured at 650°C for two phason mode wave vectors.

The time decay of the correlation function has been fitted by an exponential law from which a characteristic time τ is extracted. The top and bottom part of the left panel of Figure 12.12 correspond to a wave vector $\mathbf{q}$ equal to 0.064 and 0.13 nm^{-1}, respectively. As can be seen in the Figure, the characteristic time is much shorter at larger wave vectors, which is in agreement with the expected diffusive

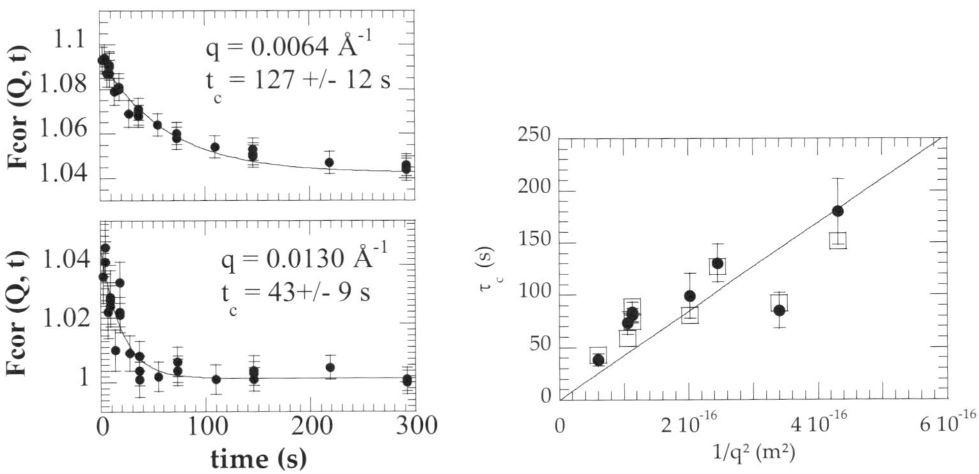

Figure 12.12. *Left: Time dependence of the intensity-intensity correlation function measured at 650°C in the i-AlPdMn phase. The solid line is an exponential fit to the data. The top and bottom Figures correspond to two different wave vectors; the decay is much faster for large wave vectors. Right: Evolution of the characteristic time τ as a function of λ² measured at 650°C along the (−τ,1,0) direction. There is a linear dependence from which the phason diffusion coefficient is extracted.*

character of phason modes. This can be understood by the hand-waving argument that says that the time needed to destroy a given fluctuation is much longer the longer the wavelength of the phason mode is. The diffusive character of the phason mode is shown in a more quantitative way in the right panel of Figure 12.12. As explained in the beginning of this section, we expect the characteristic time τ to be inversely proportional to q^{-2}. Figure 12.12 displays the characteristic time τ measured along the direction ($-\tau$ 1 0) as a function of $1/q^2$. From the linear dependence, a phason diffusion constant defined as $q^{-2} = D_{phason}\tau$ is found to be equal to $2.2.10^{-18}\,m^2s^{-1}$ at 650°C.

A few measurements were also carried out at 600°C. At this temperature, phason fluctuations are much slower than at 650°C, the characteristic τ being five times larger. An accurate estimate of the diffusion constant is thus difficult because those characteristic times are too close to the maximum time that can be experimentally achieved. Considering that phason modes are thermally activated, we estimated that the activation energy is of the order of 2 (±1) eV [63,64].

Such results are compatible with what is known about atomic diffusion in these systems. Mn, which is the slow diffuser in the i-AlPdMn phase, has a diffusion constant equal to $6 * 10^{-14}\ m^2s^{-1}$ at 650°C and an activation energy equal to 2 eV [65]. We find a much smaller phason diffusion constant; this might be understood as being the result of the atomic species exchange, which is a slow process limiting the kinetics of the dynamics of phason modes.

Phason fluctuations are believed to play an important role in the growth process and in the mechanical properties of QCs. In the i-AlPdMn QC, plastic deformation is only possible for temperatures greater than 600°C and occurs via dislocation motions [66]. Because of the quasiperiodic long-range order, the distortion field around a dislocation contains both a phonon and a phason part. When a dislocation is moving, a highly localized phason strain (called a phason wall) is left behind [67]. In situ high-resolution transmission electron microscopy (TEM) images have shown that these phason walls annealed out very rapidly above 650°C. A detailed study of these dynamics has been carried out using in-situ TEM images [68]. There is a contrast in the images that is due to the phason wall. This contrast vanishes exponentially with time, from which a characteristic time is extracted. The time scale at 640°C is of the order of 140 s for a sample thickness of the order of 200 nm; that is, the time necessary to anneal the phason wall on 100 nm (half the sample thickness) is about 140 s. This is in very good agreement with the value of the phason diffusion constant reported above at 650°C, where for a wavelength of 100 nm one finds a time scale of the order of 130 s. The authors have also determined the temperature dependence of the characteristic time, which follows an Arrhenius law. The activation energy extracted from this dependence is equal to 4 eV, which is of a same order of magnitude as the activation energy obtained with coherent X-rays. The high activation energy of the phason modes explains why the i-AlPdMn QC is brittle below 600°C; in this case, the time necessary to relax the localized phason strain is too long, and dislocations are pinned. This set of experiments on phason strain relaxation in the vicinity of a dislocation thus agrees nicely with those obtained by coherent X-ray diffraction.

5 Other Icosahedral Phases

Diffuse scattering studies have been carried out on some other quasicrystalline phases. The main result is that for all known QCs, even the so-called perfect ones, PDS is observed. The shape anisotropy (related to the ratio of the K1 and K2 phason elastic constants) and the amount of diffuse scattering (related to the magnitude of the phason elastic constants) is, however, strongly sample dependent. This is, for instance, the case in isostructural QCs i-AlPdMn, i-AlCuFe [69], and i-AlPdRe [70]. Although they share a similar atomic structure, the diffuse scattering shape anisotropy is quite different, with a fivefold type diffuse scattering, for instance, in the i-AlCuFe phase, which is in agreement with a phason-driven phase transition toward the rhombohedral phase with an intermediate phase presenting a fivefold modulation phason wave [71–73].

The binary CdYb phase [6] and related family in the ZnMgSc system [74] were also investigated. PDS has been observed in the CdYb icosahedral phase [70].

Of particular interest is the comparison carried out between the i-ZnMgSc and the 1/1 cubic ZnSc approximant. Whereas the diffuse scattering observed in the ZnSc cubic approximant can be fully accounted for considering the TDS contribution only, in the i-ZnMgSc QC there is a supplementary contribution to the diffuse scattering arising from phason modes [75,76]. We stress here again the importance of having carried out absolute scale measurements, which allowed for an exact computation of the TDS contribution. This result demonstrates that the PDS is related to the long-range quasiperiodic order and not to the local order, because both the QC and its approximant share the same large atomic clusters (see section 2). Moreover, in both cases, the inner tetrahedron corresponds to a site where the local "phason" jump occurs [5], experimentally illustrating the known result that one needs a quasiperiodic long-range order for the long wavelength phasons to take place. Finally, it was found that the amount of diffuse scattering is much smaller in the i-ZnMgSc phase than in the i-AlPdMn one. As a consequence, K1 phason elastic constant is larger and the phason Debye-Waller factor much smaller in the i-ZnMgSc phase than in the i-AlPdMn one. As a result, a much larger number of high-Q_{per}, weak reflections are observed in the diffraction pattern of the i-ZnMgSc phase than in the i-AlPdMn phase. In fact, the i-ZnMgSc diffraction pattern can only be indexed by considering that reflections have a perpendicular component of their six-dimensional reciprocal wavector up to 7, whereas the highest value of the Q_{per} component experimentally observed in i-AlPdMn is 2.5. Indeed, a larger perpendicular Debye-Waller factor will cause all weak reflections to vanish in the diffuse scattering. This is illustrated Figure 12.13, which displays a twofold Q-scan in both phases.

In this chapter we did not consider the phonon-phason coupling term K3 in the computing of the PDS [20,32]. In i-AlPdMn, K3 = 0 is a good assumption as supported by the perfect match between experiment and simulation in Figure 12.8. On the contrary, in i-CdYb, considering only the K1 and K2 phason elastic constants does not suffice in accounting for the whole measured PDS, and the phonon-phason coupling diffuse scattering term must be considered [70]. Recently, Edagawa proposed an interesting method for determining K3 [77–79] consisting of comparing the lattice parameters of the QC and of its parent crystalline approximant. Using this method, the phonon-phason coupling constant could be evaluated in a few systems.

6 Conclusion

We have shown in this chapter that the diffuse scattering measurement techniques are powerful tools for evidencing phason modes in QCs and that a convincing demonstration that diffuse scattering is indeed related to phason modes requires a detailed comparison with the hydrodynamic theory. Measuring the diffuse scattering intensity on an absolute scale is the only way to compare the amount of diffuse scattering between different samples and allows for a determination of the phason elastic constants.

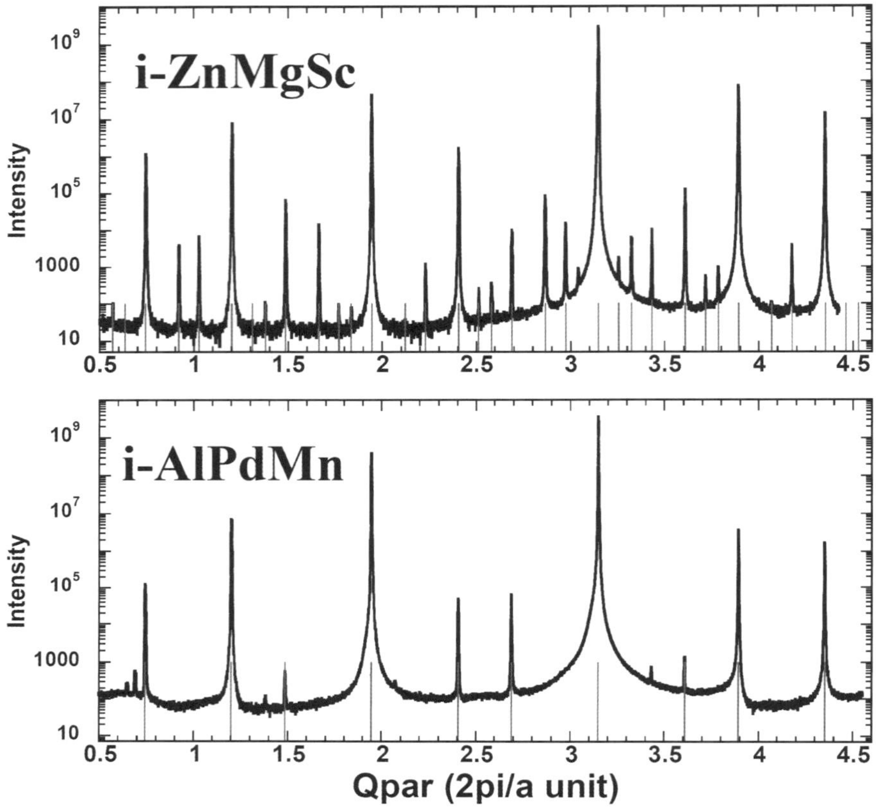

Figure 12.13. *Comparison of the scattered-intensity signal measured in the i-ZnMgSc and i-AlPdMn phases along a twofold axis. The amount of diffuse scattering is lower in i-ZnMgSc, and the number of large Q_{per} weak reflections larger. Bars stand for the calculated Bragg peak positions considering a maximum Q_{per} value of 7 in i-ZnMgSc and 2.5 in i-AlPdMn. (From de Boissieu et al. [75].)*

Phason diffuse scattering has been observed in all QCs, although the ratio and the magnitude of the phason elastic constants depend on the sample chemistry. The wavelength range of phason modes, extracted from the $1/q^2$ decay of the diffuse scattering, goes from 10 to 200 Å. In i-AlPdMn, the phason fluctuations are pretransitional fluctuations toward a threefold instability, and the diffuse scattering amount increases as the temperature is lowered, in agreement with an entropy stabilization scenario. Using coherent X-rays, the dynamics of phasons was proved to be diffusive, as predicted by the hydrodynamic theory. It was also experimentally demonstrated that phason modes originate from the long-range quasiperiodic order, because only the TDS related to phonons is observed in periodic 1/1 approximants. Among all icosahedral quasicrystalline phases studied thus far, i-ZnMgSc is the one with the smallest amount of PDS and the larger number of high Q_{per} reflections in its diffraction pattern.

In this chapter we did not give any atomic description of the buildup of phason modes. This remains one of the most challenging problems in the field of QCs. From the experimental results and from the high-dimensional picture of QCs, a few trends can nevertheless be given. The shape anisotropy, associated with the sign of the ratio of the phason elastic constants K2 and K1, is changing even between phases that are isostructural. This suggests that the tile geometry does not suffice to explain the PDS and that the details of the chemistry of the system have to be taken into account. Indeed, when a phason wave distortion is introduced in the ideal QC structure, it generates several correlated atomic "flips." There are two kinds of such flips: the first one corresponds to the analog of the LS to SL permutation. In that case, an atom goes from one position to a vacant one located nearby (on distances of the order of 0.1 nm) that has the same local environment. This usually occurs for atoms generated by the outer part of the atomic surfaces. The second type is one in which an atom A is "exchanged" with an atom B, A and B being different chemical species. The distance between the A and B sites may be of several Å, and both sites must be located at specific positions in terms of the local environments. This corresponds to atoms generated by the inner part of the atomic surface in general. The latter flips might be more auspicious at driving the phason dynamics because they involve long-range diffusion.

The i-CdYb QC, for which there is a detailed understanding of the atomic structure and for which the pair potentials are in good agreement with the lattice dynamics data [80], is a promising system in the purpose of gaining insight into the microscopic realization of phason modes. Such an understanding would certainly benefit from an interdisciplinary approach encompassing all aperiodic phases so as to address the fascinating question of why and how long-range aperiodic order sets in.

References

1. D. Shechtman, I. Blech, D. Gratias, and J. W. Cahn, *Phys. Rev. Lett.* 53 (1984): 1951.
2. T. Janssen, G. Chapuis, and M. de Boissieu, *Aperiodic Crystals: From Modulated Phases to Quasicrystals* (Oxford: Oxford University Press, 2007).
3. M. de Boissieu, R. Currat, and S. Francoual, "Phason Modes in Aperiodic Crystals," in *2007 in Physics of Quasicrystals*, ed. T. Fujiwara and Y. Ishii (Amsterdam: Elsevier Science), 107.
4. I. R. Fisher, Z. Islam, A. F. Panchula, K. O. Cheon, M. J. Kramer, P. C. Canfield, and A. I. Goldman, *Phil. Mag. B* 77 (1998): 1601.
5. H. Takakura, C. P. Gómez, A. Yamamoto, M. de Boissieu, and A. P. Tsai, *Nat. Mater.* 6 (2007): 58.
6. A. P. Tsai, J. Q. Guo, E. Abe, H. Takakura, and T. J. Sato, *Nature* 408 (2000): 537.
7. C. P. Gómez and S. Lidin, *Angew. Chem. Int.* 40 (2001): 4037.
8. C. P. Gómez and S. Lidin, *Phys. Rev. B* 68 (2003): 024203\1.
9. A. W. Overhauser, *Phys. Rev. B* 3 (1971): 3173.
10. R. Currat and Janssen T. *Sol. Stat. Phys.* 41 (1988): 201.
11. P. A. Kalugin, A. Y. Kitaev, and L. S. Levitov, *JETP Lett.* 41 (1985): 145.
12. P. Bak, *Phys. Rev. Lett.* 54 (1985): 1517.
13. T. C. Lubensky, S. Ramaswamy, and J. Torner. *Phys. Rev. B* 32(11): (1985): 7444.
14. V. Elser. *Phys. Rev. Lett.* 55 (1985): 1730.
15. T. C. Lubensky, "Symmetry, elasticity and hydrodynamics of quasicrystals," in *Aperiodicity and order: Introduction to quasicrystals*, ed. M. V. Jarić (Boston: Academic Press, 1988), 199.
16. P. Bak, *Phys. Rev. B* 32, no. 9 (1985): 5764.
17. P. A. Kalugin, A. Y. Kitaev, and L. S. Levitov, *J. Phys. (France), Lett.* 46 (1985a): L601.
18. E. Zeyher and W. Finger. *Phys. Rev. Lett.* 49 (1982): 1833.
19. T. Dotera and P. J. Steinhard, *Phys. Rev. Lett.* 72, no. 11: (1994): 1670.
20. M. V. Jarić and D. R. Nelson, *Phys. Rev. B* 37 (1988): 4458.

21. P. M. Chaikin and T. C. Lubensky, *Principles of Condensed Matter Physics* (Cambridge: Cambridge University Press, 2000).
22. Y. Ishii, *Phys. Rev. B* 39(16): (1989): 11862.
23. Y. Ishii, *Phil. Mag. Lett.* 62 (1990): 393.
24. Y. Ishii, *Phys. Rev. B* 45 (1992): 5228.
25. J. Laval, *C. R. Acad. Sci.* 207 (1938): 169.
26. J. Laval, *Bull. Soc. franc Miner* 62 (1939): 137.
27. P. Olmer, *Acta Crystallogr.* 1 (1948): 57.
28. H. Curien, *Acta Crystallogr.* 5 (1952): 393.
29. C. B. Walker, *Phys. Rev. B* (1956): 547.
30. B. E. Warren, *X-ray Diffraction* (New York: Dover Editions, 1990).
31. R. W. James, *The Optical Principles of the Diffraction of X-Rays* (Woodbridge: Ox Bow Press, 1962).
32. S. B. Rochal, *Phys. Rev. B* 64, no. 14 (2001): 144204/1.
33. L. H. Tang, *Phys. Rev. Lett.* 64 (1990): 2390.
34. C. L. Henley, "Random tiling models," in *Quasicrystals: The State of the Art*, ed. D. P. DiVicenzo and P. J. Steinhardt (Singapore: World Scientific, 1991), 429.
35. M. Mihalkovič and C. L. Henley, *Phys. Rev. B* 70 (2004): 09202–1.
36. M. Mihalkovič and M. Widom. *Phys. Rev. Lett.* 93 (2004): 095507/1.
37. C. L. Henley, V. Elser, and M. Mihalkovič. *Z. Krist.* 215 (2000): 553.
38. K. Ingersent, "Matching rules for quasicrystalline tilings," in *Quasicrystals: The State of the Art*, ed. D. P. DiVincenzo and P. J. Steinhardt (Singapore: World Scientific, 1991), 185.
39. C. Janot, *Quasicrystals: A Primer* (New York: Oxford University Press, 1992).
40. M. Wollgarten, M. Beyss, K. Urban, H. Liebertz, and U. Koster, *Phys. Rev. Lett.* 71 (1993): 549.
41. A. P. Tsai, A. Inoue, Y. Yokoyama, and T. Masumoto, *Mat. Trans. JIM* 31 (1990): 98.
42. M. M. De Boissieu Durandcharre, P. Bastie, A. Carabelli, M. Boudard, M. Bessiere, S. Lefebvre, C. Janot, and M. Audier, *Phil. Mag. Lett.* 65 (1992): 147.
43. Y. Yokoyama, A. Tsai, A. Inoue, and T. Masumoto, *Mater. Trans. JIM* 32 (1991): 1089.
44. M. Boudard, E. Bourgeat-Lami, M. de Boissieu, C. Janot, M. Durand-Charre, H. Klein, M. Audier, and B. Hennion, *Phil. Mag. Lett.* 71 (1995): 11.
45. J. Gastaldi, S. Agliozzo, A. Létoublon, J. Wang, L. Mancini, H. Klein, J. Hartwig, J. Baruchel, I. R. Fisher, T. Sato et al., *Phil. Mag. B* 83 (2003): 1.
46. S. W. Kycia, A. I. Goldman, T. A. Lograsso, D. W. Delaney, D. Black, M. Sutton, E. Dufresne, R. Bruning, and B. Rodricks, *Phys. Rev. B* 48 (1993): 3544.
47. T. Jach, Y. Zhang, R. Colella, M. de Boissieu, M. Boudard, A. I. Goldman, T. A. Lograsso, D. W. Delaney, and S. Kycia, *Phys. Rev. Lett.* 82 (1999): 2904.
48. M. de Boissieu, P. Stephens, M. Boudard, C. Janot, D. Chapman, and M. Audier, *Phys. Rev. Lett.* 72, no. 22 (1994): 3538.
49. M. de Boissieu, P. Stephens, M. Boudard, C. Janot, D. Chapman, and M. Audier, *J. Phys.: Condens. Matter* 6 (1994): 10725.
50. J. W. Cahn, D. Shechtman, and D. Gratias, *J. Mater. Res.* 1 (1986): 13.
51. M. M. De Boissieu Boudard, B. Hennion, R. Bellissent, S. Kycia, A. I. Goldman, C. Janot, and M. Audier, *Phys. Rev. Lett.* 75 (1995): 89.
52. A. Létoublon, M. de Boissieu, M. Boudard, L. Mancini, J. Gastaldi, B. Hennion, R. Caudron, and R. Bellissent, *Phil. Mag. Lett.* 81 (2001): 273.
53. M. Widom, *Phil. Mag. Lett.* 64, no. 5 (1991): 297.
54. M. Boudard, M. de Boissieu, A. Létoublon, B. Hennion, R. Bellissent, and C. Janot, *Europhys. Lett.* 33, no. 3 (1996): 199.
55. M. de Boissieu, M. Boudard, T. Ishimasa, J. P. Lauriat, A. Létoublon, M. Audier, M. Duneau, and A. Davroski, *Phil. Mag. A* 78 (1998): 305.
56. M. Audier, M. Duneau, M. de Boissieu, M. Boudard, and A. Létoublon, *Phil. Mag. A* 79 (1999): 255.
57. A. Létoublon, T. Ishimasa, M. de Boissieu, M. Boudard, B. Hennion, and M. Mori, *Phil. Mag. Lett.* 80 (2000): 205.
58. L. J. Shaw, V. Elser, and C. L. Henley, *Phys. Rev. B* 43, no. 4 (1991): 3423.

59. G. Coddens, S. Lyonnard, B. Hennion, and Y. Calvayrac, *Phys. Rev. B* 62, no. 10 (2000): 6268.
60. M. Sutton, S. G. J. Mochrie, T. S. E. N. Greytak, L. E. Berman, G. A. Held, and G. B. Stephenson, *Nature* 352 (1991): 608.
61. A. Létoublon, F. Yakhou, F. Livet, F. Bley, M. de Boissieu, L. Mancini, R. Caudron, C. Vettier, and J. Gastaldi, *Europhys. Lett.* 54 (2001): 753.
62. F. Livet, F. Bley, J. Mainville, R. Caudron, S. G. Mochrie, G. Dolino, D. Abernathy, G. Grüebel, and M. Sutton, *Nucl. Inst. Meth.* 451 (2000): 596.
63. S. Francoual, F. Livet, M. de Boissieu, F. Yakhou, F. Bley, A. Létoublon, R. Caudron, and J. Gastaldi, *Phys. Rev. Lett.* 91 (2003): 225501/1.
64. S. Francoual, F. Livet, M. de Boissieu, F. Yakhou, F. Bley, A. Létoublon, R. Caudron, J. Gastaldi, and R. Currat, *Phil. Mag.* 86, nos. 6–8 (2006): 1029.
65. R. Bluher, P. Scharwaechter, W. Frank, and H. Kronmuller, *Phys. Rev. Lett.* 80, no. 5 (1998): 1014.
66. M. Feuerbacher, B. Baufeld, R. Rosenfeld, M. Bartsch, G. Hanke, M. Beyss, M. Wollgarten, U. Messerschmidt, and K. Urban, *Phil. Mag. Lett.* 71 (1995): 91.
67. D. Caillard, G. Vanderschaeve, L. Bresson, and D. Gratias, *Phil. Mag. A* 80, no. 1 (2000): 237.
68. M. Feuerbacher and D. Caillard, *Acta Mater.* 54 (2006): 3233.
69. M. Boudard, A. Létoublon, M. de Boissieu, T. Ishimasa, M. Mori, E. Elkaim, and J. P. Lauriat, *Mater. Sci. Eng. A* 294–296 (2000): 217.
70. M. de Boissieu, H. Takakura, M. Bletry, J. Q. Guo, and A. P. Tsai, *J. Alloy Comp.* 342 (2002): 265.
71. P. A. Bancel, *Phys. Rev. Lett.* 63 (1989): 2741.
72. P. A. Bancel, *Phil. Mag. Lett.* 67 (1993): 43.
73. N. Menguy, M. de Boissieu, P. Guyot, M. Audier, E. Elkaim, and J. P. Lauriat, *J. Phys. I France* 3 (1993): 1953.
74. Y. Kaneko, Y. Arichika, and T. Ishimasa, *Phil. Mag. Lett.* 81 (2001): 777.
75. M. de Boissieu, S. Francoual, Y. Kaneko, and T. Ishimasa, *Phys. Rev. Lett.* 95 (2005): 105503\1.
76. S. Francoual (PhD diss., Université Joseph Fourier, 2006).
77. K. Edagawa, *Phil. Mag. Lett.* 85 (2005): 455.
78. K. Edagawa and Y. G. So, *Phil. Mag.* 87 (2007): 77.
79. K. Edagawa, *Phil. Mag.* 87 (2007): 2789.
80. M. de Boissieu, S. Francoual, M. Mihalkovič, K. Shibata, A. Q. R. Baron, Y. Sidis, T. Ishimasa, D. Wu, T. Lograsso, L. P. Regnault, F. Gähler et al., *Nat. Mater.* 6 (2007): 977.

13

Structural Disorder in Quasicrystals

Thomas Weber

Walter Steurer

Laboratory of Crystallography, Department of Materials, ETH Zurich, Wolfgang-Pauli-Strasse 10, CH-8093 Zurich, Switzerland

Abstract

Quasicrystals (QCs) may show all types of disorder phenomena observable in periodic crystals (e.g., substitutional or displacive disorder). Other types of disorder are special for QCs (e.g., phasonic disorder) or are better understood in higher dimensional description (e.g., out-of-phase domains, disordered superstructures). Contrary to periodic crystals, observation of diffuse scattering in diffraction experiments is not necessarily a proof for the presence of disorder. Even in cases of perfectly ordered QCs, limited experimental resolution leads to a superposition of Bragg reflections that might appear as a continuous diffuse background. Distinguishing such artificial diffuse intensities from structural diffuse scattering is possible by interpretation of Bragg intensities. Further, several examples of diffuse scattering from decagonal and icosahedral quasicrystals are discussed. It is shown that pair-distribution function analysis based on single-crystal diffuse scattering is a powerful tool for investigation of complicated disorder phenomena in QCs.

1 Introduction

All QCs studied so far show diffuse scattering—some more, some less. Does this mean that they all are disordered? What does disorder mean in the case of QCs anyway? Is it possible to apply the concepts used for disordered periodic crystals to QCs as well? In other words, can the structure of any real QC be uniquely separated into a strictly long-range ordered average structure giving rise to Bragg reflections and into a disordered difference structure causing diffuse scattering? How does the definition and description of disorder in QCs show up in their higher dimensional (nD) description, and what kind of disorder can be deduced from it? Phasonic disorder, for instance, quite naturally results from the *n*D approach. Are the experimental techniques developed for the study of disorder in periodic crystals directly applicable to QCs, and can the diffuse scattering be interpreted analogously? Understanding the relationships between disorder and diffuse scattering is crucial for answering the still open question about the factors responsible for the stabilization of QCs. Are QCs high-temperature phases, is entropy the crucial stabilizing factor, and can a quasiperiodic structure be a ground state of matter?

In the past two decades more than 150 publications have appeared on diffuse scattering in QCs. For a review summarizing the state of the art until 1998, see Steurer and Frey [1]. For later papers, see, for instance, Wang et al. [2] (on thermal diffuse scattering [TDS] in QCs), De Boissieu [3] (on phason diffuse scattering [PDS]), Francoual et al. [4] (on phason dynamics), Kobas et al. [5,6] (on general structural disorder in decagonal Al-Co-Ni), Letoublon et al. and Francoual et al. [7,8] (on coherent X-ray speckle patterns caused by phason modes), and Stewart et al. [9] (on magnetic diffuse scattering). In our present contribution, we focus on static structural diffuse scattering and skip the discussion of TDS and PDS, which will be dealt with in another part of this book. We discuss some fundamental concepts of disorder and diffuse scattering in QCs and illustrate them with several examples.

2 Theory

2.1 Disorder and Diffuse Scattering

Structural diffuse scattering reflects the deviation of a disordered ideal crystal structure from its periodic average structure, that is, the nonperiodic disordered structure taken modulo one unit cell of the underlying basic structure. Assuming the validity of kinematic scattering theory, the diffuse diffraction intensity of a disordered crystal can be written as

$$I_{diff}(\mathbf{h}) = I_{total}(\mathbf{h}) - I_{Bragg}(\mathbf{h}) = \left|FT\left[\rho_{dis}(\mathbf{x})\right] - FT\left[\rho_{aver}(\mathbf{x})\right]\right|^2 = \left|FT\left[\Delta\rho(\mathbf{x})\right]\right|^2, \qquad (13.1)$$

where FT denotes the Fourier-transform operation, $\rho_{dis}(\mathbf{x})$ is the scattering density distribution of the disordered crystal structure, $\rho_{aver}(\mathbf{x})$ is the density obtained by averaging the crystal structure over all unit cells, and $\Delta\rho(\mathbf{x})$ is its deviation from the average structure. By definition, $\Delta\rho(\mathbf{x})$ is not long-range ordered, and its average over all unit cells, $\Delta\rho_{aver}(\mathbf{x})$, is zero. The average density $\rho_{aver}(\mathbf{x})$ can be obtained by standard methods based on Bragg intensities alone. Reconstruction of $\Delta\rho(\mathbf{x})$, and thus of $\rho_{dis}(\mathbf{x})$, by assigning phases to structure amplitudes of diffuse intensities is only possible in the case of fully coherent diffraction. In general, coherent disorder scattering does not have a smooth intensity distribution; it shows a so-called speckle pattern. In case of conventional experiments with incoherent radiation, it is incoherently broadened by the experimental resolution function, leading to a smooth diffuse diffraction pattern. If we assume the resolution is constant in reciprocal space, the observed diffraction pattern can be described by the convolution product

$$I_{obs}(\mathbf{h}) = I_{speckle}(\mathbf{h}) * R(\mathbf{h}), \qquad (13.2)$$

with observed intensities, $I_{obs}(\mathbf{h})$; intensities of a fully coherent diffraction experiment, $I_{speckle}(\mathbf{h})$; and the reciprocal space resolution function, $R(\mathbf{h})$. According to the convolution theorem, the effect of limited reciprocal space resolution on direct space can be described by the product

$$P_{obs}(\mathbf{x}) = P_{dis}(\mathbf{x})\ r(\mathbf{x}), \tag{13.3}$$

where $P_{obs}(\mathbf{x})$ is the observable autocorrelation function (= Patterson function or single-crystal pair distribution function [PDF]) obtained by Fourier transformation of observed intensities, $P_{dis}(\mathbf{x})$ is the Fourier transform of $I_{speckle}(\mathbf{h})$, and $r(\mathbf{x})$ is the Fourier transform of the resolution function. If $R(\mathbf{h})$ is a Gaussian with given standard deviation, σ_R, then $r(\mathbf{x})$ is also a Gaussian with $\sigma_r = 1/(2\pi\sigma_R)$. The resolution function can be directly determined from the profiles of the Bragg reflections. Typical full widths at half-maximum (FWHM) of Bragg reflections measured in synchrotron experiments with area detectors are on the order of 0.001 Å^{-1}, giving $\sigma_R = \frac{0.001}{2\sqrt{2\ln 2}}\,\text{Å}^{-1}$. (*Note:* All reciprocal space units used in this chapter are defined by $d^* = \frac{2\sin\theta}{\lambda}$.) The consequence of limited reciprocal space resolution is that the PDF densities are damped by 50% at a distance from the origin of $\frac{1000\text{Å}\cdot 8\ln 2}{4\pi} \approx 440\text{Å}$, while those with a distance of 100 Å are weakened by only 3.5%.

2.2 Order and Disorder in Quasicrystals

2.2.1 Higher Dimensional Embedding of Ideal Quasicrystals

The structure of QCs is best represented in *n*D space, which allows a periodic description of an actually quasiperiodic *d*D structure (Figure 13.1). The *n*D lattice is decorated by hyperatoms, the (*n-d*) perpendicular-space (perp-space) component, which is called the atomic surface or occupation domain, and the parallel-space (par-space) component is just the *d*D atom (electron density distribution in the case of X-ray diffraction). The *d*D crystal structure in par-space is obtained as an irrational section of the *n*D hypercrystal structure. According to the rules of Fourier transformation, the par-space diffraction pattern then results from the projection of the hypothetical reciprocal *n*D space to the experimentally accessible *d*D reciprocal par-space.

The *n*D embedding is not unique due to additional degrees of freedom along the perp-space. In practice, however, this is no problem because only a few settings are useful and rules for the determination of a proper basis are known [10]. Information about the perp-space component of diffraction phenomena is lost by the projection onto the par-space. Since, by definition, Bragg reflections are located at the nodes of the *n*D hyperlattice, a unique *n*D indexing of Bragg reflections is possible once a proper basis has been found. Generally this method cannot be used for *n*D indexing of diffuse intensities because diffuse scattering is not restricted to lattice nodes and therefore perp-space components cannot be reconstructed from their experimentally observed par-space positions.

2.2.2 Definition of Disorder in Quasiperiodic Structures

A strictly quasiperiodic structure is purely point diffractive; that is, it shows Bragg reflections only. The set of diffraction vectors forms a *Z* module of rank *n* in *d*D space ($n > d$ for QCs, $n = d$ for periodic structures). For a given *n*D representation, there are infinitely many different (incongruent) realizations of the actual quasiperiodic structure, depending on the relative position of the cutting par-space.

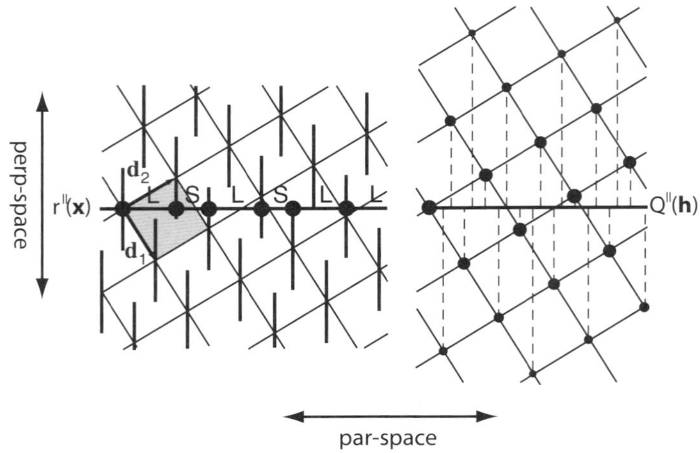

Figure 13.1. *nD embedding of a quasiperiodic structure illustrated on the example of the Fibonacci sequence (at left: direct space; at right: reciprocal space). The lattice nodes of the two-dimensional hyperlattice (unit cell shaded gray) are decorated by hyperatoms, that is, line segments with the length of a unit cell projected onto the perp-space. The point atoms in one-dimensional par-space appear wherever the par-space cuts a hyperatom. The Fibonacci sequence has two different vertex distances, L and S, with $L = \tau S$ (τ is the golden mean 1.61803...). The intensity-weighted reciprocal lattice is obtained by Fourier transformation of the hyperstructure. Its projection onto par-space gives the one-dimensional diffraction pattern. Since the relative orientation of the two-dimensional lattice to the par-space is irrational, Bragg reflections are never projected onto the same position.*

They all belong to the same local isomorphism class; that is, all structures are homometric and have the same energy and entropy. The structures have the same discrete Patterson function and, consequently, the same point diffraction pattern without any continuous contribution (diffuse scattering).

Phasons, that is, excitations of a quasiperiodic structure with polarization along the perp-space, are the origin of PDS. They are an analogy to phonons, which are excitations in par-space causing TDS. PDS (TDS) is strong beneath and around strong Bragg reflections with large perp- (par-)space components, respectively. Phason modes correspond to correlated atomic jumps in well-defined double-well potentials. In the *n*D picture, they result from perp-space fluctuations. If we exclude this kind of structural disorder from our discussion, what kind of disorder is left? First, there is chemical disorder, particularly in those cases where the compositional phase field has a certain width or, generally, at high temperatures. Second, there is displacive disorder, either as a consequence of chemical disorder due to size effects or due to symmetry mismatch. Symmetry mismatch takes place, for instance, if the symmetry of some cluster shells is lower than the average symmetry of a cluster; such is the case for the two innermost shells of the Tsai cluster. Third, there is the formation of domain structures needed for the release of strain or resulting from transformations from high-symmetry, high-temperature phases. Fourth, there is disorder due to the limited correlation length or imperfect realization of ordering parameters.

2.2.3 Models of Disorder in Higher Dimensional Description

In the following, some principles of the *n*D description of disorder are illustrated by examples derived from Fibonacci sequence-based structures. We will concentrate on cases that are special for QCs (e.g., phason flips, quasiperiodic approximant domains) or that are more easily understood in *n*D than in physical space (e.g., out-of-phase domains).

2.2.3.1 Phason Flips

Disorder due to fluctuations of hyperatoms along perp-space is called *phasonic disorder*. Analogous to phononic excitations in par-space and corresponding TDS, correlated phasonic fluctuations along perp-space are called *phason modes* leading to PDS. We will restrict our attention to uncorrelated local random phason fluctuations.

If a hyperatom, cut by par-space close to its boundary, is shifted along perp-space, it may no longer be cut, and the atom "disappears." Conversely, a perp-space shift of a hyperatom somewhere else in the structure, already close to par-space, could "create" a new atom if it is cut then. Obviously, such a scenario is not possible in a real crystal. The problem does not occur if the *n*D structure fulfills the so-called *closeness condition*. The closeness condition says that hyperatoms must be arranged in such a way that any shift of physical space along perp-space directions does not change the total number of atoms; that is, when one hyperatom is leaving par-space, there must be another close by entering it at the same time. Then we have just an atomic jump over a short distance (phason flip) (Figure 13.2). From a crystal chemical point of view, phason flips represent jumps between the minima of double-well potentials. Uncorrelated phason flips lead to broad diffuse scattering features.

Phason flips are not necessarily a signature of disorder. A continuous shift of the cutting space along the perp-space results in cascades of phason flips, which are deterministic, contrary to the random phason flips mentioned above. In the Fibonacci sequence, there are never two phason flips at exactly the same time when moving the cutting space along perp-space. All structures obtained by such a cutting-space shift belong to the same local isomorphism class. Their diffraction patterns do not differ because the two-dimensional hypercrystal structure remains unaffected by the cutting-space shift. It is a special property of quasiperiodic structures that strict long-range order can be preserved, even though they are modified at particular atomic positions. In the case of periodic structures, modifications must take place simultaneously in all unit cells; otherwise periodic long-range order is broken, leading to diffuse scattering.

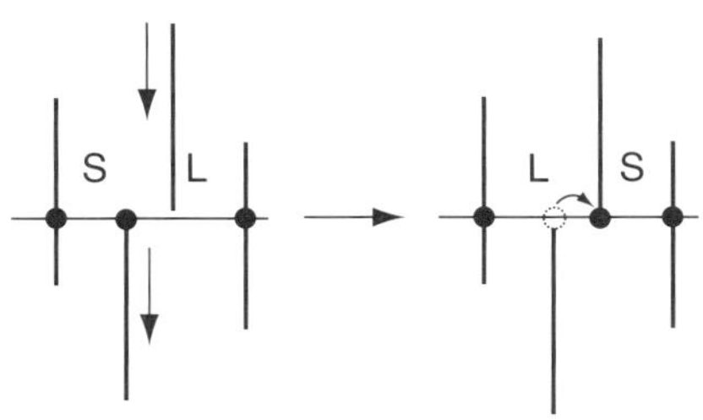

Figure 13.2. *Phason flip in a Fibonacci sequence. A correlated shift of two hyperatoms along perp-space results in an atomic jump over a distance length L-S.*

2.2.3.2 Quasiperiodic Approximant Domain Structures

Piecewise rational par-space cuts of an *n*D hypercrystal, separated by discommensurations of multiples of *n*D lattice vectors, lead to domain structures of rational approximants. If the par-space fluctuations average to zero, then the domain structure is quasiperiodic on average (Figure 13.3). Besides Bragg reflections resulting from the quasiperiodic average structure, diffuse intensities appear with maxima beneath and in the vicinity of Bragg reflections. Geometrically this kind of domain structure can be seen as a precursor of a lock-in transition of a "soft phason mode."

If the average domain size significantly exceeds the coherence length of the incident beam (depending on the source between several 100 Å up to several 10000 Å), then we mainly observe the diffraction pattern of a usually multiply twinned approximant structure accompanied by diffuse scattering from the domain boundaries.

2.2.3.3 Out-of-Phase Domains in Higher Dimensions

In the case of out-of-phase domain structures, the internal structure of neighboring domains is the same, but their lattice origins are shifted by a fraction of a lattice vector. In case of antiphase domains, it is just half a lattice vector. An example of an *n*D antiphase domain structure is shown in Figure 13.4. Locally the hypercrystal forms a 1×2 superstructure of the Fibonacci sequence due to two different types of hyperatoms (black and gray). Neighboring domains have the same hypercrystal structure; only their origins are shifted by a half lattice vector along the doubled axis. In the case of a constant domain size, one obtains a modulated *n*D structure. In our example, the commensurate modulation leads to a 6×2 superstructure of the Fibonacci sequence with additional Bragg reflections that can be indexed by integers in higher dimensional description. Averaging over the domain structure gives the 1×1 unit cell of the underlying Fibonacci sequence. The related Bragg reflections will be unaffected by domain-size fluctuations contrary to those of the 6×2 superstructure and the original 1×2 structure, which all become diffuse.

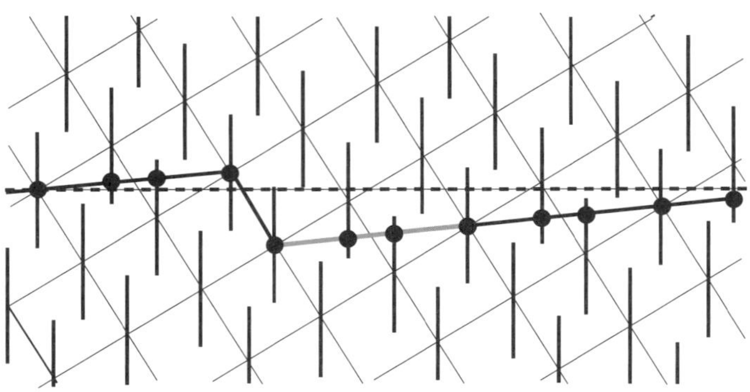

Figure 13.3. *Coherently intergrown 2/1-approximant domain structure with quasiperiodic average structure. The Fibonacci sequence ...LSLSLLSLLSL... is replaced by ...|LSL|S|LSL|LSL|..., that is, by ...|LSL|... approximant domains separated by discommensurations |S| or |L|. The cutting space of the approximant is shown by a solid line and that of the Fibonacci sequence by a dashed line. The one-dimensional approximant unit cell, |LSL|, is indicated by the gray line segment.*

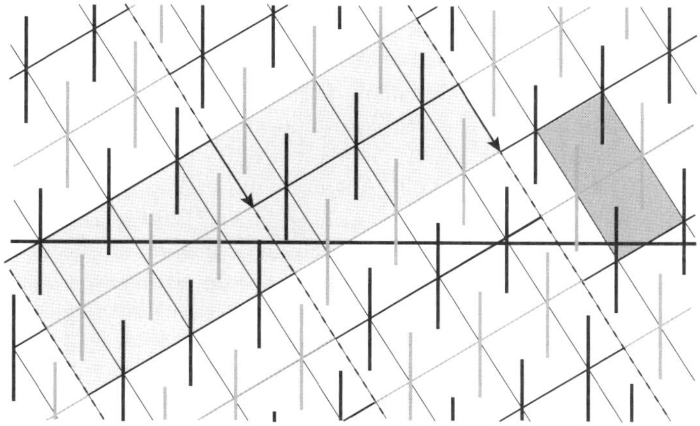

Figure 13.4. *Out-of-phase domain structure demonstrated on the example of a two-color Fibonacci sequence, which locally forms a twofold superstructure (dark gray rectangle) of a square lattice. Domain boundaries are shown by dashed lines and origin shifts between out-of-phase domains are indicated by arrows. The 6 × 2 superstructure (see text) is shown by the light gray rectangle.*

2.3 Analyzing Disorder by Pair Distribution Function (PDF) Analysis

The Fourier transform of the total scattering intensity (Bragg and diffuse scattering) gives the PDF of the real structure, that is, the density-weighted probability distribution function that two scatterers are separated by a given vector. The Fourier transform of Bragg scattering yields the PDF of the average structure, that is, the Patterson function. The possibility of higher dimensional indexing of Bragg reflections allows the calculation of Bragg scattering based PDF maps in nD as well as in dD space. The Fourier transform of the diffuse scattering results in the PDF of the difference structure $\Delta\rho(\mathbf{x})$, which will be called "Δ-PDF" in the following. It can only be obtained in physical space because diffuse intensities cannot be indexed uniquely in higher dimensions. A positive (negative) Δ-PDF peak means that the real structure shows the corresponding interatomic vector with a higher (lower) frequency than the Bragg structure. The Δ-PDF approach allows inclusion of knowledge about the average quasicrystalline structure and other a priori crystal chemical data as complementary information. However, it can also be used for the analysis of disordered structures based on diffuse scattering data alone. This is, in particular, valuable if disorder is investigated as a function of temperature, pressure, composition, and so on. As will be shown below, selective Fourier transformation of distinguishable classes of diffuse scattering as well as selective modeling of Δ-PDF densities allow a formidable reduction of complexity and the investigation of even very complicated diffuse scattering phenomena.

2.4 Experimental Aspects

According to theory, the reciprocal space of a QC is densely filled by Bragg reflections. However, only a discrete set of Bragg reflections is observable for a given dynamic range of a diffraction experiment because reflections close to each other in par-space are widely separated in perp-space. Due to the steep falloff of intensity with the increasing perp-space component of the diffraction vector, their intensities will strongly differ and only those that are in the detectability range will be observed. Strong reflections have small perp-space components and are well separated in par-space (Figure 13.1). Weak reflections with high perp-space components cannot be resolved below a certain threshold value and will be experimentally observed as a weak continuous diffuse background, as demonstrated in the following example. Figures 13.5a and 13.5b show a diffraction pattern of a perfectly ordered one-dimensional Fibonacci sequence of 10000 Al atoms with interatomic distances S = 2.40 Å and L = $\tau \times 2.40$ Å ≈ 3.88 Å The total size of the model crystal is ~33200 Å. The diffraction pattern was calculated in par-space by the squared modulo of the Fourier transform of the one-dimensional structure; that is, no higher dimensional embedding was used. For simulating experimental resolution, the diffraction pattern was convoluted by a Gaussian of half-width 0.001 Å^{-1} (Figures 13.5c and 13.5d). As a consequence, the spiky background pattern seen in Figures 13.5a and 13.5b changes into a smooth diffuse background. The maximum of the diffuse background intensity is ~4×10^{-5} of the peak intensity of the strongest Bragg reflection, which is weak but detectable with low intrinsic background detectors like image plates or pixel detectors. The ratio $I^{max}_{diffuse} / I^{max}_{Bragg}$ strongly increases if the resolution function gets broader. It is ~2×10^{-4} if the FWHM of $R(\mathbf{h})$ is doubled to 0.002 Å^{-1} and ~3×10^{-3} for a reciprocal space resolution of 0.01 Å^{-1}, a resolution typical for in-house experiments. In addition to the broad diffuse scattering, relatively narrow diffuse maxima can be observed beneath strong Bragg reflections, which from their appearance could be misinterpreted as TDS or PDS.

For comparing diffuse background induced by limited experimental resolution and diffuse scattering coming from disorder, the diffraction pattern of a Fibonacci chain affected by uncorrelated phason flips was investigated (Figures 13.5e and 13.5f). Structure simulation and Fourier transform were done similarly as for the simulations seen in Figures 13.5a,b, but now L-S sequences were randomly replaced by S-L sequences and vice versa. These phason flips were introduced without any spatial correlation, and the probability for a particular phason flip was 0.5. The diffraction patterns shown in Figures 13.5e and 13.5f represent fully coherent experiments. Interestingly, the intensity ratio between the strongest Bragg peaks and diffuse background in Figures 13.5c–f are very similar; that is, distinguishing disorder diffuse scattering from unresolved Bragg reflections is hardly possible in the shown example. Consequently, observation of diffuse scattering in diffraction experiments with QCs cannot be seen as a proof for the presence of disorder; resolution effects have to be considered as well. It is striking that the patterns shown in Figures 13.5d and 13.5f differ significantly when restricting attention to *resolved* Bragg reflections. In Figure 13.5d, Bragg reflections still form a dense set (as is the case in Figure 13.5b), but in Figure 13.5f only a few well-separated Bragg reflections are visible above the background, which all have a low perp-space component. Reflections with high perp-space values are weakened heavily such that they cannot be resolved from the diffuse background. This is a typical sign of the presence of phasonic disorder, and thus—ironically—the presence of disorder can be concluded from the behavior of Bragg reflections rather than from observation of diffuse scattering.

The diffuse background seen in Figures 13.5c and 13.5d is mainly the result of overlaps of weak reflections. The consequence of neglecting weak reflections in the Fourier transform of a Fibonacci sequence was discussed by Ma et al. and Li et al. [12,13]. It was shown that a par-space PDF calculated from strong reflections alone introduces artificial peaks at short PDF vectors. This means that limited reciprocal space resolution affects densities not only at long PDF vectors, as shown by Equation 13.3, but also at short distances when the kind of diffuse scattering seen in Figures 13.5c and 13.5d is ignored. It was further shown by Ma et al. and Li et al. [12,13] that artifacts introduced at short PDF

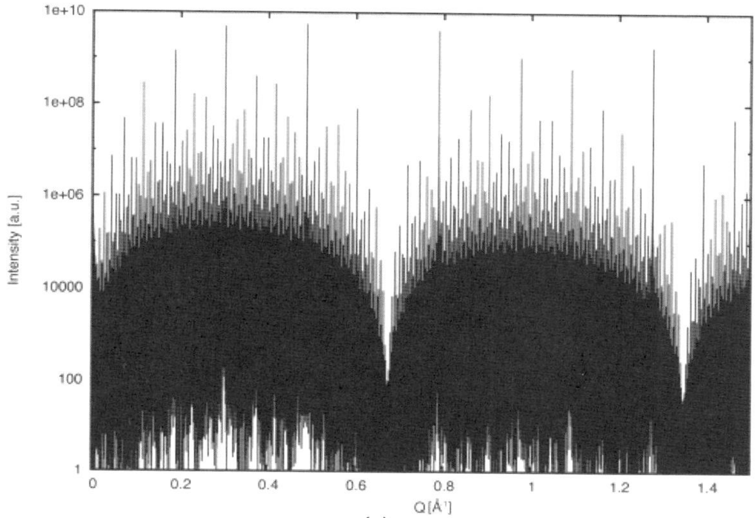

(a)

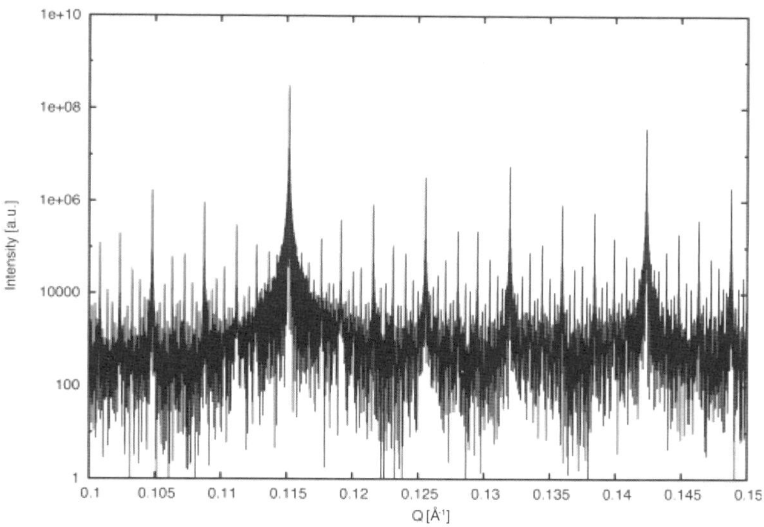

(b)

Figure 13.5. *Simulated diffraction intensities (logarithmic scale) of a Fibonacci sequence decorated with Al atoms. A fully coherent diffraction pattern is shown in (a). The same diffraction pattern convoluted by a Gaussian of FWHM 0.001 Å^{-1} is depicted in (c).* (continued on next page)

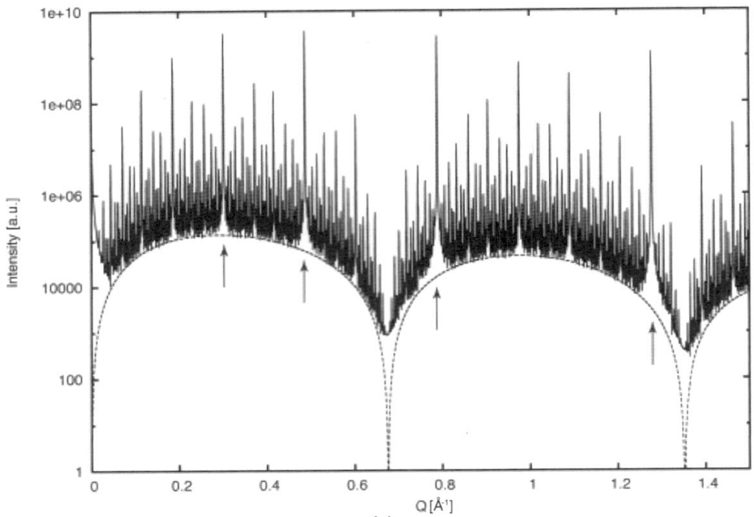

(c)

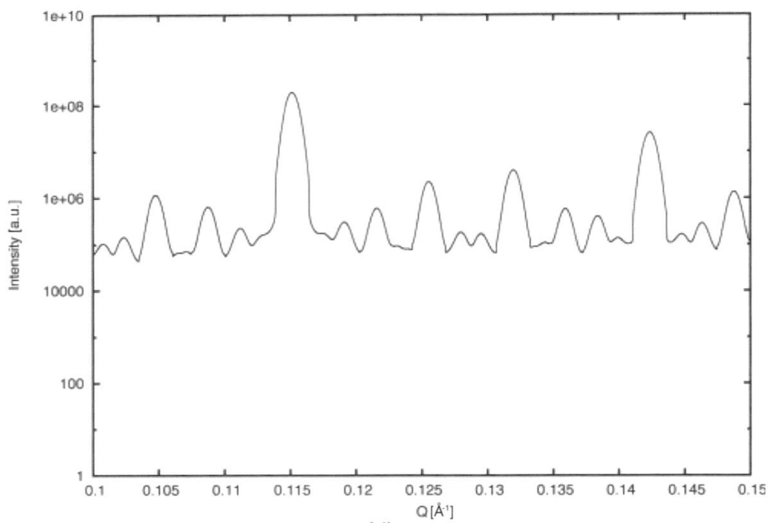

(d)

Figure 13.5. (continued) *The lower envelope line in (c) represents a continuous diffuse background that can be described by the relation* $I_{diffuse}(Q) = a \cdot f_{Al}^2(Q)\left\{1 - \cos\left[2\pi(L - S)Q\right]\right\}$ *(a is a scale factor,* $f_{Al}(Q)$ *is the atomic form factor of Al, and L and S are the long and short interatomic distances in a Fibonacci sequence).* (continued on next page)

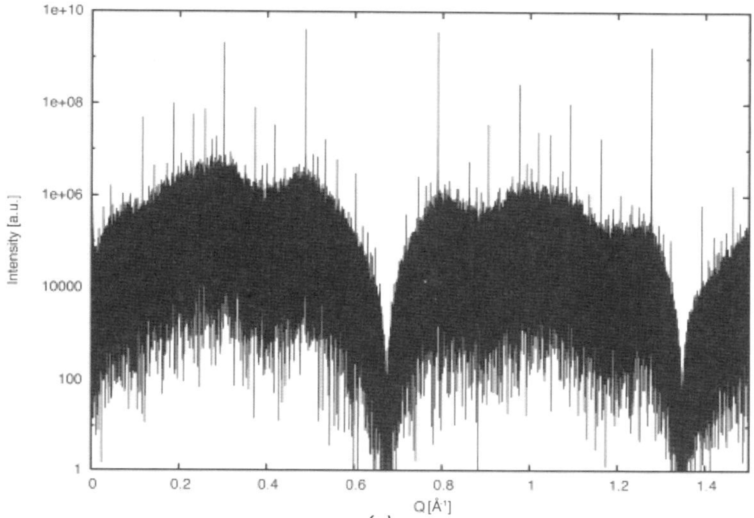

(e)

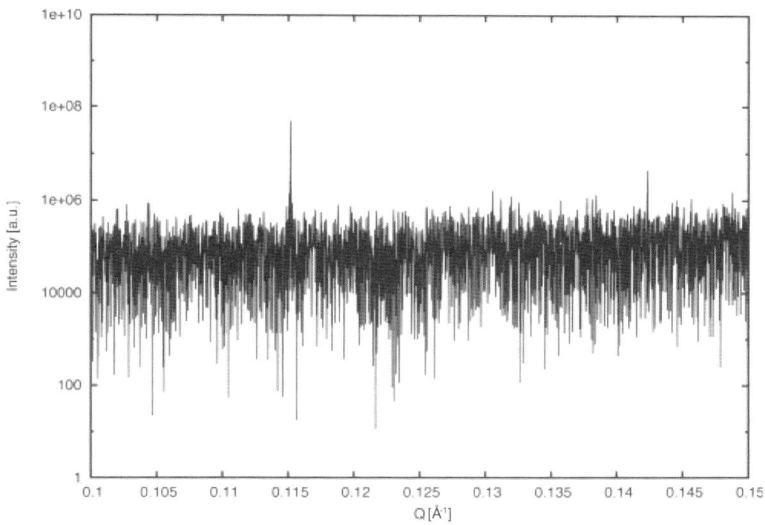

(f)

Figure 13.5. (continued) *The coherent diffraction pattern of a Fibonacci chain disordered by random phason flips is shown in (e). Magnifications of the ranges 0.1 Å^{-1} < Q < 0.15 Å^{-1} of (a), (c), and (e) are shown in (b), (d), and (f), respectively. The periodic upper envelope of the intensities is a property of the periodic average structure of a Fibonacci sequence ([11]). Some diffuse maxima having profiles similar to phononic/phasonic diffuse scattering are marked by arrows in (c).*

vectors are completely eliminated if weak reflections (i.e., diffuse scattering in the present example) are included. Note that the ambiguity in the separation of Bragg and diffuse scattering is expected to be even more serious in the case of real experimental data. First, our simulations neglect experimental factors such as background scattering or statistical noise, which would make the separation of weak reflections and background even more difficult. Second, the hyperatoms of real QCs are usually more complex than those of a Fibonacci sequence. As a consequence of complexity, Bragg reflections with higher perp-space components may gain intensity, and therefore par-space overlaps of Bragg reflections with relatively strong intensities may become more likely.

3 Examples

3.1 *Diffuse Scattering from d-$Al_{70}Co_{12}Ni_{18}$*

Figure 13.6 shows diffraction patterns of decagonal $Al_{70}Co_{12}Ni_{18}$ as a function of temperature. The crystal was taken from a sample that was quenched from 1170 K by switching off the high-vacuum furnace. In situ diffraction experiments were done at the Swiss-Norwegian beamline (SNBL) at the European Synchrotron Radiation Facility (ESRF) in Grenoble, France. Data were collected at room temperature, 1070 K, and 1120 K (in this sequence) using a high-temperature environment furnace [14] attached to a *mar345* image plate system. For further experimental details, see Steurer et al. [15]. Three types of diffraction phenomena, which are typical for many decagonal QCs, can be identified. Even-numbered layers along the periodic tenfold axis contain Bragg reflections as well as diffuse scattering, whereas odd-numbered layers ("interlayers") show diffuse scattering only. The distance between Bragg layers corresponds to a real-space periodicity of about 4 Å, whereas the diffuse layers represent a twofold (i.e., ~8 Å) periodic superstructure along the tenfold axis. The structure consists of a four-layer periodic arrangement ...ABA′B′ABA′B′..., along the tenfold axis (in the following also referred to as *c* direction or *z* coordinate) where every second layer is flat and the others are puckered [16]. Flat layers act as mirror planes for off-plane displacements in the puckered layers. Local columnar displacements of 4 Å along the periodic direction lead to a 4 Å periodic superposition $\ldots \frac{1}{2}(A{+}A') \frac{1}{2}(B{+}B') \frac{1}{2}(A{+}A') \frac{1}{2}(B{+}B') \ldots$ seen in Bragg layers. Diffuse scattering in interlayers is governed by the differences (A – A′) and (B – B′) as well as by lateral correlations of the 4 Å displacements.

Diffuse scattering in even- and odd-numbered layers shows not only a different appearance but also a different temperature behavior. Diffuse scattering in Bragg layers is relatively broad at room temperature and 1070 K but becomes streaky at high temperatures. The interlayer diffuse scattering is rather complex at ambient temperature. At 1070 K, diffuse scattering condenses to almost Bragg peak-like maxima and, finally, at 1120 K, it becomes broad. The observation of different appearance and temperature dependency indicates that diffuse scattering in Bragg layers and interlayers are of different origin and can thus be investigated separately.

3.1.1 *Diffuse Scattering in Bragg Layers*

Δ-PDF maps calculated from diffuse scattering in Bragg layers at different temperatures are shown in Figure 13.7. There is only a little variation in the PDF densities as a function of temperature. In

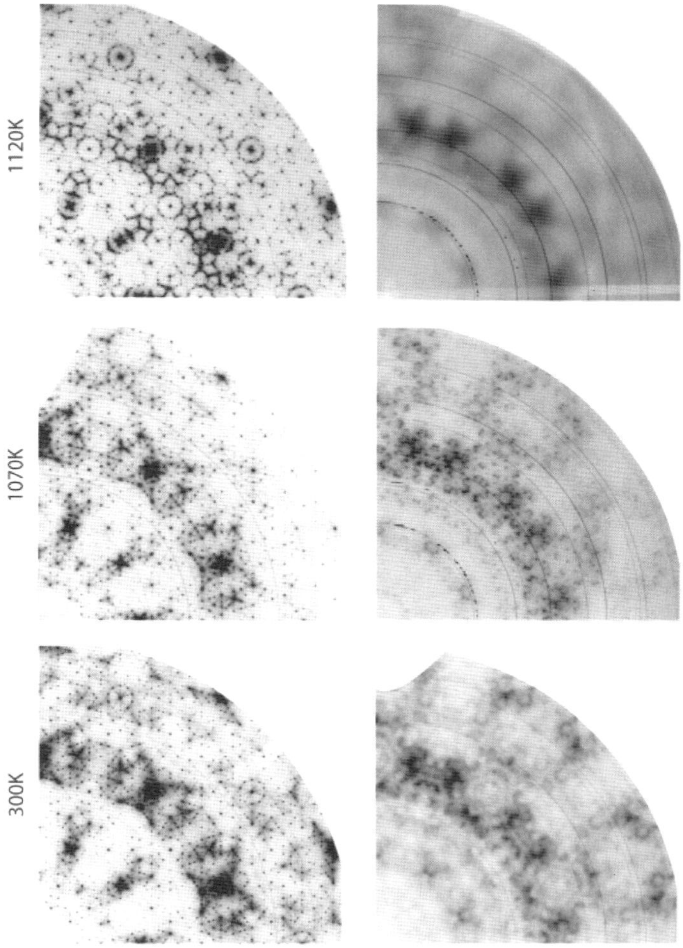

Figure 13.6. *X-ray diffraction patterns of $Al_{70}Co_{12}Ni_{18}$ taken at different temperatures with synchrotron radiation. The left column shows the zeroth Bragg layer, and the right column shows the first diffuse interlayer. The powder rings are caused by the sample holder.*

particular, the peak distribution does not change with temperature. Only densities of peaks that are far from the origin of PDF space increase slightly with increasing temperature; that is, correlations increase as a function of temperature.

In order to test whether the diffuse scattering observed in the Bragg layers of d - $Al_{70}Co_{12}Ni_{18}$ (Figure 13.7) is related to phononic/phasonic disorder, TDS/PDS was simulated employing an nD extension of the hydrodynamic theory [2,17]. An advantage of this approach is that no atomic structure model is required for calculating diffuse intensities. The input needed are the Bragg intensities,

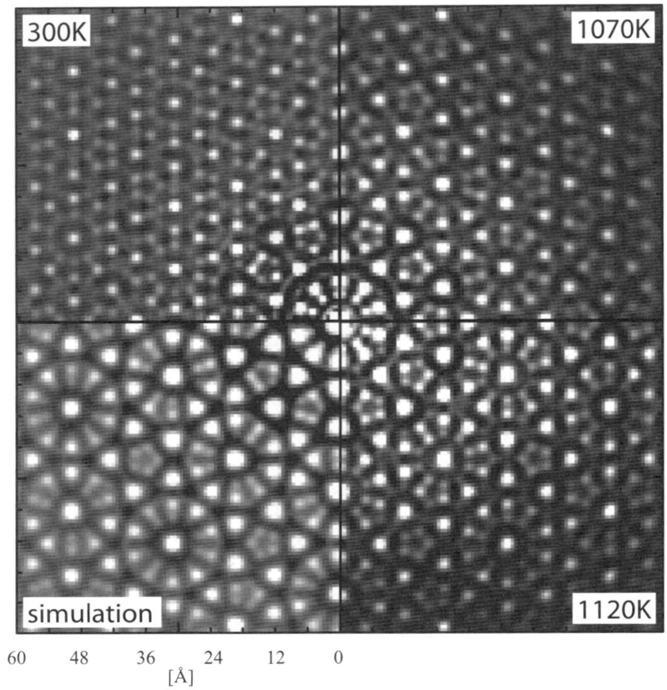

Figure 13.7. *Zero-layer Δ-PDF maps calculated from the diffuse scattering in d-$Al_{70}Co_{12}Ni_{18}$ as a function of temperature compared to a PDF map calculated from simulated PDS.*

a set of five par-space elastic constants governing the appearance of TDS, two perp-space constants describing PDS, and one phonon-phason coupling constant. Experimental values of par-space constants obtained by resonant ultrasound spectroscopy experiments were taken from Chernikov et al. [18]. Perp-space elastic constants cannot be measured directly and were found by trial and error [5,6]. Figure 13.8 shows a comparison of the 1120 K experimental diffraction intensities and the calculated TDS/PDS pattern.

The agreement between the patterns is satisfying in the sense that weak and strong diffuse scattering show a similar general distribution in experiment and simulation. The fine structure of the diffuse features, however, differs significantly, so it is difficult to decide to what extent the underlying model is a valid representation of the structure. A clearer answer can be given by inspection of the Fourier transform (i.e., the Δ-PDFs) of measured and calculated diffuse diffraction patterns (Figure 13.7). The agreement is very good for short interatomic vectors (<30 Å) and gets increasingly worse at longer distances. Separate calculation of TDS and PDS using the elastic constants employed in the simulations has shown that the diffuse diffraction patterns are highly dominated by phasonic contributions. We can therefore conclude that, at least for short interatomic distances, diffuse scattering in the Bragg layers of *d*-AlCoNi results from phasonic disorder.

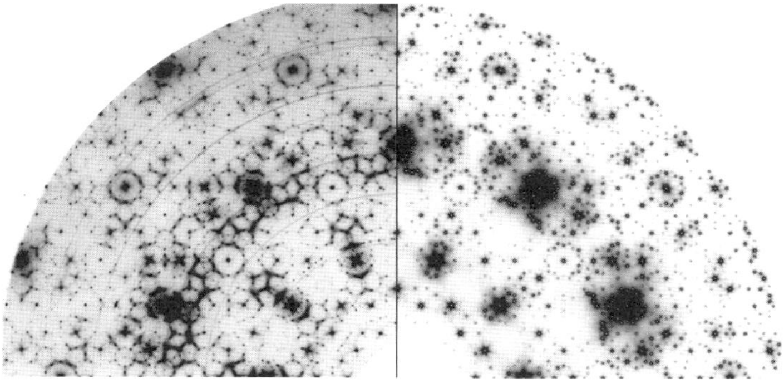

Figure 13.8. *Observed (left) and calculated (right) TDS/PDS of d-$Al_{70}Co_{12}Ni_{18}$.*

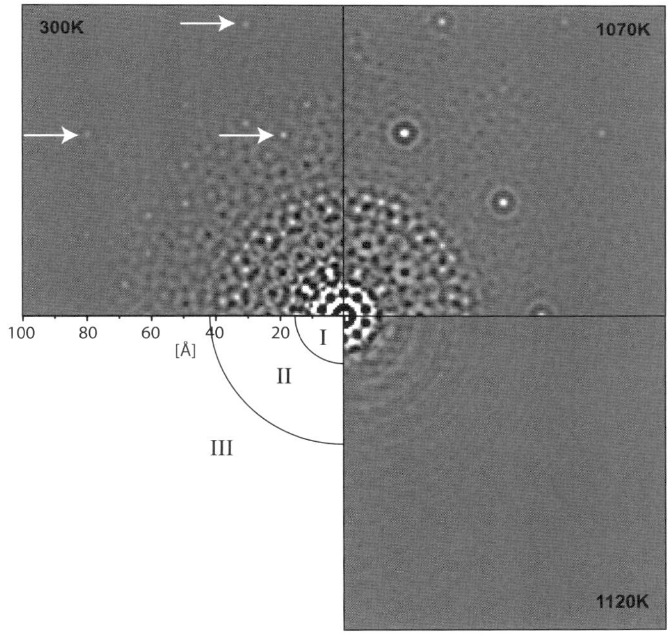

Figure 13.9. *Temperature-dependent Δ-PDF maps (z = 0) of d- $Al_{70}Co_{12}Ni_{18}$ calculated from diffuse interlayers only. Arrows indicate some superstructure peaks present in 300 K and 1070 K maps.*

3.1.2 Diffuse Scattering in Interlayers

Δ-PDF maps calculated from the diffuse interlayers alone are shown in Figure 13.9. The temperature-dependent Δ-PDF maps can be split into three zones. Zone I, which covers interatomic distances up to about 14 Å, is almost unchanged as a function of temperature; that is, there is a stable 14 Å-size structure motif ("cluster") that is unaffected by temperature. Zone II covers interatomic distances to about

42 Å. Δ-PDF densities in this zone are very similar at room temperature and at 1070 K but vanish completely at 1120 K. This can be explained by the presence of a 42 Å-size supercluster, which exists at least up to 1070 K. Finally, interatomic distances larger than 42 Å (zone III) show strong correlation peaks at room temperature as well as in the 1070 K pattern but no correlations at 1120 K. The strong peaks at 1070 K are a subset of the RT peaks, which at low temperatures are weaker than at 1070 K (see arrows). Since the room-temperature experiment was done with a quenched sample, we can conclude that the peaks present at RT and 1070 K represent a high-temperature superstructure that is not yet existing at solidification of the melt but is formed at slightly lower temperatures. The superstructure may not be stable at room temperature in this form and relaxes during cooling only partially, for kinetic reasons, decreasing long-range order. When heating up, the high-temperature superstructure is restored at 1070 K and finally completely eliminated at 1120 K. Due to limitations of the furnace, measurement of high-angle diffraction data was not possible, and thus the PDF resolution (~1.25 Å) is not sufficient for developing an atomistic model of the 14 Å cluster.

Figure 13.10. *X-ray diffraction pattern of d-$Al_{65}Co_{20}Ni_{15}$ showing the first interlayer. Data were collected at room temperature at SNBL ESRF using a mar345 image-plate detector system. Parasitic background was subtracted and the image was averaged according to 10/mmm Laue symmetry. Note that the diffuse diffraction pattern (almost) perfectly fulfills extinction rules of a five-dimensional glide plane (dotted lines) with the glide component along the periodic direction (c-glide plane).*

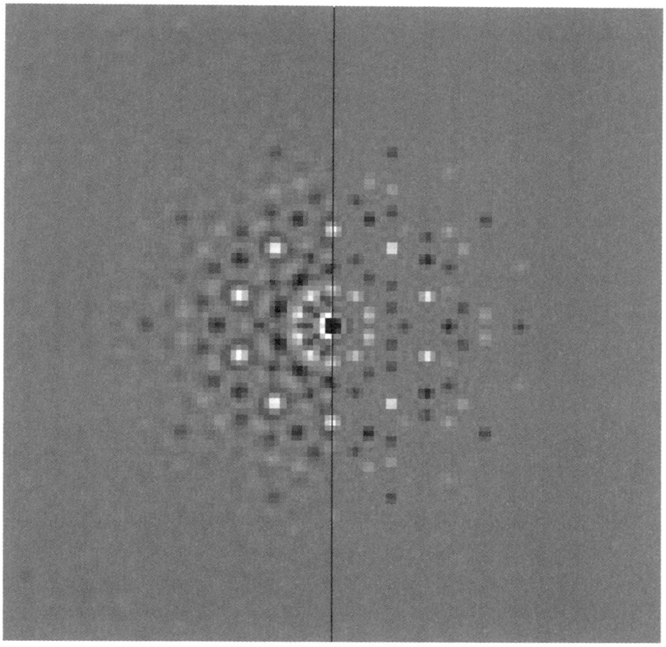

Figure 13.11. *Experimental (left) and simulated (right) Δ-PDFs of d-$Al_{65}Cu_{20}Co_{15}$ at z = 0.8 Å. The edge length of the map is ±25 Å along each direction.*

3.2 Diffuse Interlayers from d-$Al_{65}Cu_{20}Co_{15}$

Decagonal $Al_{65}Cu_{20}Co_{15}$ [19] and *d*-$Al_{70}Co_{12}Ni_{18}$ have similar diffraction patterns for both Bragg and diffuse scattering. For *d*–$Al_{65}Cu_{20}Co_{15}$, high-quality room-temperature diffraction data are available (Figure 13.10), yielding a PDF resolution of about 0.84 Å. A Δ-PDF section calculated from the diffuse interlayers is shown in Figure 13.11.

Similar to *d*-$Al_{70}Co_{12}Ni_{18}$, strong Δ-PDF peaks are observable up to a distance of ~14 Å. At longer distances, only weak correlations can be found. A good agreement with the experimental PDF pattern is achieved based on the following model (Figure 13.12): a tenfold double ring, which is present in the average structure, is puckered such that the atoms are shifted out of plane by about ±0.4 Å along the periodic direction, whereby next-neighboring atoms are shifted along opposite directions.

Although this simple model neglects disorder in the flat layers as well as other correlations within the puckered layers, its Δ-PDF map is in good agreement with the Δ-PDF from experimental data (Figure 13.11). A more complete description of the underlying disorder will be published elsewhere [20]. This example demonstrates an interesting feature of the PDF technique. The Δ-PDF map shown in Figure 13.11 is a section of three-dimensional Δ-PDF space at z = 0.8 Å; that is, it represents only correlations between atoms that have a difference of 0.8 Å along the periodic direction. This is true for atoms in the puckered layers, whereas all atoms in the flat layers have the same z coordinates, and for interlayer correlations, Δz is much larger than 0.8 Å. The PDF map shown in Figure 13.11 is therefore

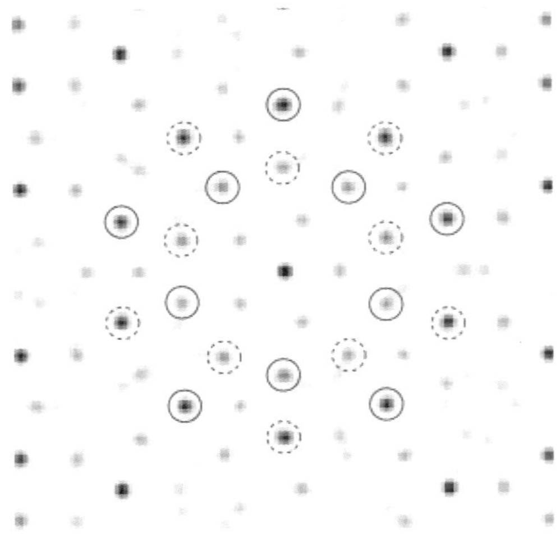

Figure 13.12. *Section from the d-$Al_{70}Co_{12}Ni_{18}$ structure as obtained from Bragg scattering [19]. Circles indicate the tenfold structure motif that was used for modeling diffuse scattering in the odd-numbered layers. Full and dashed lines indicate opposite out-of-plane displacements. The diameter of the motif is about 14 Å.*

only sensitive to atomic correlations within puckered layers. This allows a separate modeling of disorder in puckered and flat layers and therefore a significant reduction of complexity. It is interesting to note that the puckered tenfold double ring is a major component of a cluster that could be identified as being a prototypic building unit in the system *d*-AlCoNi for a broad range of quasicrystalline and approximant phases [21]. This analogy supports the assumption that interlayer diffuse scattering in *d*-AlCoNi and *d*-AlCuCo basically reflect the same kind of disorder.

3.3 Disordered Superstructure in i-$Al_{64}Cu_{23}Fe_{13}$

An example in which diffuse scattering can be interpreted as caused by a short-range ordered superstructure is shown in Figure 13.13.

The icosahedral QC *i*-$Al_{64}Cu_{23}Fe_{13}$ shows Bragg reflections that fulfill the condition that indices $h_1 \ldots h_6$ are all even or odd. Further, diffuse maxima but no sharp reflections can be observed at Bragg positions with mixed indices. The interpretation of the diffraction pattern is analogous to three-dimensional periodic crystals. Bragg reflections represent a six-dimensional face-centered lattice, but on a local scale *F*-centering is violated, leading to diffuse maxima at positions where Bragg reflections should be systematically extinct. The local structure is best described by a primitive six-dimensional unit cell; that is, hyperatoms located at edges and face centers of the six-dimensional unit cell are not identical on a local scale, but due to disorder they are indistinguishable on average. The correlation length of the primitive unit cell domains can be estimated from the FWHM of diffuse maxima to about 100 Å.

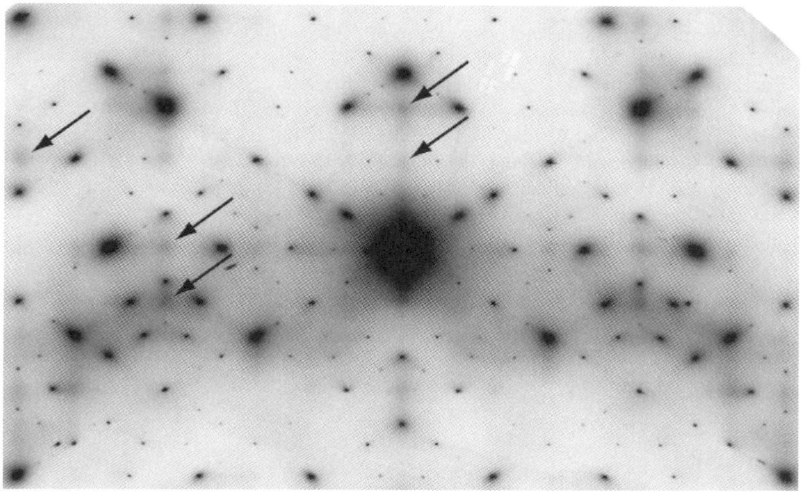

Figure 13.13. *Section from the twofold layer of i-$Al_{64}Cu_{23}Fe_{13}$. Some of the diffuse maxima occupying Bragg positions with mixed indices are indicated by arrows. Strong diffuse maxima beneath Bragg reflections are TDS/PDS-like diffraction phenomena. The data were collected at the PX beamline at Swiss Light Source (SLS) in Villigen, Switzerland, using a PILATUS 6M pixel detector. For further experimental details, see Weber et al. [22].*

4 Conclusions

Disorder, that is, structural modifications with only short-range correlations, gives rise to diffuse scattering in both periodic and quasiperiodic structures. The main difference lies in the interpretation of scattering intensities that appear diffuse in a diffraction experiment. In the case of quasiperiodic structures, this may just be caused by a lack of resolution, which smears the theoretically dense set of Bragg reflections.

References

1. W. Steurer and F. Frey, *Phase Trans.* 67 (1998): 319–361.
2. R. Wang, C. Hu, and J. Lei, *Phys. Rev. B* 61 (2000): 5843.
3. M. De Boissieu, *Phil. Mag.* 86 (2006): 1115.
4. S. Francoual, F. Livet, M. de Boissieu, F. Yakhou, F. Bley, A. Letoublon, R. Caudron, J. Gastaldi, and R. Currat, *Phil. Mag.* 86 (2006): 1029–1035.
5. M. Kobas, T. Weber, and W. Steurer, *Phys. Rev. B* 71 (2005): 224205.
6. M. Kobas, T. Weber, and W. Steurer, *Phys. Rev. B* 71 (2005): 224206.
7. A. Letoublon, F. Yakhou, F. Livet, F. Bley, M. de Boissieu, L. Mancini, R. Caudron, C. Vettier, and J. Gastaldi, *Europhys. Lett.* 54 (2001): 753–759.
8. S. Francoual, F. Livet, M. de Boissieu, F. Yakhou, F. Bley, A. Letoublon, R. Caudron, and J. Gastaldi, *Phys. Rev. Lett.* 9122 (2003): 5501–5501.
9. J. R. Stewart, K. H. Andersen, R. Cywinski, and A. P. Murani, *J. Appl. Phys.* 87 (2000): 5425–5430.
10. A. Cervellino, T. Haibach, and W. Steurer, *Phys. Rev. B* 57 (1998): 11223.

11. J. Wolny and L. Pytlik, *Acta Crystallogr.* A56 (2000): 11.
12. Y. Ma, M. A. Stern, X. Li, and J. Janot, *Phys. Rev. B* 40 (1989): 8053.
13. X. Li, A. E. Stern, and Y. Ma, *Phys. Rev. B* 43 (1991): 1371.
14. M. A. Estermann, H. Reifler, W. Steurer, F. Filser, P. Kocher, and L. J. Gaukler, *J. Appl. Cryst.* 32 (1999): 833.
15. W. Steurer, A. Cervellino, K. Lemster, S. Ortelli, and M. Estermann, *Chimia* 55 (2001): 528.
16. A. Cervellino, T. Haibach, and W. Steurer, *Acta Crystallogr.* B58 (2002): 8.
17. J. Lei, R. Wang, C. Hu, and D. H. Ding, *Phys. Rev. B* 59 (1999): 822.
18. M. A. Chernikov, H. R. Ott, A. Bianchi, A. Migliori, and T. W. Darling, *Phys. Rev. Lett.* 80 (1998): 321.
19. W. Steurer and K. H. Kuo, *Acta Crystallogr.* B46 (1990): 703.
20. P. Schaub, T. Weber, and W. Steurer (forthcoming 2009).
21. S. Deloudi and W. Steurer, *Phil. Mag.* 87 (2007): 2727.
22. T. Weber, S. Deloudi, M. Kobas, Y. Yokoyama, A. Inoue, and W. Steurer, *J. Appl. Cryst. 41*. (2008): 669.

14

DIFFUSE SCATTERING FROM QUASICRYSTALS

Short-Range Order, Atomic Size Effect, and Phason Fluctuations

Hiroshi Abe

Department of Materials Science and Engineering, National Defense Academy, Yokosuka 239-8686, Japan

Abstract

Atomic short-range order (SRO) is significant to resolve phase stability in quasicrystals. By anomalous X-ray scattering, diffuse scattering around Bragg reflections was observed in single decagonal quasicrystals of $Al_{70}Ni_{15}Co_{15}$ and $Al_{72}Ni_{20}Co_{8}$. Intensity modulations of the diffuse scattering occurred at each incident energy. By quantitative measurements and analysis, a contrast of the diffuse scattering shows the presence of SRO in three kinds of pair-correlation functions in ternary alloy systems. On the other hand, anisotropic distributions of diffuse scattering are observed in $Al_{70}Ni_{15}Co_{15}$. In addition, nonuniform deformation, which is caused by random phason strain, provides peak broadening of Bragg reflections. Both on Co concentration and temperature, atomic size effect, which is induced by SRO, influences the sensitivity of phasons in quasicrystal.

1 Introduction

Recently, functional materials have been manufactured for industrial applications. Nanotechnology in materials science is developed by self-organization. For instance, there are colloidal suspensions as nanosegregated structures, zeolites as frameworks for templates, porous material as hyperstructures, and copolymers as hybrid structures, which are self-assembled. At the same time, computer-aided material design is applied to composite materials. Large-scale simulations in complex materials are feasible using self-similarity. In the twenty-first century, great progress in experiments, theory, and simulation can enlarge complexity in science.

In principle, complexity is classified into *quantitative complexity* and *qualitative complexity*. The former is obtained by various kinds of atoms. The latter is described by an intrinsic property such as self-similarity. Self-similarity is represented mathematically based on the concept of fractals. Quasicrystals are a prototype of *qualitative complexity*. Discovery of quasicrystals had a major impact on crystallography [1]. In fact, stable phases of quasicrystals are synthesized easily in a binary or ternary alloy system [2]. By introducing hyperspace mathematics, quasiperiodicity is visualized perfectly. Hexagon-boat-star (HBS) tiling involves the smaller HBS tiling (Figure 14.1). The scaling law of HBS tiling is described by τ^2 enlargement. τ is the golden mean, which is given by $\frac{1+\sqrt{5}}{2}$.

In this chapter, atomic SRO on quasiperiodic lattices is examined by using anomalous X-ray scattering. Also, peak broadening of Bragg reflections and anomalous Debye-Waller (DW) factors are discussed in order to analyze static and dynamic fluctuations in quasicrystals.

2 Phason Models and Diffraction Patterns

The concept of phasons is introduced theoretically in quasicrystals [3]. For simplicity, two kinds of phasons are explained using a Fibonacci lattice. To express quasiperiodicity, the perpendicular space, $E_\perp$, is extended virtually (Figure 14.2a) [4]. $E_\perp$ is defined as perpendicular to the parallel space, $E_\parallel$, and the hyperspace consists of $E_\parallel$ and $E_\perp$. The Fibonacci chain is obtained by section method onto the parallel space, $E_\parallel$, when the periodic lattice in hyperspace tilts by $1/\tau$ on $E_\parallel$. $E_\perp$ is connected with a phason degree of freedom that controls quasiperiodicity while phonon fluctuations exist in $E_\parallel$. Phonon and phason displacements are given by $u_\parallel$ and $u_\perp$, respectively. If the tilt value changes to a rational number from an irrational one such as $1/\tau$, long periodicity appears in $E_\parallel$, as shown in Figure 14.2b. Therefore a quasicrystal-approximant (aperiodic-periodic) phase transition is described by tilt values. This type of phason is caused by a linear shift and is called *phason strain*. In contrast, *phason modulation* corresponds to an order-disorder phase transition in quasicrystals. If sinodal phason

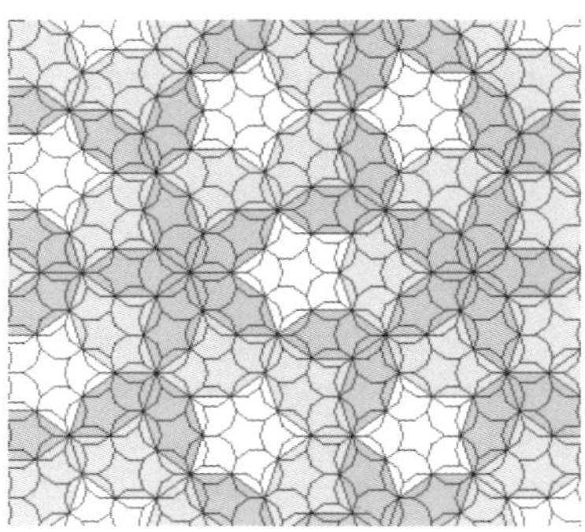

Figure 14.1. *HBS tiling involves smaller tiling. Quasiperiodicity is described by self-similarity.*

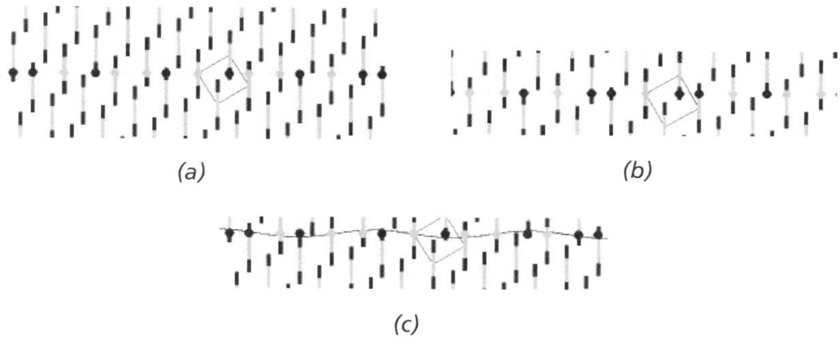

Figure 14.2. *(a) Fibonacci lattice obtained by section method on hyperlattice. (b) Periodicity appears by a tilt of the physical space, $E_{||}$, in the high-dimensional periodic space. (c) Atomic flips are caused by phason modulation. $E_{\perp}$ is perpendicular space, which is extended virtually. $E_{||}$ is conventional parallel space.*

fluctuation along $E_{\perp}$ appears, as shown in Figure 14.2c, atomic flips occur without atomic exchange. In the atomic flip, a degree of freedom is permitted only at the two atomic positions on $E_{||}$ under the condition that atomic distance is restricted to the specific values, such as *L* or *S*. The peculiar atomic flip on a quasiperiodic lattice is called a phason flip. For instance, a local sequence on a Fibonacci lattice (*LSLSLLSLLSL*…) changes to *LS* from *SL*.

By phason theory in quasicrystals, a variety of phason fluctuations are distinguished by diffraction patterns. The relationships between phason and diffraction are summarized in Table 14.1. In decagonal quasicrystals, **G** refers to the five-dimensional reciprocal lattice vector, which consists of two components, $\mathbf{G}_{||}$ and $\mathbf{G}_{\perp}$. $\mathbf{G}_{||}$ is the three-dimensional reciprocal lattice vector in the physical space. $\mathbf{G}_{\perp}$ is the two-dimensional component in the complementary perpendicular space. Phason fluctuations are provided by the $\underline{M}$ matrix, which is defined by an elastic modulus tensor. There are four kinds of phasons classified as follows:

1. *Phason strain* is represented by a tilt of the physical space in the high-dimensional periodic space. Bragg reflections shift from ideal positions, and the peak shifts are linked to a quasicrystal-approximant (aperiodic-periodic) phase transition.
2. In *phason modulation,* local atomic flip satisfying quasiperiodicity is obeyed by sinodal fluctuations along $E_{\perp}$. Supertiling imposed on ideal quasiperiodicity corresponds to an order-disorder phase transition in quasicrystals.
3. In *random phason strain*, peak broadening of Bragg reflections is caused by nonuniform strains. A phason freezes randomly along the equivalent directions.
4. *Phason modes* are explained by the hydrodynamic theory. For instance, a *random tiling model* [5] was proposed to explain the diffuse scattering, which contains some kind of disorder in the long-range geometry.

TABLE 14.1

A variety of quasicrystals is induced by phasons.

MODEL	PHASONIC FLUCTUATION	PECULIAR FEATURE ON DIFFRACTION PATTERN
Quasiperiodic Lattice	No (constant)	δ function
Phason strain (Approximant)	Linear $\mathbf{u}_{\perp} = \underline{M}\,\mathbf{r}$	Peak shift ($\Delta\mathbf{G} = \underline{M}^{t}\,\mathbf{G}_{\perp}$)
Phason modulation (Order-disorder)	Periodic $\mathbf{u}_{\perp} = \mathbf{u}_0 \sin \mathbf{q}\,\mathbf{r}$	δ function + superstructure reflection ($\Delta\mathbf{G} = \pm\mathbf{q}$)
Random phason strain	Random $<\vert\mathbf{u}_{\perp}(\mathbf{r}) - \mathbf{u}_{\perp}(\mathbf{0})\vert^2> = \vert\mathbf{r}\vert^n$	Peak broadening FWHM $\propto \vert\mathbf{G}_{\perp}\vert^{2/n}$
Phason modes (Random tiling)	Random	δ function + diffuse scattering

The shape of phason diffuse scattering (PDS) depends on *phason modes*. Some of them are related to phason softening. Slow relaxation of a phason remains with some displacements. Randomness of a freezing phason, whose displacement is not divergent, reflects the anisotropic distributions of the diffuse scattering. This is stabilized by entropy effect. As an example, the diffuse scattering studies of icosahedral $Al_{68.2}Pd_{22.8}Mn_9$ were performed using neutron diffraction [6]. Using high-energy resolution, phonon contributions to the diffuse scattering were completely separated from the phason contributions. The experimental results were consistent with the calculated PDS, which is regarded as a precursor phenomenon of a quasicrystal-approximant phase transition.

3 Scattering Theory

3.1 Diffuse Scattering Theory for a Crystal

Let $\mathbf{Q}$ be a scattered wave vector defined as $\mathbf{Q} = \mathbf{G} + \mathbf{q}$, where $\mathbf{G}$ is a reciprocal lattice vector. Several kinds of diffuse scattering are considered for conventional crystals: atomic SRO, atomic size effect (SE) from displacements induced by chemical ordering of atoms of disparate sizes, thermal diffuse scattering (TDS) from phonons, and Huang diffuse scattering (HDS) from defects are the main contributions. The total diffuse scattering is thus given by [7]

$$I^{\mathrm{diff}} = I^{\mathrm{SRO}} + I^{\mathrm{SE}} + I^{\mathrm{TDS+HDS}} . \tag{14.1}$$

The SRO diffuse intensity in a binary alloy system is given by

$$I^{SRO}(\mathbf{Q}) = x_A x_B \left|f_A - f_B\right|^2 \sum_m \alpha_m \exp\left(i\mathbf{Q}\cdot\mathbf{R}_m\right) = x_A x_B \left|\Delta f\right|^2 \alpha(\mathbf{q}), \tag{14.2}$$

where $\alpha(\mathbf{q})$ is the Fourier transform of the Warren-Cowley SRO parameters and x_A is the concentration of A atoms ($x_B = 1 - x_A$). If we put an A atom at the origin, the SRO parameter on the *m*th site, α_m, becomes

$$\alpha_m = \frac{\left\langle \sigma_0^A \sigma_m^B \right\rangle - x_A x_B}{x_A\left(\delta_{AB} - x_B\right)} = 1 - \frac{P_m^{AB}}{x_B}, \alpha_0 \equiv 1 \tag{14.3}$$

$P_m^{AB} = \left\langle \sigma_0^A \sigma_m^B \right\rangle / x_A$ is the conditional probability of a B atom on the *m*th site given an A atom at the origin. One can obtain nearly pure thermal and static quadratic and higher order displacement scattering, $I^{\text{TDS+HDS}}$, which is proportional to the square of the average structure factor, $|\overline{f}|^2$, where $|\overline{f}|$ is given by $|x_A f_A + x_B f_B|$. In general, the expression for the elastic diffuse scattering from a disordered binary alloy is given by

$$I^{\text{diff}}(\mathbf{Q}) = \text{N}\left|\Delta c(\mathbf{q})\right|^2 \times \left|\Delta f - \overline{f}\mathbf{Q} \bullet \mathbf{A}(\mathbf{q})\right|^2, \tag{14.4}$$

where $\Delta c(\mathbf{q})$ is the Fourier transform of the concentration fluctuations and $\mathbf{A}(\mathbf{q})$ is the displacement field. The SE is represented by the coupling of the atomic displacements to the concentration. In addition, the linear coupling between concentration and displacement in Equation 14.4 is also important in particular cases. This term is the cross-term of the amplitude product that contributes to I^{SRO} and $I^{\text{TDS+HDS}}$ and is regarded as both chemical and topological ordering. Therefore this term, referred to as I^{SE}, is proportional to $\Delta f \times \overline{f}$; the actual equation is more complicated [8]. In general, both TDS [9] and HDS [10] are proportional to q^{-2}. The equations are given by

$$I^{\text{TDS}} \propto k_B T \frac{|\mathbf{Q}|^2\left(C_{ijk\ell} g_k g_\ell\right)^{-1}}{q^2}, \tag{14.5}$$

and

$$I^{\text{HDS}} \propto \frac{|\mathbf{Q}_i (C_{ijk\ell} g_k g_\ell)^{-1} P_{jk} g_k|^2}{q^2}, \tag{14.6}$$

where g_i is the unit vector of q_i and P_{ij} is the elastic dipole derived from a defect.

A "null Laue" experiment is important in order to clarify SRO between two atomic species having a small difference in atomic numbers [11]. By selecting the incident X-ray energy so that $|\Delta f| = |f_A - f_B| \cong 0$, one can then estimate and remove pure $I^{\text{TDS+HDS}}$ by the null Laue and thereby obtain I^{SRO} and I^{SE}. This is because $\overline{f}$ does not depend much on the incident energies (Figure 14.3a), whereas Δf is very sensitive to them (Figure 14.3b). In a ternary alloy system, the SRO diffuse intensity of Equation 14.2 is rewritten as

$$I^{\text{SRO}} \propto x_A x_B |f_A - f_B|^2 \alpha_{AB}(\mathbf{q}) + x_B x_C |f_B - f_C|^2 \alpha_{BC}(\mathbf{q}) + x_C x_A |f_C - f_A|^2 \alpha_{CA}(\mathbf{q}) \tag{14.7}$$

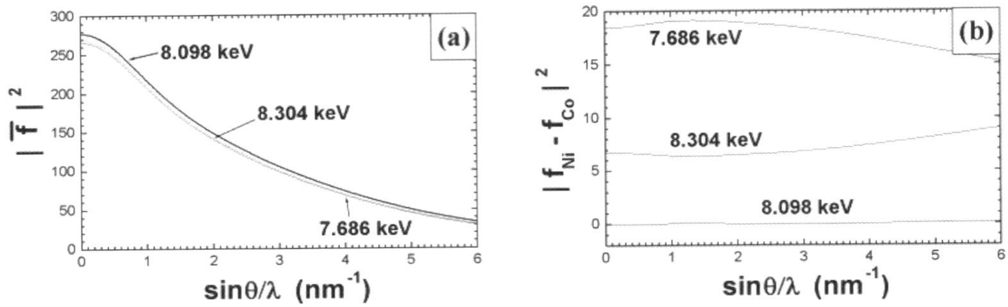

Figure 14.3. *(a)* $|\bar{f}|^2$ *and (b)* $|\Delta f|^2$ *as a function of* $(\sin\theta)/\lambda$ *for the three energies, where* $|\Delta f|$ *is equal to* $|f_{Ni} - f_{Co}|$.

3.2 Diffuse Scattering Theory for a Decagonal Quasicrystal

We focus on the phonon and phason elastic energies in a decagonal quasicrystal. Icosahedral notation for phason softening is described in detail elsewhere [12]. In a decagonal quasicrystal, the structural fluctuations of a disordered state are described by introducing a phonon displacement field, $\mathbf{u} = (u_1, u_2, u_3)$, and a phason displacement field, $\mathbf{v} = (v_1, v_2)$. Phonon displacements are influenced by phason displacements such as $\mathbf{u} \to \mathbf{u} + [C^{-1}]_{\|,\|} C_{\|,\perp}\ \mathbf{v}$, where $\hat{C}$ is the hydrodynamic elastic matrix given by [13]

$$\hat{C}(\mathbf{q}_{\|}) = \begin{pmatrix} C_{\|,\|} & C_{\|,\perp} \\ C_{\perp,\|} & C_{\perp,\perp} \end{pmatrix}. \tag{14.8}$$

The diffuse scattering is given by

$$I^{\text{diff}}(\mathbf{Q}_{\|}) \propto ({}^{t}\mathbf{G}_{\|}, {}^{t}\mathbf{G}_{\perp})\ \hat{C}^{-1}(\mathbf{q}_{\|}) \begin{pmatrix} \mathbf{G}_{\|} \\ \mathbf{G}_{\perp} \end{pmatrix} \propto F_{\text{thermal}} + F_{\text{phason}}, \tag{14.9}$$

where

$$F_{\text{thermal}} = {}^{t}\mathbf{G}_{\|}\ [C^{-1}]_{\|,\|}\ \mathbf{G}_{\|} \tag{14.10}$$

and

$$F_{\text{phason}} = {}^{t}\Delta\mathbf{G}_{\perp}(C_{\perp,\perp} - {}^{t}C_{\|,\perp}[C^{-1}]_{\|,\|}C_{\|,\perp})^{-1}\Delta\mathbf{G}_{\perp}, \tag{14.11}$$

with $\Delta\mathbf{G}_{\perp}$ equal to $\mathbf{G}_{\perp} - {}^{t}C_{\|,\perp}\ [C^{-1}]_{\|,\|}\mathbf{G}_{\|}$. Similar to I^{TDS} and I^{HDS}, Equation 14.9 shows a q^{-2} dependence. The actual components of the *C*-matrix that include the periodic direction are expressed as,

$$C_{\|,\|} = (\lambda + 2\mu)\begin{pmatrix} q_1^2 & q_1q_2 & 0 \\ q_2q_1 & q_2^2 & 0 \\ 0 & 0 & 0 \end{pmatrix} + \mu\begin{pmatrix} q_2^2 & -q_1q_2 & 0 \\ -q_2q_1 & q_1^2 & 0 \\ 0 & 0 & 0 \end{pmatrix} + C_1\begin{pmatrix} q_3^2 & 0 & 0 \\ 0 & q_3^2 & 0 \\ 0 & 0 & q_1^2+q_2^2 \end{pmatrix} + C_2\begin{pmatrix} 0 & 0 & 0 \\ 0 & 0 & 0 \\ 0 & 0 & q_3 \end{pmatrix} + C_3\begin{pmatrix} 0 & 0 & q_1q_3 \\ 0 & 0 & q_2q_3 \\ q_3q_1 & q_3q_2 & 0 \end{pmatrix},$$

$$C_{\perp,\perp} = [K_1(q_1^2+q_2^2)+K_2q_3^2]\begin{pmatrix} 1 & 0 \\ 0 & 1 \end{pmatrix}, \quad C_{\|,\perp} = K'\begin{pmatrix} q_1^2-q_2^2 & -2q_1q_2 \\ -2q_2q_1 & q_2^2-q_1^2 \\ 0 & 0 \end{pmatrix}, \tag{14.12}$$

where parallel components are rewritten as $(q_\|)_i \rightarrow q_i$ for simplicity. Here, λ and $\mu(>0)$ are the conventional Lamé coefficients, $C_1\ (>0)$, $C_2\ (>0)$, and C_3 are the phonon stiffness constants, K_1 and K_2 (>0) are the phonon stiffness constants, and K' is a coupling constant of the phonon-phason mixing. For PDS calculations, the following conditions are also required:

$$\lambda + 2\mu > 0,\ \ \mu K_1 > (K')^2,\ \ (\lambda + 2\mu)K_1 > (K')^2. \tag{14.13}$$

The phason elastic matrix $C_{\perp,\perp}$ is isotropic for decagonal quasicrystals. However, anisotropic distributions of the diffuse scattering are realized if phonon-phason coupling is taken into account.

4 Experimental Methodology

Anomalous X-ray scattering experiments were performed on the Oak Ridge National Laboratory beamline X14A of the National Synchrotron Light Source at Brookhaven National Laboratory [14] and the BL-4C of the Photon Factory at the High Energy Accelerator Research Organization in Japan [15]. A vertically focusing mirror was placed in front of a Si(111) double monochromator. The specimen was mounted on a four-circle diffractometer (Huber 5010). Air scattering was minimized by Helium-filled beam paths. X-ray fluorescence was removed by a curved highly oriented pyrolytic graphite (HOPG) (002). The incident energies were calibrated to within 1 eV using both Ni and Co foils. Here, we selected the following incident energies: 7.686 keV (near K edge of Co), 8.098 keV $(\Delta f = | f_{Ni} - f_{Co} | \cong 0)$ for the "null Laue" case, and 8.304 keV (near K edge of Ni). Figures 14.3a and 14.3b show the calculated $| \Delta f |^2$ between Ni and Co and $|\bar{f}|^2$ ($|\bar{f}| = |x_{Al}f_{Al}+x_{Ni}f_{Ni}+x_{Co}f_{Co}|$) at each energy, respectively. By anomalous X-ray scattering, we can even obtain the contrast between Ni and Co, whose difference of atomic number is only one. *Q* resolution of these beam optics is estimated to be around 0.01 nm^{-1} using a Si single crystal. Here, the unit of reciprocal space is $4\pi(\sin\theta)/\lambda(nm^{-1})$. In order to analyze the diffuse scattering quantitatively (electron units per atom), we measured several integrated intensities of a standard powder sample of Ni at each energy.

Alloy ingots with nominal composition of $Al_{70}Ni_{15}Co_{15}$ and $Al_{72}Ni_{20}Co_8$ were prepared by melting mixtures of pure Al (99.99%), Ni (99.99%), and Co (99.99%) metals under an Ar atmosphere in an arc furnace. This ingot was crushed into powder, put into an alumina crucible, and then sealed in a quartz tube. The powder sample was melted at 1423 K, slowly cooled to 1073 K at the rate of 50 K h^{-1}, kept at 1073 K for one day, and then quenched in water (T_q = 1073 K). The specimens were roughly rod shaped with a length of about 0.1 mm and a diameter of about 0.030 mm. Mosaicity of $Al_{70}Ni_{15}Co_{15}$ [16] and $Al_{72}Ni_{20}Co_8$ [17] are estimated to be 0.0988 and 0.0223 nm^{-1}, respectively.

5 Experimental Results and Discussion

The Al-Ni-Co (ANC) system is well known to form decagonal quasicrystals, which have two-dimensional quasiperiodic planes. The structures of ANC critically depend both on the Ni and Co concentrations and on temperature. The complex phase diagram of ANC was reported in a previous study [18]. The phase diagram contains a great variety of phase transitions, that is, quasicrystal-approximant phase transitions, phase transitions from two-dimensional to one-dimensional quasicrystals, order-disorder phase transitions, and so on.

5.1 *Diffuse Scattering in $Al_{70}Ni_{15}Co_{15}$*

In general, SRO diffuse scattering on a crystal appears at the zone center (clustering type) or zone boundary (ordering type). Analysis of the zone-centered diffuse scattering is accompanied by difficulties because TDS and HDS also appear at the zone center. The intensity ratio of SRO, SE, TDS, and HDS at the zone center is not determined experimentally using X-ray diffraction. In neutron diffraction, pure TDS from phonons is extracted from total diffuse scattering [6]. Therefore theoretical calculations of TDS, HDS, and PDS are required to analyze the zone-centered diffuse intensities obtained by X-ray diffraction. PDS is to quasicrystals what HDS is to crystals.

Before analyzing diffuse scattering, phason contributions are examined by the peak broadening of the Bragg reflections. In $Al_{70}Ni_{15}Co_{15}$, the full width at half-maximum (FWHM) of the Bragg reflections increases with increasing $|\mathbf{G}_{\perp}|$ values, as shown in Figure 14.4. It is clear that random phasons appear in $Al_{70}Ni_{15}Co_{15}$. In contrast, there is no $|\mathbf{G}_{\perp}|$ dependence of the FWHM of Bragg reflections in $Al_{72}Ni_{20}Co_{8}$. It is clear that no "random phason strain" exists in the as-quenched sample of $Al_{72}Ni_{20}Co_{8}$. Certainly, the structure of $Al_{72}Ni_{20}Co_{8}$ is interpreted as an atomic decoration of ideal Penrose tiling (HBS tiling) [19].

Q_x, Q_y, and Q_z are set to measure three-dimensional diffuse scattering, as shown in Figure 14.5. By measurement along the periodic direction (Q_z), we found diffuse scattering distributed only on quasiperiodic planes, but not in the periodic direction. The diffuse scattering of $Al_{70}Ni_{15}Co_{15}$

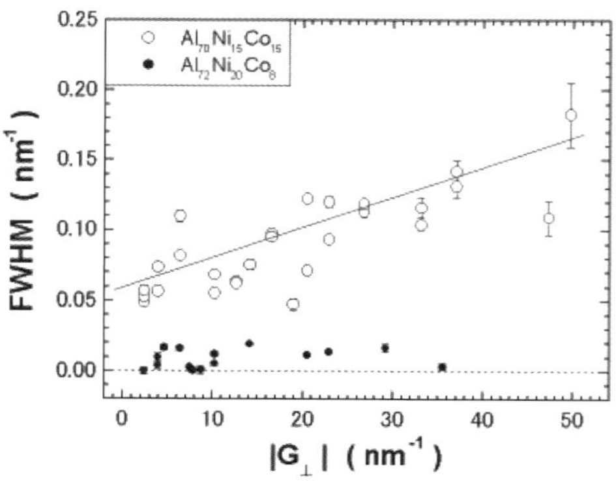

Figure 14.4. *$|G_{\perp}|$ dependence of FWHM of Bragg reflections. In comparison with that of $Al_{70}Ni_{15}Co_{15}$, no "random phason strain" is seen in $Al_{72}Ni_{20}Co_{8}$.*

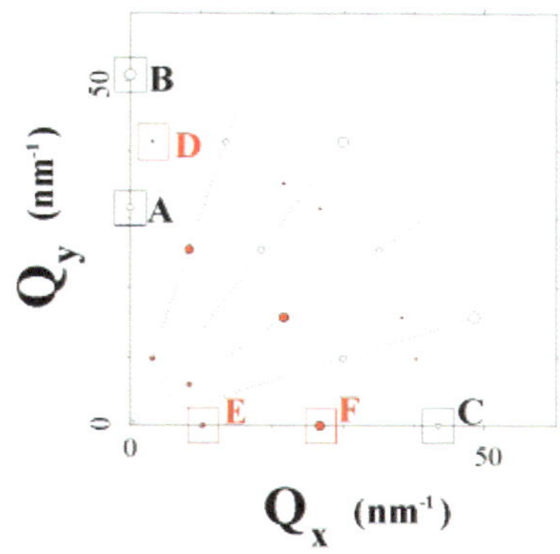

Figure 14.5. *Reciprocal space perpendicular to the periodic direction of the decagonal $Al_{70}Ni_{15}Co_{15}$ phase. The area of the spot is proportional to the intensity of the reflection. The diffuse scattering has been measured around the labeled reflections. The labeled reflections of A, B, C, D, E, and F stand for $10\bar{3}\bar{3}0$, $30\bar{4}\bar{4}0$, $1\bar{2}\bar{5}\bar{4}0$, $20\bar{4}\bar{3}1$, $10\bar{1}01$, and $1\bar{1}\bar{3}\bar{2}1$, respectively.*

quasicrystals was measured around $10\bar{3}\bar{3}0$, $30\bar{4}\bar{4}0$, $1\bar{2}\bar{5}\bar{4}0$, $20\bar{4}\bar{3}1$, $10\bar{1}01$, and $1\bar{1}\bar{3}\bar{2}1$ Bragg reflections, where the indexing of Bragg reflections was used in a previous study [20]. Here, main, S1 (superstructure 1), and S2 (superstructure 2) reflections are classified as $\sum_i h_i = 0$ (mod 5), $\sum_i h_i = \pm 1$ (mod 5), and $\sum_i h_i = \pm 2$ (mod 5), respectively. Figures 14.6a–f show the distributions of the diffuse scattering on quasiperiodic planes around $10\bar{3}\bar{3}0$ (A in Figure 14.5), $30\bar{4}\bar{4}0$ (B in Figure 14.5), $1\bar{2}\bar{5}\bar{4}0$ (C in Figure 14.5), $20\bar{4}\bar{3}1$ (D in Figure 14.5), $10\bar{1}01$ (E in Figure 14.5), and $1\bar{1}\bar{3}\bar{2}1$ (F in Figure 14.5) at 8.304 keV. The distributions of observed diffuse scattering are quite anisotropic and depend strongly on the reciprocal lattice vectors. Diffuse scattering from $Al_{70}Ni_{15}Co_{15}$ is located at the zone center.

First, SRO diffuse scattering in $Al_{70}Ni_{15}Co_{15}$ is investigated. We analyzed three kinds of pair correlations by taking advantage of anomalous X-ray scattering. Figure 14.7 shows diffuse scattering along

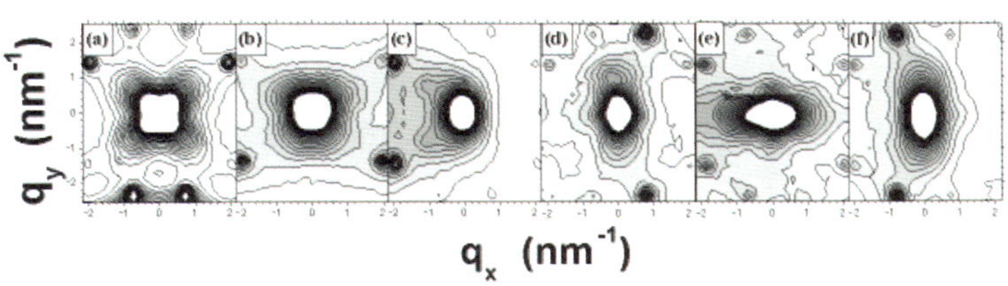

Figure 14.6. *The observed diffuse scattering around (a) $10\bar{3}\bar{3}0$, (b) $30\bar{4}\bar{4}0$, (c) $1\bar{2}\bar{5}\bar{4}0$, (d) $20\bar{4}\bar{3}1$, (e) $10\bar{1}01$, and (f) $1\bar{1}\bar{3}\bar{2}1$ Bragg reflections at 8.304 keV in $Al_{70}Ni_{15}Co_{15}$.*

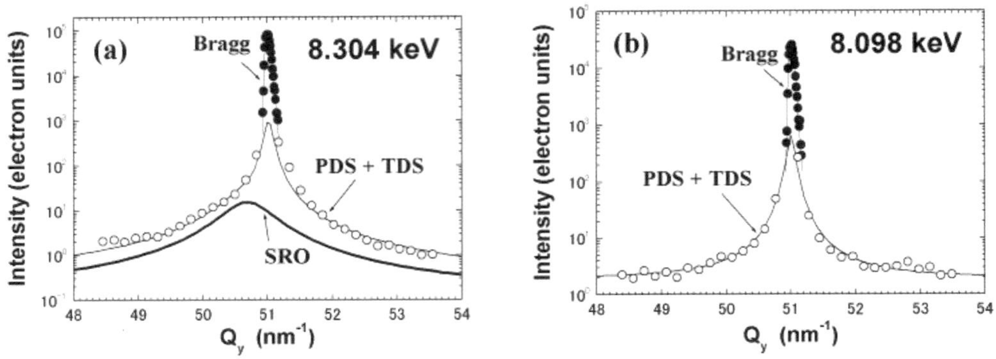

Figure 14.7. *Diffuse scattering along the longitudinal directions around* $3 0 \bar{4} \bar{4} 0$ *Bragg reflection at (a) 8.304 keV and (b) 8.098 keV. Asymmetric peak profile is caused by atomic SRO, which is modulated by atomic SE. At 8.098 keV* ($\Delta f \cong 0$)*, asymmetric SRO disappears. There are two components of the diffuse peak profile clearly at 8.304 keV.*

the longitudinal (L-) direction around $3 0 \bar{4} \bar{4} 0$ Bragg reflection at (a) 8.304 keV ($| f_{Ni} - f_{Co} | \neq 0$) and (b) 8.098 keV ($| f_{Ni} - f_{Co} | \cong 0$). Here, intensity in electron units is used for the quantitative unit. Obviously, asymmetric SRO, whose correlation length is 2.5 nm, vanishes completely at 8.098 keV. Asymmetric profiles are originated from atomic SE. The significant results imply that $\alpha_{Al-Ni}(\mathbf{q})$ and $\alpha_{Co-Al}(\mathbf{q})$ are much smaller than $\alpha_{Ni-Co}(\mathbf{q})$ in Equation 14.7. Consequently, only SRO between Ni and Co can exist in $Al_{70}Ni_{15}Co_{15}$; both $|f_{Al}-f_{Ni}|$ and $|f_{Co}-f_{Al}|$ are greater than 0 at 8.098 keV. Almost similar partial diffuse intensities were observed in $Al_{72}Ni_{18}Fe_{10}$, which has smaller mosaicity [21]. The Al-Ni-Fe (ANF) alloy system also forms decagonal quasicrystals. By the quantitative inverse matrix method, the observed diffuse scattering was decomposed into three kinds of partial diffuse intensities. It was established that $\alpha_{Ni-Fe}(\mathbf{q})$ is much larger than $\alpha_{Al-Ni}(\mathbf{q})$ and $\alpha_{Fe-Al}(\mathbf{q})$. Hence, a pair correlation between transition metals (TMs)—that is, a Ni-Fe pair—is most effective. Moreover, $\alpha_{Ni-Fe}(\mathbf{q})$ provides almost symmetric distribution due to little SE, whereas apparent SE was seen in partial diffuse intensities of Al-Ni and Fe-Al pairs.

Second, TDS and PDS contributions are evaluated by theoretical calculations. It should be noticed that only the null-Laue method can determine pure TDS and PDS contributions experimentally, as shown in Figure 14.7b. Distributions of TDS are calculated with Equations 14.10 and 14.12. The calculation results are shown in Figures 14.8a–e. In the actual calculation, we use λ (=0.5736) and μ (=0.8845) values for $Al_{71}Ni_{16}Co_{13}$ single quasicrystals [22]. TDS distributes along the transverse (T-) directions, and Q dependence of anisotropic diffuse shapes is not seen. The calculated distributions of TDS cannot reproduce the observed ones. Thus it is predicted that TDS contributions to total diffuse scattering are negligible. Next, we investigate a possibility of PDS characterized by quasiperiodicity. We calculated the theoretical PDS described by Equations 14.11 and 14.12. On $Q_z = 0$, elastic constants are reduced to λ, μ, K_1, and K'. Figures 14.9a–e exhibit the theoretical PDS distributions around $1 0 \bar{3} \bar{3} 0$ Bragg reflection that satisfy the inequalities (Equation 14.13). Only in combination with K_1 (=0.1) and K' (=0.2) is the calculated PDS distribution similar to the observed one, though

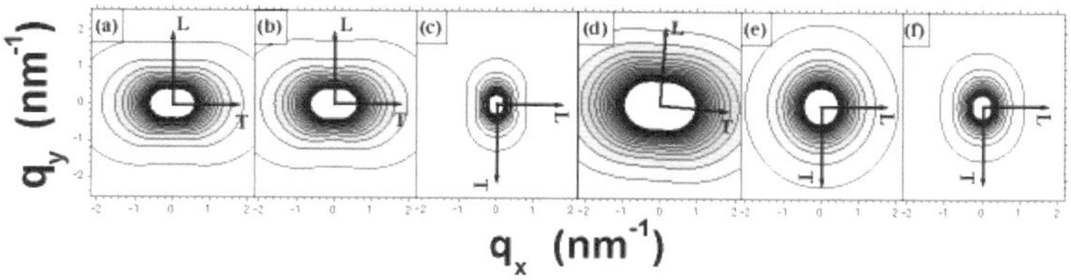

Figure 14.8. *The calculated TDS around (a)* $10\bar{3}\bar{3}0$, *(b)* $30\bar{4}\bar{4}0$, *(c)* $1\bar{2}\bar{5}\bar{4}0$, *(d)* $20\bar{4}\bar{3}1$, *(e)* $10\bar{1}01$, *and (f)* $1\bar{1}\bar{3}\bar{2}1$ *Bragg reflections. TDS distributes along the transverse (T-) directions, where longitudinal (L-) direction is perpendicular to T direction.*

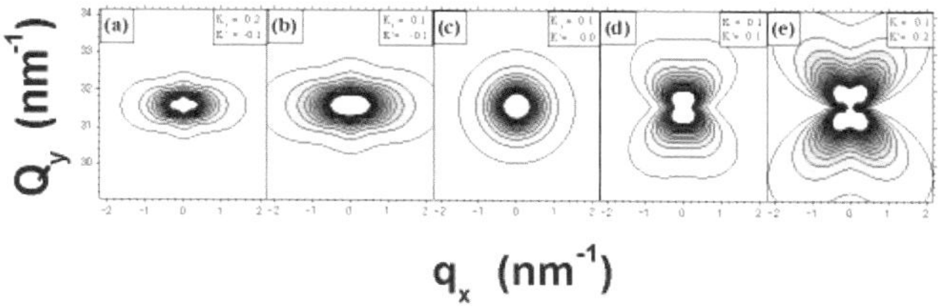

Figure 14.9. *Theoretical PDS around* $10\bar{3}\bar{3}0$ *Bragg reflections using (a)* $K_1 = 0.2$ *and* $K' = -0.1$, *(b)* $K_1 = 0.1$ *and* $K' = -0.1$, *(c)* $K_1 = 0.1$ *and* $K' = 0$, *(d)* $K_1 = 0.1$ *and* $K' = 0.1$, *and (e)* $K_1 = 0.1$ *and* $K' = 0.2$. K_1 *is the phason elastic constant. Anisotropic distributions are enhanced by the coupling constant between phonon and phason,* K'.

distributions around other Bragg reflections are quite different from the observed ones. Apart from discrepancies between experiments and the theoretical PDS calculations, diffuse intensities in electron units are proportional to $|\mathbf{G}_\perp|$ values: $|\mathbf{G}_\perp|$ of $1\bar{2}\bar{5}\bar{4}0$ is 2.42 nm^{-1}, that of $30\bar{4}\bar{4}0$ is 4.60 nm^{-1}, and that of $10\bar{3}\bar{3}0$ is 7.44 nm^{-1}. At least, it is deduced that anisotropic diffuse scattering is related to phasons. In principle, the diffuse scattering derived from pure phason contributions is isotropic for a decagonal quasicrystal; however, a coupling between phonon and phason can modify the distributions anisotropically.

5.2 *Diffuse Scattering in* $Al_{72}Ni_{20}Co_8$

At first, "random phason strain" is investigated in $Al_{72}Ni_{20}Co_8$. By an approach to phason analysis, no peak broadening of the Bragg reflections is found, as shown in Figure 14.4. $Al_{72}Ni_{20}Co_8$ has no random phason strain at room temperature. In previous studies, $Al_{72}Ni_{20}Co_8$ was regarded as a perfect quasicrystal. In fact, $Al_{72}Ni_{20}Co_8$ has quite small mosaicity in this study. Figure 14.10 shows the $Q_x - Q_y$ region for measurements in $Al_{72}Ni_{20}Co_8$. Diffuse scattering was observed around $10\bar{3}\bar{3}0$, $30\bar{4}\bar{4}0$, $0\bar{1}\bar{2}\bar{2}0$, $1\bar{2}\bar{5}\bar{4}0$, $20\bar{4}\bar{3}1$, and $1\bar{1}\bar{3}\bar{2}1$ Bragg reflections at 8.304 keV and diffuse scattering was distributed only on quasiperiodic planes. Compared with anisotropic distributions of diffuse scattering in $Al_{70}Ni_{15}Co_{15}$, the zone-centered diffuse scattering of $Al_{72}Ni_{20}Co_8$ is isotropic around the center of the Bragg reflections. Since the observed diffuse shapes are not obviously modified by a strain, the

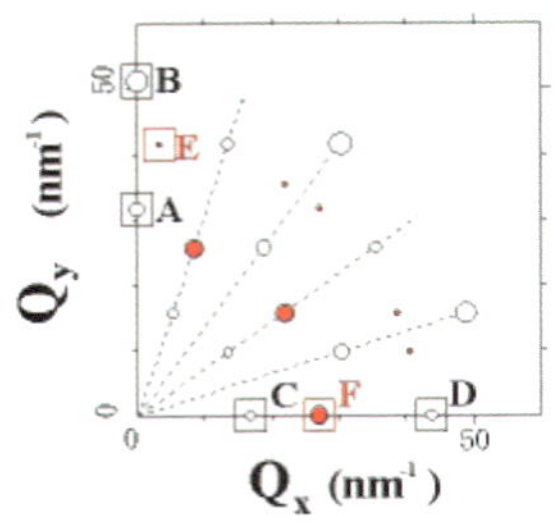

Figure 14.10. *Reciprocal space in quasiperiodic planes of the decagonal $Al_{72}Ni_{20}Co_8$ phase. The area of the spot is proportional to the intensity of the reflection. The diffuse scattering has been measured around the labeled reflections. The labeled reflections of A, B, C, D, E, and F stand for 10330, 30440, 01220, 12540, 20431, and 11321, respectively.*

existence of any phasons is not considered in $Al_{72}Ni_{20}Co_8$. Theoretical calculations of TDS and PDS are applied to the observed distributions of diffuse scattering in $Al_{72}Ni_{20}Co_8$. Neither TDS nor PDS fit the observed ones well using any combination between K_1 and K'. Therefore the experimental results in $Al_{72}Ni_{20}Co_8$ as an ideal quasicrystal refuse the existence of phasons completely. Other characteristic features of diffuse scattering in $Al_{72}Ni_{20}Co_8$ are additional broad peaks around the center of ideal positions of S1-type superstructure reflections. The ideal positions are indicated by closed stars in Figures 14.11a–f. Generally, diffuse scattering at the positions of superstructure reflections suggests an order-disorder phase transition. Analysis of diffuse scattering is standard, but indispensable, to resolve ordering processes occurring during order-disorder phase transition. In SRO analysis using anomalous X-ray scattering, diffuse scattering along the L-direction around 30440 Bragg reflection was observed at 8.304 keV and 8.098 keV (null Laue), as shown in Figures 14.12a and 14.12b. In contrast to $Al_{70}Ni_{15}Co_{15}$, asymmetric SRO diffuse scattering never disappears in spite of the null Laue condition. Therefore all atomic pair-correlation functions are effective in $Al_{72}Ni_{20}Co_8$. Recently, the total diffuse scattering in $Al_{72}Ni_{20}Co_8$ was successfully decomposed into the three kinds of partial diffuse intensities—$\alpha_{Al-Ni}(\mathbf{q})$, $\alpha_{Co-Al}(\mathbf{q})$, and $\alpha_{Ni-Co}(\mathbf{q})$—using the inverse matrix method [23]. This is because the additional diffuse scattering at the ideal positions of superstructure reflections was analyzed without

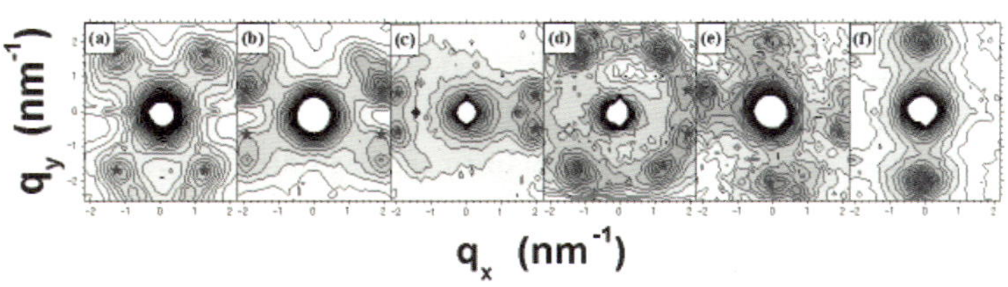

Figure 14.11. *The observed diffuse scattering around (a) 10330, (b) 30440, (c) 01220, (d) 12540, (e) 20431, and (f) 11321 Bragg reflections at 8.304 keV in $Al_{72}Ni_{20}Co_8$. Broad peaks appear at the ideal positions of superstructure reflections. The positions are indicated by closed stars.*

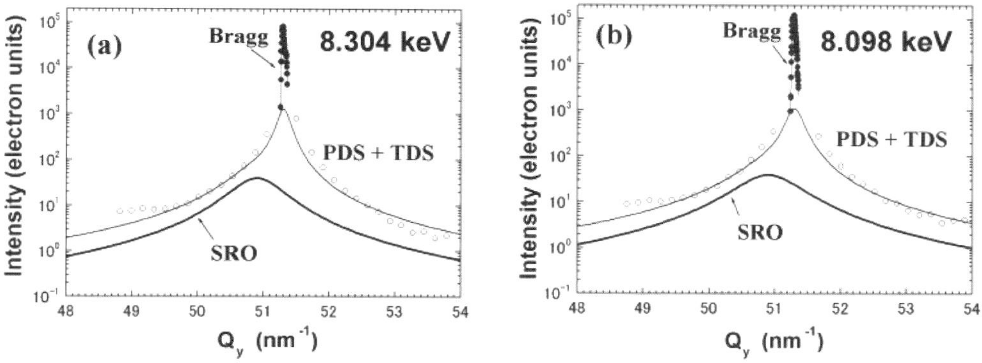

Figure 14.12. *Diffuse scattering along the longitudinal directions around $30\bar{4}\bar{4}0$ Bragg reflection at (a) 8.304 keV and (b) 8.098 keV in $Al_{72}Ni_{20}Co_8$.*

considering the influence of TDS and PDS. Moreover, it was also found that $\alpha_{Ni-Co}(\mathbf{q})$ is larger that $\alpha_{Al-Ni}(\mathbf{q})$ and $\alpha_{Co-Al}(\mathbf{q})$. By the above quantitative analysis of diffuse scattering, reverse Monte Carlo modeling based on the decomposed partial diffuse intensities was carried out to visualize the local structure in $Al_{72}Ni_{20}Co_8$. At the same time, SRO parameters were obtained from Monte Carlo simulations.

Atomic SE contains important information about local distortions in quasicrystals. By systematic measurements using $Al_{70}Ni_{15}Co_{15}$, $Al_{72}Ni_{20}Co_8$, and $Al_{72}Ni_{18}Fe_{10}$, there is a tendency that SRO between TMs is much more dominant than between other Al-TM pairs, whereas asymmetric distributions of SRO diffuse scattering are enhanced between Al-TM pairs because of the difference in atomic sizes. In addition to the entropy effect, atomic SE, which is influenced by different atomic sizes, contributes to phase stability in quasicrystals.

5.3 Order-Disorder Phase Transition in $Al_{72}Ni_{20}Co_8$

An order-disorder phase transition at high temperature was reported in $Al_{72}Ni_{20}Co_8$ [24]. Figure 14.13a shows the temperature dependence of the FWHM for the radial scan of the $1\bar{2}\bar{5}\bar{4}0$ fundamental reflection. Here, the observed FWHM were deconvoluted using the resolution function, which was estimated by the Bragg reflections in a Si single crystal. There is a clear sharp bend in the line at T_c (=992 K). The abrupt increase of the FWHM of the fundamental reflection below T_c reveals the nonuniform distortion generated by the S1-type microdomain. The FWHM of the Bragg reflections are generally proportional to the $|\mathbf{G}_{\parallel}|$ values under a nonuniform strain for conventional crystals, whereas a uniform strain provides peak shifts. In quasicrystals, the formation of microdomains was proposed theoretically in Honal, Haibach, and Steurer [25] to explain quasicrystal transformations, where each microdomain has no smooth boundary. The integrated intensities of the $1\bar{2}\bar{5}\bar{4}0$ fundamental

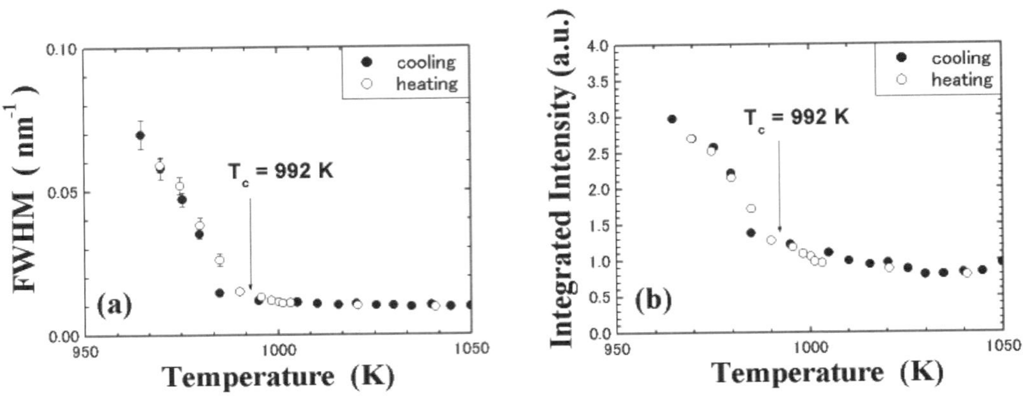

Figure 14.13. *Temperature dependences of (a) FWHM and (b) the integrated intensity for the radial scan of the* $\bar{1}\bar{2}\bar{5}40$ *fundamental reflection in* $Al_{72}Ni_{20}Co_8$.

reflection also changed drastically at T_c, as shown in Figure 14.13b. Note that a hysteresis of both the FWHM and the integrated intensity of the fundamental reflection was estimated to be 10 K at approximately 992 K, considering experimental errors. This indicated that the order-disorder phase transition in $Al_{72}Ni_{20}Co_8$ was a nearly second-order phase transition. In order to separate the contributions of the nonuniform deformation and the random phason strain, we measured the FWHM of the various fundamental reflections at 300 K (quenched above T_c), 965 K (< T_c), and 1096 K (> T_c). In many cases, peak broadening as a function of $|\mathbf{G}_{\|}|$ and $|\mathbf{G}_{\perp}|$ corresponds, respectively, to the nonuniform deformation and the random phason strain. Small-peak broadening as a function of $|\mathbf{G}_{\|}|$ of the fundamental reflections was found at room temperature in the as-quenched sample ($T_q > T_c$), as shown in Figure 14.14a. This means that only small deformation was induced by the quenching process. No increase of the FWHM of the fundamental reflections was seen at 1096 K (> T_c). In contrast, large $|\mathbf{G}_{\|}|$ dependence of the fundamental reflections at 965 K (< T_c) was observed. This can be interpreted by the nonuniform deformations introduced by the S1 microdomains below T_c, as mentioned earlier. If we look at the details, it should, however, be noted that some data points do not coincide exactly with the linear increase in the $|\mathbf{G}_{\perp}|$ axis. We believe that these disagreements with the $|\mathbf{G}_{\perp}|$ linearity imply the existence of a random phason strain in the $|\mathbf{G}_{\perp}|$ axis below T_c. The nonuniform deformation and the random phason strain occur simultaneously in an order-disorder phase transition. Thereby the observed FWHM, $\eta^{obs}\left(|\mathbf{G}_{\|}|,|\mathbf{G}_{\perp}|\right)$, was separated into the $|\mathbf{G}_{\|}|$ and $|\mathbf{G}_{\perp}|$ components. The observed FWHM of the fundamental reflections has a simple relationship in the $|\mathbf{G}_{\|}|$ and $|\mathbf{G}_{\perp}|$ components given by

$$\eta^{obs}(|\mathbf{G}_{\|}|,|\mathbf{G}_{\perp}|) = \eta_{\|} + \eta_{\perp} = \alpha|\mathbf{G}_{\|}| + \beta|\mathbf{G}_{\perp}|. \tag{14.14}$$

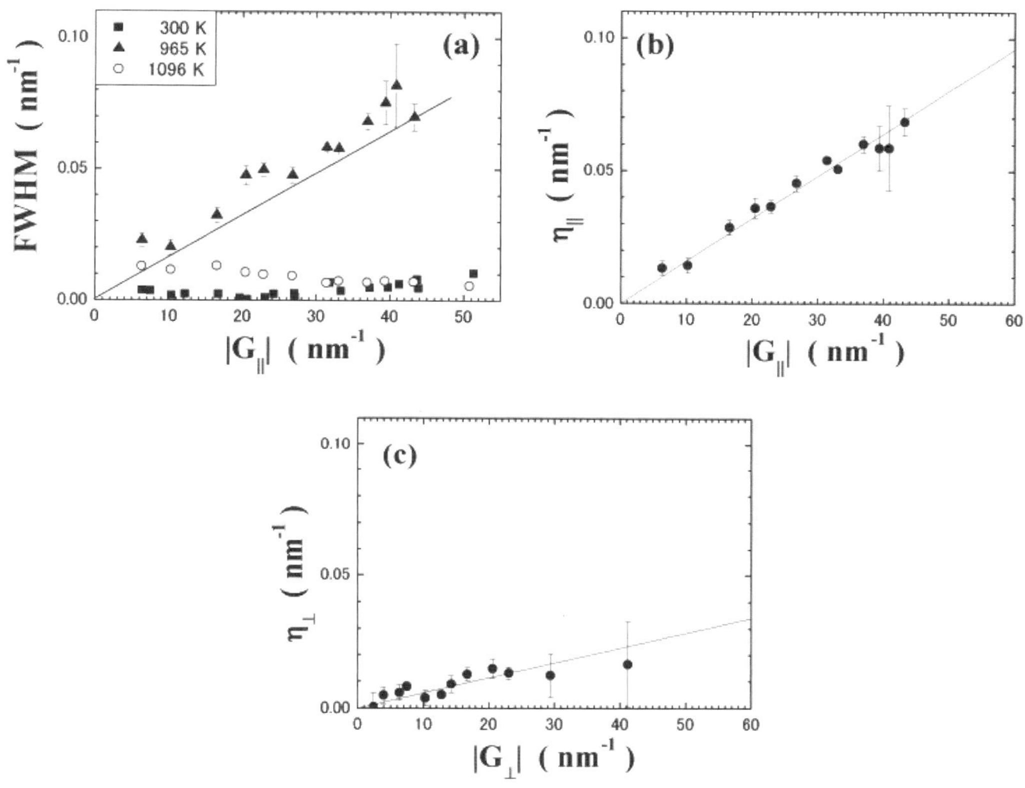

Figure 14.14. *(a) $|G_{\parallel}|$ dependences of FWHM of the fundamental reflections at 300 K (quenched above T_c), 965 K ($T < T_c$), and 1096 K ($T > T_c$). Apparent peak broadening with increasing $|G_{\parallel}|$ values at 965 K is observed. Peak broadening at 965 K ($T < T_c$) is decomposed into (b) $|G_{\parallel}|$ and (c) $|G_{\perp}|$ components.*

The closed triangles in Figure 14.14a show the $|\mathbf{G}_{\parallel}|$-dependence of $\eta^{\mathrm{obs}}(|\mathbf{G}_{\parallel}|, |\mathbf{G}_{\perp}|)$, where the observed FWHM of the fundamental reflections is projected onto the $|\mathbf{G}_{\parallel}|$ axis. The solid line in Figure 14.14a shows the lower limit of the observed FWHM as a function of $|\mathbf{G}_{\parallel}|$, that is, $\alpha\,|\mathbf{G}_{\parallel}|$. After deconvolution of the $|\mathbf{G}_{\parallel}|$ component, the deconvoluted FWHM, $\eta_{\perp}\left(\left|\mathbf{G}_{\perp}\right|\right)$, is plotted as a function of $|\mathbf{G}_{\perp}|$ in Figure 14.14c. A linear fit of the deconvoluted data is displayed by the solid line shown in Figure 14.14c. The fitted slope of $\eta_{\perp}\left(\left|\mathbf{G}_{\perp}\right|\right)$ is equal to β $(= d\eta_{\perp} / d\left|\mathbf{G}_{\perp}\right|)$. Using this β value, we remove the $|\mathbf{G}_{\perp}|$ component, that is, $\eta_{\perp}\left(\left|\mathbf{G}_{\perp}\right|\right)$, from the observed FWHM. The results show good agreement with the $|\mathbf{G}_{\parallel}|$ linearity in Figure 14.14b. The relation expressed in Equation 14.14 is appropriate because two kinds of the deconvoluted FWHM are self-consistent with each other within the experimental errors. The actual α and β values are 0.00160 and 0.000565, respectively. The β values both at room temperature (the as-quenched sample) and at 1096 K (> T_c) are almost negligible. The slope of FWHM along $|\mathbf{G}_{\perp}|$, β, is a measure of the magnitude of the random phason strain. Therefore a random phason contributes to an order-disorder phase transition because β below T_c is larger than at 1096 K (> T_c) by in situ observations. Compared with other ANC systems, the β values are shown in Table 14.2. Obviously, the β value for $Al_{72}Ni_{20}Co_8$ at 965 K (< T_c) is much smaller than for other ANC systems (as-quenched samples). Moreover, the different densities among ANC systems reflect the presence of different kinds of defects at room temperature. For example, the ratio of atomic weights between $Al_{70}Ni_{15}Co_{15}$ and $Al_{72}Ni_{20}Co_8$ is 1.02, and the ratio of densities between them is 1.14. It is clear that the Co concentration can control the phason strain on the phase transition in ANC. Figure 14.15 shows the distributions of the diffuse scattering around the $\bar{1}\bar{2}\bar{5}40$ fundamental reflection at 300 K (quenched from above T_c), 965 K ($T < T_c$), and 1096 K ($T > T_c$), respectively. The SRO diffuse scattering around the S1-superstructure positions at high temperature has a significant relationship with an order-disorder phase transition. The SRO diffuse scattering around the S1 positions disappeared completely at 1096 K ($T > T_c$), although the SRO diffuse scattering was observed around the same positions below T_c. For a comparison, the FWHM of the diffuse peak at the center of line P ($\bar{2}\bar{1}\bar{5}20$ superstructure reflection) was found to be 0.713 nm^{-1} at room temperature. The diffuse peak width at the same position was approximately 0.20 nm^{-1} around T_c. Correlation length is estimated by the Ornstein-Zernike correlation function. It is found to be 2.8 nm at room temperature. On the other hand, the correlation length of the S1-type diffuse peaks increased to 11.0 nm at

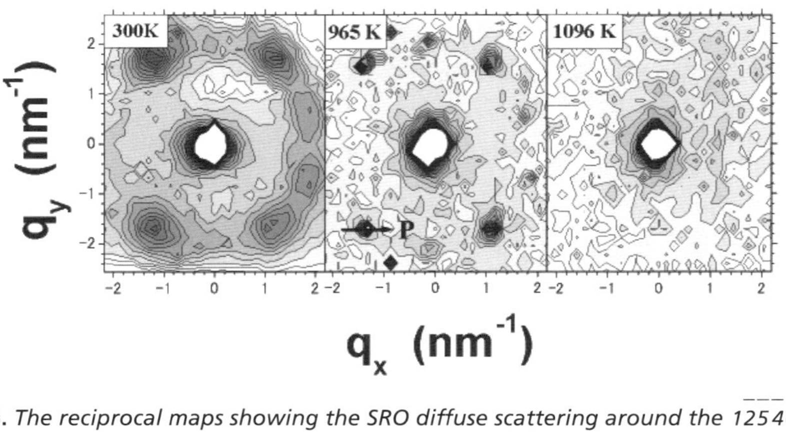

Figure 14.15. *The reciprocal maps showing the SRO diffuse scattering around the $\bar{1}\bar{2}\bar{5}40$ fundamental reflection at (a) room temperature (quenched above T_c), (b) 965 K ($T < T_c$), and (c) 1096 K ($T > T_c$). The diffuse intensity profile of the diffuse peak labeled P was measured during the temperature cycle in Figure 14.16. The diffuse scattering around the S1-superstructure positions disappeared completely above T_c + 10 K.*

965 K. The SRO analysis of the diffuse scattering enables us to determine both the atomic arrangements and the correlation length. More detailed temperature dependence of the diffuse intensity at the peak maximum is shown in Figure 14.16a. The SRO diffuse intensity around the S1-superstructure reflection $(2\bar{1}\bar{5}\bar{2}0)$ was obtained by making a line scan (line P), as drawn in Figure 14.15. According to Equation 14.2, SRO parameters correspond to the diffuse intensities in the reciprocal space. The small SRO diffuse intensity around the S1-superstructure position did not vanish, even above T_c, but it disappeared completely above $T_c + 10$ K. On the other hand, the FWHM of the SRO diffuse scattering had a minimum value around T_c (Figure 14.16b). In general, the diffuse peak width is inversely proportional to the correlation length connected with the ordering process. On cooling, the correlation length continuously increases toward T_c, indicating ordering. However, the diffuse peaks do not sharpen monotonically on cooling in $Al_{72}Ni_{20}Co_8$ because the SRO diffuse peaks are also modified by the nonuniform strain, such as peak broadening of the fundamental reflections below T_c. We estimate

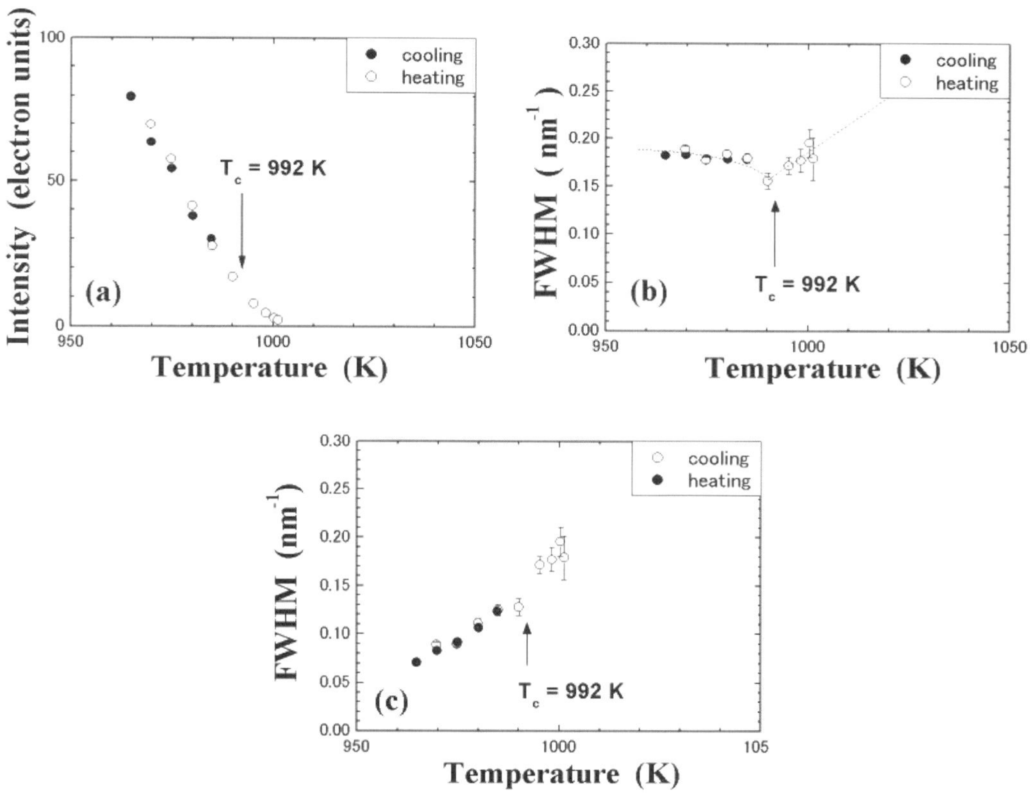

Figure 14.16. *Temperature dependences of (a) the diffuse intensity and (b) the FWHM of the diffuse peak along line P in Figure 14.15. The dotted line is a guideline for the reader's eye. A little hysteresis was shown in the temperature cycle. (c) The deconvoluted FWHM in the S1-superstructure reflection* $(2\bar{1}\bar{5}\bar{2}0)$ *decreases with decreasing temperature.*

peak broadening on the S1-superstructure ($\overline{2}\overline{1}\overline{5}20$) position by Equation 14.14. Then we introduce the reduced temperature, t (= $T/T_c - 1$). η ($-t$, $\overline{1}\overline{2}\overline{5}40$) below T_c is fitted as $\eta(-t) = |-t|^{\gamma}$. From the previous relationship, the deconvoluted FWHM at the $\overline{2}\overline{1}\overline{5}20$ superstructure position is displayed in Figure 14.16c. Temperature dependence of the FWHM at this position shows a decrease on cooling. The correlation length develops gradually with decreasing temperature.

TABLE 14.2

Fundamental data for ANC alloys. a_0 is the lattice constant and $2a_0/\sqrt{5}$ is the edge length of the Penrose pattern. c_0 is the lattice constant along the periodic direction. Some data are referred from previous studies of $Al_{72.7}Ni_{8.5}Co_{18.8}$ [33] and $Al_{70}Ni_{10}Co_{20}$ [34].

	$AL_{72}NI_{20}CO_8$	$AL_{70}NI_{20}CO_{10}$	$AL_{70}NI_{15}CO_{15}$	$AL_{72.7}NI_{8.5}CO_{18.8}$	$AL_{70}NI_{10}CO_{20}$
Diffuse scattering	isotropic	isotropic	anisotropic	anisotropic and peak shifts	—
$d/\eta_\perp / d\lvert\mathbf{G}_\perp\rvert$	−0.00001 (300 K) 0.00056 (965 K) 0.00012 (1096 K)	0.00001	0.00345	0.00424	0.00980
a_0 (nm)	0.2723	0.2733	0.2744	0.2753	0.2768
c_0 (nm)	0.4081	0.4087	0.4078	0.4080	—
density (g/cm^3)	3.94	—	4.5	—	—

5.4 *Anomalous Debye-Waller Factors in $Al_{72}Ni_{20}Co_8$*

Our next step is to detect the dynamic behavior of phasons at high temperature. Recently, a local anomaly of the DW factor was detected by in situ observations using high-angle annular dark-field scanning transmission electron microscopy (HAADF-STEM) in $Al_{72}Ni_{20}Co_8$ [26]. It is thought that local dynamic fluctuations, which correspond to atomic flips, are excited at high temperature. Using transmission geometry in X-ray diffraction, we estimate the DW factor at high temperature quantitatively in $Al_{72}Ni_{20}Co_8$ [27]. It should be noted that X-ray experiments provide information about the average of the local fluctuations in the bulk state, whereas electron microscopy experimental results correspond to direct local fluctuations at the microscopic scale. To estimate the increase in $\langle u^2 \rangle$ at high temperature by a different method, we measured the intensities of various Bragg reflections at room temperature (T_{rt}), 965 K (= $T_1 < T_c$), and 1096 K (= $T_2 > T_c$). $\langle u^2 \rangle$ is the mean square displacement from the average lattice sites. The relationship between the DW factor and integrated intensities of Bragg reflections is given by

$$I^{obs} = I_0 \exp(-2M) = I_0 \exp\{-2B(\sin\theta/\lambda)^2\} = I_0 \exp\{-2(B_{\parallel} |\mathbf{G}_{\parallel}/4\pi|^2 + B_{\perp} |\mathbf{G}_{\perp}/4\pi|^2)\} \quad (14.15)$$

where B is equal to $8\pi^2\langle u^2\rangle$. Since the reciprocal lattice vectors consist of $\mathbf{G}_{\|}$ and $\mathbf{G}_{\perp}$ in quasicrystal, B is also divided into $B_{\|}$ and $B_{\perp}$ components. In quasicrystals, the DW factor is derived from M= $\int \mathbf{G}[C(\mathbf{q}_{\|})]^{-1}\mathbf{G}d^3\mathbf{q}_{\|}$. The difference in $\langle u^2\rangle$ between T and T_{rt} is introduced by

$$-\ln\left\{\frac{I(T)}{I(T_{rt})}\right\} = 2\left\{\ \Delta B_{\|}\left|\mathbf{G}_{\|}/4\pi\right|^2 + \Delta B_{\perp}\left|\mathbf{G}_{\perp}/4\pi\right|^2\right\}. \tag{14.16}$$

Here we assume that the B factor is isotropic. This is because the elastic constants of ANC were found to be almost isotropic by ultrasonic measurements [22]. Taking account of two components, $\Delta B_{\|}|\mathbf{G}_{\|}|^2$ and $\Delta B_{\perp}|\mathbf{G}_{\perp}|^2$, we decomposed them by a Monte Carlo calculation. $|\mathbf{G}_{\perp}|^2$ dependence of $-\ln\{I(T)/I(T_{rt})\}$ are seen in Figure 14.17b. While there is no $|\mathbf{G}_{\|}|^2$ dependence within accuracy (Figure 14.17a), $\Delta\langle u^2\rangle_{\perp}$ between T_1 and T_{rt} is found to be 1.90×10^{-3} nm^2. We confirm that the anomalous attenuation of the Bragg intensities by the anomalous DW factor at high temperature is associated with the local vibration anomaly, as pointed out in HAADF-STEM experiments. On the other hand, phonon dumping at high temperature is not remarkable in quasicrystals. By in situ observation using HAADF-STEM, phason-related α and β sites are observed on HBS tiling. The distance between them is 0.095 nm, as shown in Figure 14.18, where only the pentagon in the Figure is located on the quasiperiodic plane. Since this distance is so small, atoms cannot occupy the α and β sites simultaneously. It is also reported that β sites are energetically similar to α sites by *ab initio* total-energy calculations [28]. We assume that only Al on α or β sites can flip between the energetically equivalent sites. The idea of phason-related atomic jumps between α and β sites at high temperature was introduced in the schematic diagram displayed in Figure 14.18, where *c* is the periodic direction. In the temperature region I ($T > T_c + 10$ K) in Figure 14.18, perfect HBS tiling is satisfied by allowing

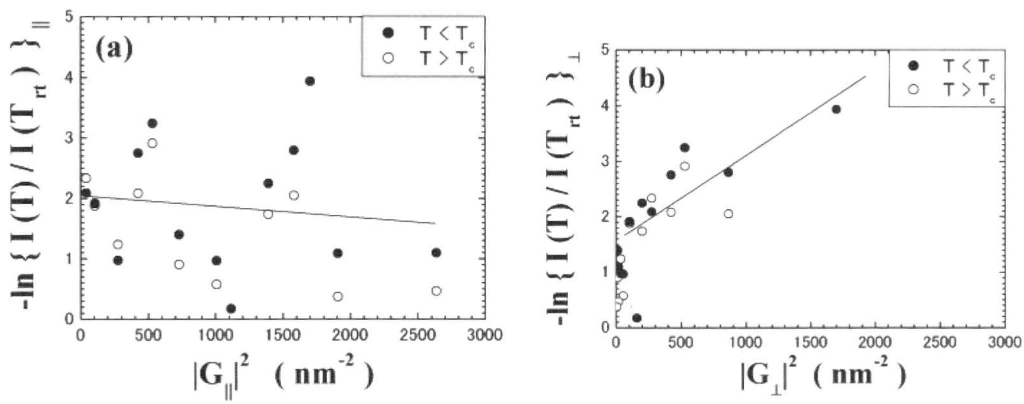

Figure 14.17. (a) $|\mathbf{G}_{\|}|^2$ dependence of the intensity ratio of Bragg reflections at room temperature (T_{rt}) and high temperature (T). (b) $|\mathbf{G}_{\perp}|^2$ dependence of the intensity ratio between room temperature (T_{rt}) and high temperature. T is 965 K (= $T_1 < T_c$) and 1096 K (= $T_2 > T_c$).

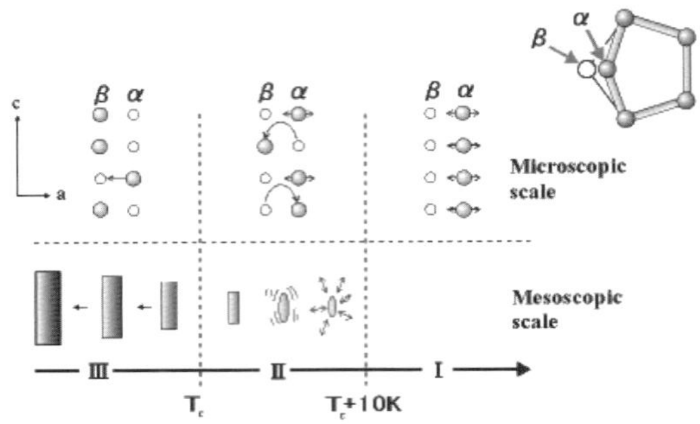

Figure 14.18. Schematic diagram of local atomic flip. Microdomain growth on mesoscopic scale is induced by microscopic atomic flip.

a slight vibration anomaly, which is associated with a phason jump. In temperature region II, some peculiar atom at the α site gradually flips to the β one by the assistance of local phason softening. Since the β site is not a ground state but a local minimum in this temperature region, β back to α flip also occurs with some probability. It is assumed that a quasiperiodic sublattice is partially formed at region II. Therefore atomic SRO appears during the ordering process as supertiling on perfect HBS tiling. On the mesoscopic scale, dynamic embryos associated with superstructure microdomains randomly appear or disappear. Further cooling to temperature region III ($T < T_c$) causes freezing at the β site cooperatively. Cooperative freezing along the periodic direction is possible. With assistance of random phasons, superstructure microdomains are formed at the mesoscopic scale. The randomly generated microdomains extensively generate nonuniform strains.

6 Summary

In this study, systematic experiments and theoretical calculations are performed in ANC decagonal quasicrystals to clarify the relationship between SRO and phasons. The results can be summarized as follows:

1. Peak broadening on $|\mathbf{G}_{\perp}|$ was observed at high Co concentrations of ANC quasicrystals (Table 14.2). Also, weak peak broadening appears below T_c at high temperature even in a perfect quasicrystal.
2. At high Co composition, anisotropic diffuse scattering appeared around Bragg reflections, which have large $|\mathbf{G}_{\perp}|$ values.
3. Intensity modulations at each incident energy were observed by anomalous X-ray scattering in ANC and ANF systems. By quantitative analysis, atomic SE is enhanced, particularly in atomic pairs between Al and TMs whose differences of atomic size are not negligible. Moreover, atomic pair correlation between TMs develops both in ANC and ANF decagonal quasicrystals compared with those between Al-TMs pairs.
4. An anomalous DW factor on $|\mathbf{G}_{\perp}|$ is detected at high temperature. The anomaly is connected with dynamic phason flip.

A competition between SRO and phasons is drawn schematically in Figure 14.19. A picture of a competition between SRO and phasons in quasiperiodic lattices will be more obvious by introducing the following assumptions. In $Al_{70}Ni_{15}Co_{15}$, "random phasons" and "phason modes" are generated in spite of relatively small mosaicity. A large density of $Al_{70}Ni_{15}Co_{15}$ implies that an adequate amount of phasons can promote efficient atomic packing by yielding partial distortions on quasiperiodic lattices. Random appearance of phasons can relax the local strains over the whole lattice, although phasons violate the matching rule locally on quasiperiodic tiling. On the other hand, SRO only between TMs that possess little SE exists in $Al_{70}Ni_{15}Co_{15}$. Other atomic pair correlations such as Al-TM pairs are excluded. One of the reasons is that $Al_{70}Ni_{15}Co_{15}$ quasicrystals avoid generating additional strains induced by atomic SE.

Atomic SE was also seen in the SRO diffuse scattering of $Al_{72}Ni_{20}Co_8$, which has no phasons and quite small mosaicity. In particular, the components of $\alpha_{Al-Ni}(\mathbf{q})$ and $\alpha_{Co-Al}(\mathbf{q})$ are extensively modified by atomic SE. Atomic SE on a perfect quasicrystal is a clue to resolve SRO mechanisms in ideal Penrose tiling. A low density of $Al_{72}Ni_{20}Co_8$ corresponds to atomic vacancies, which are predicted by precise X-ray structure analysis [29]. Some atomic sites on Penrose tiling show low occupation probability. Satisfying the matching rule on Penrose tiling, local strains induced by atomic SE can be compensated around atomic vacancies. At high temperature, dynamic behavior such as phason flip should be considered in addition to SRO and SE. Further experiments about dynamics are required to clarify dynamic SRO on quasiperiodic lattices because present data cannot explain the complicated SRO associated with phason dynamics.

Randomness on quasiperiodic lattices is essentially different from randomness on periodic ones. In ANC and ANF experiments, we found an important property of quasicrystals from SRO and SE analysis:

1. On ideal Penrose tiling, a quasiperiodic lattice satisfying the matching rule is not sensitive to local strains generated by atomic SE.
2. On a partially distorted Penrose tiling caused by phasons, atomic SE directly influences lattice instability. It is predicted that the influence of atomic SE is enhanced locally by violation of the matching rule. We insist that a trivial balance between SRO and a phason is reflected by complicated phase diagrams in quasicrystal-forming alloys.

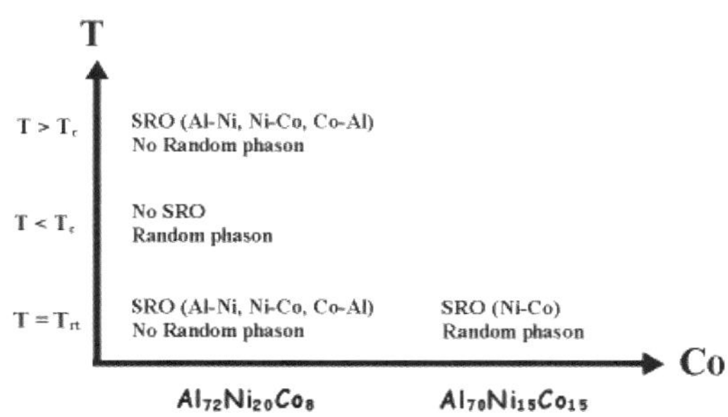

Figure 14.19. *A competition between atomic SRO and phasons varies both along Co concentration and temperature.*

Considering the electronic structure of quasicrystals, the existence of SRO is important to explain the absence of theoretically predicted spikiness in the density of states (DOS). For example, a pseudogap, as predicted theoretically, in the electronic structure of $Al_{70}Ni_{15}Co_{15}$ has been reported [30], although the DOS spikiness was washed out. Full *ab initio* calculations were carried out using modified short-range pair interactions to predict the structure of a decagonal quasicrystal of ANC [31].

In complex quasicrystal materials, it is reconfirmed that the diffuse scattering technique is much more powerful to study essential properties associated with phase stability. In particular, hidden information on chemical disorder in quasicrystals is clarified by anomalous X-ray scattering. For further study, not only SRO but also medium-range order (MRO) is important to explain order-disorder phase transition in quasicrystals. Recently, MRO parameters were proposed for inverse Monte Carlo and reverse Monte Carlo refinements of local atomic structure [32]. Highly modulated diffuse scattering from crystals has been successfully analyzed quantitatively. Nevertheless, atomic SE on quasiperiodic lattice is unresolved. More work will be needed to address MRO analysis, including SE in quasicrystals.

Acknowledgments

The author appreciates helpful discussions with Professor Y. Ishii of Chuo University, Professor E. Abe of the University of Tokyo, Professor H. Takakura of Hokkaido University, and Dr. K. Saitoh of Nagoya University. Also, the author would like to thank Dr. J. Bai of Oak Ridge Associated Universities, Dr. H. Saitoh of the Japan Atomic Energy Research Institute, Dr. H. Nakao of Tohoku University, and Dr. N. Tamura of Advanced Light Source for experimental support. The author wishes to acknowledge useful discussions with and comments and encouragement from Professor Y. Matsuo of Nara Women's University, Professor K. Ohshima of University of Tsukuba, and Professor S. C. Moss of the University of Houston.

References

1. D. Shechtman, I. Blech, D. Gratias, and J. W. Cahn, *Phys. Rev. Lett.* 53 (1984): 1951.
2. H. Takakura, C. P. Gomez, A. Yamamoto, M. de Boissieu, and A. P. Tsai, *Nat. Mater.* 6 (2007): 58.
3. T. C. Lubensky, J. E. S. Socolar, P. J. Steinhardt, P. A. Bancel, and P. A. Heiney, *Phys. Rev. Lett.* 57 (1986): 1440.
4. A. Yamamoto, *Acta Crystallogr. A* 52 (1996): 509.
5. C. L. Henley, *J. Phys. A* 21 (1988): 1649.
6. M. de Boissieu, M. Boudard, B. Hennion, R. Bellissent, S. Kycia, A. Goldman, C. Janot, and M. Audier, *Phys. Rev. Lett.* 75 (1995): 89.
7. S. C. Moss, *Mat. Res. Soc. Symp. Proc.* 376 (1995): 675.
8. S. Hashimoto, *Acta Crystallogr. A* 56 (2000): 85.
9. W. A. Wooster, *Diffuse X-ray Reflections from Crystals* (New York: Dover, 1997).
10. P. H. Dederichs, *J. Phys. F* 3 (1973): 471.
11. G. E. Ice, C. J. Sparks, A. Habenshuss, and L. B. Shaffer, *Phys. Rev. Lett.* 68 (1992): 863.
12. Y. Ishii, *Phys. Rev. B* 45 (1992): 5228.
13. Y. Ishii, *Mater. Sci. Eng.* 294–296 (2000): 377.
14. G. E. Ice and C. J. Sparks, *Nucl. Inst. Meth. A* 291 (1990): 110–116.
15. H. Abe, H. Saitoh, H. Nakao, K. Ito, and K. Ohshima, *J. Appl. Cryst.* 39 (2006): 767.
16. H. Abe, N. Tamura, D. Le Bolloc'h, S. C. Moss, Y. Matsuo, Y. Ishii, and J. Bai, *Mater. Sci. Eng. A* 294–296 (2000): 299.

17. H. Abe, Y. Matsuo, H. Saitoh, T. Kusawake, K. Ohshima, and H. Nakao, *Jpn. J. Appl. Phys.* 39 (2000): L1111.
18. S. Ritsch, C. Beeli, H.-U. Nissen, T. Godecke, M. Scheffer, and R. Luck, *Philos. Mag. Lett.* 78 (1998): 67.
19. E. Abe, K. Saitoh, H. Takakura, A. P. Tsai, P. J. Steinhardt, and H.-C. Jeong, *Phys. Rev. Lett.* 84 (2000): 4609.
20. K. Edagawa, H. Sawa, and S. Takeuchi, *Phil. Mag. Lett.* 69 (1994): 4227.
21. H. Abe, K. Yamamoto, S. Matsuoka, and Y. Matsuo, *J. Phys.: Condens. Matter* 19 (2007): 466201.
22. M. A. Chernikov, H. R. Ott, A. Bianchi, A. Migliori, and T. W. Darling, *Phys. Rev. Lett.* 80 (1998): 321.
23. H. Abe, *Jpn. J. Appl. Phys.* (forthcoming).
24. H. Abe, H. Saitoh, T. Ueno, H. Nakao, Y. Matsuo, K. Ohshima, and H. Matsumoto, *J. Phys.: Condens. Matter* 15 (2003): 1665.
25. M. Honal, T. Haibach, and W. Steurer, *Acta Crystallogr. A* 54 (1998): 374.
26. E. Abe, S. J. Pennycook, and A. P. Tsai, *Nature* 421 (2003): 347.
27. H. Abe, H. Saitoh, T. Ueno, H. Nakao, Y. Matsuo, K. Ohshima, and H. Matsumoto, *J. Phys. Soc. Jpn.* 72 (2003): 1828.
28. C. L. Henley, M. Mihalkovic, and M. Widom, *J. Alloy Comp.* 342 (2002): 221.
29. H. Takakura, A. Yamamoto, and A. P. Tsai, *Acta Crystallogr. A* 57 (2001): 576.
30. Z. M. Stadnik, D. Purdie, M. Garnier, Y. Baer, A. P. Tsai, A. Inoue, K. Edagawa, S. Takeuchi, and K. H. J. Buschow, *Phys. Rev. B* 55 (1997): 10938.
31. M. Mihalkovic, I. Al-Lehyani, E. Cockayne, C. L. Henley, N. Moghadam, J. A. Moriarty, Y. Wang, and M. Widom, *Phys. Rev. B* 65 (2002): 104205.
32. H. Abe, *J. Phys. Soc. Jpn.* 76 (2007): 094601.
33. Y. Matsuo, Y. Nishimura, K. Yamamoto, and Y. Ishii, *Mater. Trans.* 42 (2001): 2008.
34. Y. Matsuo, N. Sugiyama, and K. Yamamoto, *Mater. Trans. JIM* 40 (1999): 137.

15

Ab Initio Test of the Hume-Rothery Electron Concentration Rule for Gamma-Brasses

U. Mizutani[1]

R. Asahi[2]

H. Sato[3]

T. Takeuchi[4]

1. *Toyota Physical & Chemical Research Institute, Nagakute, AICHI, 480–1192, Japan*
2. *Toyota Central Research Laboratories, LTD, Nagakute, AICHI, 480–1192, Japan*
3. *Aichi University of Education, Kariya, Aichi-ken, 448–8542, Japan*
4. *Ecotopia Science Institute, Nagoya University, Nagoya 464–8603, Japan*

Abstract

The gamma-brasses containing fifty-two atoms in the unit cell are classified into three groups: (1) those stabilized at an electron-to-atom ratio (e/a) of 21/13 and known as being typical of obeying the Hume-Rothery electron concentration rule; (2) those involving the transition metal (TM) element as a partner to either Al or Zn and posing the difficulty not only as to whether or not the Hume-Rothery electron concentration rule holds but also as to what e/a value the TM element bears; (3) those like Ag_5Li_8, whose e/a value does not coincide with 21/13 in apparent contradiction to the Hume-Rothery electron concentration rule. First, the stability mechanism for Cu_5Zn_8 and Cu_9Al_4 gamma-brasses in group 1 is reviewed. The central issue of the present chapter is the stability mechanism and e/a determination of the TM element for the TM_2Zn_{11} (TM = Fe, Co, Ni, and Pd) and Al_8V_5 gamma-brasses in group 2 by means of first-principles full-potential linearized augmented plane wave (FLAPW) and linear muffin-tin orbital-atomic sphere approximation (LMTO-ASA) electronic structure calculations. We stress the importance of the 3d states mediated resonance of electrons with different sets of lattice planes. The 3d states mediated resonance is proved to result in a significant suppression of the sp-partial density of states (DOS) over the energy range from –2 eV through the Fermi level up to +2.2 eV. A gain in the electronic energy has been attributed to the formation of highly condensed bonding states below the Fermi level caused by the 3d states mediated resonance. We also show that all the TM elements—V, Fe, Co, Ni, and Pd—bear positive e/a values less than unity and that all the gamma-brasses in group 2 studied in this chapter tend to be stabilized at the e/a value, which increases from 21/13 = 1.605 for Cu_5Zn_8 and Cu_9Al_4 gamma-brasses to about 1.94 with a decrease in the atomic number of the TM element involved.

1 Introduction

The structure of gamma-brass, known as being typical of structurally complex alloy phases, can be described by arranging either two identical twenty-six-atom clusters to form a bcc lattice or two different ones to form a CsCl-type structure. The former belongs to the space group $I\bar{4}3m$, while the latter belongs to the space group $P\bar{4}3m$, both containing fifty-two atoms in its cubic unit cell [1]. The twenty-six-atom cluster consists of four atoms in the first shell formed by four vertices of inner tetrahedron (IT), four atoms in the second shell by those of outer tetrahedron (OT), six atoms in the third shell by those of octahedron (OH), and twelve atoms in the fourth shell by those of cube octahedron (CO) [2]. The atomic structure of many gamma-brasses has been experimentally determined. In general, chemical disorder exists in each shell of the twenty-six-atom cluster. The exceptions are for Cu_5Zn_8 and Cu_9Al_4, in which a specific atom species is filled into each shell [3–7], and hence the experimentally determined atomic structure could be directly employed for first-principles band calculations, in which chemical disorder in each shell must be eliminated.

The gamma-brass phase is found in twenty-four binary alloy systems [1,8]. In many cases, it has a finite solid-solution range. Since first-principles band calculations are made for a chemical disorder-free structure, it is important to have a stoichiometric composition without involving any chemical disorder in each shell in the twenty-six-atom cluster. The stoichiometric composition of the ideal gamma-brass should fall in one of A_2B_{11}, A_4B_9, A_5B_8, A_7B_6, and $A_{10}B_3$ with the space group $I\bar{4}3m$ and a combination of these two with the space group $P\bar{4}3m$. A maximum solid-solution range and the ideal stoichiometric composition expected in the solid-solution range for twenty-four gamma-brasses are listed in Table 15.1. Among twenty-four binary gamma-brasses, only Al_8V_5 and Mn_3In apparently exist as line compounds [1]. The composition for Mn_3In can be realized only when chemical disorder is distributed over shells in the twenty-six-atom cluster.

We have classified twenty-four gamma-brasses into three groups, as listed in Table 15.1. Group 1 consists of a combination of monovalent noble metals and polyvalent elements whose valency is well defined. They are formed in the vicinity of e/a = 21/13, as the Hume-Rothery electron concentration rule claims for gamma-brass. Among them, the fact that the prototype Cu_5Zn_8 and Cu_9Al_4 gamma-brasses are stabilized at e/a = 21/13 in spite of different solute concentrations has been known since the late 1920s [9–11]. Its stabilization mechanism has been discussed by performing first-principles FLAPW band calculations [7]. The pseudogap formed across the Fermi level at e/a = 21/13 for both systems has been proved to originate from the resonance of electron waves near the Fermi level with the sets of lattice planes {330} and {411}.

Group 2 includes a total of eleven gamma-brasses that consist of a 3d transition metal (TM) element and either divalent elements Be, Zn, and Cd or trivalent elements Al and In. Since the e/a

TABLE 15.1

Gamma-brasses in binary alloy systems

GROUP I			SPACE GROUP	GROUP II			SPACE GROUP
COMPOSITION	*e/a*	SOLUBILITY RANGE		COMPOSITION	*e/a*	SOLUBILITY RANGE	
Ag_5Cd_8	21/13	$57 \leq Cd \leq 62.5$	$I\bar{4}3m$	Fe_2Zn_{11}	?	$17 \leq Fe \leq 31$	$I\bar{4}3m$
Ag_5Zn_8	21/13	$58 \leq Zn \leq 64.7$	$I\bar{4}3m$	Ni_2Zn_{11}	?	$15 \leq Ni \leq 30$	$I\bar{4}3m$
Cu_5Cd_8	21/13	$52.2 \leq Cd \leq 66$	$I\bar{4}3m$	Pd_2Zn_{11}	?	$14.5 \leq Pd \leq 24$	$I\bar{4}3m$
Cu_5Zn_8	21/13	$57 \leq Zn \leq 70$	$I\bar{4}3m$	Pt_2Zn_{11}	?	$19 \leq Pt \leq 23$	
Au_5Cd_8	21/13	$61.2 \leq Cd \leq 67.6$	$I\bar{4}3m$	Mn_2Zn_{11}	?	$15.2 \leq Mn \leq 23$	$I\bar{4}3m$
Au_5Zn_8	21/13	$62.5 \leq Zn \leq 76$	$I\bar{4}3m$	Ir_2Zn_{11}	?	$15.3 \leq Ir \leq 15.7$	$I\bar{4}3m$
Cu_9Al_4	21/13	$31.5 \leq Al \leq 37.5$	$P\bar{4}3m$	Ni_2Be_{11}	?	$11.5 \leq Ni < ?$	$I\bar{4}3m$
Cu_9Ga_4	21/13	$29.5 \leq Ga \leq 34.7$	$P\bar{4}3m$	Al_8V_5	?	Al_8V_5	$I\bar{4}3m$
Cu_9In_4	21/13	$27.7 \leq In \leq 31.3$	$P\bar{4}3m$	Mn_3In	?	Mn_3In	$P\bar{4}3m$
$Ag\,In_4$	21/13	$31.1 \leq In \leq 33.6$	$P\bar{4}3m$	Ni_2Cd_{11}	?	$12 \leq Ni \leq 19.5$	$P\bar{4}3m$
Au_9In_4	21/13	$28.8 < In < 31.4$	$P\bar{4}3m$	Ag_5Li_8	1.0	$63.5 \leq Li \leq 76$	$I\bar{4}3m$
Co_2Zn_{11}	?	$14.6 \leq Co \leq 31$	$I\bar{4}3m$	*$Li_{10}Pb_3$	22/13	$22.5 \leq Pb \leq 23.5$	$P\bar{4}3m$

* We recently confirmed that $Li_{10}Pb_3$ does not crystallize into the gamma-brass structure. More details about this will be reported elsewhere.

value for the TM element is not a priori known, it is not clear whether the gamma-brass in group 2, having an ideal stoichiometric composition, is stabilized at e/a = 21/13 or not. This has been for a long time regarded as one of the most challenging themes in the electron theory of metals [12]. In the present chapter, we discuss the stability mechanism of TM_2Zn_{11} (TM = Fe, Co, Ni, or Pd) and Al_8V_5 gamma-brasses and the determination of the e/a value of the TM element involved. Group 3 includes gamma-brasses composed of two components whose valencies are well defined. However, a simple manipulation of e/a does not yield 21/13. Its typical example is Ag_5Li_8 gamma-brass consisting of only monovalent elements. It is of great interest to examine whether or not different stabilization mechanisms exist in different groups. This will shed more light not only on the limit of the applicability of the Hume-Rothery electron concentration rule but also on the long-lasting theme of the electron theory of metals.

2 Crystal Structure

2.1 TM_2Zn_{11} (Fe, Co, Ni, and Pd) Gamma-Brasses

The gamma-brass phase extends over 15 to 30 at.%TM in the TM-Zn (TM = Mn, Fe, Co, Ni, Pd, and Pt) alloy systems [1]. Among them, the atomic structure data are available in the literature for TM = Mn [9], Fe [5,13], Co [14], Ni [13], and Pd [15]. Chemical disorder is always present in all cases studied. To eliminate chemical disorder, we filled the majority atom in each shell of the twenty-six-atom cluster and ended up with a structure with the composition TM_2Zn_{11}, in which the atom TM is filled only into the OT shell and Zn atoms filled into all other shells: IT, OH, and CO. The lattice constant and atom positions experimentally determined with chemical disorder were used in the present band calculations for gamma-brasses TM_2Zn_{11} (TM = Fe, Co, Ni, and Pd) [5,13–15]. Fortunately, the composition of 15.38 at.% TM corresponding to TM_2Zn_{11} is found within the stable composition range in the respective phase diagrams [1].

2.2 Al_8V_5 Gamma-Brass

The Al_8V_5 gamma-brass contains the TM element V with 38.46 at.%, being more than twice as much as that in the TM_2Zn_{11} gamma-brasses. According to the structure analysis on a single crystal by Brandon et al. [16], the Al_8V_5 gamma-brass crystallizes into the same structure as the prototype Cu_5Zn_8, that is, the structure with the space group $I\bar{4}3m$ and containing fifty-two atoms in its unit cell with a lattice constant of 9.223 Å. As shown in Figure 15.1a, two Al and two V atoms are filled into the IT shell, four V atoms into the OT shell, four V atoms and two Al atoms into the OH shell, and twelve Al atoms into the CO shell. To avoid the chemical disorder in IT and OH, we exclusively filled Al atoms into IT and V atoms into OH. This can be done without changing the stoichiometric composition Al_8V_5. The results are shown in Figure 15.1b and employed in the band calculations [17].

3 LMTO-ASA and FLAPW Band Calculations

3.1 LMTO-ASA Method

The LMTO-ASA electronic structure calculations provide decisive information about the role of orbital hybridizations with computational efficiency and are performed in the periodic zone scheme in combination with local density functional theory. The potential parameters are self-consistently determined at 285 independent **k** points in the irreducible wedge of the Brillouin zone for the Al_8V_5 gamma-brass [18,19]. The hybridization terms between the V 3d and Al 3p states are intentionally deleted to examine how orbital hybridizations between the neighboring constituent elements affect the band structure.

3.2 FLAPW Method

The details of the present FLAPW electronic structure calculations have been described elsewhere [7]. Briefly, the wave function is expressed as

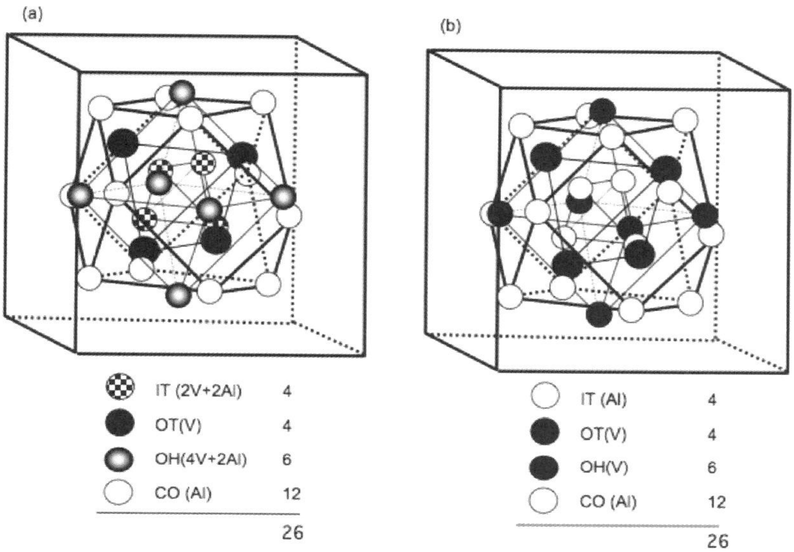

Figure 15.1. *(a) Experimentally determined atomic structure of Al_8V_5 gamma-brass [16] and (b) chemical disorder-free model employed in the electronic structure calculations [17].*

$$\psi_i(\mathbf{r},\mathbf{k}) = \sum_{\mathbf{G}} C^i_{\mathbf{k}+\mathbf{G}} \phi(\mathbf{r},\mathbf{k}+\mathbf{G}) , \tag{15.1}$$

where $\mathbf{k}$ is an arbitrary wave vector in the irreducible Brillouin zone, $\mathbf{G}$ is a reciprocal lattice vector, and i is the band index. Here the basis functions are given as [20,21]

$$\phi(\mathbf{r},\mathbf{k}+\mathbf{G}) = \begin{cases} \Omega^{-1/2} e^{i(\mathbf{k}+\mathbf{G})\cdot\mathbf{r}} & r \in \text{interstitial} \\ \sum_{lm}\left[A^{\alpha}_{lm}(\mathbf{k}+\mathbf{G})u_l(E^{\alpha}_l, r_\alpha) + B^{\alpha}_{lm}(\mathbf{k}+\mathbf{G})\dot{u}_l(E^{\alpha}_l, r_\alpha)\right]Y_{lm}(\hat{r}_\alpha) & r \in \text{sphere} \end{cases} , \tag{15.2}$$

where $u(E^{\alpha}_l, r_\alpha)$ and $\dot{u}_l(E^{\alpha}_l, r_\alpha)$ are solutions of the radial Schrödinger equation solved at a fixed energy E^{α}_l and their derivatives, respectively; $Y_{lm}(\hat{r}_\alpha)$ is a spherical harmonic; and the coefficients $A^{\alpha}_{lm}(\mathbf{k}+\mathbf{G})$ and $B^{\alpha}_{lm}(\mathbf{k}+\mathbf{G})$ are determined by the requirement that the plane waves and their radial derivatives are continuous at the surface of the atomic spheres. The number of plane waves was about 2900 in the case of the Al_8V_5 gamma-brass.

4 Stability Mechanism of Gamma-Brasses In Group 2

4.1 TM_2Zn_{11} (TM = Fe, Co, Ni, and Pd) Gamma-Brasses

Figure 15.2 shows the E–**k** relations and corresponding DOS for the TM_2Zn_{11} gamma-brasses with TM = Fe, Co, Ni, and Pd. One can immediately identify the Zn 3d band in the energy range –7.5 to –8 eV with dispersionless E–**k** relations. Since the amount of the TM element is limited to 15.38 at.%, the TM d band is small and is centered at –4.5, –2.5, –1, and –1 eV in the sequence of Pd, Ni, Co, and Fe. In the cases of TM = Pd and Ni, their d band is essentially located below the Fermi level and the pseudogap is seen immediately below it in the same manner as those in Cu_5Zn_8 and Cu_9Al_4 gamma-brasses [7]. However, the d band in TM = Co and Fe is extended above the Fermi level so that the pseudogap is apparently fully pushed above it. The presence of the pseudogap above the Fermi level would no longer contribute to reducing the electronic energy to stabilize a system.

We now focus on the Fourier component of the FLAPW function extending outside the muffin-tin (MT) sphere at the symmetry point N of the bcc Brillouin zone for all gamma-brasses studied. Note that the FLAPW function in this region is given by a sum of plane waves over the permissible reciprocal lattice vectors. At the symmetry point N (or M in the case of Cu_9Al_4 with the space group $P\bar{4}3m$), the square of the reciprocal lattice vectors in units of $2\pi/a$ is ordered as $|\mathbf{G}|^2 = 2, 6, 10, 14, 18, 22, 26, 30, 34,\ldots$, where two of the Miller indices are odd. We have shown that, in the cases of Cu_5Zn_8 and Cu_9Al_4, the square of the Fourier coefficient summed over equivalent zones $\sum|C_{\mathbf{k+G}}|^2$ is extremely large when $|\mathbf{G}|^2 = 18$ [7]. This means that the pseudogap across the

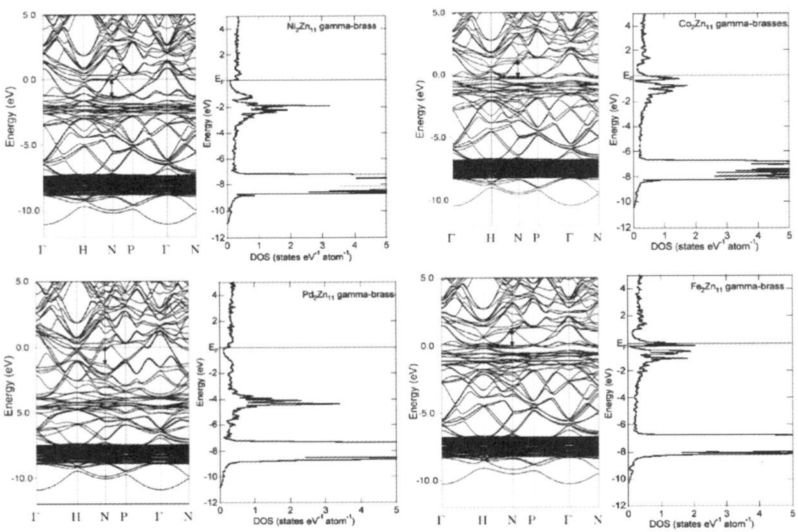

Figure 15.2. *Dispersion relations and DOS derived from FLAPW electronic structure calculations for the TM_2Zn_{11} gamma-brasses [14].*

Fermi level is caused by the resonance of electron waves having nearly the Fermi wavelength with the set of lattice planes associated with {330} and {411} zones.

To examine if the same mechanism holds for the TM_2Zn_{11} gamma-brasses, we show in Figure 15.3 the ratio of the Fourier coefficient $\sum|C_{\mathbf{k+G}}|^2$ at $|\mathbf{G}|^2 = 18$ over the next intense one as a function of energy eigenvalues near the Fermi level in comparison with the data for Cu_5Zn_8 and Cu_9Al_4. The ratio I_{18}/I_{next} in Cu_5Zn_8 and Cu_9Al_4 is obviously large, indicating that the resonance with {330} and {411} zones is responsible for the formation of the pseudogap. The same situation holds true in Ni_2Zn_{11} and Pd_2Zn_{11} gamma-brasses. We are thus led to conclude that these two gamma-brasses are stabilized in the same mechanism as that in Cu_5Zn_8 and Cu_9Al_4. However, the ratio I_{18}/I_{next} is small in the vicinity of the Fermi level for Co_2Zn_{11} and Fe_2Zn_{11}, indicating the failure of the resonance mechanism solely involving {330} and {411} zones.

4.2 Al_8V_5 Gamma-Brass

Figure 15.4 shows the FLAPW-derived E–**k** relations and corresponding DOS for the Al_8V_5 gamma-brass. First of all, the dispersionless E–**k** relations and a high DOS in the energy range from –3 to +3 eV are safely attributed to the V 3d band. A deep pseudogap found at about +0.5 eV above the Fermi level obviously separates the V 3d band into bonding and antibonding bands. Figures 15.5a and 15.5b show the LMTO-ASA-derived DOS before and after deleting the V 3d/Al 3p hybridization terms, respectively. The disappearance of the deep pseudogap at +0.5 eV in Figure 15.5b indicates that the orbital hybridizations between the V 3d and Al 3p states are responsible for the split into bonding and

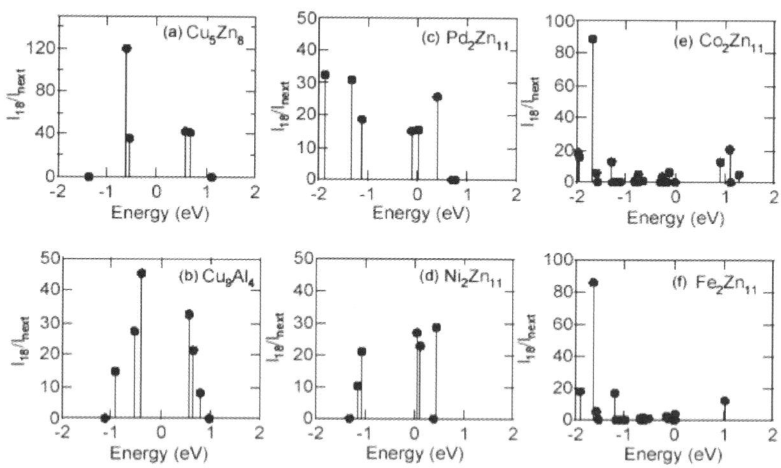

Figure 15.3. *Energy dependence of the ratio of the Fourier coefficient $I = \Sigma|C_{k+G}|^2$ of the FLAPW function at $|G_0|^2$ (= Σh^2) = 18 over the next largest one, I_{18}/I_{next}, for (a) Cu_5Zn_8, (b)Cu_9Al_4, (c) Pd_2Zn_{11}, (d) Ni_2Zn_{11}, (e) Co_2Zn_{11}, and (f) Fe_2Zn_{11} gamma-brasses [14].*

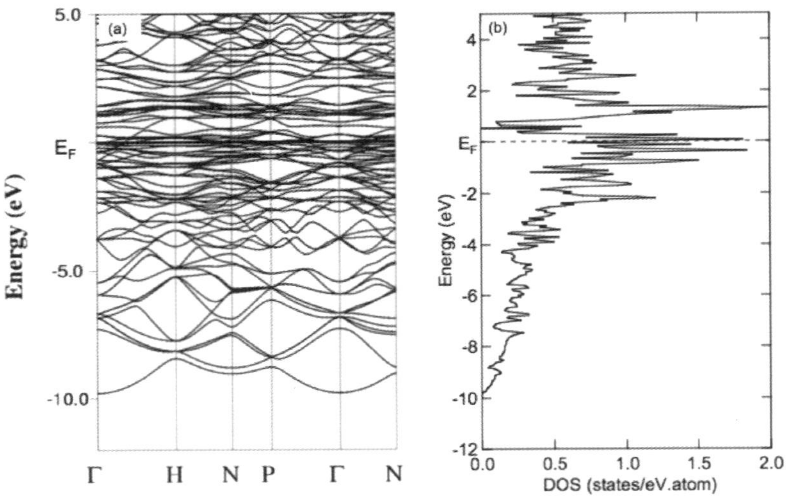

Figure 15.4. *(a) Dispersion relations and (b) DOS derived from FLAPW electronic structure calculations for the Al_8V_5 gamma-brass [17].*

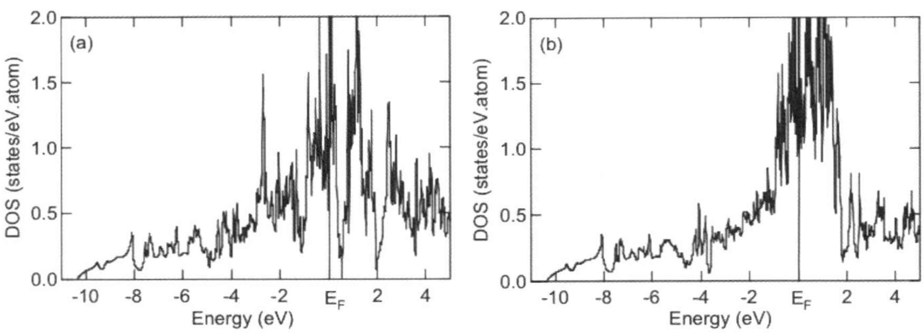

Figure 15.5. *(a) LMTO-ASA-derived total DOS and (b) the DOS after deleting hybridization terms between the V 3d and Al 3p states in the LMTO-ASA basis functions for the Al_8V_5 gamma-brass. The deep pseudogap at 0.5 eV above the Fermi level has disappeared after zeroing orbital hybridizations [17].*

antibonding states. However, the pseudogap existing above the Fermi level is not able to reduce the electronic energy of the system. Interestingly, there is no pseudogap below the Fermi level in this system.

Figure 15.6a shows the E–**k** relations along the ΓN direction in the reduced zone scheme calculated within the free-electron model for the Al_8V_5 gamma-brass. The symmetry point N is the center of the bcc Brillouin zone. All zones consisting of the set of Miller indices, two of which are odd like {330} and {411}, pass through the point N upon the reduction to the first zone [18]. The two parabolic bands intersect at –9.5 eV at the point N in Figure 15.6a. The electronic state is identified as the point

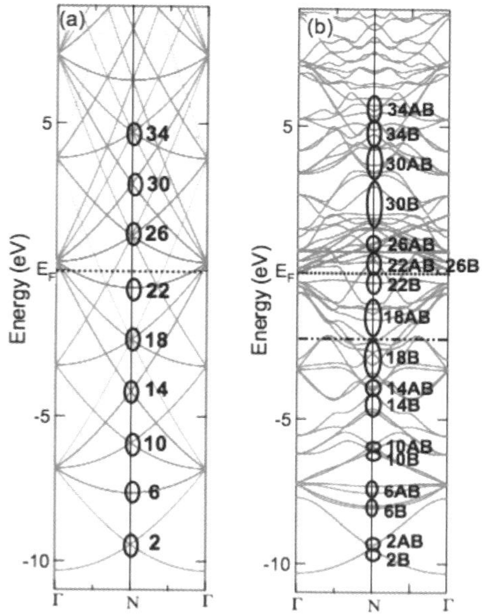

Figure 15.6. *Energy dispersion relations along ΓNΓ directions in (a) the free-electron model and (b) the nearly free-electron model, where hybridization terms of V 3d with all other states are neglected in the LMTO-ASA method for the Al_8V_5 gamma-brass [17].*

N on the {110} zones or $|\mathbf{G}|^2 = 2$. As shown in Figure 15.6a, the free-electron states at the point N can be indexed as $|\mathbf{G}|^2 = 2$, 6, 10, 14, 18, 22, 26, 30, 34,... in an increasing order of energy. Figure 15.6b indicates the E–**k** relations derived after deleting all hybridization terms between V 3d and all other states in the LMTO-ASA calculations. As a result of the isolation of the V 3d states, it is reduced to a level at about 1 eV above the Fermi level. The E–**k** relations in Figure 15.6b may be viewed as the "nearly free-electron model" for the Al_8V_5 gamma-brass because all other orbital hybridizations remain considered. The electronic states at the point N in Figure 15.6b also can be easily identified by using the assignment in Figure 15.6a as a guide. For example, the states with $|\mathbf{G}|^2 = 2$ open a small energy gap and split into bonding (denoted as "B") and antibonding states ("AB") as a result of resonance of electrons with the set of lattice planes {110}.

As can be seen from Figure 15.6b, electrons near the Fermi level are interacting with the sets of lattice planes due to $|\mathbf{G}|^2 = 22 \sim 26$. Instead, electronic states marked as 18A and 18AB are located about 2 eV below the Fermi level. This is the interaction with the set of {330} and {411} zones, that is, $|\mathbf{G}|^2 = 18$. The DOS calculated in the nearly free-electron model is shown in Figure 15.7. A pseudogap caused by the resonance with $|\mathbf{G}|^2 = 18$ is indeed found at –2 eV. The integration of the DOS from the bottom of the valence band up to the center of the pseudogap yields the electron concentration close

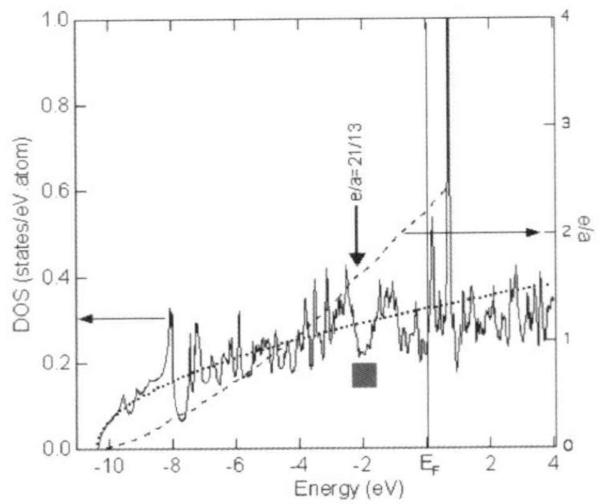

Figure 15.7. *The total DOS calculated from the dispersion relations shown in Figure 15.6b for the Al_8V_5 gamma-brass. The pseudogap centered at –2 eV is easily identified as originating from the resonance with the set of lattice planes specified by $|\mathbf{G}|^2 = 18$ and the e/a value of 21/13 is obtained by integrating the DOS up to the middle of this pseudogap [17].*

to 1.6, in good agreement with the Mott and Jones model [11]. However, the electron concentration obtained by integrating the DOS up to the Fermi level turns out to be 2.15, suggesting that the Al_8V_5 gamma-brass is no longer an e/a = 21/13 compound. But the discussion above is less accurate because the V 3d band is fully ignored.

We now study the resonance condition of electrons with sets of lattice planes in the presence of the V 3d states by making full use of the Fourier expansion technique for the FLAPW function outside the MT sphere [7,14,17]. Figure 15.8 shows the energy dependence of the Fourier coefficient $|C_{\mathbf{k+G}}(E)|^2$ characterized by $|\mathbf{G}|^2$ of the FLAPW function outside the MT sphere at the point N for the Al_8V_5 gamma-brass. As can be understood from inspection of Figure 15.4a in comparison with Figure 15.6, the electronic states at the point N at energies around –7.4 eV are indexed as $|\mathbf{G}|^2 = 6$ and are essentially free-electron-like because they are well below the V 3d band. Indeed, the energy spectrum at $|\mathbf{G}|^2 = 6$ is narrow and centered at about –7.4 eV. When $|\mathbf{G}|^2$ is increased in the region $10 \le |\mathbf{G}|^2 \le 34$, the V 3d states come into play and widen the energy spectra, as shown in Figure 15.8. More important is the fact that the electronic states are apparently split into bonding and antibonding states across the Fermi level as if they avoid the V 3d states. The characteristic feature of the splitting above can be understood as arising from the V 3d mediated resonance of mobile electrons extending outside the MT sphere with sets of lattice planes in the range $10 \le |\mathbf{G}|^2 \le 34$. We hereafter call it the V 3d mediated resonance and consider below how the resonance can contribute to reducing the electronic energy of the systems.

We calculated the $|\mathbf{G}|^2$ dependence of two different electronic energies defined below at symmetry points N, Γ, and H of the bcc Brillouin zone:

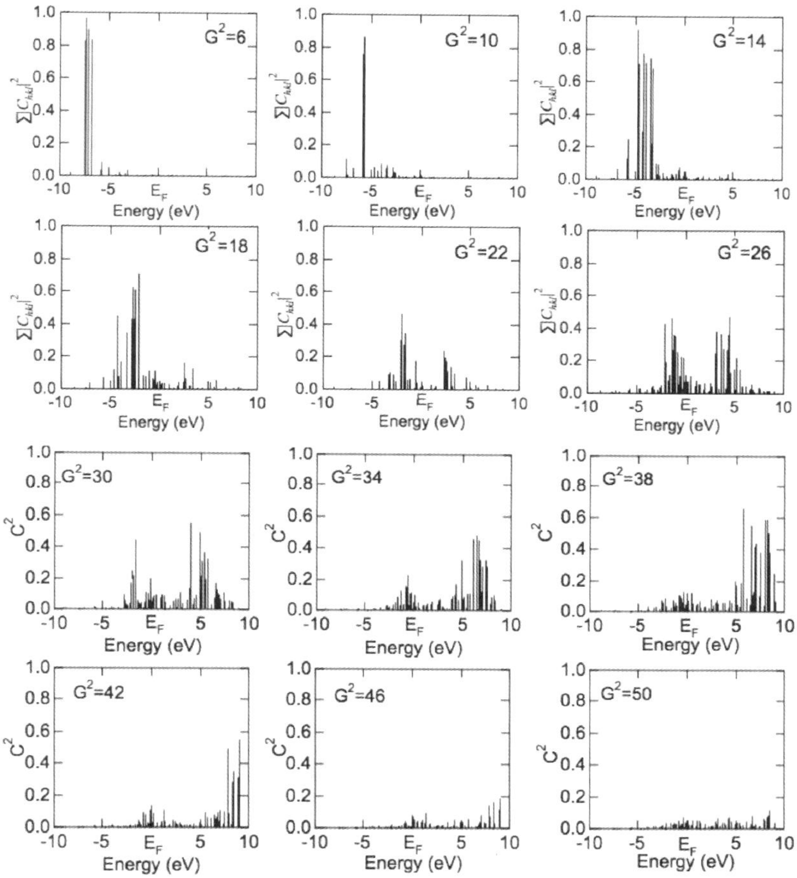

Figure 15.8. *Energy dependence of* $\sum_{\{hkl\}} \left|C_{hkl}\right|^2$ *at selected values of* $|\mathbf{G}|^2$ *at the point N for the* Al_8V_5 *gamma-brass. Note that the splitting into bonding and antibonding states is observed for all* $|\mathbf{G}|^2$*s over the range 14 to 50 [17].*

$$U_{av}(\mathbf{G}) = \frac{\sum_{E=E_0}^{\infty} \left|C_{\mathbf{k+G}}(E)\right|^2 E}{\sum_{E=E_0}^{\infty} \left|C_{\mathbf{k+G}}(E)\right|^2} \tag{15.3}$$

and

$$U_{oc}(\mathbf{G}) = \frac{\sum_{E=E_0}^{E_F} \left|C_{\mathbf{k+G}}(E)\right|^2 E}{\sum_{E=E_0}^{E_F} \left|C_{\mathbf{k+G}}(E)\right|^2}, \quad (15.4)$$

where E_0 refers to the energy at the bottom of the valence band. The electronic energy in Equation 15.3 is evaluated by taking into account both bonding and antibonding states across the Fermi level, whereas that in Equation 15.4 only takes into account the bonding states below the Fermi level. The results are shown in Figure 15.9. It can be seen that both $U_{av}(\mathbf{G})$ and $U_{oc}(\mathbf{G})$ fall on respective universal curves irrespective of the choice of symmetry points N, Γ, and H. This means that, although both $U_{av}(\mathbf{G})$ and $U_{oc}(\mathbf{G})$ are calculated only at special symmetry points of the Brillouin zone, they may well represent the electronic energies calculated over whole electronic states in the Brillouin zone. The value of $U_{av}(\mathbf{G})$ is close to the line drawn on the basis of the free-electron model, whereas $U_{oc}(\mathbf{G})$ is consistently lower than $U_{av}(\mathbf{G})$. We are therefore led to conclude that the electronic energy of the present system is definitely lowered by the V 3d states mediated resonance effect.

We further pursue how the V 3d states mediated resonance contributes to the stability of the Al_8V_5 gamma-brass. Figure 15.10 shows the energy dependence of $\left|C_{\mathbf{G}}(E)\right|^2$ by dividing the spectra into three regimes in terms of the magnitude of $\left|\mathbf{G}\right|^2$ at the symmetry point N: (a) $6 \le \left|\mathbf{G}\right|^2 \le 18$, (b) $22 \le \left|\mathbf{G}\right|^2 \le 30$, and (c) $34 \le \left|\mathbf{G}\right|^2 \le 50$. The resonance in regime (a) is found to produce the bonding

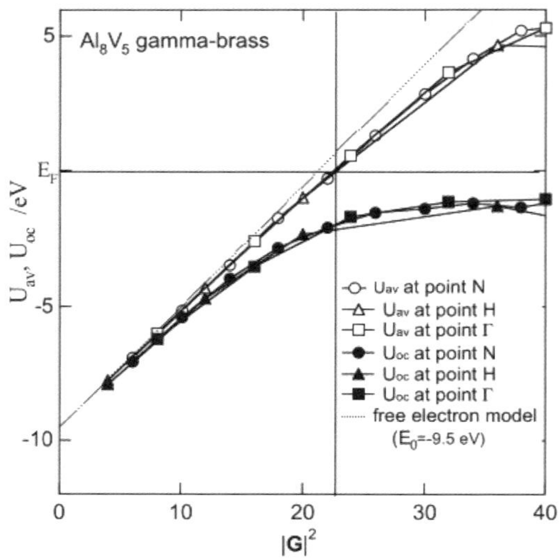

Figure 15.9. *$|G|^2$ dependence of the electronic energies U_{av} and U_{oc} calculated from Equations 15.3 and 15.4 at the points N, H, and Γ for the Al_8V_5 gamma-brass, respectively [17].*

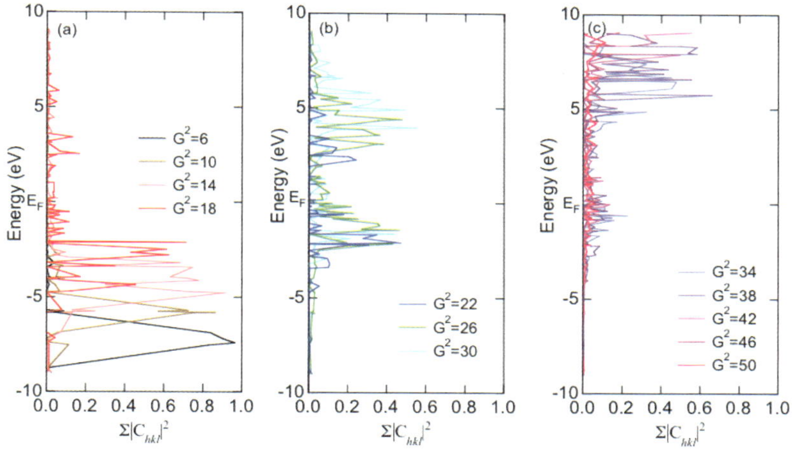

Figure 15.10. *Energy dependence of $\Sigma|C_{k+G}|^2$ in three different regimes: (a) $6 \le |\mathbf{G}|^2 \le 18$, (b) $22 \le |\mathbf{G}|^2 \le 30$, and (c) $34 \le |\mathbf{G}|^2 \le 50$ at the point N for the Al_8V_5 gamma-brass.*

states mainly below –2 eV and contributes most effectively to lowering the electronic energy of the system. In regime (b), the bonding states remain substantial below the Fermi level but decrease rapidly toward the Fermi level. The states are very sparse from the Fermi level up to about +2 eV, resulting in a deep gap in this energy range. The resonance in regime (c) yields negligibly small bonding states below the Fermi level but substantial antibonding states above about +5 eV. Hence, we consider the resonance in regimes (a) and (b) to play a key role but that in regime (c) to be of minor importance in contributing to the reduction in the electronic energy in the system. If we superimpose all spectra in the three regimes, we immediately realize that the V 3d states mediated resonance substantially reduces the DOS of mobile electrons from –2 eV toward the Fermi level and causes essentially the formation of a gap from the Fermi level up to +2 eV.

We cannot find any pseudogap in the range below the Fermi level down to –2 eV in the total DOS shown in Figure 15.4b. This is because the V 3d states, which mostly reside inside the MT sphere, dominate the total DOS in the energy range –3 to +3 eV and mask the structure exhibited by mobile electrons discussed above. The V 3d states mediated resonance would be reflected in the sp-partial DOS. Figure 15.11 shows the FLAPW-derived sp-partial DOS for the Al_8V_5 gamma-brass. One can clearly see the deep pseudogap over the range –2 to +2 eV, reflecting the characteristic feature discussed above. We therefore conclude that the V 3d states mediated resonance is the stabilization mechanism for the structurally complex Al_8V_5 gamma-brass. Although we have not yet explored the d states mediated resonance in the Co_2Zn_{11} and Fe_2Zn_{11} gamma-brasses, we tend to believe that the TM 3d mediated resonance would serve to stabilize these gamma-brasses, although the TM 3d states dominate the total DOS across the Fermi level.

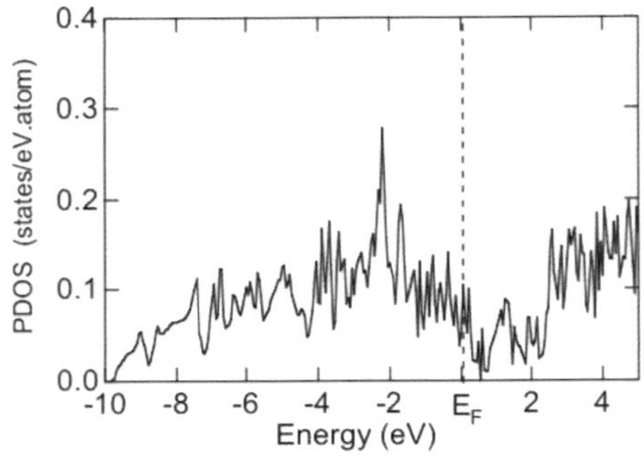

Figure 15.11. *sp-partial DOS calculated from FLAPW for Al_8V_5 compounds.*

5 Determination of e/a for Gamma-Brasses in Groups 1 and 2

The parameter e/a represents an average number of valence electrons per atom and serves a key role in the Hume-Rothery electron concentration rule. Let us consider the case of noble metals Cu, Ag, and Au. The Fermi level is determined by totally filling eleven electrons per atom (ten d electrons and one s electron per atom) into their valence band. But noble metals are known to be treated as monovalent elements with e/a = 1.0. This means that more localized d-electrons occupying the total DOS a few eV below the Fermi level are not counted. The value of e/a for polyvalent elements situated on the right-hand side of the noble metals in the periodic table is simply given by the number of outermost electrons in the state of a free atom. However, the Fermi level in TMs on the left-hand side of the noble metals in the periodic table is located in the middle of their d band. As a result, their effective e/a value is not self-evident. The determination of the e/a value for TM elements in an alloy has long been controversial in spite of a large number of discussions available in the literature [12].

Raynor [22] pointed out that the Al-richest Al-TM binary compounds existing in the equilibrium phase diagram decrease the Al content with an increase in the atomic number of the TM element—$CrAl_7$, $MnAl_6$, $FeAl_3$, Co_2Al_9, and $NiAl_3$—and that all these Al-TM compounds are centered at the e/a value of 2.1 and satisfy the Hume-Rothery electron concentration rule, provided that Cr, Mn, Fe, Co, and Ni possess negative values of e/a equal to –4.66, –3.66, –2.66, –1.66, and –0.66, respectively. Raynor explained the assignment of negative valencies to the TM element on the assumption that the TM element absorbs mobile electrons into a more localized d band. Tsai et al. [23] discovered Al-Cu-TM (TM = Fe, Ru, and Os) thermally stable quasicrystals and pointed out that they are stabilized in the neighborhood of e/a = 1.8 if the negative valencies proposed by Raynor [22] are adopted for the TM elements. Since Tsai and his group [24] further discovered new quasicrystals in Al-Pd-TM (TM = Mn, Re) systems by using the Hume-Rothery electron concentration rule with the Raynor's e/a scheme, a number of researchers have frequently discussed the stability range of thermally stable quasicrystals by employing Raynor's e/a scheme for TM elements. The possession of a negative valency for the TM element has been discussed even in the framework of first-principles band calculations [25,26].

Literature also exists reporting a positive e/a value for TM elements. For example, Haworth and Hume-Rothery [27] assigned e/a = 1.8, 1.0, 0.8, and 0.6 to Mn, Fe, Co, and Ni, respectively, so as to decrease the maximum solubility of the α phase into the same e/a value. Mizutani et al. [28]

interpreted the origin of the pseudogap in the Al-Cu-Ru-Si approximant in terms of the resonance of electrons near the Fermi level with sets of {543}, {550}, and {710} lattice planes and proposed that Ru possesses a positive e/a value of 0.76. More recently, Ishimasa et al. [29] pointed out that the Sc-TM quasicrystals they discovered are stabilized at e/a = 2.1, provided that the positive e/a values for the TM elements proposed by Haworth and Hume-Rothery [27] are employed.

As described above, we could extract the resonance effect of electrons with a set of lattice planes characterized by the magnitude of $|\mathbf{G}|^2$ by focusing on the motion of mobile electrons extending outside the MT sphere for the Al_8V_5 gamma-brass. By extending the Fourier spectrum analysis of the FLAPW function, we have proposed a new method to evaluate the e/a value for TM elements involved in TM_2Zn_{11} (TM = Fe, Co, Ni, and Pd) [14] and Al_8V_5 gamma-brasses [17]. The method allows us to determine the e/a value by constructing the Fermi sphere from momentum distribution of mobile electrons propagating outside the MT sphere.

First, we calculate the FLAPW function with an energy E at the wave vector $\mathbf{k}_i$ at the center of an element obtained by partitioning into two hundred elements an irreducible wedge of 1/48th of the bcc Brillouin zone and extract the plane-wave component $2(\mathbf{k}_i+\mathbf{G})$ with the largest Fourier component outside the MT sphere. The quantity of $2(\mathbf{k}_i+\mathbf{G})$ is calculated at 200 $\mathbf{k}_i$ points in the wedge, and its average $2k_G(E)$ is calculated by using equation

$$2k_G(E) \equiv 2\sum_{i=1}^{200} \omega_i \left|\mathbf{k}_i+\mathbf{G}\right|_E , \tag{15.5}$$

where ω_i is a weighting factor representing the degeneracy. Since the energy eigenvalues are not equally spaced, the energy E, at which $2(\mathbf{k}_i+\mathbf{G})$ is calculated, is obtained by linearly interpolating the neighboring energy eigenvalues. In this way, we can achieve $\{2k_G(E)\}^2-E$ relations, which represent a single-branch isotropic dispersion relation in the extended-zone scheme and describe the motion of mobile electrons extending outside the MT sphere. Because the value of $\{2k_G(E)\}^2$ at the Fermi level provides the square of the diameter of the Fermi sphere, we can immediately deduce the total number of electrons per atom $(e/a)_{total}$.

However, one must be aware of the fact that the value of $\{2k_G(E)\}^2$ is a quantity obtained by an averaging process through Equation 15.5. The variance must be small, otherwise the $2k_G(E)$ derived from Equation 15.5 becomes physically meaningless. The variance $\sigma^2(E)$ is defined as

$$\sigma^2(E)=\left[2\{k_G(E)+\sigma_G(E)\}\right]^2-\{2k_G(E)\}^2$$

$$=8k_G(E)\sigma_G(E)+4\sigma_G(E)^2, \tag{15.6}$$

where the standard deviation $\sigma_G(E)$ is given by

$$\sigma_G(E) \equiv \sqrt{\sum_i \omega_i \left(\left|\mathbf{k}_i + \mathbf{G}\right|_E - k_G(E)\right)^2}. \tag{15.7}$$

Figure 15.12 shows the energy dependence of $\{2k_G(E)\}^2$ or dispersion relations for (a) Cu_5Zn_8, Cu_9Al_4, (b) Ni_2Zn_{11}, Pd_2Zn_{11}, and (c) Co_2Zn_{11}, Fe_2Zn_{11} gamma-brasses [14]. The energy dependence of the corresponding variance $\sigma^2(E)$ is plotted in Figure 15.13. One can see that the variance in the neighborhood of the Fermi level in (a) Cu_5Zn_8 and Cu_9Al_4 and (b) Ni_2Zn_{11} and Pd_2Zn_{11} is small, and hence these data are reliable. The value of $\{2k_G(E)\}^2$ at the Fermi level, which is hereafter denoted as $(2k_F)^2$, is found to be close to 18 to 19. On the other hand, the variance across the Fermi level is significant in (c) Co_2Zn_{11} and Fe_2Zn_{11} gamma-brasses because of the presence of their 3d bands. The value of $(2k_F)^2$ is determined by ignoring the data in this energy range and extrapolating the data in the energy range, where the variance is small. This is done by using a line in Figure 15.12c.

The energy dependences of both $\{2k_G(E)\}^2$ and $\sigma^2(E)$ for the Al_8V_5 gamma-brass are shown in Figure 15.14 [17]. The variance in the energy range –3 to +3 eV is large because of the presence of the V 3d band. The two lines are drawn, as shown in Figure 15.14, by using the data below –3 eV and above +3 eV, where the variance is sufficiently low. The intercepts at the Fermi level and its average are denoted as A, B, and C, respectively. The value of $(2k_F)^2$ is determined from the point C. All the results for the TM_2Zn_{11} (TM = Fe, Co, Ni, and Pd) and Al_8V_5 gamma-brasses are summarized in Table 15.2. An uncertainty in determining the value of $(2k_F)^2$ for Co_2Zn_{11}, Fe_2Zn_{11}, and Al_8V_5 is larger than that for others because of the unavoidable location of the d band across the Fermi level.

The total number of electrons per atom, $(e/a)_{total}$, is accommodated in the Fermi sphere with the radius k_F. Its value is easily calculated by inserting the radius k_F in units of $2\pi/a$ into the well-known relation $(e/a)_{total} = 8\pi k_F^3/3N$, where N is the number of atoms in the unit cell (equal to fifty-two). As listed in Table 15.2, the values for Cu_5Zn_8 and Cu_9Al_4 are 1.60, in good agreement with 21/13, while those for Ni_2Zn_{11} and Pd_2Zn_{11} are increased to 1.7. The value of $(e/a)_{total}$ further slightly increases in the Fe_2Zn_{11} and Al_8V_5 gamma-brasses. This is indicative of a gradual increase with a decrease in the atomic number of the TM element.

The value of $(e/a)_{TM}$ is also easily calculated, provided that valencies of Al and Zn are assigned the values of 3 and 2, respectively. As listed in Table 15.2, the valency of Cu is indeed very close to unity. The valency of Ni and Pd is very close to zero but slightly positive in sign. The valencies of Fe, Co, and V also turn out to be positive in sign. The fact that all TM elements studied in the present work possess positive values of $(e/a)_{TM}$ indicates that they contribute to enhancing the charge density outside the MT sphere, in contrast to the proposal by Raynor.

6 Conclusions

We have classified twenty-four gamma-brasses existing in binary phase diagrams into three groups. Group 1 refers to those subjected to the Hume-Rothery electron concentration rule and stabilized at

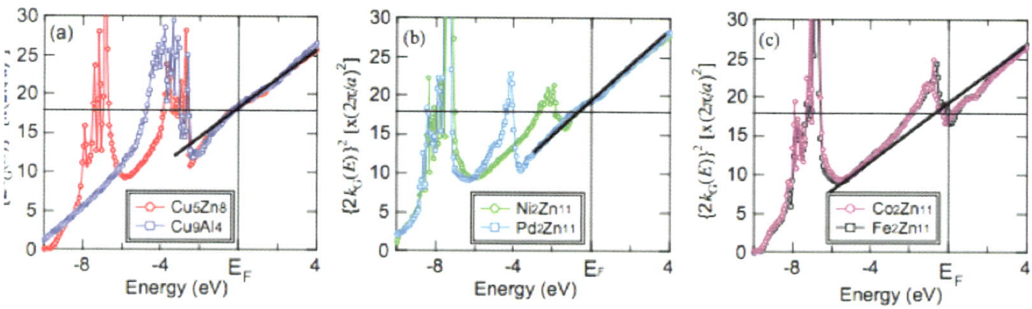

Figure 15.12. *The energy dependence of* $\{2k_G(E)\}^2$ *in units of* $(2\pi/a)^2$*, where a is the lattice constant, is plotted over the energy range –10 to +4 eV across* E_F *for (a)* Cu_5Zn_8 *and* Cu_9Al_4*, (b)* Ni_2Zn_{11} *and* Pd_2Zn_{11}*, and (c)* Co_2Zn_{11} *and* Fe_2Zn_{11} *gamma-brasses [14].*

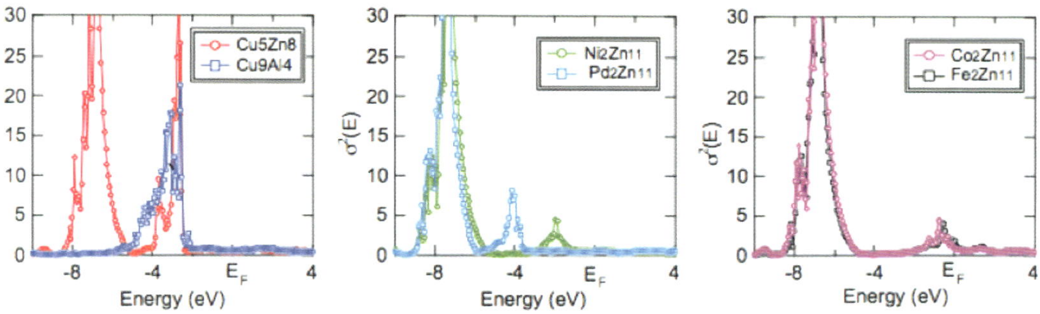

Figure 15.13. *Energy dependence of variance* $\sigma^2(E)$ *in units of* $(2\pi/a)^2$ *defined by Equations 15.6 and 15.7 for (a)* Cu_5Zn_8 *and* Cu_9Al_4*, (b)* Ni_2Zn_{11} *and* Pd_2Zn_{11}*, and (c)* Co_2Zn_{11}*, and* Fe_2Zn_{11} *gamma-brasses [14].*

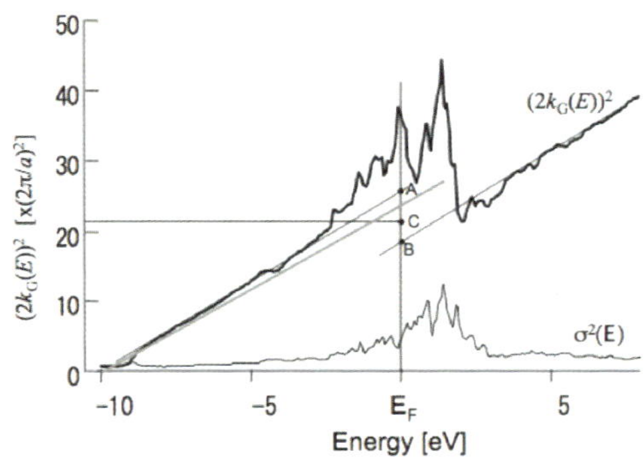

Figure 15.14. *Energy dependence of {2kG(E)}2 and its variance σ2(E) calculated from Equations 15.5 to 15.7 for the Al8V5 gamma-brass. The square of the Fermi diameter is determined by taking a mean value of A and B corresponding to the intersection of the two straight lines with the Fermi level. The gray line refers to the free-electron model [17].*

TABLE 15.2

$(2k_F)^2$, $(e/a)_{total}$, and $(e/a)_{TM}$ for gamma-brasses in groups I and II

	GAMMA-BRASS						
	Al_8V_5	Fe_2Zn_{11}	Co_2Zn_{11}	Ni_2Zn_{11}	Pd_2Zn_{11}	Cu_5Zn_8	Cu_9Al_4
$(2k_F)^2$	21.0	20.0	19.5	19.36	19.27	18.47	18.45
Zones	$14 \leq G^2 \leq 30$	Not studied	Not studied	$G^2 = 18$ {411}, {330}	$G^2 = 18$ {411}, {330}	$G^2 = 18$ {411}, {330}	$G^2 = 18$ {411}, {330}
$(e/a)_{total}$	1.94	1.80	1.73	1.72	1.70	1.60	1.60
$(e/a)_{TM}$	0.23	0.70	0.26	0.15	0.07	0.96	0.97
Ref	[17]	[14,17]	[14,17]	[14]	[14]	[7]	[7]

e/a = 21/13. Typical examples are Cu_5Zn_8 and Cu_9Al_4. Gamma-brasses involving the transition metal element TM as a partner element to polyvalent elements like Zn and Al are classified into group 2.

We studied the stability mechanism and the e/a value for gamma-brasses in group 2 in comparison with those in group 1 by making full use of first-principles LMTO-ASA and FLAPW band calculations. In the case of Al_8V_5 gamma-brass, mobile electrons extending outside the MT sphere resonate sets of lattice planes in the range of $10 \leq |\mathbf{G}|^2 \leq 30$ through the mediation of the V 3d states and generate a deep pseudogap in the energy range –2 to +2 eV in the sp-partial DOS. However, this is masked in the total DOS as a result of the superposition of the V 3d band in the range –3 to +3 eV. We assume that the d states mediated resonance is equally responsible for the stability of Fe_2Zn_{11} and Co_2Zn_{11}. The Ni_2Zn_{11} and Pd_2Zn_{11} gamma-brasses are classified into group 2. But as discussed above, they are stabilized by the pseudogap formed immediately below the Fermi level. Their stability mechanism is attributed to the resonance of electron waves with the set of {330} and {411} lattice planes in the same way as in Cu_5Zn_8 and Cu_9Al_4 in group 1. Indeed, they obey the Hume-Rothery electron concentration rule if the e/a value close to zero is employed for Ni and Pd. Therefore, they must be included in group 1 from the viewpoint of the stability mechanism.

Acknowledgment

One of the authors (U. M.) wishes to thank the Grant-in-Aid for Scientific Research (C) (contract No: 20560620) from the Japan Society for the Promotion of Science for financial support.

References

1. P. Villar, ed., *Pearson痴 Handbook*, vol. 2. (Materials Park, OH: ASM International, 1997).
2. A. J. Bradley and P. Jones, *J. Inst. Met.* 51 (1933): 131–162.
3. S. Westman, *Acta Chem. Scand.* 19 (1965): 1411–1419.
4. O. V. Heidenstam, A. Johansson, and S. Westman, *Acta Chem. Scand.* 22 (1968): 653–661.

5. J. K. Brandon, R. Y. Brizard, P. C. Chieh, R. K. McMillan, and W. B. Pearson, *Acta Crystallogr. B* 30 (1974): 1412–1417.
6. L. Arnberg and S. Westman, *Acta Crystallogr. A* 34 (1978): 399–404.
7. R. Asahi, H. Sato, T. Takeuchi, and U. Mizutani, *Phys. Rev. B* 71 (2005): 165103.
8. L. Arnberg and S. Westman, *Acta Chem. Scand.* 26 (1972): 513–517.
9. J. K. Brandon, H. S. Kim, and W. B. Pearson, *Acta Crystallogr. B* 35 (1979): 1937–1944.
10. A. F. Westgren and G. Phragmen, *Trans. Farad. Soc.* 25 (1929): 379.
11. N. F. Mott and H. Jones, *The Theory of Properties of Metals and Alloys* (New York: Dover, 1958), 168–174 (first edition: 1936).
12. C. S. Barrett and T. B. Massalski, *Structure of Metals* (New York: McGraw-Hill, 1966): 308–310.
13. A. Johansson, H. Ljung, and S. Westman, *Acta Chem. Scand.* 22 (1968): 2743–2753.
14. R. Asahi, H. Sato, T. Takeuchi, and U. Mizutani, *Phys. Rev. B* 72 (2005): 125102.
15. V.-A. Edström and S. Westman, *Acta Chem. Scand.* 23 (1969): 279–285.
16. J. K. Brandon, W. B. Pearson, P. W. Riley, C. Chieh, and R. Stokhuyzen, *Acta Crystallogr. B* 33 (1977): 1088–1095.
17. U. Mizutani, R. Asahi, H. Sato, and T. Takeuchi, *Phys. Rev. B* 74 (2006): 235119.
18. H. Sato, T. Takeuchi, and U. Mizutani, *Phys. Rev. B* 64 (2001): 094207.
19. H. Sato, T. Takeuchi, and U. Mizutani, *Phys. Rev. B* 70 (2004): 024210.
20. E. Wimmer, H. Krakauer, M. Weinert, and A. J. Freeman, *Phys. Rev. B* 24 (1981): 864.
21. M. Weinert, E. Wimmer, and A. J. Freeman, *Phys. Rev. B* 26 (1982): 4571.
22. G. V. Raynor, *Prog. Met. Phys.* 1 (1949): 1–76.
23. A. P. Tsai, A. Inoue, and T. Masumoto, *Jpn. J. Appl. Phys.* 27, L1587; A. P. Tsai, A. Inoue, Y. Yokoyama, and T. Masumoto, 1990, *Mater. Trans. JIM.* 31 (1988): 98.
24. Y. Yokoyama, A. P. Tsai, A. Inoue, T. Masumoto, and H. S. Chen, *Mater. Trans. JIM.* 32 (1991): 421.
25. G. Trambly de Laissardiere, D. N. Manh, L. Magaud, J. P. Julien, F. Cyrot-Lackmann, and D. Mayou, *Phys. Rev. B* 52 (1995): 7920.
26. E. S. Zijlstra and S. K. Bose, *Phil. Mag.* 86 (2006): 717.
27. J. B. Haworth and W. Hume-Rothery. *Phil. Mag.* 43 (1952): 613.
28. U. Mizutani, T. Takeuchi, and H. Sato, *Prog. Mat. Sci.* 49 (2004): 227–261.
29. T. Ishimasa, Y. Kaneko, and H. Kaneko, *J. Non-Cryst. Solids* 334–335 (2004): 1–7.

16

The Role of Chemical and Displacement Pair Correlation in the Determination of Higher order Correlation

D. M. C. Nicholson[1]

R. I. Barabash[1,2]

Y. S. Puzyrev[1]

C. Y. Gao[3]

D. J. Keffer[3]

G. E. Ice[1]

1. *Oak Ridge National Laboratory, Oak Ridge, TN 37831, USA*
2. *Department of Materials Science and Engineering, University of Tennessee, Knoxville, TN 37996, USA*
3. *Department of Chemical Biomechanical Engineering, University of Tennessee, Knoxville, TN 37996, USA*

Abstract

Scattering experiments are routinely used to measure atomic pair correlations while theoretical models of materials include pair and higher order correlations. Only in special cases can model structures that reproduce the pair correlations be used to unambiguously determine higher order correlations because there may exist a range of higher correlations that are consistent with the pair correlation. Constraints can reduce this ambiguity. We discuss the role of geometry, thermodynamic equilibrium, displacement short-range order, and thermodynamic functions in constraining higher order correlations constructed from a measured pair correlation. Strategies aimed at further specification of higher order correlations by augmenting measured pair correlations with thermodynamic measurements and first-principles calculations are explored.

1 Introduction

Usually all that is known about atomic-scale structure are the pair correlations (PCs) determined by scattering experiments. However, correlations to all orders are needed to completely define the structure [1]. Additional constraints may sometimes result in relationships between pair and higher order correlations that remold this requirement into the requirement that the constraints together with lower-order correlations completely determine the structure. In the context of periodic, piecewise uniform structures, for example, Rozman [2] showed that for several structures with smooth boundaries, the PCs were sufficient to define the structure. In that study, structures were found that could be reconstructed from their PCs, but the particular constraints that allowed this were left as a matter of speculation.

Here we explore the role of constraints in extending pair correlation information to higher orders. This is a long-standing problem in many fields [3–5]. Possible constraints include geometry (structure), thermodynamic, and model interactions. For example, diagrammatic methods have been used to express the grand potential for liquids in terms of the point and pair correlation functions [6]. In the tradition of the study of short-range order (SRO) and diffuse scattering, early contributions were by Krivoglaz [7] and others [8–16]. Henderson [17] used the minimum condition on the free energy attributed to Bogoliubov [18] to show that for pair-potential Hamiltonians (PPHs) the PC was unique. Evans [19] commented on Henderson's work and asserted—without proof—that a many-body Hamiltonian (MBH) of order n is uniquely determined by correlations to order n. Stillinger and Torquato [20] studied short linear chains in which they could enumerate all configurations. They found atom chains with lengths between 11 and 15 for which a range of many-body correlations was consistent with a single PC. They point out the need for caution when interpreting higher order correlations based on reverse Monte Carlo. Wolverton et al. [21] showed that a PPH and an MBH could give the same SRO. Schweika and Carlsson [22] derived an expression for an effective pair potential that is guaranteed to reproduce the SRO resulting from an MBH to order $(1/T^2)$. Nicholson et al. [23] discussed the relationship between pair and higher order correlations for pair and higher order interactions.

Our discussion is limited to homogeneous materials. Although some progress has been made, even the determination of higher order correlations in homogeneous bulk systems remains a daunting task. We focus on the virtues and limitations of extracting structural information from the scattered intensity from bulk homogeneous materials. Because the scattering intensity provides the Fourier transform of the pair correlation function, the discussion centers on the relationship between the pair correlation functions and higher order correlations.

Because PCs restrict but do not generally determine higher order correlation, additional constraints are needed. We begin by reminding the reader of some strictly geometric constraints on the higher order correlations that are true independent of the thermodynamic state. Thermodynamic equilibrium is such a powerful concept that we will quickly adopt it as the guiding principle for the remainder of the chapter. The crux of the arguments will depend on the assumption of equilibrium or metastability. We will assume that the task of reducing the X-ray, electron, or neutron scattering data to PCs has been accomplished so that the type of scattering is no longer an issue. It has been shown algebraically that if the Hamiltonian is of pair-potential form, then all the higher order correlations can be completely determined from the PCs [23]. Furthermore, it has been shown numerically by counterexample that, at least for some MBH, the higher order correlations are not determined by the PCs. This unsettling result leaves us in a quandary; scattering experiments give the PCs, but we must justify strategies for constructing higher order correlations via, for example, reverse Monte Carlo. The resolution of this quandary may come from the use of auxiliary information. Historically, for ordered crystals with small unit cells, the constraints of crystal symmetry together with pair information permits confident specification of structures such as that of NaCl. We propose that first-principles calculations can provide complementary information so that, together with measured pair information, light can be shed on higher order correlations. This possibility is illustrated for the simple case of an "iso-g" distribution with a simple rule for environment-dependent displacement of atoms. A final possibility is that although it is unlikely that any material has strictly pair interactions, it may nonetheless be possible that the pair interactions dominate the determination of the higher order correlations. How can we judge the relative importance of the pair interactions? We make one step toward addressing this issue by considering another aspect of the unique relationship between the PC and the pair potential; for a pair potential system, the free energy is a functional of the PCs. Hence there is a direct relationship between the thermodynamic functions and the PCs. Thermodynamic data can help clarify the role of many-body interactions.

2 Geometric Constraints

Gragg [24] used the conservation of probability and the positive definiteness of probabilities to find restrictions on specific higher order correlations for a given set of PCs. These constraints do not depend on the system being in equilibrium. The constraints are dependent on the type of lattice; it is for this reason that we refer to them as geometric. If these constraints were tight enough to satisfy our need for accuracy in the higher order correlations, our task would be complete once the PCs were measured. We illustrate Gragg's results for an fcc lattice with a concentration, c, of 0.5. This will simplify a comparison (see following section) with the further constraints imposed by specifying that the system is in equilibrium. For the $c = 0.5$ case, an interesting argument by Clapp [25] proves that for a pair-potential system, the three-particle correlations are exactly zero. We illustrate in Figure 16.1 the geometry of the three-particle correlations upon which our discussion is centered. If the three atoms are labeled 1, 2, and 3 with pairs (1, 2), (1, 3), and (2, 3), then the pairs (1, 2) and (1, 3) are nearest-neighbors pairs.

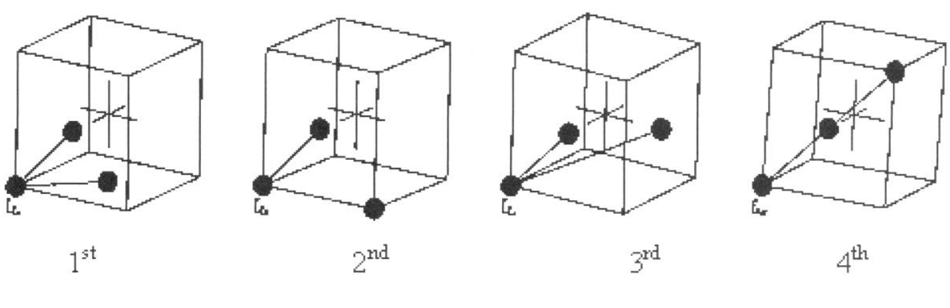

Figure 16.1. *Kinds of triplets: first: (000), (101), (110); second: (000), (200), (101); third: (000), (2,1,1), (101); and fourth: (0,0,0), (2,2,0), (1,0,1)*

The third pair is either a first-, second-, third-, or fourth-neighbor pair. Gragg's constraints will involve the near-neighbor Warren-Cowley short-range-order (SRO) parameter, α and α_n, where $n = 1, 2, 3,$ or 4 ($\alpha_1 = \alpha$) for the nth neighbor. The probability that any triangle (1,2,3) is occupied as (A,A,A) is P^{AAA}, (A,A,B) is P^{AAB},....

Following Gragg we find that

$$\begin{aligned}
&P^{AAA} + P^{AAB} + P^{ABA} + P^{ABB} + P^{BAA} + P^{BAB} + P^{BBA} + P^{BBB} = 1\\
&P^{AAA} + P^{AAB} + P^{ABA} + P^{ABB} = x_A\\
&P^{AAA} + P^{AAB} + P^{BAA} + P^{BAB} = x_A\\
&P^{AAA} + P^{ABA} + P^{BAA} + P^{BBA} = x_A\\
&P^{AAA} + P^{AAB} = x_A\left(x_A + x_B\alpha\right)\\
&P^{AAA} + P^{ABA} = x_A\left(x_A + x_B\alpha\right)\\
&P^{AAA} + P^{BAA} = x_A\left(x_A + x_B\alpha_n\right).
\end{aligned} \tag{16.1}$$

Defining the site occupation $\sigma = 1(-1)$ for A(B), the three-particle correlation is given by

$$\begin{aligned}
&\sigma = \tfrac{1}{N}\sum_{i=1}^{N}\langle\sigma_i\rangle = 2c-1\\
&\sigma_j^{(2)} = \tfrac{1}{N}\sum_{i=1}^{N}\langle\sigma_i\sigma_{i+j}\rangle = 4(c^2 - c(1-c)\alpha_j) - 2c - 1\\
&\sigma_{j,k}^{(3)} = \tfrac{1}{N}\sum_{i=1}^{N}\langle\sigma_i\sigma_{i+j}\sigma_{i+k}\rangle = 8P_{jk}^{AAA} - \sigma_j^{(2)} - \sigma_k^{(2)} - \sigma_{j-k}^{(2)} - 3c - 1,\\
&\sigma^3 = \langle\sigma_1\sigma_2\sigma_3\rangle = P^{AAA} - P^{AAB} - P^{ABA} + P^{ABB} - P^{BAA} + P^{BAB} + P^{BBA} - P^{BBB}.
\end{aligned} \tag{16.2}$$

If P^{AAA} is specified, then all other probabilities can be determined from Equation 16.1. However, depending on the specified value of P^{AAA}, one or more of the other probabilities may be negative or greater than one. Only those P^{AAA} that result in probabilities that are all in the proper range are allowable. The range of allowable P^{AAA} depends on the α and α_n. Figure 16.2 shows the allowable range of P^{AAA} for the first nearest-neighbor triangle as a function α. The range of the closely related quantity σ^3 is centered on zero. For a random alloy, σ^3 can fall anywhere between ± 1; it is completely unrestricted by geometric considerations.

2.1 Constraints of Equilibrium

It was shown in the previous section that geometric constraints do not significantly restrict three-particle correlations. The power of thermodynamic equilibrium constraints can be simply demonstrated utilizing the $c = 0.5$ alloy that was the subject of our discussion of geometric constraints. We place two additional conditions: (1) the alloy is in thermodynamic equilibrium at temperature, T, and (2) the interactions are pairwise. The key to the simplicity of the argument is that for any configuration of the

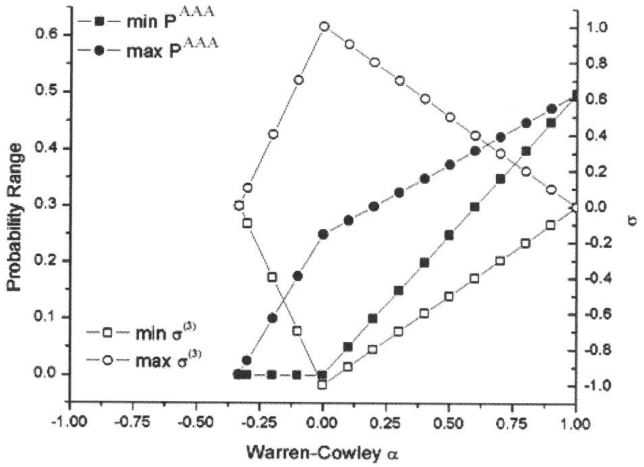

Figure 16.2. *The geometric constraint of average fcc structure sets the limits on the possible three-body probability P^{AAA}.*

alloy, the configuration with As switched to Bs (and visa versa) has exactly the same distribution of atomic pairs and hence the same energy. Thus each configuration can be paired with its occupation-flipped partner. The two partners have the same energy and therefore the same Boltzmann weight but contribute oppositely to all odd products of σ, such as the product appearing in the three-body correlation function. The three-body correlations—and, in fact, all odd-correlation functions—are therefore zero. This result is independent of lattice type and hinges entirely on the absence of interactions beyond pairs, and on thermal equilibrium. Whereas geometric constraints provided minimal constraint of higher order correlation, the equilibrium constraints for this particular pair-potential problem completely constrained all odd higher order correlations to be exactly zero.

In general, for PPH, the higher order correlations are not necessarily zero, but they are completely determined by the PCs. A corollary is that for any nth order interaction, all correlations higher than the nth order are determined by the correlations of order n and lower. Note also that if the pair potential is known to be of the central force variety, the higher order correlations are completely determined by the "spherical" pair distribution function (PDF), that is, the average of atoms of a given type within a spherical shell around atoms of a given type. Model Hamiltonians for crystals often specify only the pair interchange potential, which can be considered to result from evaluating pair potentials at lattice separations. Typically, the underlying pair potentials would not be central force potentials. For example, in fcc if the $V_{2,1/2,1/2}$ and $V_{3/2,3/2,0}$ interchange potentials are different, they cannot be based on central force potentials. If the potentials underlying the pair interchange potentials are not central force potentials, then there are difficulties in consistently treating alternative structures such as the liquid, glass, hcp, or bcc. In fact, whenever the interchange potentials cannot be based on central force potentials that are concentration, structure, and temperature independent, there is an implied significant many-body contribution to the interactions. Therefore, reconstruction techniques justified by theorems based on a chosen pair-potential interaction usually assume that for a particular range of temperature, concentration, and structure, the many-body interactions do not significantly impact the local structure.

In order to assess the limitations of reverse Monte Carlo estimates of higher order correlations, we analyze numerical simulations using two realistic models for NiV [22]. One Hamiltonian is pair only (PPH); the other is pair + many body (MBH). Earlier Monte Carlo simulations generated two sets of equilibrium PCs, $\left\{\sigma_j^{(2)}\right\}_{PPH}$ and $\left\{\sigma_j^{(2)}\right\}_{MBH}$, respectively, as well as many-body correlations, some of which are depicted in Figures 16.3 and 16.4. We refer to these correlations from Monte Carlo as "equilibrium" correlations. Configurations with PCs statistically identical to $\left\{\sigma_j^{(2)}\right\}_{PPH}$ and $\left\{\sigma_j^{(2)}\right\}_{MBH}$ were generated from a random configuration by repeatedly introducing pair interchanges until agreement was reached out to 150 atomic shells. We refer to these generated structures as "constructed." By construction, "equilibrium" and "constructed" values of PCs are equal, so the relevant comparison is between the many-body correlations in the "equilibrium" and "constructed" samples. The probability of AAA neighbor triangles is shown for the various cases in Figure 16.3. We observe that for the PPH, the equilibrium and constructed samples have almost equal P^{AAA}. We determined that they are equal to within the limitations imposed by our sample size (64000 for the equilibrium structures and 108000 for the constructed structures).

However, we note further that for the MBH, the P^{AAA} of the equilibrium and constructed structure differ appreciably. Such differences could impact physical properties such as electron transport, which can depend on the probability of linear paths of a particular species [26]. The difference in three-body correlations shown in Figure 16.3b supports the analytic results that for MBH, PCs do not constrain the values of many-body correlations.

In Figure 16.4, the distribution of first nearest neighbors is shown. This is, in a sense, a less fundamental set of quantities than the probability of triangles, in that the distribution is composed of sums over thirteen-body correlations having the same numbers of B atoms. However, the number and distribution of nearest neighbors is so fundamental to descriptions of materials behavior that it warrants detailed consideration. For PPH, the equilibrium and constructed samples have "identical" distributions, whereas for MBH, the differences are significant. This result cautions again of the dangers

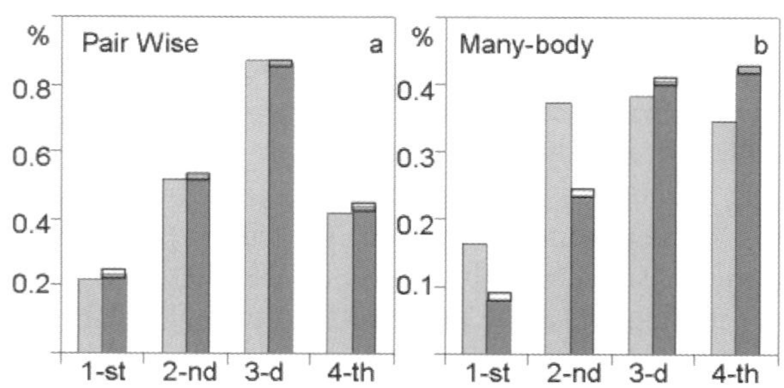

Figure 16.3. *Percent of triplets that are all A atoms in four different kinds of triplet configurations. Light gray: Equilibrium lattice. Dark gray: Constructed lattice with the same PCs as the equilibrium lattice. For the constructed MBH, lattice error bars are indicated by empty boxes.*

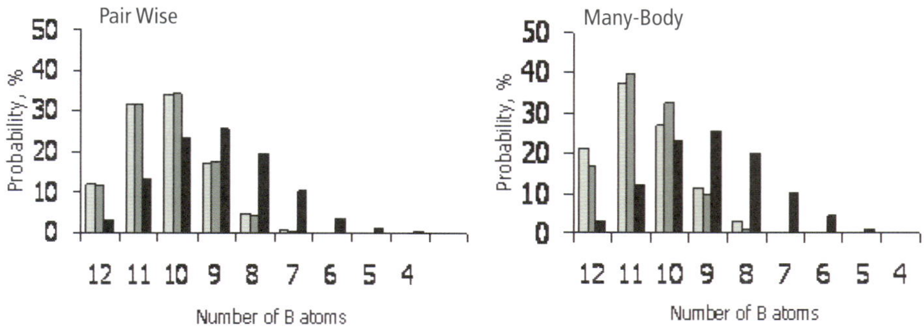

Figure 16.4. *Percent A sites with n B atoms as nearest neighbors. Light gray: Random lattice. Gray: Equilibrium lattice. Dark gray: Constructed lattice with the same PCs as the equilibrium lattice.*

of using higher order correlations obtained from fitting measured PCs without auxiliary information indicating that the Hamiltonian is well approximated by a pair-potential model (which is rarely the case for real physical systems).

The analysis above emphasizes the connection between higher order correlation and the response of lower-order correlation to external fields and to the fluctuations of lower-order correlations. If external fields could be controlled at the atomic scale, it would be straightforward to determine higher order correlations by measuring the PC in the presence of the appropriate external field; typical fields, such as magnetic and stress fields, have a global effect. Changes in the higher order interactions do give new information but usually require knowledge of the relationship between the interactions and the applied field. These relationships can be determined from first-principles calculations or from measurements with some caveats. The change in ordering temperature with strain could indicate the dependence on near-neighbor interactions on strain.

2.2 Constraints from Displacement SRO

We begin the discussion of displacement SRO by recapping the work of Stillinger and Torquato [20], who studied short linear chains with periodic boundary conditions (see Figure 16.5). They were considering the chain as a lattice liquid model in which they sought a unique Hamiltonian that would

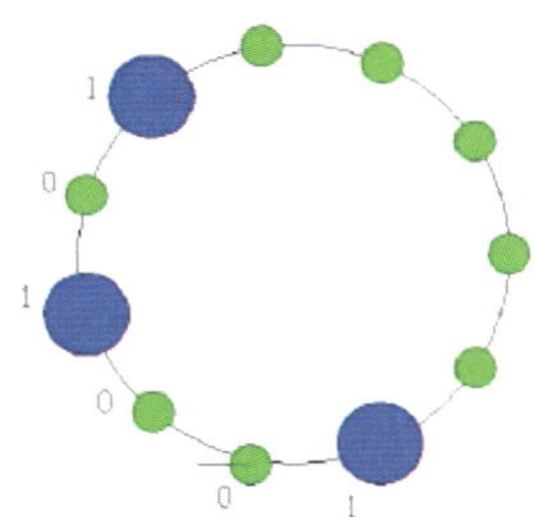

Figure 16.5. *An eleven-site chain with periodic boundary conditions. Blue atom "A" is kept unmoved, while green atom "B" is pushed to the left by neighbors.*

yield a PDF that was zero at small separations (first neighbor in their model) and was a constant beyond the first neighbor. The symbolic representation of such a configuration is a row of ones and zeros. Because of the short length of the chains considered and because no nearest-neighbor occupation was allowed, they could enumerate all configurations. To clarify the restrictions that displacements place on the configuration, we reconsider Stillenger's example of an eleven-site chain with $c = 3/11$ but allow for displacement of the atoms. The possible configurations can be reduced by symmetry to five groups of "identical" arrangements; members of a group have the same energy. Specifying the five energies, $(\varepsilon_1, \varepsilon_2, \varepsilon_3, \varepsilon_4, \varepsilon_5)$ completely defines the Hamiltonian. Each group also makes a particular contribution to the PDF at each atomic separation. If the system is in thermodynamic equilibrium, the weight of each group to the PDF is proportional to its multiplicity (number of members in the group m_n) and its Boltzmann weight, $w_n = e^{-\frac{\varepsilon_n}{k_B T}} \Big/ \sum e^{-\frac{\varepsilon_n}{k_B T}}$. The number of minority second neighbors to the right (equals number of second nearest-neighbor bonds) of the minority atoms is listed in column 4 of Table 16.1 for each configuration. We assume that an atom with one and only one second nearest neighbor is displaced away from its equilibrium position by δ for minority atoms. The five energies can be determined (within an arbitrary constant) from the five weights, and therefore the Hamiltonian is completely specified once the weights are known. For simplicity, the target PDF is again chosen to be constant beyond the first nearest neighbors.

TABLE 16.1

Second near neighbor pairs 1-1 for each configuration and their contribution to the statistical sum

i	CONFIGURATION	MULITIPILCITY, *m*	SECOND NEAR NEIGHBOR 1-1 PAIRS, *n*	*n*/*N**MULTIPLICITY
1	10101000000	11	2	2/11*11
2	10100100000	22	1	1/11*22
3	10100010000	22	1	1/11*22
4	10010010000	11	0	
5	10010001000	11	0	

The requirement that the model PDF is zero for nearest neighbors and a constant for all other neighbors results in the following equations:

$$\begin{aligned} 2w_1 + 2w_2 + 2w_3 &= 3/44 \quad \textit{2nd near neighbor} \\ 2w_2 + 2w_4 + 1w_5 &= 3/44 \quad \textit{3rd near neighbor} \\ 1w_1 + 2w_3 + 2w_5 &= 3/44 \quad \textit{4th near neighbor} \\ 2w_2 + 2w_3 + 1w_4 &= 3/44 \quad \textit{5th near neighbor} \end{aligned} \tag{16.3}$$

If there is only one choice (solution to Equation 16.3) for the five numbers $(w_1, w_2, w_3, w_4, w_5)$ that give the iso-*g* PDF, then the Hamiltonian is uniquely determined by the PDF, and therefore all higher order correlations are determined by the PDF. However, if there is a range of values for $(w_1, w_2, w_3, w_4, w_5)$, then neither the Hamiltonian nor the higher order correlations are determined by the PDF. The solution to Equation 16.3 is

$$\begin{aligned} w_2 &= 1/44 - 3w_1/2 \\ w_3 &= 1/88 + w_1/2 \\ w_4 &= 2w_1 \\ w_5 &= 1/44 - w_1 \end{aligned} \quad . \tag{16.4}$$

There were only four equations and five weights, so one weight remains as an independent variable, w_1. However, the requirement that all the weights be positive restricts the possible value of w_1:

$$0 \le w_1 \le 1/66 \,. \tag{16.5}$$

Depending on the length of the chain and the target PDF, there could be no solution, a single solution, or multiple solutions. If there are multiple solutions, each solution will typically have different higher order correlations. When we analyze a measured PDF, our main concern is to resolve the multiple solutions so that we can recover the higher order correlations. Having reviewed the results of Stillinger and Torquato [20], we move on to consider strategies for resolving the multiple solutions. We propose to utilize the fact that the lattice is strained differently at the atomic level depending on the local configuration. The richest source of this information is the displacement SRO, which has been measured in several systems [27–29]. The local displacements can reflect the higher order correlations by limiting the possible configuration weights that are consistent with iso-*g*(*r*), but a model is needed in order to extract information about the correlation from the displacements. Models can be based on measurements of elastic constants as a function of concentrations and SROs (different SROs can be induced by quenching from different temperatures) or they can be based on first-principles calculations.

To illustrate how this might work, we propose several simple models of displacement SROs as "stand ins" for a more elaborate model based on first principles. In Table 16.2, the "1-1" bond expansion, *d*, is listed in units of δ for each model and each class of configurations.

In our first model, we assume that the displacement of atoms labeled "1" will move away from a second neighbor that is also a "1" unless it falls between two second neighbors that are "1s," in which case it cannot shift due to symmetry. The shift from the average lattice site will be δ. The average displacement is

$$\Delta_{ave} = \left(2w_1 + 4w_2 + 4w_3\right)\delta \,. \tag{16.6}$$

Because δ is assumed to be known from the model, if Δ_{ave} is known from measurement, then we have a fifth equation; therefore, the weights and higher order correlations are completely determined by the PDF. This model is manifestly different from the conventional perception of displacements as a

TABLE 16.2

Displacement correlation contribution, $\delta_j^{(2)} = \frac{1}{N}\sum_{i=1}^{N}\left\langle \delta_{i+j} - \delta_i \right\rangle$, from each configuration to the statistical average

i	CONFIGURATION	MULITIPILCITY, *m*	MODEL 1	MODEL 2	MODEL 3
1	10101000000	11	2	4	0.78
2	10100100000	22	2	2	0.62
3	10100010000	22	2	2	0.62
4	10010010000	11	0	0	0
5	10010001000	11	0	0	0

pairwise effect. Consider the situation where one would use more springlike interaction between atoms labeled "1" and second nearest-neighbor "1." This will cause second-neighbor atoms in the symmetric case to move additional δ from the lattice site. The average displacement is

$$\Delta_{ave} = \left(4w_1 + 4w_2 + 4w_3\right)\delta . \tag{16.7}$$

Unlike the previous case, Equation 16.7 either brings no new information (for $\Delta_{ave} = 6\delta/44$) or makes the requirement of iso-*g*(*r*) unrealizable.

Imagine a ring in which the near neighbors interact via springs with force constant $k = 1$ and natural length $l = 1$, while the second near neighbor's interaction is described by springs with $k = k_0$ and $l = 2 + \delta$ for the pairs with both minority-kind atoms and $k = 0$ and $l = 2$ for other pairs. For $k_0 = 1$, the additional equation provides enough information to specify the configuration and determine the higher order correlations:

$$\Delta_{ave} = \left(0.78w_1 + 0.62w_2 + 0.62w_3\right)\delta . \tag{16.8}$$

Thus different local interaction models can impose additional constraints and introduce additional requirements for PDF in systems with well-defined long-range interaction.

This approach utilizes the strengths of both measurement and calculation. Measurement includes the exact interaction and can equilibrate over long times, whereas calculation can determine atomic positions with subangstrom precision. The accuracy of calculated displacements will also improve as theories that go beyond the local density approximation mature. Figure 16.6 shows two configurations with the same Warren-Cowley parameters but very different three-particle correlations. The samples are of appropriate size for first-principles calculation. The calculated displacement SRO for a series of structures of this type with different three-particle correlations could be compared to measured displacement SRO to determine which choice of the three-particle correlations was consistent with the measured displacement SRO.

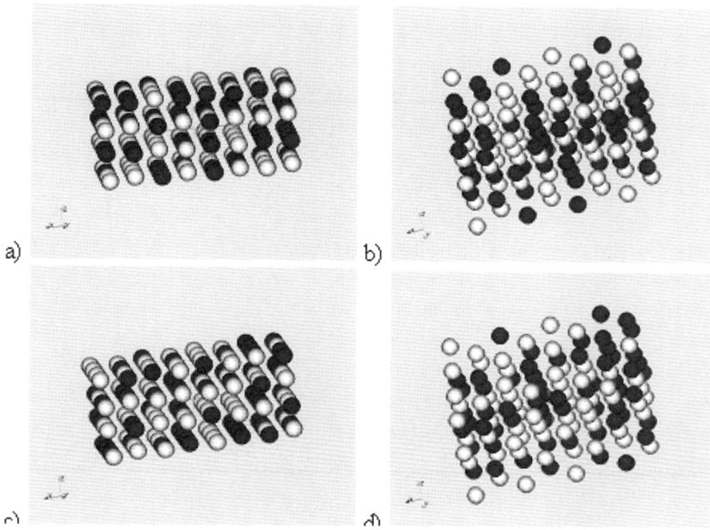

Figure 16.6. *Two configurations of $Fe_{46.5}Ni_{53.5}$—in which black spheres represent Ni atoms, and light spheres represent Fe atoms—have identical pair SROs but different three-body correlations. (a) and (b) are projections for configuration $P^{FeFeFe} = 0.1$ and (c) and (d) are projections for configuration $P^{FeFeFe} = 0.4$, respectively.*

2.3 Constraints from Thermodynamic Functions

For PPH, the uniqueness of the pair potential ensures that the thermodynamic functions, such as entropy, are functionals of the PDF. For MBH, the thermodynamic function requires additional information from higher order correlations. However, if reliable functionals were known for PPH, then the thermodynamic functions could be used as an indicator of missing higher order effects. Furthermore, direct evaluation of the thermodynamic functions based on measured PCs would be a valuable tool for interpreting phase stability.

It is natural to assume that a knowledge of pair information over a wide temperature range could provide information about higher order correlation beyond that available from a single temperature, T. We might be led astray in this regard, however, if the interactions change rapidly with T. An unpredictable T dependence of the interactions would leave us with separate independent problems at each T. Nonetheless, it is physically reasonable to judge the continuity with T of the interactions from the continuity of the PDF; thus there is hope that the higher order interactions will be further revealed in the gradual variation of the PDF with T.

As a first step in this direction, and in hope of stirring interest in this line of attack, we look at the PDF of Fe simulated with a PPH at several temperatures and relate the PDF to the entropy (see Schweika [22] for a discussion of the enthalpy of a solid solution). The simulations were performed using 16000 atoms interacting via the Johnson potential [30] in NVT conditions at the room-temperature density of Fe. Figure 16.7 shows the PDFs for bcc at 750 K, glass at 750 K, supercooled liquid at 2250 K, and liquid at $T = 9000$. We expect that the entropy associated with this list proceeds accordingly

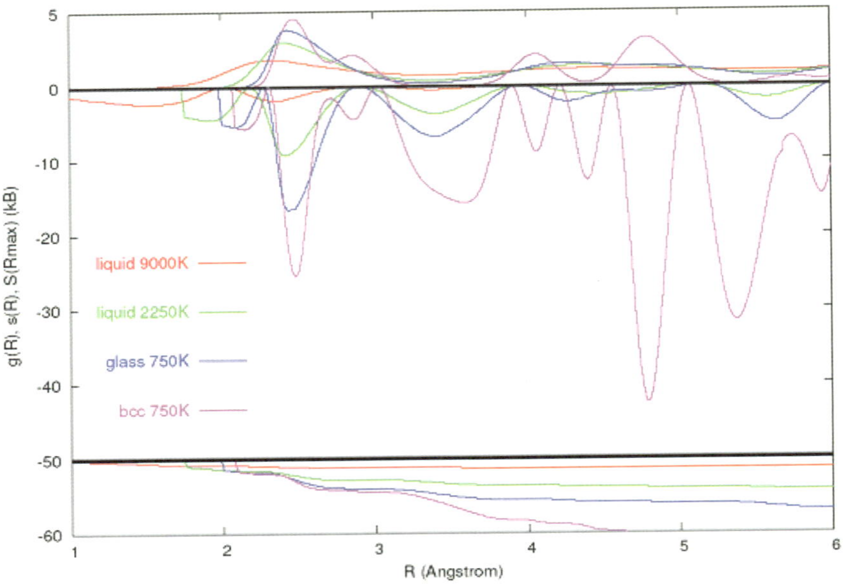

Figure 16.7. *For each PDF shown at the top of the Figure, the contribution to the excess entropy from each spherical shell, s(R), is shown. At the bottom (shifted by –50) is the integral of s(R) up to R.*

from lowest to highest. But can we say anything more quantitative on the basis of the PDF? Indeed, we can relate specific structures in the PDF to the entropy. Although it is known that the entropy is a functional of the PDF, the form of the functional is unknown. Here we use a particularly simple approximate form that we refer to as the pair-product approximation [31], which is based on the Kirkwood approximation to the N-body correlation function. For the liquid phase it becomes

$$S[n,g]-S^{ideal}(n,T)=-\frac{1}{2}k_B n\left[\int_0^{\infty} 4\pi R^2\, dR g(R)\ln(g(R))-\{g(R)-1\}\right]-\frac{1}{2}k_B. \qquad (16.9)$$

At this point the accuracy of this approximate expression for the entropy is not a concern; the point is that the expression depends only on n and $g(r)$ and does not involve the Hamiltonian. This expression gives physically reasonable results; moreover, it can be improved. There is an unknown but exact functional relationship between the entropy and $g(r)$ that can be approached with further effort.

Results from Equation 16.9 are illustrated in Figure 16.7, where the excess entropy is plotted as a function of R. The contribution to the integral is zero when $g(R)$ is one. Therefore a $g(R)$ that rapidly approaches one as R increases has a smaller contribution to the integral; this results in higher entropy. Deviation of the PC function either below one or above one results in lower entropy. Consequently, we note three structural features in $g(R)$ associated with reduced entropy: (1) at shells where atoms are depleted from a region, for example, interior to the first-neighbor shell; (2) when atomic concentrations are increased, for example, at the location of the nearest and subsequent neighbor shells; and

(3) at large distances due to the persistence of correlation to large separations in the crystal structure. The entropy difference between the ideal gas and the high-T liquid is primarily due to the exclusion of atoms of short separations; that is, the atoms of an ideal gas are volumeless points, whereas the simulated atoms have volumes due to repulsive centers. The entropy difference between high- and low-T liquids results primarily from sharpness of the first-neighbor peak and the depth of the trough beyond the first-neighbor shell. The difference between the glass and the supercooled liquid has only a small contribution from the split second peak but a large contribution from the trough beyond the split second peak. The crystal differs from the glass mainly at large separations. In fact, the crystal entropy does not converge in the pair-product approximation. Improved functionals [32], which are beyond the scope of this discussion, modify the Kirkwood approximation for the crystal and yield an entropy consistent with physically reasonable reduction of entropy seen in Figure 16.7 at small distances and add a supplemental convergent contribution from the long-range correlation.

3 Conclusions

We have discussed the relationship between PCs and higher order correlations and the possibility that constraints together with the PCs can yield useful information about the higher order correlations. If a straightforward connection exists between the atomic structure and a measured or calculated property, then the property provides a possible constraint on the structure. We have focused on a few possibilities. As modeling improves, the connection between structure and properties will become more direct and provide more possibilities for constraints. All measurements and calculations are therefore sources of possible constraints, for example, magnetic properties. We have explored basic geometric and thermodynamic properties that are fundamentally connected to correlations. In addition, we have proposed displacement SROs in connection with first-principles calculations as a particularly suitable source of constraints on structure. We question inferences about higher order correlations based purely on PCs that do not evaluate the ranges of possible higher order correlations that are consistent with the PCs.

Acknowledgment

This research used resources of the Center for Computational Sciences at Oak Ridge National Laboratory through the University of Tennessee Computational Sciences Initiative. Additional support for one author (C. Y. G) was provided by the University of Tennessee Computational Sciences Initiative. This work, sponsored by the Department of Energy, Office of Science (DOE-OS) through the Office of Basic Energy Sciences (BES) and by the Laboratory Directed Research and Development Program, was performed at Oak Ridge National Laboratory, which is operated for the Department of Energy by UT-Battelle, LLC, under contract number DE-AC0500OR22725.

References

1. S. Torquato, *Random Heterogeneous Materials* (New York: Springer, 2001).
2. M. G. Rozman and M. Utz, *Phys. Rev. Lett.* 89 (2002): 135501.
3. A. L. Paterson, *Phys. Rev.* 65 (1944): 195.
4. J. I. Yellot Jr., *J. Opt. Soc. Am. A* (1993): 777.

5. C. Chubb and J. I. Yellot, *Vision Research*, 40 (2000): 485.
6. C. de Dominicis, *J. of Math. Phys.* 3 (1962): 983.
7. M. A. Krivoglaz, *Diffuse Scattering of X-rays and Neutrons by Fluctuations* (New York: Springer, 1996).
8. D. de Fontaine, in *Solid State Physics*, ed. H. Ehrenreich, F. Seitz, and D. Turnbull (New York: Academic, 1979), 34.
9. S. C. Moss, *Phys. Rev. Lett.* 22 (1969): 1108.
10. B. L. Gyorffy and G. M. Stocks, *Phys. Rev. Lett.* 50 (1983): 374.
11. W. Schweika, *Disordered Alloys: Diffuse Scattering and Monte Carlo Simulations* (Berlin: Springer-Verlag, 1998).
12. S. Dietrich and W. Fenzl, *Phys. Rev. B* 39 (1989): 8873.
13. B. Schonfeld, "Local atomic arrangements in binary alloy," *Progress in Materials Science* 44 (1999): 435–543.
14. H. Dosch, *Critical Phenomena at Surfaces and Interfaces: Evanescent X-ray and Neutron Scattering Springer Tracts in Modern Physics* (Berlin: Springer, 1992).
15. J. S. Faulkner and G. M. Stocks, *Phys. Rev. B* 21 (1980): 8.
16. B. L. Gyorffy, D. D. Johnson, F. J. Pinski, D. M. Nicholson, and G. M. Stocks, *Alloy Phase Stability*, ed. G. M. Stocks and G. Gonis (1980), 421.
17. R. L. Henderson, *Phys. Lett. A* 49 (1974): 197–198.
18. H. Falk, *Am. J. Phys.* 38 (1970): 858.
19. R. Evans, *Mol. Simul.* 4 (1990): 409–411.
20. F. H. Stillinger and S. Torquato, *J. Phys. Chem. B*, 108 (2004): 19589.
21. C. Wolverton, A. Zunger, and B. Schonfeld, *Solid State Commun.* 101 (1997): 519–523.
22. W. Schweika and A. E. Carlsson, *Phys. Rev. B* 40 (1989): 4990.
23. D. M. C. Nicholson, R. I. Barabash, G. E. Ice, C. J. Sparks, J. Lee Robertson, and C. Wolverton, *J. Phys. Cond. Mat.* 18 (2006): 11585.
24. J. Gragg, J. Bardham, and J. Cohen, in *Critical Phenomena in Alloys, Magnets and Superconductors*, ed. R. Mills, E. Ascher, and R. Jaffee (New York: McGraw-Hill, 1977), 309–337.
25. P. C. Clapp, *Phys. Rev.* 164, no. 1018 (1967); *Phys. Rev. B* 4, no. 2 (1971): 255.
26. D. M. C. Nicholson and R. H. Brown, *Phys. Rev. Lett.* (1993): 70.
27. X. Jiang, G. E. Ice, C. J. Sparks, L. Robertson, and P. Zschack, *Phys. Rev. B* 54, no. 5 (1996): 3211.
28. G. E. Ice, C. J. Sparks, A. Habenschuss, and L. B. Shaffer, *Phys. Rev. Lett.* 68, no. 6 (1992): 863.
29. J. L. Robertson, G. E. Ice, C. J. Sparks, X. Jiang, P. Zschack, F. Bley, S. Lefebvre, and M. Bessiere, *Phys. Rev. Lett.* 82, no. 14 (1999): 2911.
30. R. A. Johnson, *Phys. Rev.* 134, no. 1330A (1964).
31. H. S. Green, *The Molecular Theory of Fluids* (Amsterdam: North Holland, 1952); D. C. Wallace, *J. Chem. Phys.* 29 (1971): 2250; A. Baranyai and D. J. Evans, *Phys. Rev. A* 40 (1989): 3817.
32. Y. Gao, D. M. Nicholson, and D. Keffer, "Entropy as a functional of the density and pair correlation," *Phys. Rev. B.* (forthcoming)

17

Recent Advances in the Study of Complex Materials by Neutron Scattering

Takeshi Egami

Joint Institute for Neutron Sciences, Department of Materials Science and Engineering, and Department of Physics and Astronomy, University of Tennessee, Knoxville, TN 37996, USA

Materials Science and Technology Division, Oak Ridge National Laboratory, Oak Ridge, TN 37831, USA

Abstract

Diffuse scattering is usually modeled in the Q (reciprocal) space, as discussed in many chapters of this book. However, there is an alternative real-space approach via the atomic pair distribution function (PDF). In this chapter we first discuss how the PDF determined through pulsed neutron diffraction provides knowledge of the local atomic structure and how it seamlessly connects the local atomic structure with the global average structure. We then turn our focus to the special nature of thermal neutrons, that their energy is comparable to the energy of phonons. This creates a difficulty in the PDF obtained by neutron powder diffraction, and we suggest a new approach to resolve this difficulty. Finally, the novel method of dynamic PDF to determine the local dynamics of atoms and spins through inelastic neutron scattering is discussed.

1 Local Structural Analysis by the PDF Approach

Structural characterization of a solid by X-ray or neutron diffraction usually means determination of the crystal structure through the analysis of the Bragg diffraction peaks. But the Bragg peaks carry information only on the periodic components of the structure, whereas the information regarding deviations from periodicity is contained in the diffuse scattering, as discussed in a number of chapters in this book. The study of diffuse scattering is more important today than ever because the modern materials being studied or developed are usually complex in structure and very often not perfectly periodic due to structural defects, mesoscopic modulations, and other complexities. The conventional approach is to model the diffuse scattering intensity as a function of the momentum transfer, $\mathbf{Q}$, or $Q = |\mathbf{Q}|$, to determine the deviations from perfect periodicity. Such modeling is not too difficult if the nature of the defects or modulation is well known. For instance, if the chemical, or compositional short-range order (CSRO), is present, the intensity of the diffuse scattering will tell us directly the pattern of modulation and the extent of CSRO.

However, in many cases we do not know the nature of the defects and the extent of structural complexity, and finding their nature often is the purpose of the structural study. But to work backward from the diffuse scattering data to find a model starting from nothing is very difficult; the Q dependence of the diffuse scattering tells us the distribution of the wavelengths of the structural modulation but rarely tells us directly what kind of modulation exists. For instance, in the case of defects with atomic displacements, the diffuse scattering intensity is the product of the square of the displacement and the number density of the displaced atoms. In order to separate these two variables, we have to determine the diffuse scattering intensities from a crystal over many Brillouin zones, which is a difficult and tedious task. A much simpler alternative is to use atomic PDF analysis on a powder sample [1] before getting involved with a single-crystal diffuse scattering study.

Usually powder diffraction measurements are used in determining the crystal structure by identifying the Bragg peaks and analyzing their position in $Q\ (= 4\pi\sin\theta / \lambda)$, where θ is the diffraction angle and λ is the wavelength of the probe), discarding the background. However, the "background" contains all the diffuse scattering intensities, and thus relevant information on the deviation from periodicity. Throwing the background away is similar to making soup out of pieces of meat and throwing away the soup to eat the meat. But in order to include the diffuse scattering, the real sample-independent background, such as air scattering, has to be determined by a direct measurement without the sample and then subtracted from the data after correction for sample absorption. After further correction for multiple scattering and Compton scattering for X-rays and incoherent scattering for neutrons, we obtain the structure function, $S(Q)$, due to interference from the atomic structure of the sample

$$S(\mathbf{Q}) = \frac{1}{N}\sum_{i,j}\left\langle \exp\left(i\mathbf{Q}\cdot\left(\mathbf{r}_i - \mathbf{r}_j\right)\right)\right\rangle, \tag{17.1}$$

where $\mathbf{r}_i$ is the position of the ith atom and <......> denotes thermal and ensemble average. The PDF, $\rho_0 g(r)$, where ρ_0 is the number density of atoms in the sample, is obtained by the Fourier transformation

$$G(r) = 4\pi r\rho_0\left[g(r) - 1\right] = \frac{2}{\pi}\int_0^\infty \left[S(Q) - 1\right]\sin(Qr)Q\,dQ\,. \tag{17.2}$$

The PDF describes the distribution of the distances between all atoms in the sample.

Since the PDF analysis does not presume lattice periodicity, it can describe periodic as well as aperiodic structures. For this reason the PDF analysis has long been used in the structural study of glasses and liquids [2]. For the PDF method to be applied successfully to crystalline materials, however, the limitation in the wavelength of the probe had to be overcome. Theoretically the Fourier transform (Equation17.2) has to be performed up to infinity in Q. In reality, however, Q is limited to be less than $4\pi/\lambda$ so that the wavelength of the probe sets the maximum in Q. Prematurely terminating the integration in Equation17.2 results in spurious oscillations known as termination errors. It was only after the development of synchrotron-based radiation sources, such as pulsed neutron sources and synchrotron-radiation sources which provide probes with short wavelengths, that the PDF method became applicable to crystalline materials and contributed to resolving difficult problems involving disorder in the lattice [1]. Furthermore, by employing a spectrometer with high resolution in Q, it is now possible to determine the PDF up to large distances, thus bridging the gap between local structure analysis and average structure analysis. It is well known that the lattice vibration results in the Debye-Waller factor, $\exp(-Q^2u^2)$, where u is the amplitude of lattice vibration. But the lattice vibration also broadens the PDF peaks by $\sqrt{2}u$. In the same manner, the uncertainty in Q, ΔQ, produces an envelope in the real space for the PDF, $\exp(-\Delta Q^2 r^2)$, which attenuates the PDF at large distances, limiting the range of the PDF. In order to overcome this difficulty, the author and collaborators have upgraded a high-resolution powder spectrometer at the Lujan Center of Neutron Scattering at the Los Alamos National Laboratory to produce the high-resolution PDF spectrometer (NPDF) [3].

An example that amply demonstrates the power of the NPDF is the structural study of $LiNiO_2$ [4]. Using the capacity of the NPDF to be able to determine not only the short-range structure but also the medium-range nanoscale structure, the PDF of $LiNiO_2$ was determined up to 200 Å. In $LiNiO_2$, Ni^{3+} has the d-electron configuration of t_{2g}^6- e_g^1, and because of the singly occupied e_g level, it is a Jahn-Teller (JT) ion. However, the average structure, $R\bar{3}m$, has no sign of macroscopic JT distortion, even though a closely related compound, $NaNiO_2$, has strong macroscopic JT distortion in the structure. Since a small amount (2% ~ 4%) of site mixture of Ni and Li always exists in the sample, it is possible that these substitutional defects are randomizing the local JT distortion. It was also proposed that the orbital moment may be in the orbital-glass state, just as in spin glasses [5], so the structure of $LiNiO_2$ has been controversial for a long time.

As shown in Figure 17.1, the first peak of the PDF, which corresponds to the Ni-O distances, is clearly split into two subpeaks at 1.91 and 2.04 Å, as expected for the local JT distortion. Furthermore, by modeling the medium-range PDF up to 5 Å, it was found that the axes of local JT distortion are not parallel as in $NaNiO_2$ but are arranged in three sublattices (Figure 17.2). The JT distortions in the three sublattices almost cancel each other, leaving the average structure without macroscopic distortion. The pattern of the local JT distortions shown in Figure 17.2, however, is the projection on the xy plane, whereas in reality these axes are not in the plane but are tilted, forming a right angle to each other. As a consequence, the trimer ordering in Figure 17.2 gives rise to a curvature of the NiO_2 plane, and to accommodate such a curvature, nanoscale domains with alternating curvatures are formed. As shown in Figure 17.1, the PDF peak height at short distances decreases with temperature, as expected from atomic vibration, but at distances larger than 90 Å, the peak height increases up to T_d = 400 K because domains are formed only below T_d [4]. Thus the NPDF spectrometer proved that the extended-range PDF can detect nanoscale domain formations, which are very difficult to detect by other methods.

2 Advances in Data Reduction: Inelastic Correction

Neutrons and X-rays are scattered by matter either elastically or inelastically. Information on the static structure is contained in the elastic scattering intensity, whereas the inelastic scattering intensity carries information about the dynamics of the system. However, isolating the elastic scattering from the inelastic scattering requires an experimental setup that results in significantly lower intensity, usually

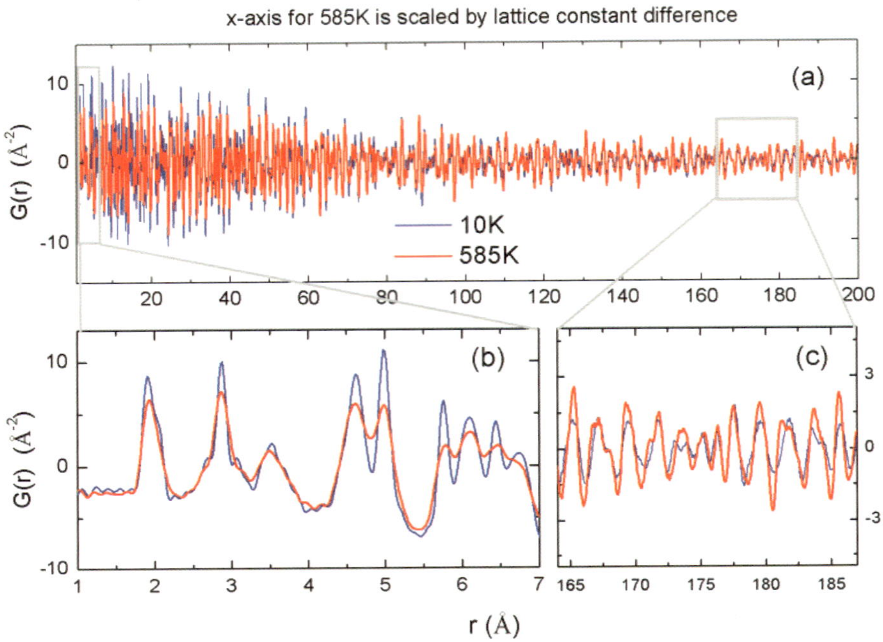

Figure 17.1. *Pulsed neutron PDF of $LiNiO_2$ at $T = 10$ K and 585 K [4]. Note that the PDF peak height decreases with temperature at short distances but increases at long distances.*

Figure 17.2. *Trimer pattern of the JT distortion, projected on the NiO_2 plane [4].*

much more than a couple of orders of magnitude. Consequently, the PDF is usually determined by the powder diffraction measurement carried out in the energy-integrated mode without separating the elastic and inelastic intensities. In the case of X-ray diffraction, this shortcutting does not cause much harm because the energy of X-rays (tens of keV) is much higher than the energy of lattice vibration (tens of meV). But in the case of neutron scattering, there are some significant penalties to pay because the energy of thermal neutrons is comparable to the energy of lattice vibration. Early on, when PDF measurements were not so accurate, owing to termination errors and other errors, such penalties did not present a real problem. However, as neutron spectrometers became more refined and accurate, this problem of inelastic scattering emerged as an important issue. In this section, I discuss how this problem can be overcome.

Through inelastic X-ray or neutron scattering measurements, it is possible to measure the dynamic structure factor

$$S(\boldsymbol{Q},\omega)=\frac{1}{N}\sum_{i,j}\int\left\langle \exp\left(i\boldsymbol{Q}\cdot\left(\boldsymbol{r}_i(0)-\boldsymbol{r}_j(t)\right)\right)\right\rangle e^{i\omega t}\,dt\,, \tag{17.3}$$

where $\boldsymbol{Q}$ and $\hbar\omega$ are the momentum and energy transfers of the scattering, $\boldsymbol{r}_i(t)$ is the position of the ith atom at time t, and <......> denotes the thermal and ensemble average [1,6]. However, in the case of inelastic X-ray scattering (IXS), the cost of obtaining the energy resolution necessary to measure phonons—say, 1 meV out of the incident energy of 20 keV, thus $\Delta E / E = 0.5\times10^{-7}$ —results in the loss of intensity by many orders of magnitude. For that reason often a much lower resolution, on the order of 1 eV, is used for actual measurement. As far as phonons are concerned, using such a low resolution means measuring the intensity integrated over the energy transfer. Thus the measured structure factor is

$$S_{total}(\boldsymbol{Q})=\int S(\boldsymbol{Q},\omega)\,d\omega=\frac{1}{N}\sum_{i,j}\left\langle \exp\left(i\boldsymbol{Q}\cdot\left(\boldsymbol{r}_i(0)-\boldsymbol{r}_i(0)\right)\right)\right\rangle, \tag{17.4}$$

or the same-time (snapshot) correlation function. This is why it is often stated that X-ray diffraction gives a snapshot picture.

Now, the situation is more complex for neutron scattering. For X-ray photons the energy transfer due to phonons, ~100 meV, is very small compared to the X-ray energy, ~10 keV. Thus the magnitude of the wave vector of the incident X-ray, $\boldsymbol{k}_i$, is almost the same as that of the scattered X-ray, $\boldsymbol{k}_f$, so that the relation for elastic scattering,

$$\left|\boldsymbol{Q}\right|=\left|\boldsymbol{k}_i-\boldsymbol{k}_f\right|=2k\sin\theta\,, \tag{17.5}$$

where $k=\left|\boldsymbol{k}_i\right|=\left|\boldsymbol{k}_f\right|$ and θ is the diffraction angle, is approximately valid even for inelastic scattering by phonons. That is why in Equation17.4 we could integrate over ω for a constant Q. For neutron scattering, however, this assumption is invalid because the kinetic energy of neutrons used in neutron scattering is on the order of kT (25 meV for room temperature) and is comparable to the energy of phonons. Consequently, $\left|\boldsymbol{k}_f\right|$ is appreciably different from $\left|\boldsymbol{k}_i\right|$, so for a given θ, Q depends significantly on ω. Thus the measured intensity is given by

$$S_{measured}(\mathbf{Q}) = \int S(\mathbf{Q}(\omega),\omega)\, d\omega\,, \tag{17.6}$$

which is not easy to evaluate and strongly depends on the experimental setup. For the PDF analysis, currently it is customary to use the so-called Placzek correction [7] to evaluate the difference between $S_{\mathrm{total}}(Q)$ and $S_{\mathrm{measured}}(Q)$. Here we use $Q = |\mathbf{Q}|$, assuming the isotropic nature of the sample. A standard PDF software package has this correction included. However, I came to recognize that the Placzek correction quite often does more harm than good to the PDF analysis. Thus it is recommended to **turn off** the Placzek correction for the PDF analysis.

The Placzek correction assumes that a neutron is scattered by the atom ballistically. In general,

$$\hbar\omega = \frac{\hbar^2}{2m}(k_i^2 - k_f^2) \tag{17.7}$$

and

$$Q^2 = k_f^2 + k_i^2 - 2k_f k_i \cos 2\theta\,, \tag{17.8}$$

where m is the neutron mass, $k_f = |\mathbf{k}_f|$, and $k_i = |\mathbf{k}_i|$. In the Placzek correction, the neutron is assumed to have a high energy, and the energy lost by a neutron is transferred entirely to the kinetic energy of one atom,

$$\begin{aligned} \hbar\omega &= \frac{\hbar^2}{2M}\left(\left|\boldsymbol{q}_f\right|^2 - \left|\boldsymbol{q}_i\right|^2\right), \\ \mathbf{Q} &= \boldsymbol{q}_i - \boldsymbol{q}_f \end{aligned} \tag{17.9}$$

where $\boldsymbol{q}_i$ and $\boldsymbol{q}_f$ are the momenta of the atom before and after scattering. Thus the energy transfer depends only on the atomic mass M and not on the restoring force, represented by the Debye temperature, for example. By assuming that the energy of the incident neutron is much larger than the energy transfer, from Equation 17.8 we obtain

$$\begin{aligned} Q^2 &= 2k_i^2 + \frac{2m\omega}{\hbar} - 2k_i^2 \cos 2\theta \left(1 + \frac{2m\omega}{\hbar k_i^2}\right)^{1/2} \\ &= 2k_i^2 + \frac{2m\omega}{\hbar} - 2k_i^2 \cos 2\theta \left(1 + \frac{m\omega}{\hbar k_i^2} + \ldots..\right) \\ &= 4k_i^2 \sin^2\theta \left(1 + \frac{m\omega}{\hbar k_i^2} + \ldots..\right) \end{aligned} \tag{17.10}$$

$$Q = 2k_i \sin\theta \left(1 + \frac{m\omega}{2\hbar k_i^2} + \ldots\right). \tag{17.11}$$

If we neglect the initial velocity of the atom, the kinetic energy gain of the atom due to the momentum transfer, Q, is given by

$$\hbar\omega = \frac{\hbar^2}{2M}Q^2, \tag{17.12}$$

where M is the atomic mass. Thus we obtain the first-order shift in Q in the Placzek correction:

$$\frac{\Delta Q}{Q_0} = \frac{m\omega}{2\hbar k_i^2} = \frac{m}{M}\sin^2\theta + \dots. \tag{17.13}$$

The Placzek correction is valid when energy and momentum transfers are large. However, it is a poor approximation at low to medium Q values, making this correction rather unsatisfactory for the analysis of a solid. This is because, in the case of scattering from a crystal, the dominant term at low to medium Q is the elastic scattering. In this scattering process, the momentum is transferred to the *whole crystal*, not to an atom. That is why the approximation of ballistic scattering by a single atom is a poor approximation. Since the mass of the whole crystal is huge compared to the mass of a neutron, the energy transfer is practically zero; thus scattering is elastic. The same applies to glasses, and also nanoparticles, because the mass of a nanoparticle is still huge compared to the mass of a single neutron. The intensity of the elastic scattering can be expressed as

$$S_{elastic}(Q) = e^{-u^2Q^2}S_0(Q), \tag{17.14}$$

where $S_0(Q)$ is the structure function of the average lattice,

$$S_0(Q) = \frac{1}{N}\sum_{i,j}\exp\left(i\boldsymbol{Q}\cdot\left(\langle \boldsymbol{r}_i\rangle - \langle \boldsymbol{r}_j\rangle\right)\right), \tag{17.15}$$

and u is the average atomic displacement,

$$u = \frac{1}{N}\sum_i \sqrt{\left\langle \left|\boldsymbol{r}_i - \langle \boldsymbol{r}_i\rangle\right|^2\right\rangle/3}. \tag{17.16}$$

The term $\exp(-u^2Q^2)$ in Equation 17.14 is the famous Debye-Waller factor. Thus up to $Q \sim 1/u$, which is on the order of 10 Å^{-1}, because $u \sim 0.1$ Å, the scattering is dominated by the elastic scattering. This shows how wrong the Placzek correction can be.

For a solid with Einstein phonons (single frequency), the dynamic structure factor, Equation 17.3, can be written in terms of the phonon amplitude, u, as

$$S(Q,\omega) = e^{-u^2Q^2}\left[S_{elastic}(Q) + u^2Q^2S_1(Q,\omega) + \frac{u^4Q^4}{2!}S_2(Q,\omega) + \dots\right], \tag{17.17}$$

where $S_n(Q,\omega)$ is the nth-order phonon dynamic structure factor [1]. Just for a moment, if we totally neglect the structure and assume

$$S_{elastic}(Q) = \int S_1(Q,\omega)\,d\omega = \int S_2(Q,\omega)\,d\omega = \ldots\ldots = 1, \tag{17.18}$$

then the terms in the square bracket in Equation 17.17 sum up to $\exp(u^2Q^2)$ and we obtain $S_{total}(Q) = 1$, as expected. The actual dynamic structure factor calculated for the system of Einstein phonons is shown in Figure 17.3. It is not difficult to calculate the phonon contribution to $S(Q,\omega)$ for Einstein phonons and actually evaluate $S(Q,\omega)$ from $S_{measured}(Q)$. Such software will soon be added to the PDF package.

3 Dynamic Pair Distribution Function Analysis

For a well-ordered single crystal sample, the phonon or magnon dispersion can be directly determined by measuring the dynamic structure factor $S(Q,\omega)$ with an inelastic neutron scattering instrument, such as a triple-axis spectrometer for the reactor source or a chopper spectrometer for the pulsed neutron source. However, if the sample contains a high degree of disorder, phonons or magnons are strongly damped and the dispersion will be diffuse in the Q and ω space, so it will be difficult to

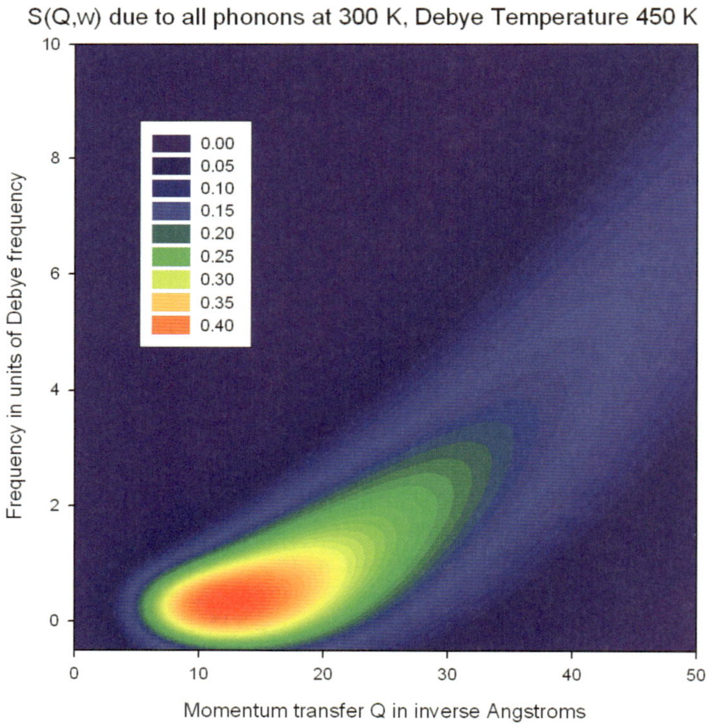

Figure 17.3. *Dynamic structure factor of a solid with Einstein phonon [8].*

determine the phonon dispersion. For instance, in strongly disordered ferroelectric solids, so-called relaxor ferroelectrics, the phonon vibrations are so strongly localized that a part of the dispersion simply disappears, resulting in a "waterfall" behavior [9]. Such localized phonons are best observed by the dynamic PDF (DPDF) technique.

Van Hove [6] pointed out that the Fourier transform of the dynamic structure factor (Equation 17.3) gives the time-dependent correlation function, now called the Van Hove function:

$$G(\boldsymbol{r},t)=\int S(\boldsymbol{Q},\omega)e^{i[\omega t-\boldsymbol{Q}\cdot\boldsymbol{r}]}d\omega\, d\boldsymbol{Q}=\frac{1}{N}\sum_{i,j}\left\langle \delta\left(\boldsymbol{r}-\left(\boldsymbol{r}_i(0)-\boldsymbol{r}_j(t)\right)\right)\right\rangle . \tag{17.19}$$

Also, the frequency Fourier transform is called the intermediate correlation function,

$$I(\boldsymbol{Q},t)=\int S(\boldsymbol{Q},\omega)e^{i\omega t}\, d\omega=\frac{1}{N}\sum_{i,j}\left\langle \exp\left(i\boldsymbol{Q}\cdot\left(\boldsymbol{r}_i(0)-\boldsymbol{r}_j(t)\right)\right)\right\rangle , \tag{17.20}$$

and is used in the analysis of neutron scattering data [10]. However, these functions are difficult to understand except for some limiting cases of overdamped vibration in liquids and are cumbersome to deal with since they usually are an extended function of time. Instead, we proposed to use the Fourier transform to the real space,

$$\rho_0 g(r,\omega)=\frac{1}{2\pi^2 r}\int S(Q,\omega)\sin(Qr)Q\, dQ , \tag{17.21}$$

and call it the DPDF [1,11]. Figure 17.4a shows the DPDF of polycrystalline Ni. The dynamic structure factor $S(Q,\omega)$ was determined by a pulsed neutron chopper spectrometer, PHAROS, at the

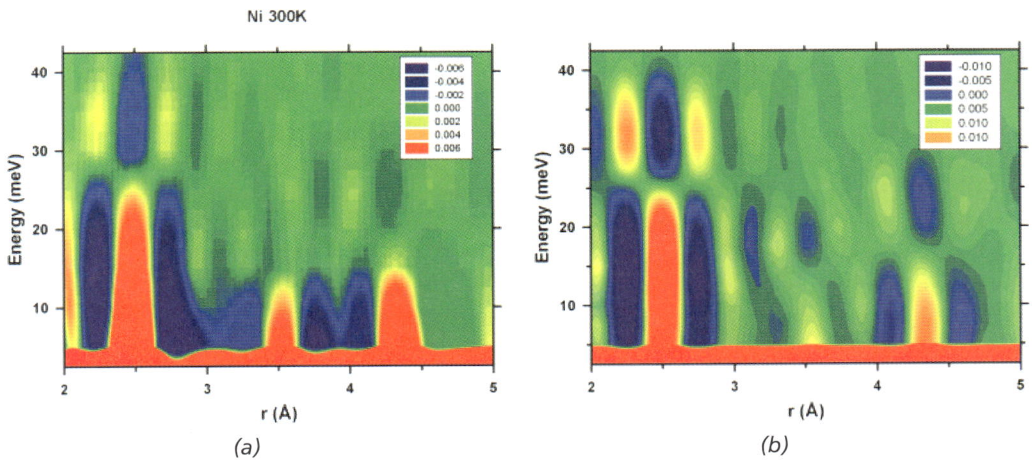

Figure 17.4. *(a) DPDF of polycrystalline Ni determined by pulsed neutron inelastic scattering. (b) Simulation for polycrystalline Ni [13].*

Lujan Center of the Los Alamos National Laboratory. In order to ensure that the Fourier transformation is accurately evaluated, $S(Q,\omega)$ has to be determined over a wide range of Q to minimize the termination errors. My colleagues and I used the initial neutron energy of 250 meV so that Q extends to 20 Å^{-1}. Figure 17.4b is the simulated DPDF for Ni using the known phonon dispersions in Ni. It is noted that the first peak at 2.5 Å splits into two above 30 meV. The magnitude of the separation between the two subpeaks, 0.5 Å, is not commensurate with the expected wavelength of the termination errors, $2\pi / Q_{max} \sim 0.3$ Å, indicating that this split is not the termination effect but is genuine. This splitting is caused by the van Hove singularity at 32 meV [12]. Essentially the DPDF shows only the local phonons. Signals from fast-dispersing phonons are canceled out by phase mixing. Only the local phonons can produce local phase coherence and a split DPDF peak.

Recently this technique was applied to the study of local lattice dynamics in disordered ferroelectrics called relaxor ferroelectrics [13]. Ferroelectric materials show high dielectric permittivity and piezoelectricity near the Curie temperature, but the response is strongly temperature dependent. Such a temperature sensitivity is obviously a major source of problems in applications. On the other hand, strongly disordered ferroelectrics with diffuse ferroelectric transition, the relaxor ferroelectrics, have a much weaker temperature dependence in response and are favored in applications [14]. $Pb(Mg_{1/3}Nb_{2/3})O_3$ (PMN) is a prototypical relaxor ferroelectric oxide. It shows frequency-dependent permittivity from 230 K to 300 K, but above 300 K permittivity is independent of frequency and only weakly dependent on temperature. Such behavior is known to originate from the formation of "polar nanoregions" (PNRs), nanoscale ferroelectric domains, but how and why they are formed has not been known at the atomistic level. In PMN, Mg^{2+} and Nb^{5+} occupy the B site of the perovskite structure semirandomly. Mg and Nb form NaCl-type sublattices, with one mostly occupied by Nb and the other occupied with (2/3)Mg and (1/3)Nb [15]. Since their valences are so much different, disorder in one of the sublattices is enough to drive the system into a glasslike relaxor behavior. According to optical responses, some polar activity appears to start as high as 600 K, known as the Barns temperature [16]. But what exactly happens at this temperature has not been known.

The recent total as well as DPDF studies of PMN using pulsed neutron scattering succeeded in shedding light on this confusing problem. The total PDF was determined with the NPDF and the DPDF with PHAROS, both at the Lujan Center, Los Alamos National Laboratory. The average crystal structure of PMN is a simple cubic perovskite. But the pulsed neutron PDF shows marked deviations from the average structure [17]. In particular, a pronounced PDF peak is seen at 2.4 Å, whereas there is no such peak in the average PDF. This peak corresponds to the short Pb-O distances in the PbO_{12} cluster [18]. The average Pb-O distance is about 2.8 Å, but because of the lone-pair *s-p* electrons, the Pb^{2+} ion is usually off-centered in the PbO_{12} cluster, resulting in short Pb-O bonds of about 2.4 Å and long bonds of about 3.2 Å [18]. Thus the presence of the PDF peak at 2.4 Å is a telltale sign of local Pb off-centering. In addition, the analysis of the total PDF [19] told us that the medium-range structure up to 2 nm or so is partly distorted into a rhombohedral structure, which is identified as PNR. The volume fraction of the distorted structure (NPR) is about 30% at $T = 0$ K, decreasing to less than 10% at $T = 300$ K, and vanishes around 600 K. So the NPR exists mostly below 300 K and completely disappears at the Burns temperature.

The total PDF, however, is determined in the energy-integrated mode so that the dynamic correlations are partially included, but the extent of the inclusion is not well understood. This was made clear by the DPDF measurement. The DPDF of PMN at $T = 450$ K is shown in Figure 17.5. It is noted that the peak at 2.4 Å is observed only at energies above 15 meV (~4 THz). This peak suggests dynamic Pb

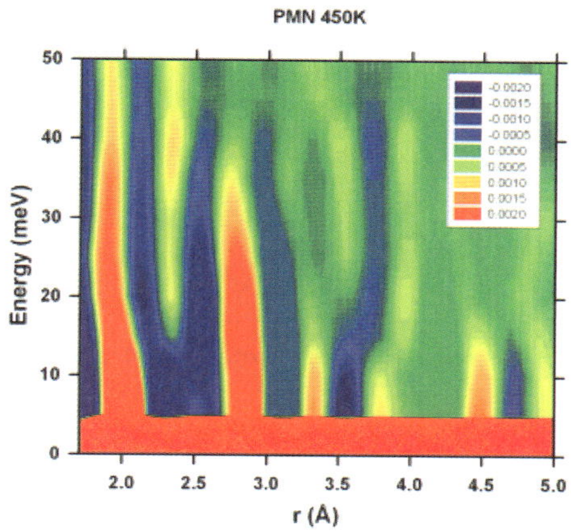

Figure 17.5. *DPDF of PMN at 450 K [13].*

off-centering with a frequency higher than 4 THz. The peak at 3.4 Å corresponding to the Pb-(Mg,Nb) distance is also split above 15 meV, again indicating dynamic polarization. Furthermore, the temperature dependence of the DPDF was found to depend strongly on frequency (energy). Figure 17.6a shows the temperature dependence of the static PDF, $\rho(\mathbf{r}, 0)$. The extra peaks at 2.4 Å and 3.2 Å disappear around 300 K. On the other hand, the DPDF at 15 meV (Figure 17.6b) shows these peaks persisting up to 600 K. These results suggest that dynamic local Pb polarizations set in around the Burns temperature and become static (below 1 THz) around 300 K. Between 300 K and 600 K, dynamic polarization clouds exist, but only below 300 K do they form quasistatic PNR with local rhombohedral distortion. Thus the total PDF and DPDF of PMN yielded a more detailed dynamic picture of the way dynamic inhomogeneous polarization in the relaxor ferroelectrics evolves with temperature.

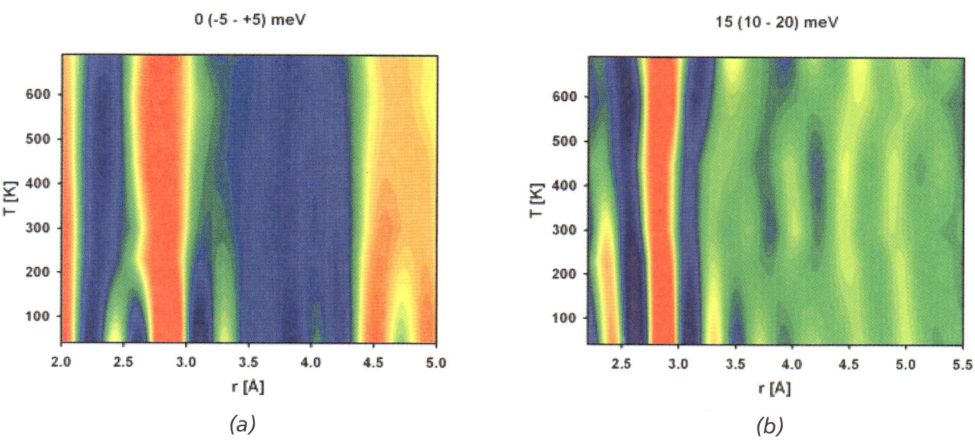

Figure 17.6. *Temperature dependence of the DPDF of PMN (a) E = 0 and (b) E = 15 meV [13].*

Acknowledgments

The author acknowledges the support of the National Science Foundation, through grants DMR04-04781 and DMR06-02876, and the Laboratory Directed Research and Development Program of Oak Ridge National Laboratory, managed by UT-Battelle, LLC, for the U.S. Department of Energy under contract no. DE-AC05-00OR22725.

References

1. T. Egami and S. J. L. Billinge, *Underneath the Bragg Peaks: Structural Analysis of Complex Materials* (Oxford: Pergamon Press, Elsevier Ltd., 2003).
2. B. E. Warren, *X-ray Diffraction* (Reading: Addison-Wesley, 1969; New York: Dover, 1990).
3. Th. Proffen, T. Egami, S. J. L. Billinge, A. K. Cheetham, D. Louca, and J. B. Parise, *Appl. Phys. A* 74 (2002): S163.
4. J.-H. Chung, Th. Proffen, S.-I. Shamoto, A. M. Ghorayeb, L. Croguennec, W. Tian, C. Sales, R. Jin, D. Mandrus, and T. Egami, *Phys. Rev. B* 71 (2005): 064410.
5. F. Reynaud, D. Mertz, F. Celestini, J.-M. Debierre, A. M. Ghorayeb, P. Simon, A. Stepanov, J. Voiron, and C. Delmas, *Phys. Rev. Lett.* 86 (2001): 3638.
6. L. van Hove, *Phys. Rev.* 95 (1954): 249.
7. G. Placzek, *Phys. Rev.* 86 (1952): 377.
8. V. A. Levashov and T. Egami, unpublished.
9. P. M. Gehring, S. Wakimoto, Z.-G. Ye, and G. Shirane, *Phys. Rev. Lett.* 87 (2001): 277601.
10. See, for example, W. Marshall and S. W. Lovesey, *Theory of Thermal Neutron Scattering* (Oxford: Clarendon Press, 1971).
11. R. J. McQueeney, *Phys. Rev. B* 57 (1998): 10560.
12. R. J. Birgeneau, J. Cordes, G. Dolling, and A. D. B. Woods, *Phys. Rev.* 136 (1964): A1359.
13. W. Dmowski, S. B. Vakhrushev, I.-K. Jeong, M. Hehlen, F. Trouw, and T. Egami, *Phys. Rev. Lett.* 100 (2008): 137602.
14. G. A. Smolensky, ed., *Ferroelectrics and Related Materials* (New York: Gordon and Breach, 1981).
15. W. Dmowski, M. K. Akbas, P. K. Davies, and T. Egami, *J. Phys. Chem. Solids* 61 (2000): 229.
16. G. Burns and B. A. Scott, *Solid State Commun.*, 13(1973): 423.
17. T. Egami, H. D. Rosenfeld, B. H. Toby, and A. Bhalla, *Ferroelectrics* 120(1991): 11.
18. S. Teslic and T. Egami, *Acta Crystallogr. B* 54 (1998): 750.
19. I.-K. Jeong, T. W. Darling, J. K. Lee, Th. Proffen, R. H. Heffner, J. S. Park, K. S. Hong, W. Dmowski, and T. Egami, *Phys. Rev. Lett.* 94 (2005): 147602.

Part IV

Scattering from Distorted Crystals

18

The Role of Diffuse Scatterings in Understanding and Controlling Radiation-Induced Swelling of Nonmetallic Materials

V. M. Kosenkov

P. P. Silantiev

Research Institute of Atomic Reaktors, Dimitrovgrad, Russia

Abstract

Diffuse scattering is a sensitive probe of materials defects and has proven essential in the study irradiated materials. Here we describe the important role of diffuse scattering in understanding and controlling radiation-induced swelling in nonmetallic materials. Diffuse X-ray scattering theory explains the appearance of near-Bragg peaks due to the presence of aggregations of defects in the crystal lattice. A criterion of radiation damage of nonmetallic materials was formulated based on experimental diffuse X-ray scattering measurements and resulted in a new approach to mitigate radiation-induced swelling by explosive shockwave treatment of the materials. This example illustrates the application of diffuse scattering to difficult materials science problems that are key to technology in the twenty-first century.

1 Introduction

X-ray characterization of reactor materials emerged in the middle of the last century as an essential tool for the investigation of materials irradiated in the first nuclear reactors. Compared to the characterization of nonirradiated materials, X-ray characterization of *irradiated* samples presents two unique challenges: (1) the instruments must deal with specimens that produce ionizing radiation with intensities greatly exceeding the intensity of the primary beam of an X-ray tube and (2) the data analysis requires special attention to treat materials with very high densities of radiation-induced defects. These two challenges place special requirements on both the hardware and the data analysis theory. Although the importance of specially shielded characterization instruments is clear with high-activity specimens, the necessity of special theoretical attention to a large defect density is also critical, as eloquently proved in a recently published monograph [1]. In this monograph, measurements are presented for the unit cell parameters of molybdenum and armko-iron foils irradiated in the SM-2 reactor with fast neutron fluences of up to $2.8.10^{20}cm^{-2}$ and $2.2.10^{21}cm^{-2}$, respectively. The angular positions of large 2θ diffraction lines were measured with a diffractometer equipped with an X-ray tube with both copper and chrome anodes. The uncertainties of the measurements were ±0.003%. After irradiation, the iron lattice parameter was observed to increase with $\Delta\mathbf{a} / \mathbf{a} = 0.013 \pm 0.003\%$ for both Cu and Cr radiation. However, for the molybdenum sample, the calculated increase was significantly different for the Cu and Cr radiation; $\Delta\mathbf{a} / \mathbf{a} = 0.010 \pm 0.003\%$ and $0.038 \pm 0.003\%$ for copper and chrome radiation, respectively. In other words, the value of the molybdenum unit cell parameter appeared to depend on the radiation type, which clearly is unphysical and requires a reevaluation of the data analysis.

2 Instrumentation

As described above, specialized equipment is required to study highly radioactive materials. For very small sample volumes—such as for monocrystals or transplutonium element compounds, which are produced in tiny amounts (several milligrams)—straightforward sample handling can be used. For example, these samples are contained in glass capillaries placed into a Debye chamber. However, for many of the most interesting samples, a special remote-controlled diffractometer with specially designed shielding is required. For example, special instrumentation is required for the characterization of specimens more than several millimeters in size, especially with reactor core items (e.g., surfaces of steel and zirconium claddings of fuel elements [containing fuel]; metallographic sections of fuel element cross sections; and reflector, absorber, and neutron moderator materials irradiated up to a neutron fluence of 10^{22} n/cm^2). Details about this specialized instrumentation are given in Kosenkov [1], Ferguson [2], and Kleykamp and Pejsa [3].

In our investigations we use a double-axis X-ray diffractometer with remote control (DARD) produced by the Scientific Production Association "Bourevestnik," St. Petersburg, Russia. The diffractometer is a double-axis design with rotating sample and source axes and with a fixed diffracted-beam monochromator and scintillation counter (Figure 18.1). The X-ray tube is fixed to the outer rotating circle and the sample is mounted to the inner axis. The scintillation counter is heavily shielded with lead. The 320 mm thick lead shielding permits diffraction studies of steel cladding of a fuel element ($\varnothing 8$ mm, L = 25–20 cm) containing fuel after 15% burnup. Integrated specimen holders and ancillary equipment extend the possible reactor material science investigations. Possible measurements include studies of surface macrostresses of cylindrical items (diameter in the range of 6.0–42.0 mm and length up to 0.5 m); studies of planar samples (e.g., hexagonal fuel assemblies); characterization of polished samples in standard metallographic mounts; and measurements of the texture of thin-walled zirconium claddings after removal of irradiated fuel. The maximum diffraction angle is 164° and the uncertainty of the measured angles is similar to the uncertainty of traditional commercial diffractometers. *Remote-controlled* DARD diffractometers have been successfully operated for ten to twelve years, and five generations of DARD diffractometers have replaced each other since 1964.

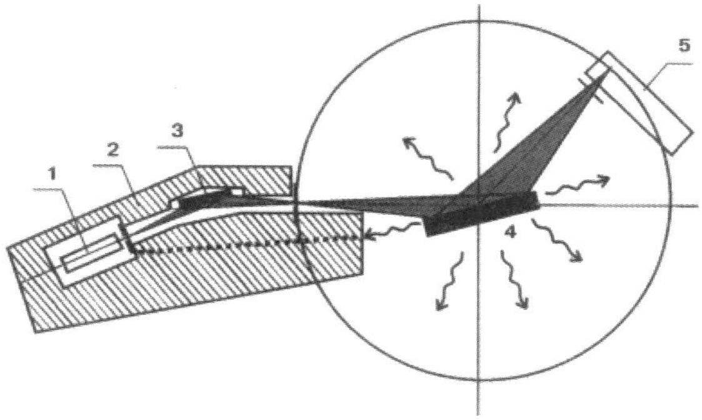

Figure 18.1. *Remote-controlled diffractometer scheme: (1) counter, (2) lead protection, (3) monochromator, (4) specimen, and (5) X-ray tube.*

3 Diffuse Scattering from Irradiated Materials

With the instrumentation described above, there have been no cases in which radiation from the specimen presented an obstacle for diffraction characterization, that is, data on the location and profile of diffraction lines. *However, difficulties associated with the interpretation of X-ray diffraction patterns from irradiated materials still impeded wide application of X-radiography in reactor material science.* In addition to the irradiated molybdenum and armko-iron examples describe above, there are numerous other cases with apparently unphysical experimental results. For example, during the measurement of lattice parameters of beryllium oxide irradiated up to a fluence of 1.0×10^{22}n/cm^2, the lattice spacing calculated at the center of gravity of (100), (110), (200), and (120) lines is equal to $\mathbf{a} = 2.701 \pm 0.005$ Å. However, **c** unit-cell dimensions calculated at the center of gravity of lines (002), (201), (102), and (121) are 4.392 Å, 4.479 Å, 4.490 Å, and 4.630 Å, respectively. In addition, diffraction lines (HKL) with $L \neq 0$ were considerably wider than lines with L = 0 (Figure 18.2).

Broadened diffraction lines (e.g., after plastic deformation of metals) can be analyzed using the method of Warren and Averbach [4]. According to this method, a crystal is represented as an assembly of mosaic blocks. Through Fourier analysis of the diffraction line profiles, the distribution of the mosaic block sizes, deformation of the blocks, and concentration of the stacking defects can presumably be characterized. However, a simple application of Warren-Averbach's method for analyzing broadened diffraction lines of irradiated materials *does not* recover physically meaningful results.

The failure of standard data analysis can be explained by the large contribution of diffuse scattering near the broadened Bragg peak. The diffuse scattering arises from radiation-induced defects. The defects can be numerous (up to 10^{17} cm^{-3}) point defect clusters. Ekstein [5] was the first to consider X-ray scattering of crystals with defects. These defects are limited in size and create static fields of atomic displacements that decrease following the law $\mathbf{U} = \mathbf{C} / \mathbf{r}^2$. The theory predicted attenuation of Bragg peaks due to the exponential factor $\mathbf{I}_b \sim \mathbf{I}_0 \exp(-2M)$, where the 2M attenuation factor depends on parameters of the defects and on the diffraction line indices. Small defect clusters do not alter the Bragg peak width but redistribute the X-ray scattering intensity, and a second component—that is, a diffuse peak located close to a Bragg peak—can be observed. Parameters of the diffuse peak (intensity, width, degree of asymmetry, angular distance from the Bragg component, etc.) depend on

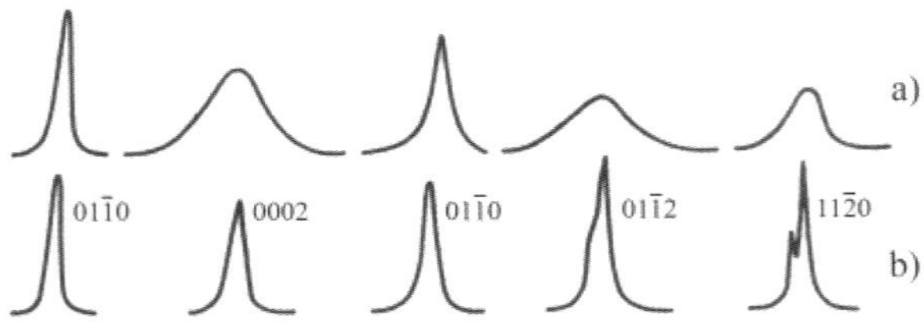

Figure 18.2. *Profiles of diffraction lines of BeO: (a) nonirradiated BeO and (b) BeO irradiated up to a fluence of 2.10^{21} n/cm^2.*

the radiation-defect parameters and density. A comprehensive bibliography on diffuse scattering (not only of irradiated materials) is given in Krivoglaz [6]. Here we present a few illustrative examples from irradiated materials.

Theoretical predictions [6,7] were entirely confirmed during investigations of irradiated molybdenum and nickel samples. As a first step in analyzing materials with high defect density, the diffraction profiles were separated into Bragg and diffuse components in order to determine their separate parameters using either a graphical method or special computer programs. Our investigation differs from previous studies in that Bragg and diffuse peaks were next separated from the diffractometer instrumental resolution by computer programs in two stages. At the first stage, an intrinsic broadening function was calculated by applying a regularization method [8]. This method allows determination of a "genuine" profile of reflexes in an X-ray analysis. The intrinsic broadening function is a theoretical intensity distribution function free from both the influence of equipment factors and *a priori* information about the physical structure of the crystal. At the next stage, the function of intrinsic broadening was separated into two components approximated by Gaussian distributions [8]. Here the choice of a Gaussian shape approximation for the diffuse component is justified by the fact that electron microscopy of irradiated metals reveals the presence of dislocation loops—clusters of radiation defects—both of vacancy type and interstitial atom type. Although the details of the defect parameters affect the shape of the diffuse peak, the shape was predicted for the case of one-type dislocation loops only. The density and average dimensions of dislocation loops calculated on the basis of characteristics of diffuse scattering (Figures 18.3 and 18.4) are in good agreement with the results of their direct observation by means of an electron microscope.

4 Overview of Swelling in Nonmetallic Materials

Following the steps described above, we have resolved the essential aspects of X-ray characterization of irradiated materials. This has enabled us to correctly measure "X-ray swelling," that is, to accurately determine the increase of unit cell volume per atom/molecule of a crystal lattice. It is well known that the vast (~20%) swelling of stainless steel is the result of an increase in the number of vacancy pores at practically steady unit cell volume. However, almost the same and often larger swelling of diamonds, beryllium oxide, graphite, and other materials irradiated at low temperature (~100°C) are associated with strong fields from atoms displaced statically due to point defects and their aggregates in the crystal lattice. As this takes place, the crystalline structure of materials does not change, as evidenced by the

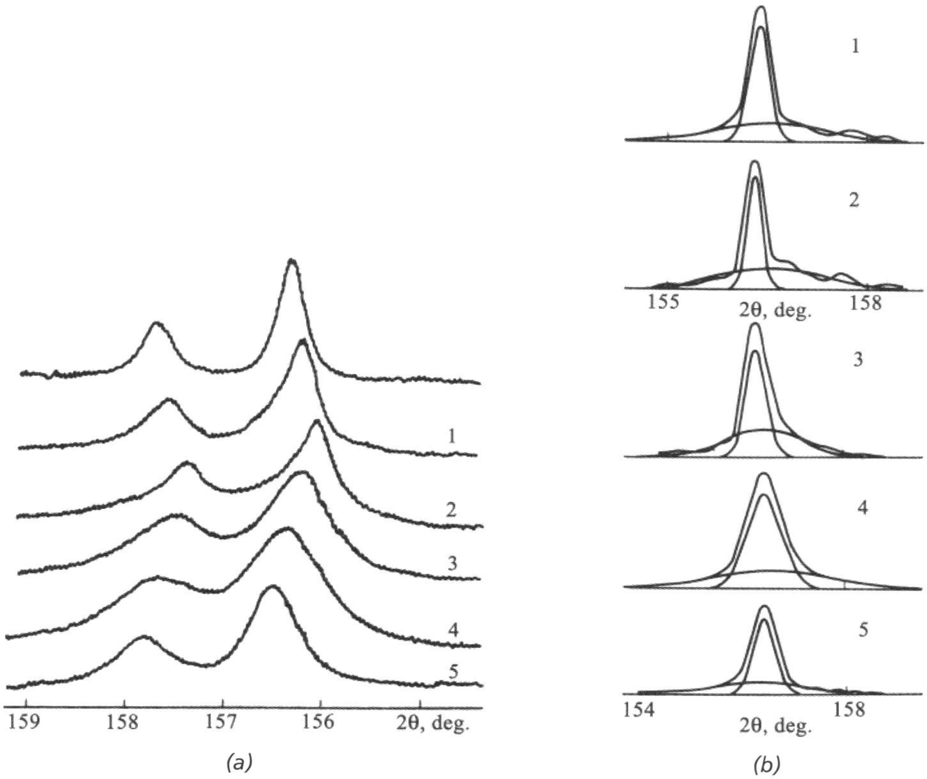

Figure 18.3. *Profiles of (a) line (400) of Mo and (b) their physical profiles after irradiation up to a fluence of (1) 1.5.10*^{20}*, (2) 2.8.10*^{20}*, (3) 1.4.10*^{21}*, (4) 2.2.10*^{21}*, and (5) 1.4.10*^{22} *n/cm*^{2}*.*

constant number of diffraction lines; the observable number of diffraction lines gradually decreases solely because of amorphization.

The presence of a diffuse component in the diffraction profiles of irradiated graphite was first demonstrated by Bacon [9]. Figure 18.5 displays distortion of the profile of the (0004) line with an increase of parameter **C** by 3.1%. Parameter **C** is the so-called strength of the defect, which characterizes the amount of displacement in the matrix around the defect. A similar distortion of the profile of the (0002) line is observed after an increase in the **C** parameter by 9.1%, and a Bragg peak is not observed near the normal (0004) line position. Kosenkov et al. [10] describe the investigation of graphite taken from the core of the Leningrad Nuclear Power Plant after irradiation to a fluence of 2 $\times$ 10^{21} fn/cm^2 at 40°C–90°C. In this case, no Bragg peak was observed near the normal (0002) line position, where a 5°-wide peak was observed; the integral breadth of a nonirradiated specimen was ~0.6°. The angle of the Bragg peak was calculated using the formula

$$2(\theta_b - \theta_d) = -0.22 \sin \Delta V \cdot 2\delta(\theta)\,. \tag{18.1}$$

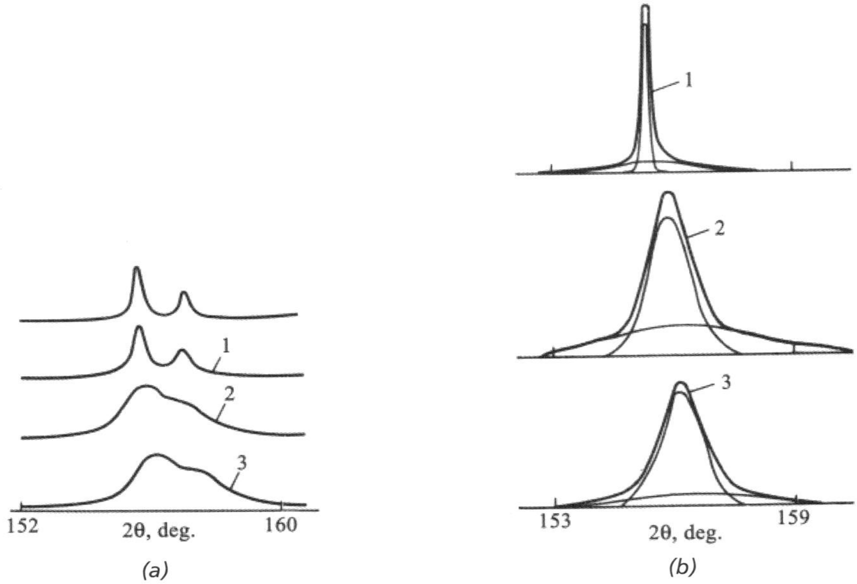

Figure 18.4. *Profiles of (a) line (420) of Ni and (b) their physical profiles after irradiation (b) up to a fluence of (1) 1.1.10^{20}, (2) 1.4.10^{21}, and (3) 2.2.10^{21} n/cm^2.*

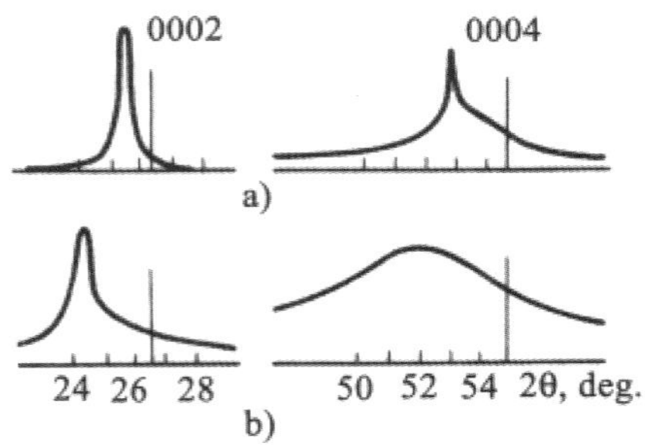

Figure 18.5. *The form of diffraction lines of graphite irradiated in the nuclear reactor: (a) ΔC/C = 3.1% and (b) ΔC/C = 9.1%.*

Here $2(\theta_b - \theta_d)$ is the angular distance between the diffuse and Bragg components, ΔV is the change in the volume per atom, and $2\delta(\theta)$ is the integral breadth of the diffuse peak. This approach allows us to determine that the average spacing **C** increased by 12% in comparison with the spacing of a nonirradiated specimen. It is in good agreement with the decrease of graphite density by 12.3%, which was measured using a picnometer method. The results demonstrate that the volume changes observed in graphite follow the mechanism of "Roentgen density" decrease.

Changes in the diffraction line profiles of diamonds were also observed for specimens irradiated up to an amorphous state. Figure 18.6 shows two lines from specimens irradiated up to a fluence of 1×10^{20} to 8.5×10^{20} n/cm^2 at a temperature of 100°C–150°C. Calculated values of the distortion factor, 2M, asymmetry, integral breadth of the diffuse peak, and difference in angular position of Bragg and diffuse components are all in good agreement with their ratios according to the theoretical formulae. The validity of Equation 18.1 was tested using specimen III. For specimen III, it was easy to separate Bragg and diffuse components. The unit cell parameter was calculated from the position of the Bragg component on line (111). For the (220) line, the Bragg component is absent and the unit cell parameter was obtained using parameters of diffuse scattering near the (220) line. The values of the unit cell parameter calculated from these two lines coincided with each other. Therefore, the measured swelling of specimen IV ($\Delta V / V = 11.4\%$), from the (111) line, is completely consistent. Remaining lines were not observed at acceptable signal-to-background noise ratios.

Another interesting phenomena was observed during a study of irradiated beryllium oxide. To determine the precise value of the parameter **C**, we had to use one of the final lines, (121) (Figure 18.7); the X-ray measurements were performed with an X-ray tube with a copper anode. The relation $(h_2 + hk + k_2)/l^2$ for the line equals 7 and the value of attenuation factor 2M allows us to observe the Bragg component even for a specimen irradiated up to a fluence of 1.0×10^{22} n/cm^2 (though with a considerable error). Further increase of parameter **C** made up $16.0 \pm 1.5\%$.

Differences in the angular positions of Bragg and diffuse components of irradiated beryllium oxide are demonstrated in Austerman and Miller [11]. This work deals with the study of monocrystals by means of a dual spectrometer and using the rotating-crystal method (Figure 18.8). The distribution of intensity in reciprocal space is shown in Kosenkov [1]. Investigations were performed by means of an X-ray diffractometer capable of producing nondistorted images of reciprocal lattice planes (Figure 18.9).

Lesser swelling is observed in another group of materials in which irradiation results in structure change, as vividly demonstrated by the change in the number of powder lines. There are a considerable

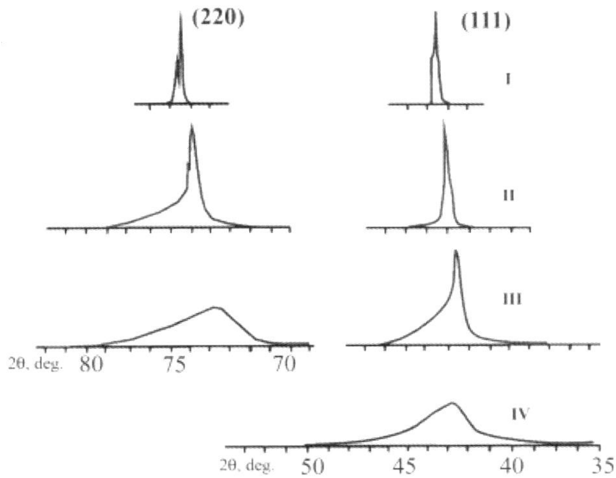

Figure 18.6. *Diffraction lines (111) and (220) of diamond specimens irradiated up to a fluence of (1) 1.10^{20}, (2) 3.5.10^{20}, (3) 5.10^{20}, and (4) 8.5.10^{20} n/cm^2.*

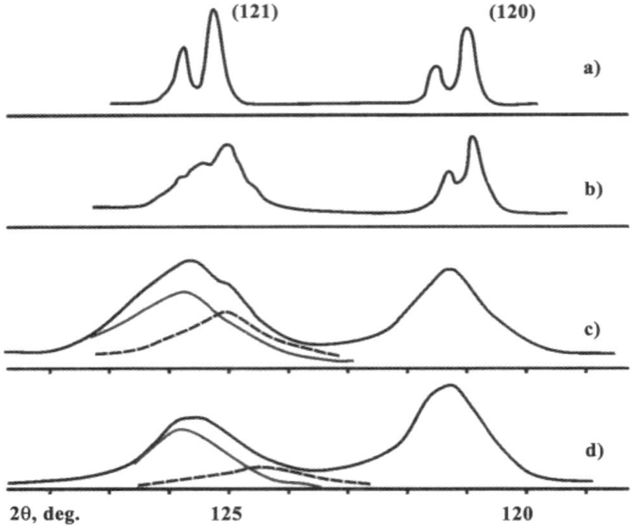

Figure 18.7. *Diffraction lines (12.0) and (12.1) of BeO specimens irradiated up to a fluence of (a) –5.5.10*[20]*, (b) –2.10*[21]*, (c) –5.10*[21]*, and (d) –1.10*[22] *n/cm*[2]

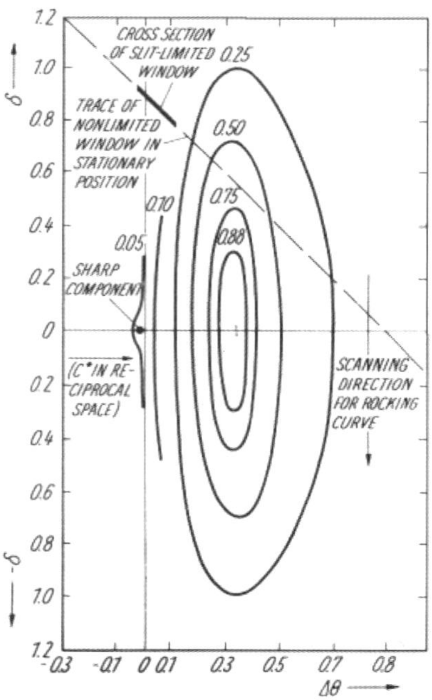

Figure 18.8. *Map of the intensity of X-rays scattering close to the 002 joint of irradiated beryllium oxide monocrystal /10/.*

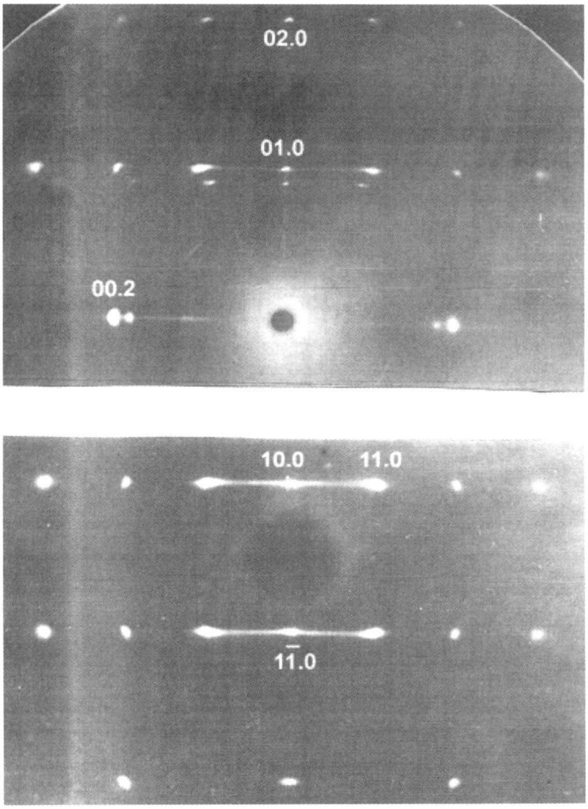

Figure 18.9. *Undistorted image of the planes of the reciprocal lattice of BeO monocrystal irradiated up to fluence 1.8.10^{21} n/cm^{2}.*

number of publications on this subject. The work of Kosenkov [1] contains results of the study of rare earth monotitanates and dititanates (Figures 18.10 and 18.11). Structural transitions in these materials are followed by an opposing change of density (irradiation decreases the density of dititanates and increases the density of Gd_2TiO_5). This is explained by the interpretation of the atomic structure of substances derived from a fluorite structure. The absence of diffuse scattering in such materials, for instance, in $GdAlO_3$ [12], and their considerably smaller swelling are typical of the substances characterized by temperature-induced phase transitions of a displacement type.

5 Swelling Can Be Defeated

In publications on the subject, there are numerous attempts to determine the dependence of radiation resistance of nonmetallic materials on their initial crystal-chemical properties. Factors proposed include bond ionicity, symmetry of the crystal lattice, and the ability of the structure to assist the displaced atom in moving from its vacancy following the mechanism of dynamic crowd ions. Based on

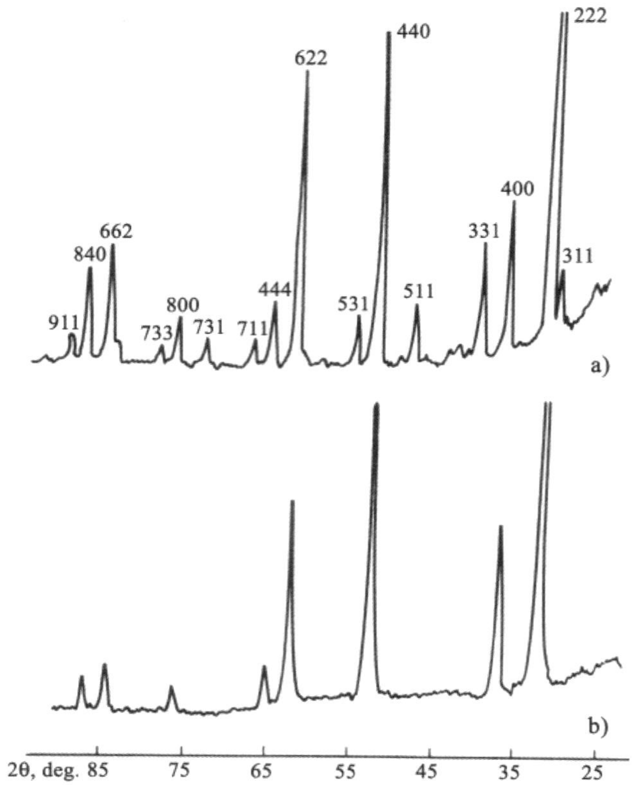

Figure 18.10. *Radiographs of (a) nonirradiated and (b) irradiated $Gd_2Ti_2O_7$ specimens.*

our investigations, we formulated a criterion of radiation damageability (namely dimension stability) of nonmetallic materials. It consists of the degree of change of the nature of atomic bonds in the area of radiation defects, which results in a change of volume per atom/molecule. Certainly, the volume change is a sum of several factors: change in the volume per atom (at some points or throughout the whole volume of the crystal) due to a change in the nature of the binding force, the appearance of gas or vacancy pores, and microcracking. We ignore the last two factors and consider the change in the nature of the binding force in detail. Swelling of nonmetallic materials depends on neutron fluence and crystal chemistry of the sample.

A natural question arises: Is it possible to change the swelling for a given neutron fluence? For example, is it possible to change the crystal chemistry by using the same substance, but in another crystallographic form? In Kosenkov [1], results are presented for simultaneous irradiation of boron nitride specimens in sphalerite, wurtzite, and graphitelike modifications. The swellings made up 1.9%, 6.5%, and 30%, respectively. However, not all substances have a set of stable phases. Materials scientists and engineers widely use alloying when it becomes necessary to change material properties. Unfortunately, this method cannot be applied to nuclear reactor core materials if it changes the essential

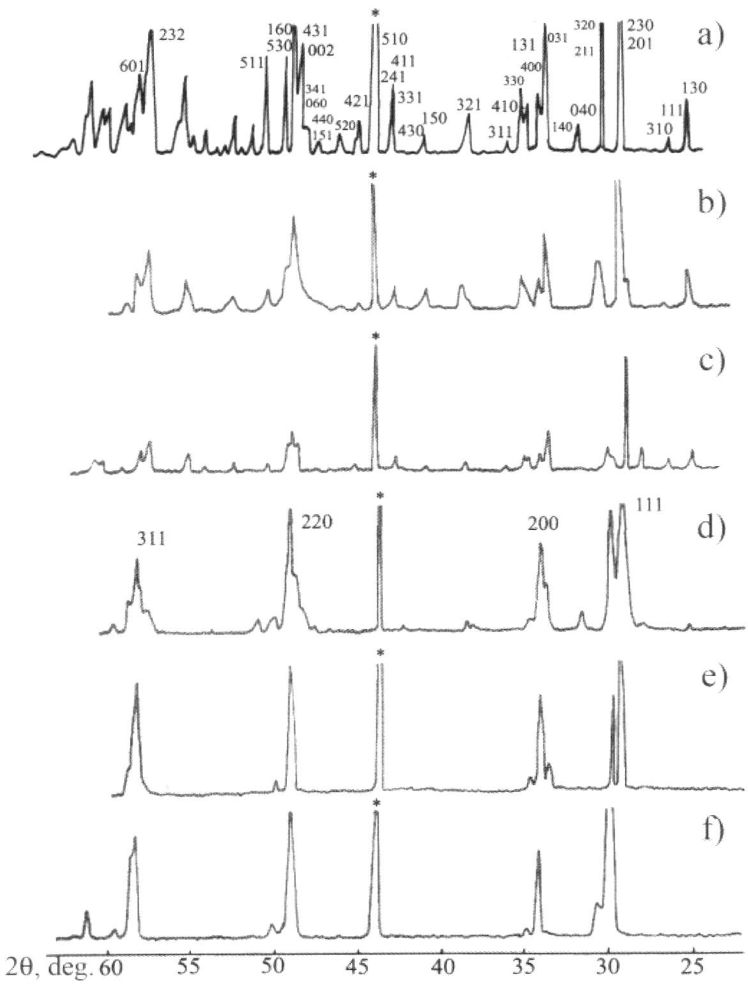

Figure 18.11. *Radiographs of (a) nonirradiated and (b) irradiated Gd_2TiO_5 specimens. An asterisk marks the diamond line (111).*

nuclear and physical properties of materials. Therefore, we propose to eliminate primary radiation defects (interstitial atoms and vacancies), which appear as a result of a fast neutron knocking an atom outside the crystal lattice site. Diffusing, the primary defects unite into immovable clusters but retain their influence on swelling.

One of the methods to decrease radiation swelling of nonmetallic materials is to treat them by explosive shock waves. The explosion-induced structural defects in the crystal lattice serve as sinks for the primary radiation-induced defects. The results of experiments on simultaneously irradiated unshocked and shocked powders are presented in Kosenkov et al. [13,14]. The calculated stress at the peak of the shock wave was 12–70 MPa. The explosion caused partial disordering, formation of multiple lattice defects, and splitting of crystallites into mosaic blocks. It was determined that the explosive treatment of beryllium oxide resulted in the appearance of a dislocation network with a dislocation density of ~10^{11} cm^{-2}. These experiments did not try to find optimal parameters of the shock wave for maximum reduction of swelling. It is evident from Table 18.1 that the maximum efficiency of swelling reduction can be achieved in materials in which irradiation creates defects leading to strong local distortions of the crystal lattice. The tenfold reduction of beryllium oxide swelling is especially impressive. It is seen in Figure 18.12 that the profile of line (002) changed along with a decrease in swelling (shifting of the angle of line (101)). It demonstrates the marked decrease in the number of radiation defects that merge into clusters.

TABLE 18.1

Comparison of radiation-induced swelling of initial materials and materials subjected to explosion

MATERIAL	F, N/CM²	$T_{IRRAD,}$ °C	$\frac{\Delta V}{Vcell}$, %
BeO, initial	$6{\cdot}10^{21}$	150	4.6
BeO, expl.			0.44
AlN, initial	$1.1{\cdot}10^{21}$	90	1.4
AlN, expl.			0.5
ZnO, initial	-//-	-//-	0.6
ZnO, expl.			0.3
$GdAlO_3$, init.	-//-	-//-	3.7
$GdAlO_3$, expl.			2.7
$Gd_2Ti_2O_7$, init.	-//-	-//-	1.3
$Gd_2Ti_2O_7$, exp.			1.1
graphite, init.	$1{\cdot}10^{22}$	475	0.48
graphite, expl.			0.16

In the twenty-first century, explosive wave treatment of nonmetallic materials will probably increase as a method for reducing radiation swelling. This new approach to materials processing will extend the range of materials that can be used in strong fields of ionizing radiation.

7 Conclusions

After the theoretical discovery of diffuse X-ray scattering in 1945, diffuse characterization became an integral part of irradiated materials science. Development of experimental methods has increased our understanding of radiation-induced statically displaced atoms and the crystal lattice distortions they

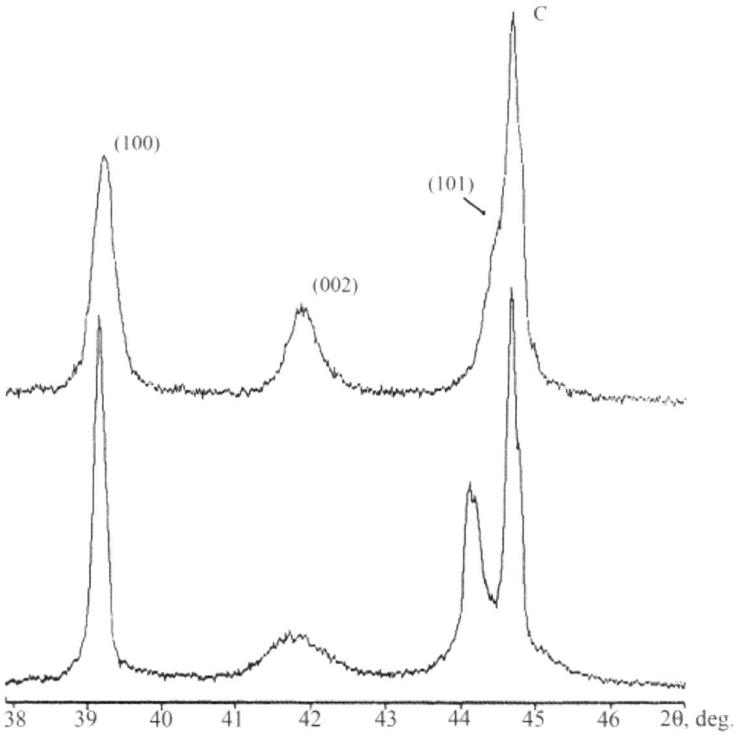

Figure 18.12. *X-ray powder patterns of irradiated BeO: initial (below) and after explosion (above). C is the diamond's line.*

cause. Specialized techniques have allowed accurate X-ray measurements of "X-ray swelling" in non-metallic materials irradiated in nuclear reactors. Based on these experimental and theoretical methods, a clearer understanding has emerged of how to limit changes in the volume of irradiated ceramics by their pretreatment by explosive wave processing.

Acknowledgment

The authors acknowledge the Russian Foundation for Fundamental Research for partial financial support of these investigations.

References

1. V. M. Kosenkov. *Radiography in Reactor Material Science.* (Ulyanovsk: UlSu, 2006), 168.
2. I. F. Ferguson. "Developments in Physical Techniques for Nuclear Materials Science." (UKAEA Diffraction analysis Conference, 30th Meeting, Oxford, Sept. 28–30 1982, London, 1983).
3. H. Kleykamp and R. Pejsa. *J. Nucl. Mat.* 124 (1984): 56–63.

4. B. E.Warren and B. L. Averbach. "The effect of cold-work distortion on x-ray pattern." *J. Appl. Phys.* 21, no. 6 (1950): 595–599.
5 H. Ekstein. "Scattering of x-ray on the local distortion." *Phys. Rav.* 68, nos. 5, 6 (1945): 120–124.
6. M. A. Krivoglaz. *Diffraction of X-rays and Neutrons in Non-Ideal Crystals* (Kiev: Naukova Dumka, 1983), 407.
7. R. I. Barabash and M. A. Krivoglaz, "Influence of anisotropy on X-ray scattering by highly distorted ageing solid solutions," *Phys. Met. Metallogr.*, 51, no. 5(1981): 1–12; "X-ray scattering by polychrystals containing defects with Coulomb displacement fields," *Phys. Met. Metallogr.* 45, no. 1 (1978): 1–10.
8. R. B. Grabova, Yu. D. Goncharenko, and A. A. Kaznelson, "Application of the regularization method for determination of a 'genuine' profile of reflexes in an X-ray analysis," *Equipment and methods of X-ray analysis. L.: Machine-building* 23 (1980): 77–86.
9. G. E. Bacon and B. E. Warren. "X-ray investigation of the graphite after neutron irradiation," *Acta Crystallogr.* 9 (1956): 2604.
10. V. M. Kosenkov, A. A. Khudyakov, and S. A. Vorobjev, "Determination of reactor graphite swelling using the methods of hydrostatic weighing and X-ray analysis," *RIAR Transactions*, SCC RIAR edition, 4 (2002): 66–69.
11. S. B. Austerman and K. T. Miller. *Phys. Stat. Sol.* 11 (1965): 241.
12. C. J. Ball, R. G. Blake, D. J. Cassidy, and J. L. Woolfrey, "Neutron irradiation effects in perovskite," *J. Nucl. Mat.* 151, no. 2 (1988): 151–161.
13. V. M. Kosenkov, S. A. Vorobjev, and A. V. Kolesnicov. "Radiation swelling decrease by means of explosive wave." *J. Nucl. Mat.* 258–263 (1998): 1809–1811.
14. V. M. Kosenkov, A. V. Kolesnikov, and S. A. Vorobjev. "Irradiation swelling of explosively shocked materials." *J. Nucl. Mat.* 273 (1999): 228–231.

19

Small-Angle Scattering from Dislocation Structures

L. E. Levine[1]

G. G. Long[2]

1. *Materials Science and Engineering Laboratory, National Institute of Standards and Technology, Gaithersburg, MD 20899, USA*

2. *Advanced Photon Source, Argonne National Laboratory, Argonne, IL 60439, USA*

Abstract

Small-angle scattering (SAS) is one of the most extensively used X-ray and neutron techniques for microstructural characterization. However, its application to dislocations has been limited by several important factors, such as an inherently low scattering contrast and complex angular dependencies. Nevertheless, SAS can provide valuable information about dislocation structures that cannot be obtained using other experimental techniques. We briefly review the theory of SAS from dislocations and dislocation walls, discuss the general experimental requirements for dislocation SAS experiments, and present results from example *in situ* studies on deformed Al single crystals.

1 Introduction

When a metal is plastically deformed, its mechanical properties change dramatically in a process known as work hardening. The underlying physical mechanism for work hardening involves the creation and evolution of complicated dislocation wall structures [1–4] such as two-dimensional "carpets" at modest strains and three-dimensional cells at larger strains. These structures impede the motion of mobile dislocations during straining [5,6]. This dislocation pattern formation [7,8], and the resulting work hardening, is very sensitive to intrinsic parameters such as the crystal structure, alloy composition, and the starting microstructure, as well as extrinsic parameters such as the orientation of the strain axis, sample temperature, applied strain, and strain rate.

Current theoretical approaches to understanding work hardening in bulk metal samples directly incorporate dislocation wall structure parameters [6,9–12]. However, experimental data on these wall structures is lacking because of inherent limitations in the available techniques. In transmission electron microscopy (TEM), for example, *in situ* studies can be conducted only on films with thicknesses less than about 0.3 μm. However, the formation of dislocation cells is a bulk process, and a minimum sample thickness of approximately 50 μm is required for *in situ* experiments. Other experimental techniques such as etch-pit studies [13], ultrasonic measurements [14], slip-line studies [15], atomic force microscopy measurements of slip-band evolution [16], resistivity measurements [17], positron annihilation spectroscopy [18] and photostimulated electron emission [19] have also provided very useful information, but these techniques have their own limitations, and TEM remains the most extensively used experimental technique for studying work-hardened metals [20–26].

Researchers have attempted to overcome some of the above experimental limitations by using X-ray and neutron scattering to study dislocation structures in plastically deformed samples. These probes are used on samples that are much thicker than TEM foils, and the experiments can be adapted to provide *in situ* data. The primary techniques that have been employed are profiling of Bragg peaks and SAS. Bragg diffraction has been the most extensively used technique, and much excellent *ex situ* work has been published on both polycrystalline and single-crystal metals. A small sampling of relevant experimental and theoretical Bragg scattering papers are referenced here [27–36]. Most recently, the rapid development of diffraction techniques using monochromatic and polychromatic synchrotron X-ray sources shows tremendous promise [37–39]. In particular, a new diffraction technique using depth-resolved submicrometer X-ray beams has provided the first quantitative elastic strain information at the individual dislocation cell level [40,41]. Although this technique is not yet directly applicable to studies of dislocation wall structure, these early microbeam experiments clearly demonstrate that some of the diffraction peak broadening observed during the plastic deformation of single-crystal Cu results from large cell-to-cell variations in the local elastic strain (and therefore stress) and not directly from the dislocation structures. Thus diffraction peak broadening analyses that neglect these cell strains may be leaving out important physics.

The primary emphasis of this review article is the use of SAS techniques for characterizing dislocations and dislocation wall structures in plastically deformed metals. In common with all of the techniques described above, the use of SAS for dislocation studies has weaknesses that limit its usefulness. Nevertheless, this technique can provide unique and valuable structural data on dislocation walls that cannot be obtained using other experimental methods. The proper role of SAS for such studies is a complementary one in which multiple techniques are used synergistically to provide the detailed dislocation wall data required to guide and validate theories and models of dislocation structure evolution.

In materials science, a number of experimental methods exist for the determination of the microstructural morphology that controls a material's properties. Among these, small-angle X-ray and neutron scattering have proven valuable for the statistically significant determination of bulk microstructures. In many physical systems, the scattering contrast is adequately large and the microstructures of interest are adequately dilute so that characterization of features on the scale of 1 nm to 100 nm can be probed quite easily. Access to ultrasmall angles raises the upper size limit to micrometers in the case of X-rays and tens of micrometers in the case of neutrons. For plastic deformation, the classical formalisms of SAS are not available, because the features of interest are single dislocations, dislocation dipoles, and dislocation wall configurations. Each involves small redistributions of electronic charge, and thus a description of the scattering contrast requires theoretical development directed at these particular types of features. In addition, the measurements are challenging since contrast is small, and a number of competing high-contrast processes must be carefully avoided.

Among the earliest SAS experiments were attempts to study the dislocation microstructure of cold-worked metals [42,43]. Although not successful, these early experiments led the way to increasingly sophisticated experiments [44–53] and the development of new theoretical descriptions [54–59] of SAS from individual dislocations and dislocation wall structures. The primary difficulties that have limited the use of SAS for dislocation microstructure studies are (1) the inherently low scattering contrast of dislocations, (2) the angular dependence of the dislocation scattering, and (3) the intrusion of higher contrast processes, such as double-Bragg diffraction. In the following sections, we will summarize the current theoretical understanding of SAS from dislocations and dislocation walls, describe the experimental conditions required for obtaining useful data from complex dislocation microstructures, and provide examples of the data analysis process for *in situ* deformation experiments on single-crystal Al.

The experiments reported in this chapter were conducted using the ultrasmall-angle X-ray scattering (USAXS) facility at the National Institute of Standards and Technology's materials science beamline (X23A3) at the National Synchrotron Light Source (NSLS) at Brookhaven National Laboratory and the USAXS instrument on the UNICAT 33-ID beamline at the Advanced Photon Source at Argonne National Laboratory.

2 Theoretical Background

Small-angle scattering by dislocations originates from small changes in the electron density (or nuclear density for neutron scattering) induced by the long-range elastic dilatation field, density changes in the dislocation core, and the nonlinear part of the elastic field. Scattering from elastic dilatation is the dominant process, and the other contributions can generally be neglected. Thus the linear-elastic approximation applies, greatly simplifying the theoretical treatment. Also, since the dilatation field of a screw dislocation is nonexistent in this approximation, only the edge components of dislocations can generally be detected using SAS techniques.

The long-range dilatation field of an edge dislocation is smoothly varying and therefore cannot be approximated using interfaces that separate regions with distinct electron or nuclear density, as is usually assumed for SAS theories. Instead, the smoothly varying dilatation field must be explicitly included in the analysis. The theoretical treatment presented below closely follows that found in the work of Long and Levine [53] and Thomson et al. [59] and is also fully consistent with earlier work [54–58].

Scattering at small angles can be written in terms of the total scattering amplitude function, A, as

$$I = A(\mathbf{q})A^*(\mathbf{q}) \tag{19.1}$$

$$A(\mathbf{q}) = a_a \int \left[1 + \phi_{tot}(\mathbf{x})\right] \exp(i\mathbf{q}\cdot\mathbf{x})\,d\mathbf{x}\,,$$

where $\mathbf{q}$ is the scattering vector, a_a is the scattering amplitude for a unit volume of perfect crystal, and the integral is the amplitude structure factor for the scattering volume. As usual, q is defined by $q = |\mathbf{q}| = (4\pi/\lambda)\sin(\theta)$, where λ is the X-ray or neutron wavelength and 2θ is the scattering angle. Dislocations contribute a total fractional density change, ϕ_{tot}, at the point $\mathbf{x}$. The first term describes scattering from the sample as a whole, which is only visible at extremely low q. This can therefore be neglected, leaving

$$A(\mathbf{q}) = a_a \int \phi_{tot}(\mathbf{x}) \exp(i\mathbf{q}\cdot\mathbf{x})\,d\mathbf{x}\,. \tag{19.2}$$

Since ϕ_{tot} is the linear sum of the fractional density changes caused by the individual defects (dislocation, dislocation pair, etc.), the scattering intensity from a set of identical defects can be written as

$$I/|a_a|^2 = |a_d|^2 c, \tag{19.3}$$

where

$$a_d = \int \phi(\mathbf{x}) \exp(i\mathbf{q}\cdot\mathbf{x})\,d\mathbf{x} \tag{19.4}$$

and

$$c = \sum_{i,j} \exp\left[i\mathbf{q}\cdot(\mathbf{R}_i - \mathbf{R}_j)\right]. \tag{19.5}$$

In these equations, a_d is the structure factor of a single defect and c is a configuration factor that describes their spatial distribution (as in dislocation wall structures). Although Equation 19.4 is only valid for configurations of identical defects, the linear sum over the fractional density changes (allowed by the linear-elastic approximation) allows more complicated systems to be handled in an analogous way. Thus the corresponding configuration factor for complex dislocation walls produces a modulation of the structure factors of the individual dislocations.

The most important defects to be treated are individual dislocations and dislocation dipoles. The coordinate system we will use to describe the dislocation and dislocation wall geometry is shown in Figure 19.1. The edge component of the Burgers vector, b_e, defines the X axis and the dislocation line direction defines the Z axis. Thus the Y axis is perpendicular to the slip plane containing the dislocation. Assuming isotropic elasticity, the density change of an individual dislocation is given by [60]

$$\phi = b_e \kappa(\sin\theta)/u \tag{19.6}$$

$$\kappa = -\frac{1-2\nu}{2\pi(1-\nu)},$$

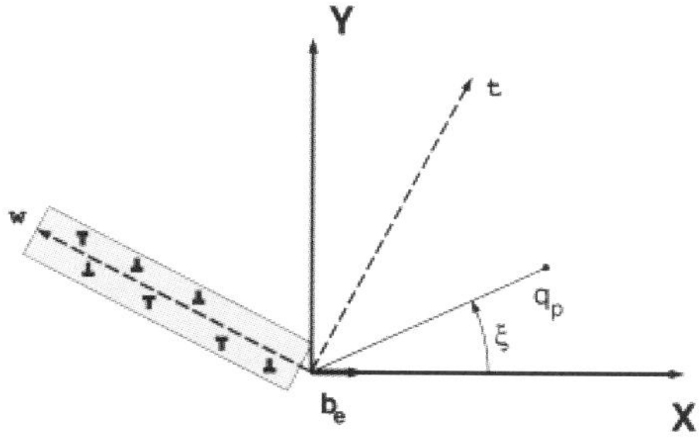

Figure 19.1. *Coordinate system for dislocations and dislocation walls. The dislocation lines define the Z axis, and the edge component of the Burgers vector defines the X axis. The projection of the scattering vector, q, in the XY plane has cylindrical coordinates (q_p/ξ). The vectors* **w** *and* **t** *are also in the XY plane with* **w** *parallel to the dislocation wall and* **t** *perpendicular to the wall.*

where θ and u are polar coordinates in the XY plane and ν is Poisson's ratio. Real dislocations are generally part of large loops that limit their extent along the Z axis. Thus Equation 19.4 is only integrated over a finite range, $-H$ to H, and this length scale plays a significant role in the long-range dilatation field of the dislocation. Thus substituting Equation 19.6 into Equation 19.4 and integrating gives

$$a_{\text{single}} = 4\pi b_e \kappa \sin\xi \left[\frac{J_0(q_p R) - J_0(q_p R_0)}{q_p} \right] \frac{\sin(q_z H)}{q_z}$$

$$\approx 4\pi b_e \kappa \frac{\sin\xi}{q_p} \frac{\sin(q_z H)}{q_z}, \qquad (19.7)$$

where a_{single} is the structure factor for a single dislocation, J_0 is the zero-order Bessel function, and q_z is the Z component of the scattering vector. The polar angle ξ and the radial distance q_p give the projection of the scattering vector onto the XY plane (see Figure 19.1). The angle is measured with respect to the edge component of the Burgers vector, which defines the X axis. R and R_0 are the upper and lower cutoffs for the dislocation strain field, respectively, and the last approximation is only valid when $q_p R >> 1$ and $q_p R_0 << 1$. These limits have practical implications that will be demonstrated in the experimental section of this chapter.

The dilatation field of a dislocation dipole is very different from that of an individual dislocation, and an analysis similar to the one shown above gives the following dislocation dipole structure factor:

$$a_{\text{dipole}} \approx \frac{8\pi b_e \kappa \sin^2\xi}{q_p^{3/2}} \left(\frac{2}{\pi d} \right)^{1/2} \frac{\sin(q_z H)}{q_z}. \qquad (19.8)$$

Here, the two dislocations comprising the dipole are displaced from one another by a distance $2d$ along the Y axis. Notice that the q_p dependencies in Equations 19.7 and 19.8 are very different, allowing scattering from individual dislocations and dislocation dipoles to be readily distinguished experimentally.

It is important to note that the factor $\sin(q_z H)/q_z$, appearing in both structure factors, has a very strong and sharp peak at $q_z = 0$ for dislocation lengths typical of deformed metals. Therefore, unless the scattering vector is within about 10° of being perpendicular to a given dislocation line, there will be no appreciable scattering. Similarly, the $\sin\xi$ terms in both structure factors describe the effect of rotating **q** with respect to the edge component of the Burgers vector (*X* axis); the scattering reaches a maximum when **q** is perpendicular to this axis. The scattering *intensity* (AA^*) for an individual dislocation falls off as $\sin^2\xi$ as compared with a much sharper $\sin^4\xi$ behavior for a dislocation dipole. In all cases, we see that the maximum scattering intensity occurs when the scattering vector is perpendicular to both the dislocation line and the edge component of the Burgers vector. This defines the slip planes of the dislocations, which, for fcc Al, are the {111} planes. The strong orientational dependence of SAS from dislocations is both a disadvantage and an advantage. Thus it is clear that SAS is not an appropriate tool for probing the total dislocation content of a deformed metal sample. However, the selective nature of dislocation SAS provides a unique experimental ability to compare the volume-averaged response to strain of dislocations on specific slip planes (latent hardening).

Dislocations in deformed fcc metals self-organize into wall structures during plastic deformation, and these dislocation arrangements greatly affect the SAS. In 1999, Thomson et al. [59] investigated the theory of SAS from dislocations arranged specifically in walls, and this analysis was later extended by Long and Levine [53]. The detailed derivations are beyond the scope of this chapter, so only the results will be presented. The general form predicted for the SAS intensity from a dislocation wall is

$$I = A^2\left[cL + F_t(q_t, L)F_w(q_w, L)\right] \tag{19.9}$$

$$A_{\text{single}} \approx \frac{4\pi\kappa}{q_p}\left(\frac{\sin(q_z H)}{q_z}\right)$$

$$A_{\text{dipole}} \approx \frac{8\pi\kappa}{q_p^{3/2}}\left(\frac{2}{\pi d}\right)^{1/2}\left(\frac{\sin(q_z H)}{q_z}\right),$$

where c is a constant, L is the total visible dislocation line length, q_t is the projection of the scattering vector perpendicular to the wall, q_w is a projection along the wall and perpendicular to the scattering dislocations, and $F_t(q_t, L)$ and $F_w(q_w, L)$ are factors that depend on the dislocation configurations in the respective directions. The ξ-dependent factors have been incorporated into c, F_w, and F_t, so the visibility criteria for single dislocations and dislocation dipoles remain unchanged.

$F_t(q_t, L)$ exhibits a relatively simple, monotonic behavior with q_t. For example, if the dislocation density in a wall is uniform with a sharp interface, then $F_t(q_t) \propto q_t^{-2}$. If the interface is diffuse, then the scattering intensity decreases more rapidly. Modeling the interface as the convolution of a sharp interface with a Gaussian function gives the approximate functional form

$$F_t(q_t) \propto q_t^{-2}\left(1 - \sigma^2 q_t^2\right), \tag{19.10}$$

where σ is the standard deviation of the Gaussian that corresponds to the effective width of the diffuse interface.

$F_w(q_w, L)$ is more complex and is best described as an interference function for the partially ordered dislocation structures. If the sample is oriented so that q_w from a given wall is appreciable, then $F_w(q_w, L)$ is highly sensitive to the ordering of the visible dislocations within the walls. Two special cases have been examined that are relevant to the experiments described below. First, if $N + 1$ dislocations with identical Burgers vectors form a periodic arrangement with spacing l/N, then $F_w(q_w, L)$ produces a simple diffraction grating function given by

$$F_w(q_w) \propto \frac{\sin^2(q_w l/2)}{\sin^2(q_w l/2N)}. \tag{19.11}$$

Second, an identically conFigured wall with alternating sign dislocations gives

$$F_w(q_w) \propto \frac{\sin^2(q_w l/2)}{\cos^2(q_w l/2N)}, \tag{19.12}$$

which just shifts the diffraction pattern by half the period. Partially disordering the dislocations in these walls introduces broadening but does not affect the overall periodic behavior described by Equations 19.11 and 19.12.

Finally, the cL term in Equation 19.9 is very important since it describes the SAS from the individual dislocations and dislocation dipoles independent of their larger-scale arrangements. As mentioned before, c includes all of the angular factors for SAS from individual dislocations and dislocation dipoles.

Depending on the specific case in question, the rich functional behavior described above can be used to analyze most dislocation SAS measurements from deformed single-crystal fcc samples. A simple example that we have found particularly useful is the case where q_t is appreciable, q_w is small, the SAS is dominated by individual dislocations rather than dislocation dipoles, and the dislocations are arranged in walls with diffuse interfaces. In that case, the SAS intensity can be written as

$$I \propto A_1 q^{-2} + B_1 q^{-4}\left(1 - \sigma^2 q^2\right), \tag{19.13}$$

where A_1 and B_1 are constants. Since the modified q^{-4} term from dislocation walls falls off much more rapidly than the q^{-2} term from individual dislocations, both behaviors can be examined independently by looking at different q ranges. Note that as the wall interface width goes to zero, the q behavior of the second term in Equation 19.13 becomes q^{-4}.

3 Experimental Considerations

Small-angle scattering from dislocations and dislocation wall structures is most significant in the q range of 10^{-4} Å^{-1} to 10^{-2} Å^{-1}, so all of the experiments described in this chapter were conducted using a modified Bonse-Hart USAXS instrument [61]. In this instrument, the photon energy is selected using a double-crystal monochromator, and higher order harmonics are rejected using a pair of grazing-incidence mirrors. A pair of diffracting crystals just before the sample serves as a collimator (beam conditioner), and the sample-incident beam intensity is continuously monitored using a transmission ionization chamber. After diffraction from a double-crystal analyzer, the scattered X-rays are detected by means of a photodiode detector with a linear response from 2000 to 2×10^{12} photons s^{-1}. This wide dynamic range enables direct measurement of the instrument profile with no sample in place as well as the scattered intensity profile from the sample. During all USAXS scans, data are acquired both with and without the sample in the beam, allowing us to explicitly remove the instrument profile from the data and determine the sample transmission with a high degree of accuracy. Scattering intensity data were converted into an absolute-scaled differential scattering cross section (in units of cm^{-1} sr^{-1}). The effective q range for these experiments was 5×10^{-4} Å^{-1} to 1×10^{-1} Å^{-1}.

As discussed above, dislocation SAS is an inherently low-contrast scattering process, and it has strong angular dependencies. Thus it is important to minimize higher contrast processes such as double-Bragg diffraction while maximizing the dislocation SAS. Accidental Bragg diffraction is the most significant problem faced by a dislocation SAS experiment. Ideally, Bragg diffraction can be eliminated using a photon energy low enough so that no Bragg diffraction is possible. Unfortunately, the resulting penetration depth (even in low electron density metals such as Al) would be smaller than the sample thickness necessary to guarantee "bulk" dislocation structure evolution. Therefore, the selected photon energy is a trade-off between minimizing the opportunity for Bragg diffraction and maximizing sample thickness. The experiments described here used a photon energy of 7.0 keV that allowed us to use Al samples with thicknesses up to 200 μm. Since the Bragg peaks from plastically deformed metal samples are very broad, it is impossible to entirely avoid double-Bragg scattering from deformed polycrystalline samples using a conventional SAS or USAXS design. Highly spatially resolved USAXS techniques (such as USAXS imaging [62]) may ultimately allow such experiments to be conducted, but for now we are restricted to using single-crystal specimens for which the geometry can be more carefully controlled. Full three-axis angular control of the scattering geometry is required for aligning **q** perpendicular to the desired slip plane, for probing the relevant angular factors described above, and for avoiding Bragg diffraction conditions. Angular control in these experiments was provided by a three-axis rotation stage using *xyz* Euler angles, referred to as ζ (yaw), θ (pitch), and χ (roll). The laboratory-frame coordinate system is shown in Figure 19.2. All sample orientations in this chapter are given as (ζ, θ, χ).

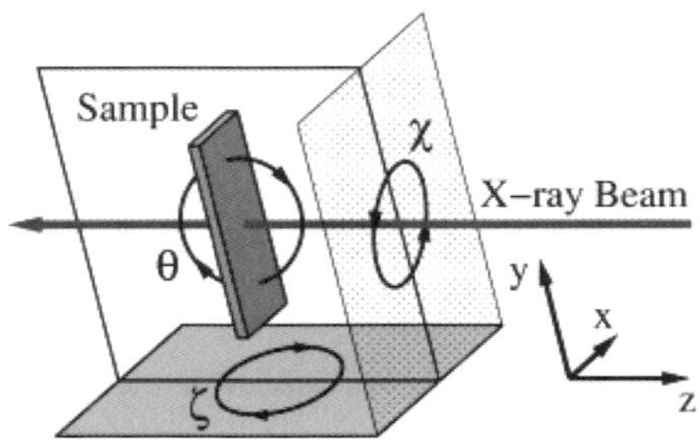

Figure 19.2. *Laboratory-frame coordinate system along with the xyz Euler angles used for sample rotations.*

4 Samples and Sample Orientations

The samples used in these experiments were ≈ 200 μm thick Al (99.99+%) single crystals in the shape of flat dog bones with gauge section dimensions of 3 mm × 3 mm. The short length was necessary to facilitate loading of the delicate samples. The sample width outside of the gauge section was 5 mm. The samples were thick enough to allow the evolution of bulk dislocation structures and thin enough for conducting USAXS measurements at a photon energy of 7.0 keV. Preshaped Al single crystals with random crystallographic orientations were grown by directional solidification in a soft mold of Al_2O_3 powder. Each crystal was sliced by an acid saw to produce several dog bone-shaped tensile specimens of identical crystallographic orientation. Samples were then chemically etched and annealed in vacuum for forty-eight hours at 350°C.

The samples were deformed incrementally in uniaxial tension during the experiment (*in situ*) using a computer-controlled tensile stage. The load and the relative displacement of the grips were monitored and recorded throughout the experiments. This allowed us to compare the USAXS data from the evolving dislocation structures with the changing mechanical properties. True stress and true strain were not calculated because the required sample geometry did not produce a completely uniform strain distribution within the sample gauge section. The final plastic strain in the sample gauge section was measured using a traveling microscope after the experiment for comparison with the *in situ* strain measurements. The elastic behavior of the tensile stage was measured *ex situ* by applying an elastic strain to a steel sample. This behavior was subtracted from the Al stress-strain data prior to analysis. All X-ray measurements were conducted while the sample was held just under the flow stress (≈0.99) because removing the load completely could affect the dislocation structures.

Since the orientation of the sample with respect to the beam is critical, a final accurate determination of the orientation was made *in situ* with the sample mounted in the USAXS instrument. To accomplish this, the 2 mm × 2 mm beam was centered on the sample gauge section and the forward-transmitted beam intensity was monitored using an X-ray video camera while the sample was rotated. When a diffracting condition occurs, part of the X-ray beam is diffracted away from the forward direction, causing a sudden decrease in the forward-transmitted intensity. The measured angles where diffraction occurs can then be used to determine the sample orientation with a one-standard-deviation uncertainty of approximately 0.2°. Figure 19.3 shows a plot of the measured sample angles where Bragg diffraction was detected from sample P (open circles), along with theoretical angles calculated using the final determined sample orientation and photon energy (dark curves).

During *in situ* deformation, lattice rotations can occur, causing the unexpected intrusion of Bragg peaks. A process similar to that described above has been shown to be effective at detecting Bragg peak contamination before it becomes appreciable [53]. Briefly, the sample transmission is monitored by measuring the forward-transmitted beam both with and without the sample present and accounting for the sample orientation and strain state. Encroaching Bragg diffraction redirects beam intensity away from the forward direction, causing an anomalous decrease in the measured transmission.

With the sample mounted and its orientation measured, sets of Euler angles are calculated for aligning the scattering vector perpendicular to specified {111} slip planes while avoiding Bragg diffractions. Generally, sample orientations are selected that are at least 3° away from the nearest Bragg diffraction. If no suitable orientations are possible for a desired slip plane, the photon energy can be fine-tuned to shift the Bragg diffractions away from the desired sample orientation.

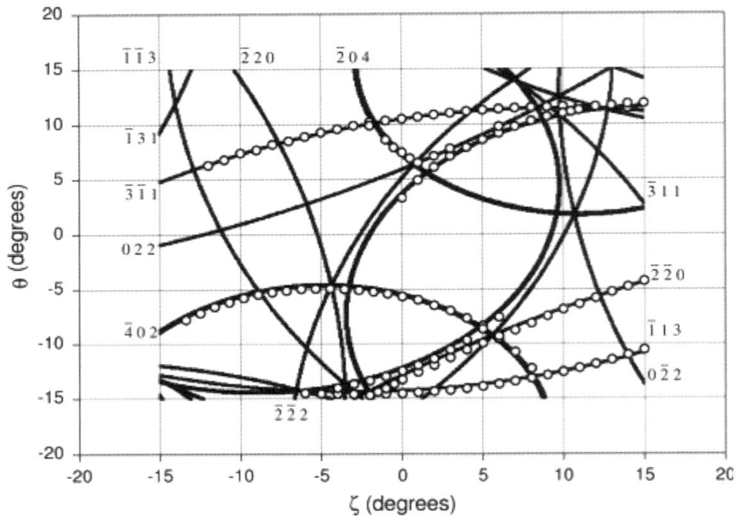

Figure 19.3. *Orientation plot for sample P with $\chi = 0°$. The open circles indicate angles where Bragg diffraction was detected and the dark lines are the corresponding theoretical calculations based on the final determined sample orientation and the photon energy. The experimental uncertainties are described in the text.*

5 Experimental Results

This section describes a series of dislocation USAXS measurements that have been selected to provide a useful overview of the kinds of data that can be obtained. Some of these USAXS data and analyses have been reported previously [53]. All of the USAXS data were obtained from four Al single crystals that were deformed incrementally *in situ* as described above. The first two crystals (P and Q) were oriented for high dislocation SAS contrast, with **q** aligned perpendicular to slip planes with active slip systems. The third crystal (G) was oriented such that **q** was still nearly perpendicular to a large population of dislocations but not perpendicular to the edge components of their Burgers vectors (medium contrast). The resulting USAXS data from these three crystals show pronounced changes in dislocation structure throughout the *in situ* experiments, and quantitative results are extracted. The results include measurements of positional correlations between dislocations in walls, the changing width of the wall/channel interface, dislocation dipoles, local inhomogeneity of the dislocation microstructure, dislocation structure changes during creep, and changes in dislocation density. Finally, data are presented from a fourth crystal (J) that was oriented such that dislocation SAS contrast was low. Table 19.1 lists information relevant to the samples for reference.

5.1 High-Contrast Orientations

Sample P was deformed uniaxially in tension along the crystallographic direction [0.66 0.42 0.62]. The slip systems with the highest resolved shear stresses (largest Schmid factors [63]) are therefore

TABLE 19.1

Sample information

SAMPLE	STRAIN AXIS	STRAIN RANGE	ORIENTATION	DESMEARED POWER LAW
P	[0.66 0.42 0.62]	10%	**F** = (0.0°, 1.7°, 60.2°) **J** = (−13.7°, 0.8°, 59.7°)	−2 and −4
			A = (10.1°, −24.4°, −20.4°)	−2.3
Q	[−0.85 −0.49 0.18]	2.3% to 6%	**H** = (10.1°, −25.4°, −20.4°)	−3
			M = (10.1°, −26.4°, −20.4°)	−2
G	[0.03 0.85 −0.53]	0.4% to 6.7%		−2
J	[0.03 0.85 −0.53]	5.8% to 11.7%		

$[011](\bar{1}1\bar{1})$ and $[110](\bar{1}1\bar{1})$, with calculated Schmid factors of 0.37 and 0.38, respectively. The slip systems with the largest Schmid factors are those expected to exhibit the greatest dislocation activity. If multiple slip systems have nearly equal Schmid factors (as with sample P), then the sample starts out in "multiple slip," where all such slip systems have roughly equivalent dislocation activity when averaged over the entire sample.

With the Euler angle χ fixed, an orientation plot using the remaining two Euler angles was produced as shown in Figure 19.4, where $\chi = 60.2°$. Sample orientations $\mathbf{F} = (0°, 1.7°, 60.2°)$ and $\mathbf{J} = (-13.7°, 0.8°, 59.7°)$ align the $(\bar{1}1\bar{1})$ slip plane perpendicular to **q**, while avoiding the intrusion of Bragg peaks.

Figure 19.5 shows absolute-calibrated slit-smeared USAXS data from orientation **F** at 10% applied strain. For comparison, the theoretical slit-smeared slopes equivalent to –2 for scattering from individual dislocations and –4 for scattering from sharp dislocation walls are also shown. The observed slopes are in excellent agreement with the theoretical predictions [53,59]. The prominent peak structure was visible in all of the scans at this level of strain.

The peak structure in Figure 19.5 demonstrates the presence of spatially ordered dislocations. To interpret this structure, we recall that the data from the USAXS double-crystal instrument are intensity-averaged (slit-smeared) over an angular range defined by the geometry. The intensity averaging is greatest at the lowest values of q and decreases monotonically with increasing q until the camera length is reached. The camera length for the present measurements was 0.028 Å^{-1}. Beyond this value of q, the data are the same as from pinhole (nonsmeared) SAS.

Figure 19.6 shows the extracted peaks from sample P at orientation **F** and 10% strain. To extract the peaks, we first fitted and subtracted the high-q background and the slit-smeared equivalent q^{-4} component. We then fitted a q^{-2} power law to the resulting data and subtracted it. The smooth curve shown in Figure 19.6 is a polynomial fit to the data. The asymmetry of the peaks and the height above the zero line are attributed to instrumental smearing as described above. The peak positions are $0.0033\ \text{Å}^{-1} \pm 0.0015\ \text{Å}^{-1}$ and $0.0069\ \text{Å}^{-1} \pm 0.0020\ \text{Å}^{-1}$. These positions are consistent with a nearly periodic sequence of like-signed dislocations as described in section 2.

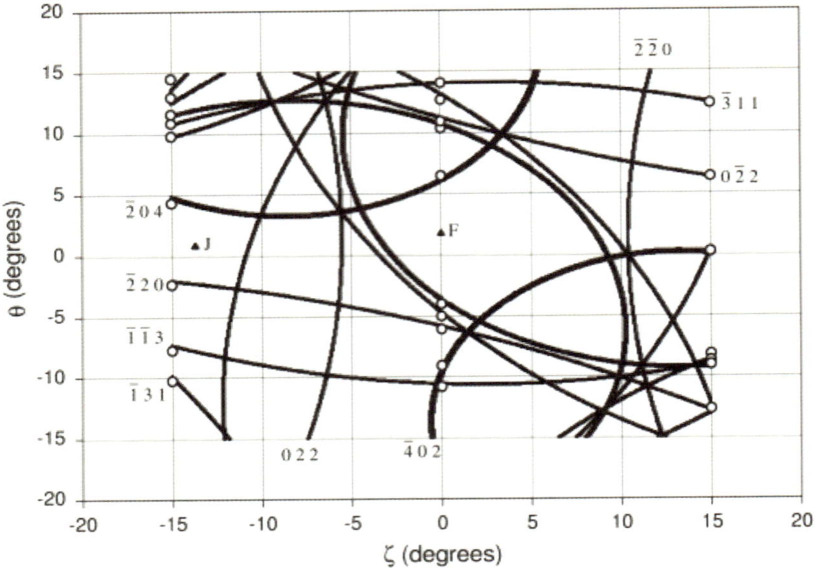

Figure 19.4. *Orientation plot for sample P with χ = 60.2°. The open circles indicate orientations where Bragg reflections were observed. The solid triangles indicate orientations from which USAXS data are presented.*

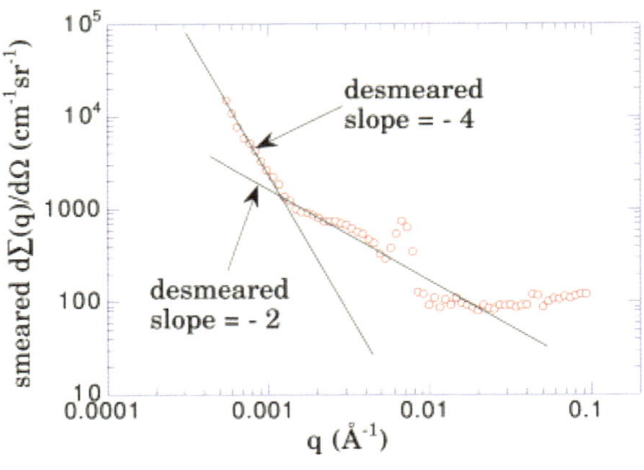

Figure 19.5. *Slit-smeared ultrasmall-angle X-ray scattering data from sample P, strained in situ to 10% strain. The straight lines indicate the equivalent desmeared slopes from dislocation wall SAS. The peak structure is produced by correlations in the dislocation spacing.*

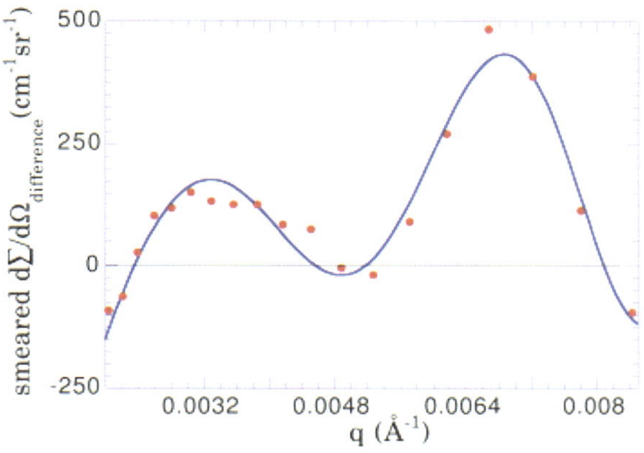

Figure 19.6. *Peak structure from Figure 19.5, with power law and background contributions subtracted. The one-standard-deviation statistical uncertainties are approximately 80 cm^{-1} sr^{-1}.*

The positions of the correlation peaks correspond to an apparent dislocation spacing of 1820 Å ± 130 Å . However, since q_w is the projection of the scattering vector within the dislocation walls, knowledge of the orientation of the walls is required to extract the true separation. Although the existing data set does not allow this orientation to be uniquely determined, walls aligned perpendicular to the $(\bar{1}1\bar{1})$ slip plane would have the largest possible q_w, resulting in a large scattering signal. Any walls with radically different orientations would not contribute significantly to such a prominent USAXS peak structure. Also, the presence of nearly periodic like-signed dislocations strongly suggests that the walls are low-angle tilt boundaries. Such tilt boundaries would be expected to be aligned approximately perpendicular to the $(\bar{1}1\bar{1})$ slip plane. Making this assumption about the orientation of the walls allows us to calculate the mean lattice rotation across these walls to be $\theta_M = 0.090° \pm 0.006°$.

The widths of the peaks shown in Figure 19.6 are too large to be caused by finite wall lengths. Instead, the observed widths originate from disorder in the dislocation spacing (either within an individual wall or between multiple walls). Thus the full-width half-maximum (FWHM) of these interference peaks corresponds to a 0.04° FWHM of the underlying misorientation distribution.

Although a single peak structure dominates the scattering, USAXS data from local regions of the gauge section demonstrate that the dislocation structures are not homogeneous across the sample. With sample P at orientation **F** at 5% applied strain, the incident X-ray beam cross section was reduced to 1.0 mm × 1.0 mm, and USAXS data were obtained from two adjacent but nonoverlapping sample volumes. These data are presented in Figure 19.7. Although the counting statistics are relatively poor in these small scattering-volume scans, the scattering profiles are clearly very different, demonstrating that the underlying dislocation structures are inhomogeneous over this length scale.

Such inhomogeneity has important implications for making meaningful comparisons between USAXS scans at different orientations and strains. Detailed comparisons can only be made when the same scattering volume is used for all of the measurements. In all of our data, the beam is positioned on the sample with an accuracy of 10 μm before each scan is made. Since changing the sample orientation changes the projection of the beam onto the sample, perfect volume matching is impossible, but we estimate the resulting volume mismatch for a 10° sample rotation to be less than 1%.

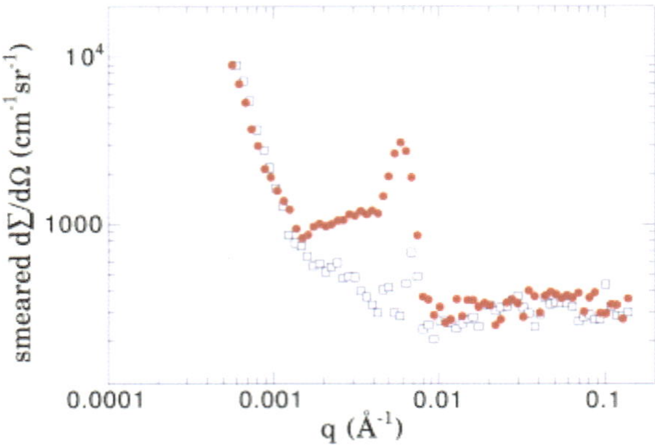

Figure 19.7. *Slit-smeared ultrasmall-angle X-ray scattering data from two adjacent nonoverlapping regions of sample P at 5% strain. The large changes in the peak structures indicate that the dislocation structures are inhomogeneous on millimeter-length scales.*

Sample Q was deformed uniaxially in tension along [–0.85 –0.49 0.18]. The slip systems with the highest resolved shear stress are therefore $[10\bar{1}](111)$, $[0\bar{1}1](111)$, and $[101](\bar{1}\bar{1}1)$, with calculated Schmid factors of 0.49, 0.32, and 0.42, respectively. The slip plane with the highest resolved shear stress was outside our rotation limits, so data were taken from the $(\bar{1}\bar{1}1)$ slip planes. Orientation $\mathbf{A} = (10.1°, -24.4°, \; -20.4°)$ aligns the $(\bar{1}\bar{1}1)$ slip plane perpendicular to $\mathbf{q}$ while avoiding interference from Bragg peaks. It must be kept in mind that single-crystal samples rotate during straining, so orientation **A** slowly shifts position with respect to the crystal axes. As we will describe below, this shifting was followed throughout the later stages of the straining process.

The USAXS data shown in Figure 19.8 were obtained from orientation **A** at a strain of 2.3%. These data show a clear q^{-2} behavior indicative of scattering from individual dislocations and a single weak peak at $0.0027\ \text{Å}^{-1} \pm 0.0002\ \text{Å}^{-1}$. The small size of this peak relative to that observed in sample P indicates that a smaller fraction of visible dislocations are periodically spaced. Straining to 5.9% shifts this peak to larger q $(0.0031\ \text{Å}^{-1} \pm 0.0002\ \text{Å}^{-1})$, reflecting a decrease in the visible dislocation spacing of approximately 13%.

At a strain of 11%, this single weak peak is replaced by a more prominent structure that changes over time. Figure 19.9 shows this scattering structure after subtraction of the power law components and background. The first scan shown in Figure 19.9 was obtained immediately after 11% strain was reached, and the second scan was taken 5.5 hours later. The two scans were made with an identical sample orientation and scattering volume. The differences between the two scans indicate that the local dislocation structures evolved while the sample was held at a nearly constant stress, just below the flow stress. Data from the tensile stage show that creep occurred during this period, resulting in an additional strain of $0.24\% \pm 0.06\%$.

The changes observed between the two scans in Figure 19.9 are consistent with a relaxation of the underlying dislocation structures. Over 5.5 hours, the peak position shifted to lower q, indicating that the visible dislocation spacing increased slightly (by about 10%); this was probably due to

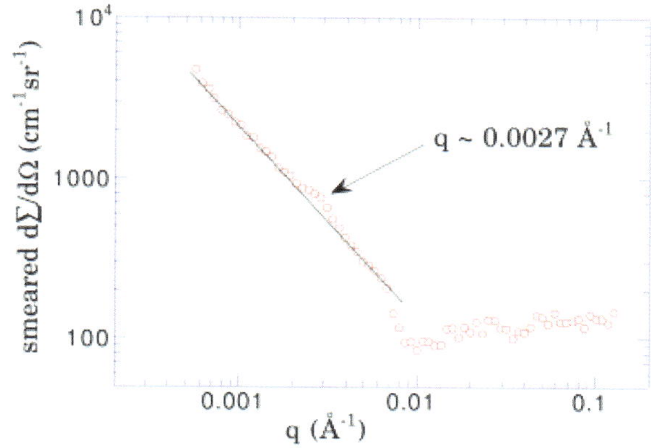

Figure 19.8. *Slit-smeared USAXS data from sample Q, strained in situ to 2.3% strain. The peak structure is produced by correlations in the dislocation spacing.*

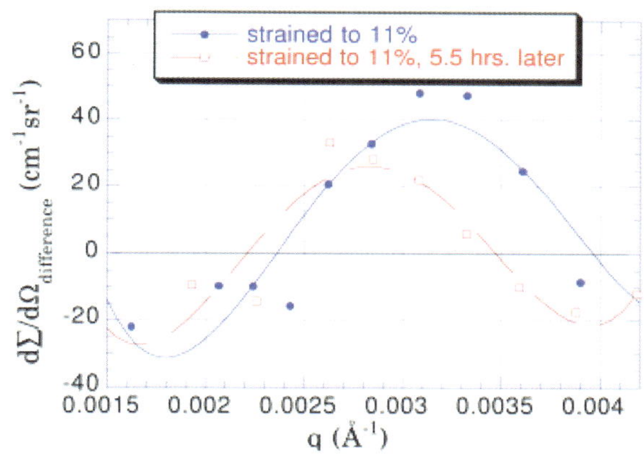

Figure 19.9. *Slit-smeared ultrasmall-angle X-ray scattering data from sample Q, strained in situ to 11% strain. The power law and background contributions have been subtracted. The differences between the two scans reflect the evolution of the dislocation structures during creep.*

thermal-activated climb and the subsequent annihilation of dislocations. The peak in the as-strained data is also broader than it is in the data taken 5.5 hours later. This narrowing suggests that the dislocations are shifting to more highly ordered positions as the dislocation structures relax to lower energy.

As shown by Equations 19.7 and 19.8, dislocation SAS theory predicts $a_{ind}^2 \propto q^{-2}$ for individual dislocations and $a_{diple}^2 \propto q^{-3}$ for dislocation dipoles. However, these equations are approximate expressions with well-defined regions of validity. For individual dislocations, Equation 19.7 is valid for the full range of q that we can reach in this experiment. For dipoles, however, Equation 19.8 is only valid when $qR_0 << 1$, where R_0 is the dipole spacing. Thus it is possible for us to detect the transition from dipole to individual dislocation behavior for dipole spacings larger than about 10 nm.

Figure 19.10 shows USAXS scans taken from orientations **A**, **H**, and **M** at 6% strain. Orientations **H** and **M** are obtained by rotating the sample away from **A** by –1° and –2°, respectively, in the θ direction. At this strain, orientation **H** aligns **q** perpendicular to the $(\bar{1}\bar{1}1)$ slip planes. Fitting a power law to the data from **H** gives a desmeared slope of -3.00 ± 0.05. This slope demonstrates that the arrangement of dislocations producing the scattering has a significant dipole moment. Rotating just ±1° from this orientation produces the lower scans shown in Figure 19.10 with fitted desmeared slopes of –2.3 for **A** and –2.0 for **M**. An additional rotation of 1° in the positive θ direction returns the orientation **A** slope to –2.0. From the predicted functional form for a^2_{dipole} in Equation 19.8, we thus obtain a lower limit of 3 μm for the length over which the dipoles satisfy the visibility requirements. This long length is consistent with dislocations trapped by geometrically necessary boundaries [64] or large cell walls. The high-*q* end of the dipole scattering exhibits a decreasing slope, starting at approximately $q = 0.004$ Å^{-1}. This gives an upper limit of about 20 nm for the dipole separation. Additional information concerning the orientation of the dipoles is required to extract further quantitative results such as the ratio of dipoles to individual dislocations. More extensive rotation data would allow these factors to be determined.

5.2 Medium-Contrast Orientations

Sample G was rotated to an orientation that was not perpendicular to a primary slip plane yet still exhibited significant excess scattering. Out of several attempts, this was the only sample where we succeeded in locating a nonslip plane orientation where **q** remained nearly perpendicular to a significant number of dislocations. No further rotations were attempted on this sample; instead, the USAXS was measured for a single orientation at applied strains of 0%, 0.2%, 0.4%, 0.7%, 0.9%, 1.3%, 1.8%, 2.2%, 2.7%, 3.6%, 4.4%, 5.3%, 6.7%, and 8.0%. The data from the 8.0% strain were not used in the

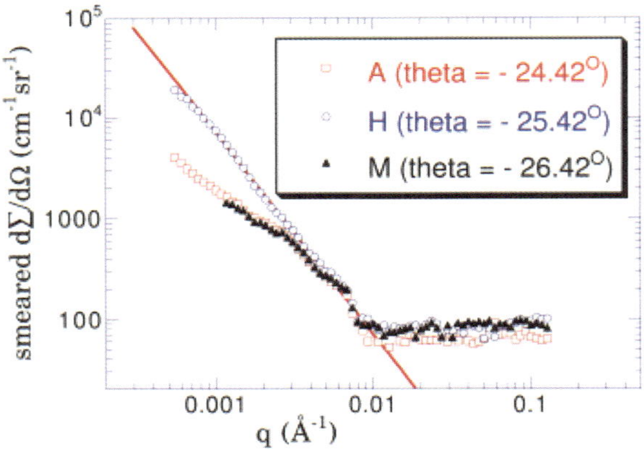

Figure 19.10. *Slit-smeared ultrasmall-angle X-ray scattering data from sample Q, strained in situ to 6% strain. The three scans were taken from orientations just 1° apart. The desmeared q^{-3} slope demonstrates the presence of dislocation dipoles in the sample.*

analysis because encroaching double-Bragg diffraction was found to be significant for this strain. The sampled volume was 2 mm × 2 mm × 151 μm.

With an angular uncertainty of approximately 0.5°, both the tensile axis and the scattering vector for sample G were [0.03 0.85 –0.53]. This scattering vector was approximately 3° away from the nearest Bragg diffraction.

Information on the orientations of the visible dislocations and dislocation walls is required to interpret the scattering results. Two facts are clear immediately. First, the requirement that the dislocation lines reside on slip planes means that they are perpendicular to <111> directions in the crystal. Second, for visibility, the dislocations must also be nearly perpendicular to **q**. Calculating the resulting $\sin^2(\xi)$ (after squaring Equation 19.7 to get the scattering intensity) for all four possible slip planes gives 0.04 for (111), 0.61 for $(\bar{1}1\bar{1})$, 0.66 for $(\bar{1}\bar{1}1)$, and 0.03 for $(1\bar{1}\bar{1})$. Thus the visible dislocations must lie on either the $(\bar{1}1\bar{1})$ or $(\bar{1}\bar{1}1)$ slip planes. These are also the slip planes with the highest (and nearly equal) resolved shear stresses; the relevant Schmid factors are 0.42 for [1 0 1] $(\bar{1}\bar{1}1)$ and 0.37 for [1 1 0] $(\bar{1}1\bar{1})$.

An earlier *in situ* high-resolution diffraction imaging experiment was conducted using monochromatic topography [65] on an identically oriented single-crystal tensile specimen [66] to determine the geometry of the wall structures at these low strains. The resulting images showed that the dislocations formed wall structures parallel to the $(\bar{1}\bar{1}1)$ and $(\bar{1}1\bar{1})$ slip planes. Such structures have been observed previously in *ex situ* transmission electron microscope studies of deformed fcc metals and are usually referred to as dislocation carpets [21].

Figure 19.11 shows the absolute-calibrated slit-smeared USAXS intensity plotted as a function of q for strain values of 0.4%, 0.9%, 2.2%, 3.6%, 4.4%, 5.3%, and 6.7%. Each scan exhibits two straight lines that intersect near q = 0.001 Å^{-1}. As before, the data at larger q (around 0.001 Å^{-1} to 0.01 Å^{-1}) have a slit-desmeared equivalent slope of –2 and correspond to scattering from individual dislocations. The changing height of the –2 slope region indicates that the visible dislocation content is increasing

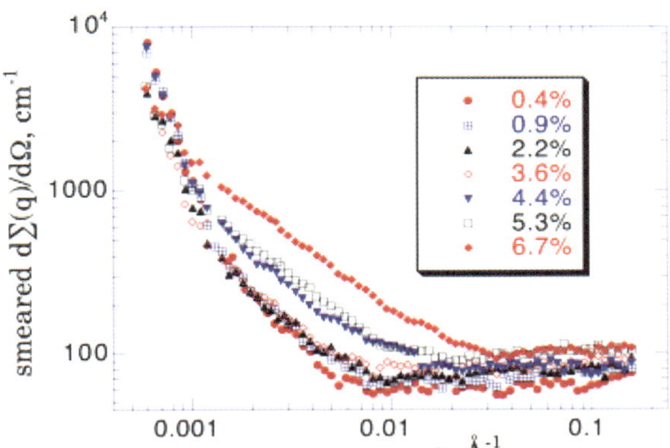

Figure 19.11. *Slit-smeared ultrasmall-angle X-ray scattering data from sample G at strains of 0.4%, 0.9%, 2.2%, 3.6%, 4.4%, 5.3%, and 6.7%.*

with strain. At smaller q, the large wall structures become visible, resulting in a different scattering behavior. This behavior will be treated quantitatively below.

To determine how the visible dislocation line length varies as a function of the strain, we integrated the larger q scattering data in the range $0.0015\ \text{Å}^{-1} \leq q \leq 0.003\text{Å}^{-1}$. This integral should be proportional to the visible line length. Figure 19.12 shows this integrated intensity plotted as a function of the plastic strain. The low dislocation density seen at 0% strain reflects the relatively low dislocation density of the starting crystal. Application of a strain of just 0.4% causes the density of visible dislocations to rise to a value that remains constant out to approximately 3.6% strain. At larger strains, the visible dislocation line length increases dramatically. Since it is expected that the total dislocation line length is increasing from 0.4% to 3.6% strain, this behavior implies that the visible dislocations do not belong to the principal slip systems (although they do reside on a plane with a high resolved shear stress). It is possible that 3.6% strain corresponds to a structural change in the dislocation structure.

The smaller q scattering is attributed to the distribution of dislocations in the thickness direction of the walls, and these structures are observed in all scans of sample G except for the first two at 0% and 0.2% strain. To extract the width, σ, of the wall/channel interface from the desmeared smaller q USAXS data, we use Equation 19.13 for the scattering intensity from dislocation walls. We first fit the larger q data with a power law and subtract this function from the smaller q data. This subtraction effectively removes the first term from Equation 19.13. Fitting the resulting data and including the appropriate projection for q_t allows us to extract the width of the interface region in the direction perpendicular to the dislocation walls. Figure 19.13 shows the resulting σ values (with one-standard-deviation uncertainties) plotted as a function of plastic strain. These data are consistent with an interface width that starts at around 1100 Å and decreases systematically with increasing strain. Beyond 4.4% strain, the number of data points in the low-q region becomes too small for reliable fitting. Additional data supporting the conclusion that the interface width decreases with increasing strain will be presented later in this section. The measured magnitude and systematic behavior of the diffuse

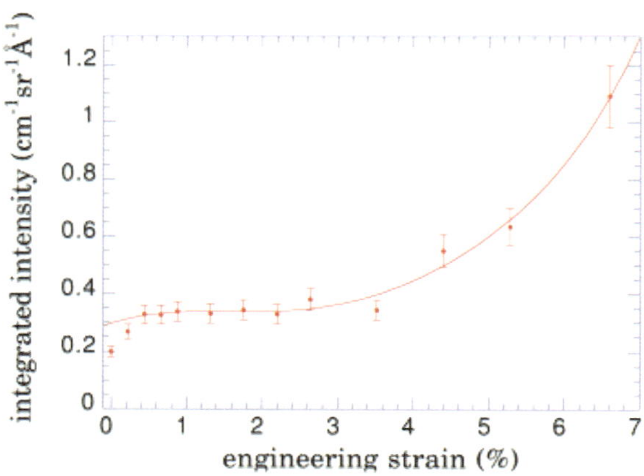

Figure 19.12. *Sample G integrated scattering intensity from $0.0015\ \text{Å}^{-1} \leq q \leq 0.0030\ \text{Å}^{-1}$ plotted as a function of the average linear strain. This integrated intensity is proportional to the total line length of the visible dislocations. The solid curve is a polynomial fit to the data above 0.2% strain.*

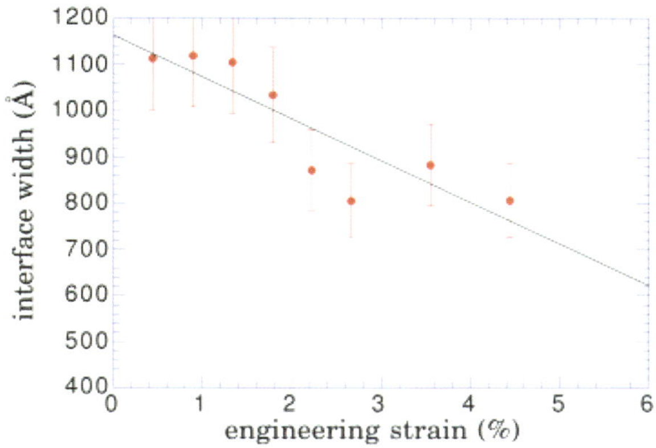

Figure 19.13. *True interface widths of dislocation wall boundaries in sample G plotted as a function of strain. These widths were extracted from the lower q scattering data using Equation 19.13. The solid line is a linear fit to the data.*

width are consistent with empirical models of dislocation structure evolution. The line in Figure 19.13 is a linear fit to the displayed data.

In the course of these studies, a number of parameters (such as the distances between ordered dislocations and the widths of the diffuse interfaces between the dislocation walls and the cell interiors) were extracted from the USAXS data. However, the parameter that has received the most attention over the years—that is, the average dislocation density—was not determined. It is clear from the discussion on dislocation visibility in section 2 that SAS is not an appropriate technique for determining this parameter. This can be considered a strength rather than a weakness, since SAS is the only available probe that can directly measure dislocation parameters from well-defined subsets of the total dislocation population. Even for these subsets, however, measuring the total visible dislocation line length, L, is a difficult undertaking. Extracting L from a q^{-2} scattering profile, for example, requires accurate estimates of various angular factors, the lengths over which the dislocations satisfy the visibility conditions, and the projection of the scattering vector along the dislocation lines. Complete determination of the appropriate parameters requires information from other experimental techniques, such as diffraction imaging and TEM. Since such information is not yet available, the extraction of L from the USAXS data is not yet possible. This is why the results in Figure 19.12 are displayed as integrated intensities that are proportional to L. Finally, it should be noted that SAS is only sensitive to the edge components of the dislocations. Thus any extracted value for L would not include the screw components of the dislocations.

5.3 Low-Contrast Orientations

As discussed in section 4, the sample preparation procedure allows us to produce multiple samples of identical crystallographic orientation. Sample J has the same crystallographic orientation as sample G, but USAXS data were obtained from a different sample orientation that still avoids Bragg scattering. Unfortunately, an electrical noise problem occurred during the early part of this experiment and data from strains below 5.8% are not available.

As expected, the scattered intensity from sample J (which is in a low-contrast orientation) was much lower than the scattered intensity from sample G, even when the strain in J was significantly higher. Nevertheless, a small strain-dependent scattering signature was detectable and is consistent with dislocation and dislocation wall scattering. The slope exhibited by sample J was not quite the theoretical desmeared slope of –2. The desmeared equivalent slopes ranged from –2.26 to –2.10, which was different from the steady –2.0 slope exhibited by all samples oriented for high dislocation scattering contrast. The deviation was probably due to a small additional scattering component such as a dislocation multipole configuration. The scattered intensity in this region slowly but systematically increased with increasing strain and strongly resembled the behavior observed in sample G.

As discussed previously, the low-*q* USAXS data from sample G exhibited changes that were consistent with a sharpening interface of the dislocation walls with increasing strain. The low-*q* USAXS data from sample J showed similar behavior, and the interface width was extracted using the same procedure. For sample G, **q** was symmetrically oriented with respect to both possible wall orientations, so the true interface width, σ_t, could be extracted. For sample J, however, the projections for the $(\bar{1}1\bar{1})$ and $(\bar{1}\bar{1}1)$ dislocation walls were slightly different (0.77 and 0.86, respectively). Figure 19.14 shows the extracted interface widths using both possible projections for sample J along with the σ_t values obtained from sample G. The data from both samples fit smoothly together, demonstrating that the same sets of dislocation walls are being sampled in the two crystallographically identical samples. Since both wall projections for sample J are consistent with the sample G data, no distinctions can be made between them. The combined data show that the width of the interface region decreases approximately linearly with increasing strain. Theoretical models of work hardening are just now starting to consider the role of such evolving dislocation density gradients [9]; at present, USAXS provides the only viable experimental technique for measuring this parameter.

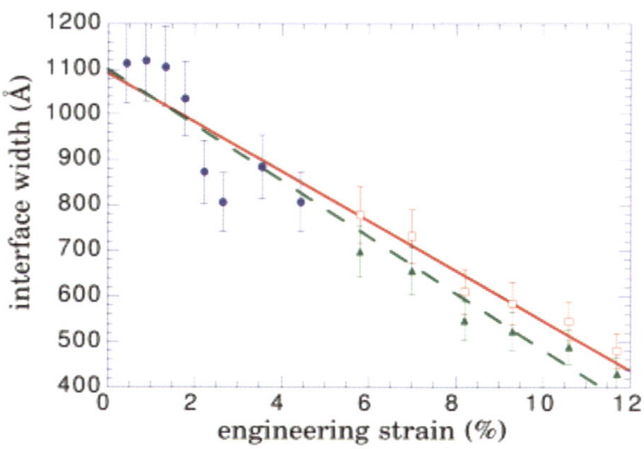

Figure 19.14. *True interface widths (with one-standard-deviation uncertainties) of dislocation wall boundaries in crystallographically identical samples G and J plotted as a function of strain. The* $(\bar{1}1\bar{1})$ *(open squares) and* $(\bar{1}\bar{1}1)$ *(solid triangles) dislocation walls have different projections, and the data cannot decide between them. The solid line is a linear fit using the* $(\bar{1}1\bar{1})$ *data along with the sample G data. The dashed line uses the* $(\bar{1}\bar{1}1)$ *data along with the sample G data.*

6 Conclusions

This chapter had three primary objectives. First, we presented the theoretical background required for the quantitative interpretation of SAS data from dislocations and dislocation walls. Second, we described the experimental conditions that we use to measure SAS from single crystal Al samples deformed *in situ*. Necessary factors to consider are the q range, the orientation of the scattering vector with respect to the slip planes, and avoidance of higher scattering contrast processes such as double-Bragg diffraction. Finally, selected dislocation SAS measurements were presented to demonstrate how dislocation-structure parameters may be extracted using *in situ* USAXS. The data were used to derive the positional correlations between dislocations, to measure the changing interface width of the dislocation walls, to detect the presence of dislocation dipoles, to examine the inhomogeneity of the microstructure, to measure the increase in the secondary dislocation density with strain, and to follow dislocation structure changes that occurred during creep at room temperature. Orienting the scattering vector perpendicular to a given slip plane maximizes the scattering from this set of dislocations while minimizing the contribution from the conjugate slip systems. This is responsible for the selective nature of the USAXS probe. For this reason, USAXS cannot be used to measure the total dislocation content of a sample. Instead, USAXS may be used for the purpose of comparing well-defined subsets of dislocations that develop and evolve as a sample deforms plastically. Experiments on latent hardening are, perhaps, the most immediately obvious application of this capability.

Although these experiments were conducted on high-purity Al samples, dislocation USAXS is applicable to a wide range of single-crystal systems (including solid-solution alloys of virtually any fcc metal). Crystal systems in which perfect screw dislocations dominate the work-hardening process (such as bcc metals at low homologous temperatures) would be less suitable since dislocation USAXS is only sensitive to the edge components of the dislocations. Dislocation USAXS could also be used to study dislocation evolution within individual large grains in polycrystalline samples. Finally, dislocation USAXS could be used to characterize dislocation structures in nonmetallic systems such as semiconductors. In this case, the dislocation structures are typically grown in rather than produced through plastic deformation.

Dislocation USAXS measurements are complementary to other experimental techniques. This is primarily due to the high sensitivity to various angular factors, including the orientation of $\mathbf{q}$ with respect to both the dislocation walls and the component dislocations. Samples G and J are good examples of this. High-resolution diffraction-imaging experiments were conducted on an identically oriented sample and showed that dislocation walls were formed parallel to the $(\bar{1}1\bar{1})$ and $(\bar{1}\bar{1}1)$ slip planes. This additional information allowed the true mean width of the dislocation wall interfaces to be extracted from the USAXS data. Data from the two samples were consistent and demonstrated that the interface width decreased approximately linearly with increasing strain. These results could not have been obtained without using complementary experimental information.

Acknowledgments

We thank H. E. Burdette for measuring the orientation of our single-crystal samples using Laue diffraction. The NSLS is operated under contract no. DE-AC02-76CH00016 with the U.S. Department of Energy (DoE). Research at the Advanced Photon Source is supported by the U.S. DoE, Office of Science, Office of Basic Energy Sciences under contract no. DE-AC02-06CH11357.

References

1. L. Kubin, "Dislocation patterning," in *Materials Science and Technology: A Comprehensive Treatment*, vol. 6, ed. H. Mughrabi (Weinheim, Germany: VCH, 1993), 137.
2. S. J. Gil, "Flow stress and work hardening," in *Treatise in Materials Science and Technology: A Comprehensive Treatment*, vol. 6, ed. H. Mughrabi (Weinheim, Germany: VCH, 1993), 19.
3. A. S. Argon, in "Mechanical Properties of single-phase crystalline media: Deformation at low temperatures," in *Physical Metallurgy*, 4th ed., vol. 3, ed. R. W. Cahn and P. Haasen (Amsterdam: Elsevier, 1996), 1877.
4. D. Kuhlmann-Wilsdorf, *Mat. Res. Innovat.* 1 (1998): 265.
5. R. Thomson and L. E. Levine, *Phys. Rev. Lett.* 81 (1998): 3884.
6. L. E. Levine and R. Thomson, *Mater. Sci. Eng. A* 400–401 (2005): 202.
7. P. Hähner, K. Bay, and M. Zaiser, *Phys. Rev. Lett.* 81 (1998): 2470.
8. D. Gomez-Garcia, B. Devincre, and L. P. Kubin, *Phys. Rev. Lett.* 96 (2006): 125503.
9. H. Mughrabi, *Mater. Sci. Eng. A* 317 (2001): 171.
10. R. Sedlacek and W. Blum, *Comput. Mater. Sci.* 25 (2002): 200.
11. F. Roters, D. Raabe, and G. Gottstein, *Acta Mater.* 48 (2000): 4181.
12. M. Zaiser, *Mater. Sci. Eng. A* 249 (1998): 145.
13. F. Y. Young Jr., *J. Appl. Phys.* 32 (1961): 192.
14. G. A Alers and K. Salama, "Interaction of dislocations with high-frequency sound waves," in *Dislocation Dynamics*, ed. A. R. Ronsenfield, G. T. Hahn, A. L. Bement Jr., and R. I. Jaffee (New York: McGraw-Hill, 1968), 211.
15. H. G. Van Bueren, *Imperfections in Crystals* (Amsterdam: North-Holland, 1961), chapter VI.
16. D. E. Kramer, M. F. Savage, and L. E. Levine, *Acta Mater.* 53 (2005): 4655.
17. H. G. Van Bueren, *Imperfections in Crystals* (Amsterdam: North-Holland, 1961), chapter XV.
18. E. Hashimoto, Y. Ueda, N. Uematsu, M. Iwami, and T. Kino, *J. Phys. Soc. Jpn.* 61 (1992): 3799.
19. M. Cai, L. E. Levine, S. C. Langford, and J. T. Dickinson, *Mater. Sci. Eng. A* 400–401 (2005): 476.
20. Z. S. Basinski, *Discuss. Faraday Soc.* 38 (1964): 93.
21. J. W. Steeds, *Proc. Roy. Soc. A* 292 (1966): 343.
22. E. Göttler, *Phil. Mag.* 28 (1973): 1057.
23. F. Prinz and A. S. Argon, *Phys. Stat. Sol. A* 57 (1980): 741.
24. C. T. Young, T. J. Headley, and J. L. Lytton, *Mater. Sci. Eng.* 81 (1986): 391.
25. D. A. Hughes, *Acta Metall. Mater.* 41 (1996): 1421.
26. B. W. Lagow, I. M. Robertson, M. Jouiad, D. H. Lassila, T. C. Lee, and H. K. Birnbaum, *Mater. Sci. Eng. A* 309 (2001), 445.
27. T. Ungár, H. Mughrabi, D. Rönnpagel, and M. Wilkens, *Acta Metall.* 32 (1984): 333.
28. H. Mughrabi, T. Ungár, W. Kienle, and M. Wilkens, *Phil. Mag. A* 53 (1986): 79.
29. T. Ungár and A. Borbély, *Appl. Phys. Lett.* 69 (1996): 3173.
30. T. Ungár, I. Groma, and M. Wilkens, *J. Appl. Cryst.* 22 (1989): 26.
31. M. Wilkens, K. Herz, and H. Mughrabi, *Z. Metallkd* 71 (1980): 376.
32. S. Straub W. Blum, H. J. Maier, T. Ungar, A. Borbély, and H. Renner, *Acta Mater.* 44 (1996): 4337.
33. M. A. Krivoglaz and K. P. Ryaboshapka, *Fiz. Metal. Metalloved.* 15, 18. Engl. transl: *Phys. Metals Metallog.* 15 (1963): 14.
34. M. Wilkens, *Phys. Stat. Sol. A* 2 (1970): 359.
35. L. E. Levine and R. Thomson, *Acta Crystallogr. A* 53 (1997): 590.
36. I. Groma, *Phys. Rev. B* 57 (1998): 7535.
37. B. Jakobsen, H. F. Poulsen, U. Lienert, J. Almer, S. D. Shastri, H. O. Sørensen, C. Gundlach, and W. Pantleon, *Science* 312 (2006): 889.
38. H. Mughrabi and T. Ungár, *Nat. Mater.* 5 (2006): 602.
39. R. I. Barabash, G. E. Ice, and F. J. Walker, *J. Appl. Phys.* 93 (2003): 1457.
40. L. E. Levine, B. C. Larson, W. Yang, M. E. Kassner, J. Z. Tischler, M. A. Delos-Reyes, R. J. Fields, and W. Liu, *Nat. Mater.* 5 (2006): 619.
41. L. E. Levine, B. C. Larson, J. Z. Tischler, P. Geantil, M. E. Kassner, W. Liu, and M. R. Stoudt, *Zeitschrift für Kristallographie* (in press).

42. J. Blin and A. Guinier, *C. R. Acad. Sci.* 233 (1951): 1288.
43. J. Blin and A. Guinier, *C. R. Acad. Sci.* 236 (1953): 2150.
44. S. Hayes and R. Smoluchowski, *Appl. Sci. Res.* B4 (1954): 10.
45. W. W. Beeman, P. Kaesberg, J. W. Anderegg, and M. B. Webb, *Handb. der Physik* 32 (1957): 440.
46. H. H. Atkinson and R. D. Lowde, *Phil. Mag.* 2 (1957): 589.
47. J. Christ, *Phys. Stat. Sol.* 7 (1964): 557.
48. E. Taglauer, *Phys. Stat. Sol.* 29 (1968): 259.
49. P. Kettunen, T. Lepistö, G. Kostorz, and G. Göltz, *Acta Metall.* 29 (1981): 969.
50. T. K. Lepistö, G. Kostorz, V.-T. Kuokkala, and P. Kettunen, *Mater. Sci. Eng. A* 131 (1991): 171.
51. B. J. Heuser, *J. Appl. Cryst.* 27 (1994): 1020.
52. G. G. Long, L. E. Levine, and R. Thomson, *J. Appl. Cryst.* 33 (2000): 456.
53. G. G. Long and L. E. Levine, *Acta Crystallogr. A* 61 (2005): 557.
54. D. Dexter, *Phys. Rev.* 90 (1953): 1007.
55. H. H. Atkinson and P. B. Hirsch, *Phil. Mag.* 3 (1958): 213.
56. A. Z. Seeger, *Naturforschg.* 14a (1959): 74.
57. A. Seeger, *J. Appl. Phys.* 30 (1959): 629.
58. A. Seeger and P. Brand, "Small-angle scattering from dislocation rings in crystals, and its relation to the depleted zones in neutron-irradiated metals," in *Small Angle Scattering*, ed. H. Brumberger (New York: Gordon and Breach, 1965), 383–99.
59. R. Thomson, L. E. Levine, and G. G. Long, *Acta Crystallogr. A* 55 (1999): 433.
60. J. P. Hirth and J. Lothe, *Theory of Dislocations* (New York: Wiley, 1982).
61. G. G. Long, P. R. Jemian, J. R. Weertman, D. R. Black, H. E. Burdette, and R. Spal, *J. Appl. Cryst.* 24 (1991): 30.
62. L. E. Levine and G. G. Long, *J. Appl. Cryst.* 37 (2004): 757.
63. J. J. Gilman, *Micromechanics of Flow in Solids* (New York: McGraw-Hill, 1969).
64. D. A. Hughes, D. C. Chrzan, Q. Liu, and N. Hansen, *Phys. Rev. Lett.* 81 (1998): 4664.
65. M. Kuriyama, W. J. Boettinger, and G. G. Cohen, *Annu. Rev. Mater. Sci.* 12 (1982): 23.
66. L. E. Levine, G. G. Long, and D. R. Black, "Complementary experimental techniques for multi-scale modeling of plasticity," in *Multiscale Phenomena in Materials: Experiments and Modeling*, ed. D. H. Lassila, I. M. Robertson, R. Phillips, and B. Devincre, MRS Proc., vol. 578 (Warrendale: Materials Research Society, 2000), 73.

20

Inhomogeneous Lattice Modulations and Fermi Surface Effects in Yttrium Barium Copper Oxide Superconductors

Zahirul Islam[1]

Xuerong Liu[2]

Sunil K. Sinha[2]

Simon C. Moss[3]

1. *Advanced Photon Source, Argonne National Laboratory, 9700 S. Cass Ave., Argonne, IL 60439, USA*
2. *Physics Department, University of California, San Diego, 9500 Gilman Drive, La Jolla, CA 92093, USA*
3. *Department of Physics and Texas Center for Superconductivity, University of Houston, Houston, TX 77204, USA*

Abstract

As a copper-oxide-based Mott insulator is doped with charge carriers ("holes"), the long-range ordered antiferromagnetic ground state is destroyed and a superconducting (SC) ground state emerges above a certain doping level. However, numerous experiments indicate that these cuprates are inhomogeneous even in their SC state. In the case of $YBa_2Cu_3O_{6+x}$ (YBCO), charge carriers are doped via oxygen stoichiometry (x) variations. Systematic X-ray diffuse scattering experiments on YBCO revealed the presence of inhomogeneous short-range ordered modulations characterized by a single wave vector of the form $\mathbf{q} = (q_x, 0, 0)$, where q_x depends on oxygen stoichiometry, x. These modulations correspond to correlated atomic displacements of Ba, Cu, and O atoms, respectively, due to oxygen-vacancy ordered "nanodomains." These superstructures (ortho phases) form well above room temperature, persist well below T_c, and at high charge-carrier concentrations appear to be susceptible to Fermi surface-induced effects at an incommensurate wave vector of the form $\mathbf{q}_1 = (0, q_{IC}, 0)$. Furthermore, "bowtie"-shaped Huang diffuse scattering indicates coherent strain at all compositions regardless of the average structure and independent of twinning, which creates an intrinsically inhomogeneous lattice throughout the phase diagram.

1 Introduction

High-temperature superconductivity in the cuprates is perhaps one of the most compelling problems in contemporary condensed matter physics and materials science. The parent compound of cuprate-based superconductors is a Mott insulator. As charge carriers (typically "holes") are introduced ("doped") into this insulator, the antiferromagnetic ground state is quickly suppressed and an SC ground state emerges beyond a certain critical concentration of "holes." In these oxides, "holes" are doped by some form of chemical substitutions or controlled variations of the oxygen stoichiometry. From numerous transport, thermodynamic, and spectroscopic measurements, there is mounting evidence that these high-T_c cuprates are inhomogeneous even in the SC phase. Although the driving force for these inhomogeneities may well be electronic instabilities, lattice strain, or a combination of both, a consensus on the nature of these inhomogeneities and their role in the overall phase diagram remains elusive. Establishing an understanding of these inhomogeneous phases is very important because they influence fundamental properties of these materials.

Recent experimental and theoretical studies on $Bi_2Sr_2CaCu_2O_{8+\delta}$ (BSCCO), one of the three main families of SC cuprates, are a case in point. Scanning tunneling microscopy (STM) work [1] on BSCCO highlights the role of inhomogeneities in electronic properties. By correlating inhomogeneous spatial distribution of local density of states (LDOS) and the SC energy gap with defect locations, the STM work has identified atomic-scale sources and mechanisms of nanoscale electronic disorder related to oxygen dopants. These defects are spatially extended in nature. Theoretical work [2] provides an explanation of how these dopant atoms make attractive coupling between charge carriers spatially modulated, giving rise to electronic inhomogeneities observed in the STM work. Such modulations are due to local lattice distortions surrounding the dopant atoms, which can modify both electron-phonon coupling and superexchange interactions. Bulk-sensitive X-ray diffraction studies [3] identified three distinct incommensurate lattice modulations on different length scales coexisting in BSCCO, one of which can be ascribed to nanoscale modulations observed in STM images [4]. What these studies suggest is that there is a clear interdependence of displacement modulation and electronic inhomogeneities in SC cuprates.

Controlled variations of oxygen stoichiometry, as is done in BSCCO, is a widely used method for doping other families of cuprates, such as YBCO, $HgBa_2CuO_{4+\delta}$ (Hg1201), and $La_2CuO_{4+\delta}$ (LCO), respectively. The very process of doping, however, invariably induces fundamental structural changes in the material. These structural changes might have profound implications for electronic properties, as discussed above. In the case of YBCO and LCO, it is well documented that some form of ordering phenomena accompanies this process. While in YBCO, oxygen vacancies tend to form quasi-one-dimensional inhomogeneous superstructures [7–9], in LCO, interstitial oxygens exhibit a remarkable phenomenon of staging [10,11]. X-ray scattering evidence of structural superstructures in Hg1201 has also been gathering [12]. A microscopic determination of the interplay of such structural effects related to dopant atoms and electronic properties is fundamental to our understanding of the overall phase diagram of cuprates. Experimental and theoretical studies using scattering techniques have made tremendous contributions to our understanding of the role of inhomogeneous phases in electronic properties of both YBCO [6,13–15] and LCO [10].

In this chapter, an attempt is made to present a unified picture of our current understanding of inhomogeneities in YBCO from an X-ray scattering point of view. The experiments, details of which can be found elsewhere [5,6], were performed on three undulator beamlines (XOR 1, XOR 4, and μ-CAT, respectively) at the Advanced Photon Source. It is important to note that in addition to the ease of observing "static" superstructures, there are distinct advantages of using synchrotron X-ray scattering in studying phenomena that may involve dynamic fluctuations (e.g., charge fluctuations or "stripes," Jahn-Teller polarons, etc.), as diffraction effects are primarily from the atomic displacements or lattice distortions that they generate. Since X-ray scattering integrates over all relevant energies at any point, **Q**, in momentum (reciprocal) space as follows,

$$S(\mathbf{Q}) = \int_0^\infty S(\mathbf{Q},\omega)d\omega, \tag{20.1}$$

where $S(\mathbf{Q},\omega)$ is the dynamical structure factor of the system, dynamical charge fluctuations with well-defined periodicity may also be observed as "static" superstructure peaks. Such superstructure peaks would appear away from the crystal Bragg points, yielding direct information of the spatial periodicity of the associated lattice modulations. Second, the high flux of extremely well-collimated photons delivered into a small spot by an undulator at a third-generation synchrotron facility makes it easier to observe scattering from such modulations in very small high-quality crystals, which can be quite weak. Third, the high energy of X-rays ensures complete sampling of the bulk of the sample and makes a large region of momentum space accessible for systematic studies. In particular, in order to observe interplanar correlations, an important part of this study, it is necessary to collect scattering intensity for as large a momentum transfer as possible so that a good real-space resolution can be obtained. Finally, the high **Q** resolution enables us to study very long-wave fluctuations in the system.

2 Overview

2.1 Superstructures and Their Correlations

The average structure of YBCO compounds with an SC ground state is orthorhombic ($a < b < c$), which is primitive. An ideal unit cell (see Figure 20.1) consists of three-dimensional layering of one-dimensional CuO chains and a pair of nearly square and slightly puckered CuO_2 planes, respectively. Ba and apical O atoms lie nearly equidistant from the neighboring CuO chain and CuO_2 planes, respectively. The crystallographic **a** axis is along the shortest Cu-Cu (or Cu-O-Cu) bond direction and the **b** axis is identified to be along the chain, respectively. The **c** axis is the perpendicular direction; in particular, it is normal to the two-dimensional CuO_2 planes. The **a** axis is the primary direction of interest because theoretical models predict the appearance of modulations [7,16] and experimental work observed charge and spin fluctuations [17], respectively, as well as phonon anomalies [18] along

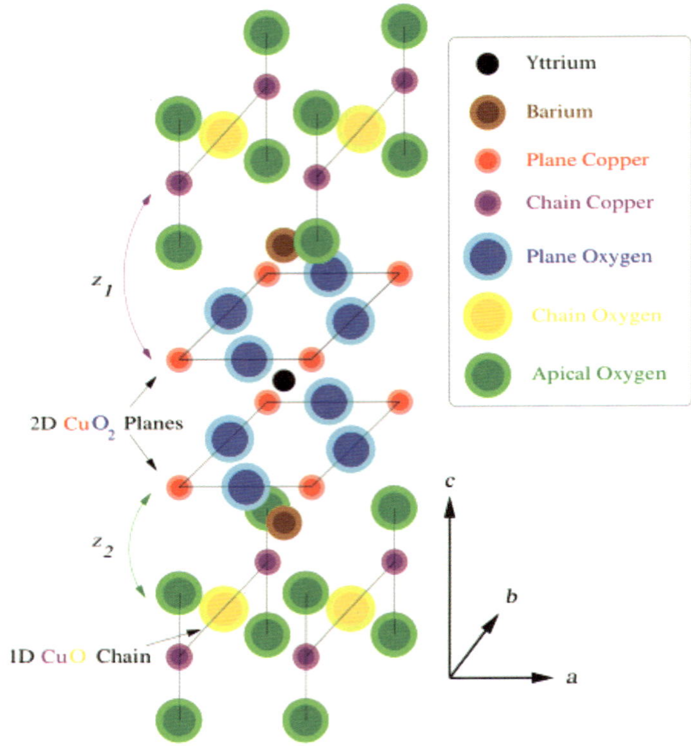

Figure 20.1. *Chemical unit cell identifying interplanar separations. One-dimensional Cu-O-Cu chains define the* **b** *axis. On the CuO-chain plane (ab plane) there are no O atoms in between the Cu atoms along* **a**. *The apical oxygen lies directly above and below each Cu atom of the chain. Note that Ba atoms just above and below the ab plane are separated from each other by* z_1 *as well. The puckering of the two-dimensional* CuO_2 *planes have not been shown for clarity.*

this direction. Measurements of periodicity, correlations, and anisotropy of modulations are done by collecting scattering intensity profiles (i.e., H and K scans) along directions **a*** and **b*** in the reciprocal space corresponding to planes orthogonal to the **a** and **b** axes, respectively. Similarly, in order to determine interplanar correlations in the **c** direction, the dependence of scattering intensity of superlattice/superstructure peaks on L (i.e., along **c***) is collected.

Superconductivity in orthorhombic YBCO arises from correlated "holes" in the CuO_2 bilayers. The hole density is controlled by tuning the oxygen stoichiometry, which changes the number of oxygen atoms, or vacancies, in the CuO_x-chain planes (or *ab*-planes). These vacancies tend to form superstructures [9] which have been calculated by de Fontaine et al. using an anisotropic Ising model [8]. These superstructures, so-called "Ortho" phases according to the nomenclature introduced in de Fontaine et al. [8], consist of different arrangements of filled and empty chains (or chain fragments) with well-defined modulations along the **a** axis. As a qualitative example, consider YBCO with $x = 0.6$. The modulation vector is $\mathbf{q} = \left(\frac{2}{5},0,0\right)$, relative to the nearest Bragg point, resulting from complex arrangements of oxygen-full and oxygen-vacant CuO_x chains (two full, one empty, one full, one empty; denoted by <11010>, where "1" means an O-full chain and "0" stands for an O-vacant chain, respectively) that run parallel to the **b** axis. Figure 20.2 depicts an overall picture of what diffuse scattering from such a superstructure in the reciprocal space looks like. A contour plot (Figure 20.2,

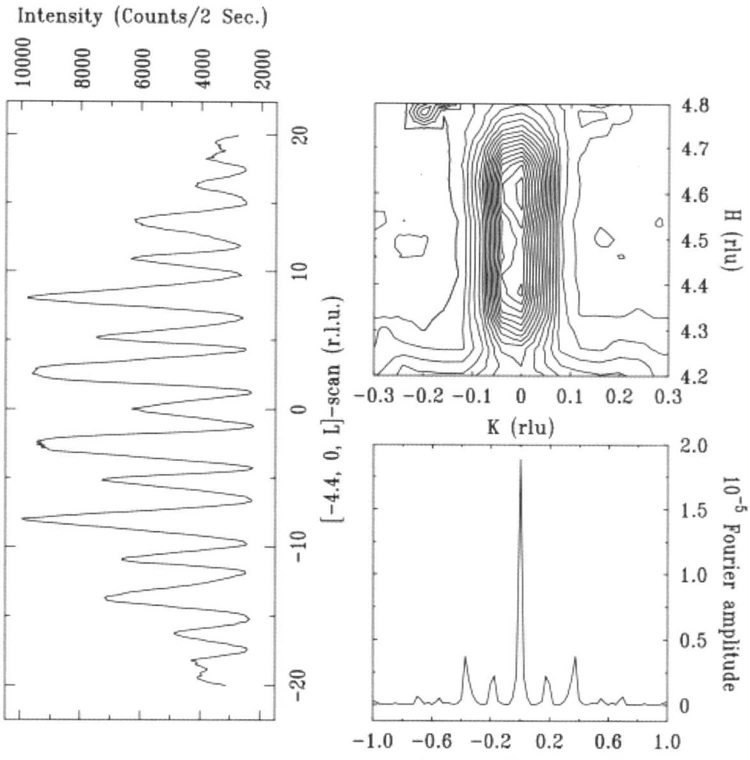

Figure 20.2. *Top right: Contour plot of scattering intensity distribution in the HK plane collected on YBCO with x = 0.63. Two diffuse superstructure peaks are clearly present away from Bragg points. Left: L scan revealing remarkable intensity oscillation as a function of L of the peak at H = 4.4, K = 0. Bottom right: Fourier transform (Patterson function) of the intensity oscillation revealing two distances along the **c** axis of planes with scattering events that are correlated with each other. (Adapted from Islam et al. [5].)*

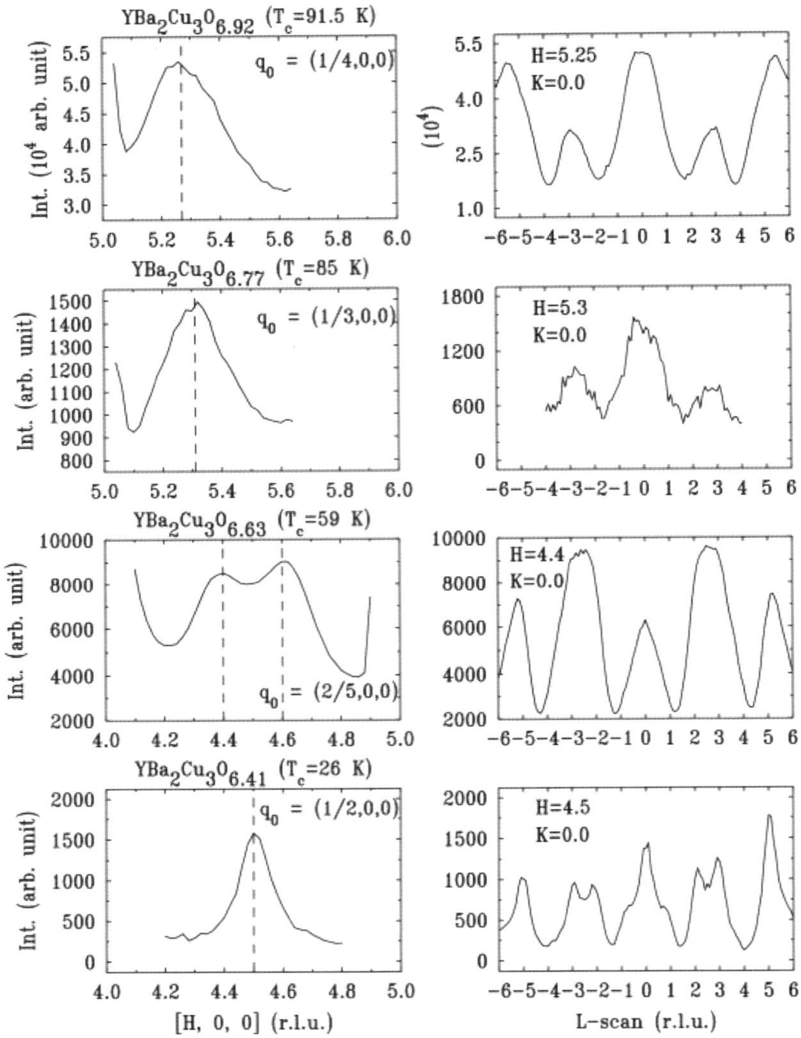

Figure 20.3. *H scans for K = 0 and L = 0 (left panels) and L scans (right panels), respectively, at various oxygen concentrations in YBCO compounds. Superlattice peaks correspond to a wave vector of the form* **q** *=* $(q_x,0,0)$*. L scans were carried out through the peak position indicated in each panel. Only the optimally doped compound was detwinned.*

top right) of intensity distribution in the HK plane shows two peaks characterized by a wave vector $\left(-\frac{2}{5},0,0\right)$ located away from the (4,0,0) and (5,0,0) Bragg points, respectively. Both peaks are diffuse with a width slightly narrower in the *K* direction than in the *H* direction. This difference in widths indicates a slightly longer correlation length along the chains (i.e., **b** axis) compared to that along the modulation direction (i.e., **a** axis). The intensity distribution along *L* of one of the two diffuse peaks shows remarkable oscillations. The bottom panel on the right in Figure 20.2 presents a Fourier transform (see below for details) of this oscillation revealing two interplanar distances within a unit cell with scattering events that are correlated with each other (see Figure 20.1).

A graphical overview of the superstructures and their correlations as a function of oxygen stoichiometry is presented in Figure 20.3. Left panels in Figure 20.3 show that as the average O concentration increases (i.e., O vacancy concentration decreases), the wave vector **q** changes as follows: $\left(\frac{1}{2},0,0\right)$ for $x = 0.41$, $\left(\frac{2}{5},0,0\right)$ for $x = 0.63$, $\left(\frac{1}{3},0,0\right)$ for $x = 0.77$, and $\left(\frac{1}{4},0,0\right)$ for $x = 0.92$, respectively. These wave vectors correspond to the first harmonic of the O vacancy ordered phases in the CuO chains referred to as Ortho-II, Ortho-V, Ortho-III, and Ortho-IV, respectively [8]. It is clear that the peak width (ΔH) increases relative to the Ortho-II phase, indicating a decreasing correlation along the **a** axis for higher oxygen concentrations. Although it is difficult to be very precise, correlation lengths extracted by fitting Gaussian line profiles to the diffuse peaks range from ~2–3a for higher order Ortho phases.

As **q** gets smaller, superstructure peaks approach the Bragg point. *K* scans (see Figure 20.8 below) through the superstructure peak, for example, (5.25, 0, 0) in optimally doped YBCO, clearly show the presence of a broader component underlying a sharper component, in addition to nearly constant background primarily due to thermal diffuse scattering (TDS). The sharp component is associated with **q**, whereas the broad structure is due to additional modulations discussed below. By fitting the *K* scan data (see Figure 20.8) with three Gaussians and a constant term to account for the background, we find the correlation length of the superstructure in optimally doped YBCO to be ~6 unit cells along the CuO-chain direction (**b**).

By measuring how the scattering intensity of the superstructure peaks depends on momentum transfer along the **c*** axis (perpendicular to the CuO_2 planes; see Figure 20.1), one can identify all the planes perpendicular to the **c** axis that contribute to the peak. If only two planes are correlated, the variation of intensity is sinusoidal, with frequency in momentum space along **c*** given by the inverse of the interplanar separation, whereas if all the planes scatter independently of each other, then scattering occurs along a rod parallel to the **c*** axis. This intensity modulation is not purely sinusoidal, suggesting that the displacements on more than two scattering planes are correlated along the **c** axis. Figure 20.3 (right panels) shows such scans. In the case of the Ortho-II phase, the peaks are narrow and are well defined on the integer values of *L*, suggesting a significant range of correlations beyond a unit cell along the **c** axis. However, for higher concentrations, the intensity of the satellites exhibits remarkable oscillations, observed in both detwinned and twinned crystals, respectively. If one describes the scattering from a given layer n ($\perp$**c**) by a complex structure factor F_n, then the intensity modulation along **c*** of the diffuse peak at $\mathbf{Q} = (Q_x,0,Q_z)$, where Q_x is the absolute position of the satellite along **a***, is given by

$$I(Q_z) \propto \sum_{n,n'} \langle F_n(Q_x) F(Q_x)^*_{n'} \rangle e^{-iQ_z(z_n - z_{n'})}, \tag{20.2}$$

where z_n is the height along the **c** axis of layer n. So the separations along the **c** axis of the correlated planes can be obtained by a direct Fourier transform (Patterson function) of the intensity modulation. In this case, there are two interplanar distances (see Figure 20.2), $z_1 \approx 0.36c$ and $z_2 \approx 0.2c$, respectively. Scattering from theses plane is correlated: z_1 corresponds to the distance between the CuO_2 and CuO-chain planes as well as to the shortest Ba-Ba distance along **c**, whereas z_2 is consistent with both the CuO_2-O_{apical} and the chain-Ba distances, respectively (see Figure 20.1).

The anisotropy of the correlation length is understandable from the kinetics of chain oxygen atoms in this material. Although the superstructures typically form above 300 K, Monte Carlo simulations [8] predict additional "branching phases" to appear as $T \rightarrow 0$ if the sample is annealed at sufficiently low temperature [8]. However, transformation between these phases and their temperature dependence have been monitored [19] with a variety of techniques (i.e., diffusion, internal friction, resistivity and superconducting T_c, Raman spectroscopy, Hall measurements, and photoconductivity). Relaxation times resulting from these measurements universally show low mobility of oxygen along **a** and an activated behavior with an activation energy of ~1 eV for an interstitial oxygen atom to attach to a chain fragment along **b**. Thus the superstructures observed at room temperature and below are "frozen" in phases that start to form at higher temperatures. In other words, the formation of longer-range order by making chain fragments longer is kinetically limited at low temperatures, accounting for correlation length asymmetry. Interplanar correlation in turn depends on the degree of order in the chains (*ab* plane) and cannot extend beyond a unit cell along **c** when the vacancy concentration in the chains gets smaller.

As a summary, Figure 20.4 displays the wave vectors as a function of oxygen nonstoichiometry x. The Ortho-II phase appears below $x \approx 0.62$ with significant correlations, whereas short-range ordered domains of higher order phases form at or above $x \approx 0.62$. Relative to the Ortho-II phase, higher order phases show a much shorter range of correlation. The decrease of correlation lengths with increasing O content has been reported in similar X-ray diffraction studies [20] as well. Note that a given Ortho phase is stable for a range of concentrations, and the Ortho-II phase appears to be the most stable.

2.2 *Long-Range Coherent Strain of the Lattice*

In addition to the presence of short-range ordered modulated regions in YBCO, the lattice is coherently strained. This strain gives rise to a strongly anisotropic "bowtie"-shaped Huang diffuse scattering (HDS) in the vicinity of Bragg points that is visible at low temperature. The first observation of such an HDS pattern was in Al-doped YBCO [14]. The bottom panels (left and middle) in Figure 20.5 display a two-dimensional scan around the (4,0,0) Bragg peak for an underdoped (x = 0.41; twinned) and optimally doped (x = 0.92; detwinned) YBCO crystals, respectively. In both cases, four lobes of diffuse scattering extend along the four [±1, ±1, 0] directions, although the symmetry is slightly distorted in the twinned crystal due to contributions from other twin domains. Note that the Ortho-II $\left(\frac{1}{2},0,0\right)$ satellites are well separated, whereas superlattice peaks at $\left(4\pm\frac{1}{4},0,0\right)$ characterizing the Ortho-IV phase are weakly discernible. What is intriguing is that the pattern of diffuse scattering is manifestly different than the TDS pattern (bottom right) calculated [21] using the elastic constants of YBCO. At room temperature the pattern (not shown) is similar to that of observed TDS [6]. The top panels (left and middle) show line scans that reveal the lobes that contrast the TDS peak more clearly. The fact that

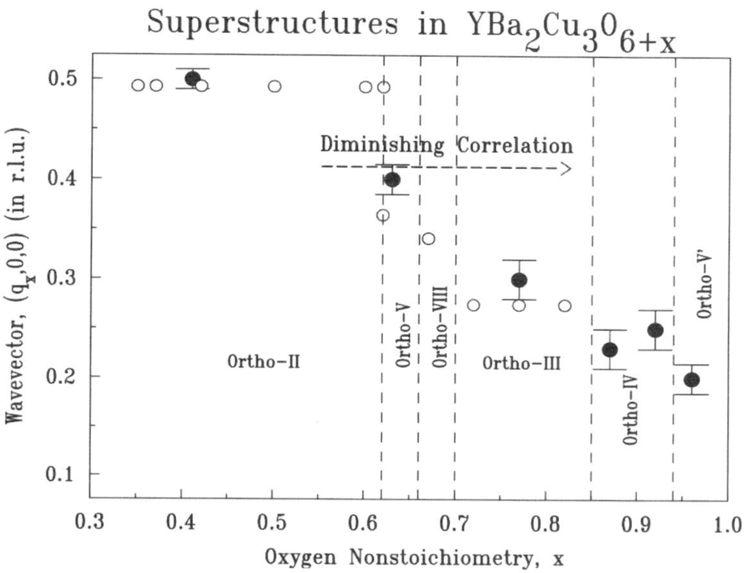

Figure 20.4. *Ordering wave vector as a function of oxygen concentration as observed at room temperature and below. At least six different phases can be distinguished. Note that there are ranges of x in which the magnitude of the wave vector remains the same within experimental uncertainty. Filled circles with error bars are from the present study. Open circles are data from Zimmermann et al. [20].*

this anisotropic pattern is observed irrespective of twinning clearly eliminates the possibility that HDS in the optimally doped material is due to stress detwinning of the crystal.

We note that earlier X-ray studies [14,22] of O-disordered *tetragonal* systems, such as aluminum-doped YBCO ($YBa_2(Cu_{0.955}Al_{0.045})_3O_7$), observed Huang scattering of the same symmetry in the vicinity of the (4,0,0) Bragg point at room temperature. Using a coupled concentration wave static displacement wave approach [23], these studies showed that Huang scattering arises from (1,1,0):($\bar{1}$,1,0) shear displacements. The disordered array of oxygens and vacancies on the CuO chain planes contains long-wave fluctuations of concentrations, which cause a static displacement wave. Such long-wave distortions of the lattice give rise to Huang scattering in the vicinity of Bragg points. It is clear that coherent lattice strain is a fundamental property of YBCO regardless of doping, electronic ground state, average structure, and twinning.

2.3 *Intensity Asymmetry and Atomic Displacements*

Although the observed **q** for the superstructure peaks is consistent with the harmonics expected of the Ortho phases discussed above, it is not conclusive proof that the peaks are due to oxygen-ordered templates. One of the salient features of the superlattice peaks at higher oxygen concentrations is the intensity asymmetry (IA) between the two satellites on either side of a Bragg point. The IA was observed for all such pairs measured in an underdoped YBCO [5,15], indicating that the origin of these superstructures is due to lattice distortions characterized by the ordering wave vector. When

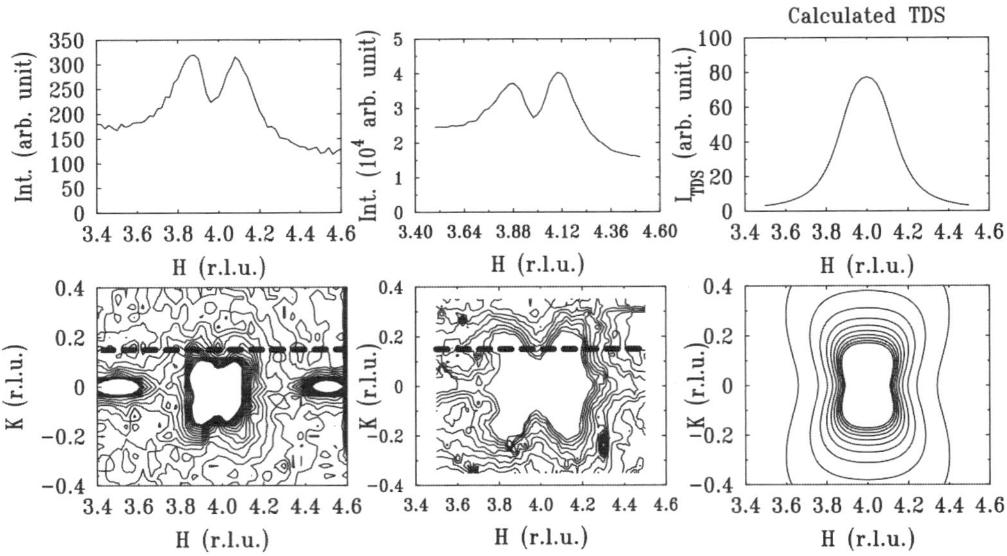

Figure 20.5. *Two-dimensional scan around the (4,0,0) Bragg point revealing an anisotropic "bowtie"-shaped HDS pattern. Left: Twinned crystal with Ortho-II phase. The* $\left(\frac{1}{2},0,0\right)$ *peak is visible on either side. Middle: Detwinned optimally doped compound with Ortho-IV phase. Right: Calculated TDS pattern. Top panels show the line scan along the direction indicated by dashes in the lower panels. Two HDS peaks, in contrast to a single TDS peak, are observed.*

the atomic displacements ($\delta\mathbf{u}_i$) are sufficiently *small*, they give rise to X-ray scattering at a point $\mathbf{Q}$ in the momentum space with an amplitude $\mathbf{Q}\cdot\delta\mathbf{u}_i$. It was found that the intensities of the satellites of the form $\left(H\pm\frac{2}{5},0,0\right)$ scaled as $\dfrac{I\left(H+\frac{2}{5}\right)}{I\left(H-\frac{2}{5}\right)}=\dfrac{\left|H+\frac{2}{5}\right|^2}{\left|H-\frac{2}{5}\right|^2}$, suggesting that the atomic displacements (|| **a**) giving rise to these superstructure peaks are *small.*

Figure 20.6 shows several scans ([H,0,0]-scan) for different integer values of K for the optimally doped material. These scans were taken at ~10K, which is well inside the superconducting state of the optimally doped YBCO. Broad superlattice peaks corresponding to $\mathbf{q}=\left(\frac{1}{4},0,0\right)$ are clearly visible between Bragg points. Unlike in the underdoped compound, a much stronger IA is observed around all the Bragg points (h,k,0) where h and k have different parity. When h and k have the same parity, strong TDS emanating from the Bragg points overwhelms the superlattice peaks. Indeed, the satellites around the even-order Bragg peaks are observed in parallel scans at a noninteger value of L [24]. Unlike in the underdoped YBCO, the IA in optimally doped YBCO does not scale as Q^2. Similar IA is also observed for higher order Ortho phases [24]. A strong IA is due to *large* atomic displacements along the **a** axis (i.e., perpendicular to the chain direction).

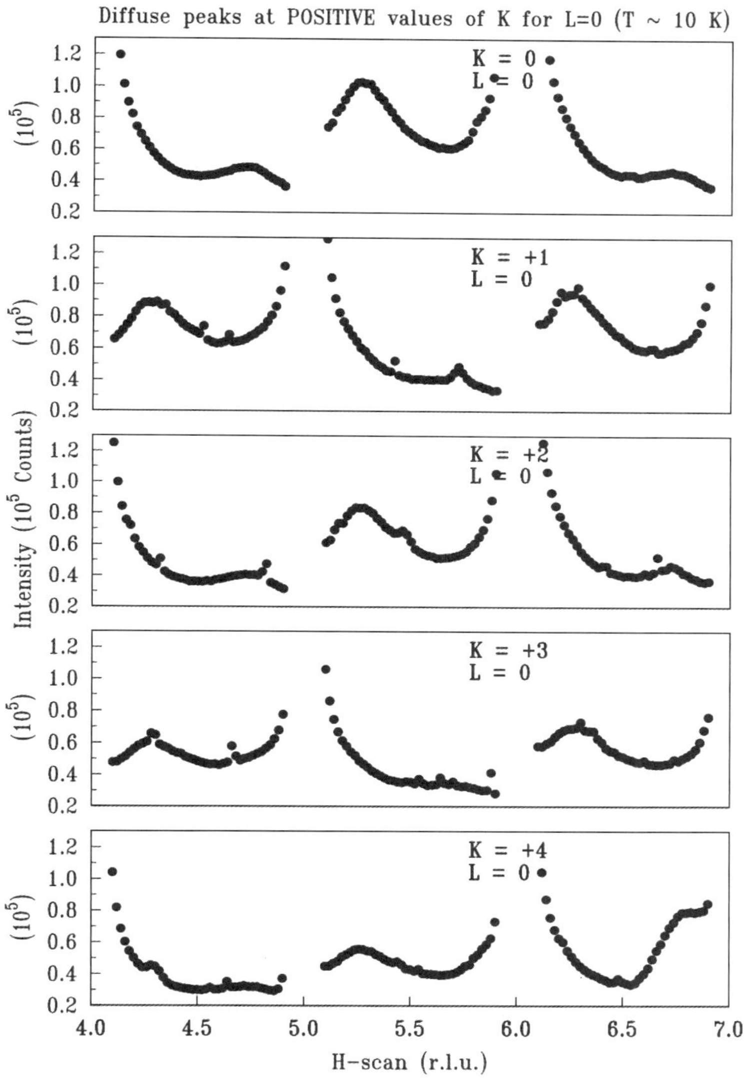

Figure 20.6. *Scans along* **a*** *for several integer values of K. Superlattice peaks corresponding to a wave vector of* **q** = $\left(\frac{1}{4}, 0, 0\right)$ *are clearly observable on either side of the (h,k,0) Bragg points where h and k have different parity. Scans for negative values of K are qualitatively the same. Data were taken on a detwinned optimally doped sample. Data for the twinned sample is very similar [24].*

To appreciate how large displacements give rise to strong IA, one can construct a simple one-dimensional model [25] in which atomic planes ⊥**a** are sinusoidally displaced along the **a** axis (i.e., longitudinal distortions) as

$$u_n = u_x * \cos(\mathbf{q}\cdot\mathbf{r}_n), \tag{20.3}$$

where $\mathbf{r}_n$ is the position of the nth plane, $\mathbf{q}$ is the modulation vector along **a**, and u_x can be large. Scattering intensity from a set of N planes with unit scattering strength that are modulated according to Equation 20.3 is given by

$$I = N^2 \sum_{m=-\infty}^{m=+\infty} J_{|m|}^2(Q_x u_x) \frac{\sin^2\{\frac{Na}{2}(Q_x + mq_x)\}}{\sin^2\{\frac{a}{2}(Q_x + mq_x)\}}, \tag{20.4}$$

where $J_{|m|}$ is a Bessel function of order m and Q_x and q_x are the components of **Q** and **q**, respectively, along **a**. The physical meaning of this expression is that there can appear an infinite number of superlattice peaks with intensities proportional to $J_{|m|}^2$ around each Bragg point, although the lattice modulation is purely sinusoidal. Note that the zeroth-order satellites are superimposed on the Bragg points, and only the first-order satellites are likely to be strong enough to be observed. This simple calculation clearly explains the IA observed, as the intensity of the positive (+m) satellite is always stronger than the corresponding negative (–m) satellite. In particular, for the first-order satellite, the ratio is given by

$$\frac{I(\mathbf{Q} = \mathbf{G} + \mathbf{q})}{I(\mathbf{Q} = \mathbf{G} - \mathbf{q})} = \frac{J_1^2\{(\mathbf{G} + \mathbf{q})\cdot u_x \mathbf{a}\}}{J_1^2\{(\mathbf{G} - \mathbf{q})\cdot u_x \mathbf{a}\}}, \tag{20.5}$$

where **G** is a Bragg point (reciprocal lattice vector). In order to explain a large IA, u_x must be a significant fraction (e.g., a few percents) of a, which is much larger than the typical displacements due to thermal vibrations at cryogenic temperatures. Such large displacements away from crystallographic sites (i.e., equilibrium positions in the lattice) can result in anharmonic forces acting on the displaced atoms, which might have a significant impact on their dynamic behavior, as discussed below.

In order to obtain a model for displacement pattern ($\delta\mathbf{u}$'s) of Ba, Cu, and O, some general features of the data should be noted. First, a strong IA between the **+q** and **–q** satellites is observed around all Bragg points. Second, for a given satellite at $(h,k,0) + \mathbf{q}$, the intensity is either very strong or weak when h and k have the same or mixed parity, respectively. This implies out-of-phase displacements of the dominant scatterers. Third, no second harmonic (2**q**) satellites were observed indicating essentially a sinusoidal modulation. Finally, scans such as $[H,-5, 0]$ ($H \in [0.1\text{–}0.9]$), which are sensitive to displacements along **b**, found no superlattice peaks suggesting the absence of any $\delta\mathbf{u}||\mathbf{b}$ associated with the **q** modulation [6]. As discussed by Islam et al. [6] in detail, the calculations and fitting of integrated intensities of a large number of peaks were performed without assuming any displacements to be small in the presence of an Ortho-IV phase in the CuO-chain plane using

$$I_{diffuse}(\mathbf{Q}) \propto \left| \sum_{n}^{Supercell} f_n(Q) e^{-W_n(Q)} e^{-i\mathbf{Q}\cdot(\mathbf{R}_n + \delta\mathbf{u}_n)} \right|^2,$$

where the displacement relative to an average lattice site ($\mathbf{R}_n$) of the nth atom is $\delta\mathbf{u}_n$, $f_n(Q)$ and $e^{-W_n(Q)}$ are the form factors and Debye-Waller factors (DWFs), respectively. While the experimentally deduced model, displayed in Figure 20.7, may not be perfect, it does account for all the systematics of the data. Furthermore, it portrays a pattern of displacements consistent with that of the Ortho-V phase in an underdoped YBCO obtained from first-principles electronic calculations [15]. In the present case of an optimally doped compound, however, the periodicity is four unit cells. The salient features of the model are large displacements along the **a** axis of Ba and correlated antiphased displacements of Cu and O atoms on both chain and two-dimensional CuO_2 planes, respectively. There is a small Ba displacement along **c** as well. The displacements are symmetric with respect to the defect center and diminish away from it. The model displays a strong lattice relaxation in the immediate neighborhood of oxygen vacancy ordering. Longer-range strain emanating from such distorted "nanodomains" must also exist in order to account for HDS, as discussed above [23].

2.4 Temperature Dependence

Superstructure peaks at all compositions—in particular, in higher order Ortho phases—weaken significantly on warming to room temperature. The top panel in Figure 20.8 shows [H,0,0] scans through the Ortho-IV superlattice peak. Scans along **b*** perpendicular to the [H,0,0] scans at the

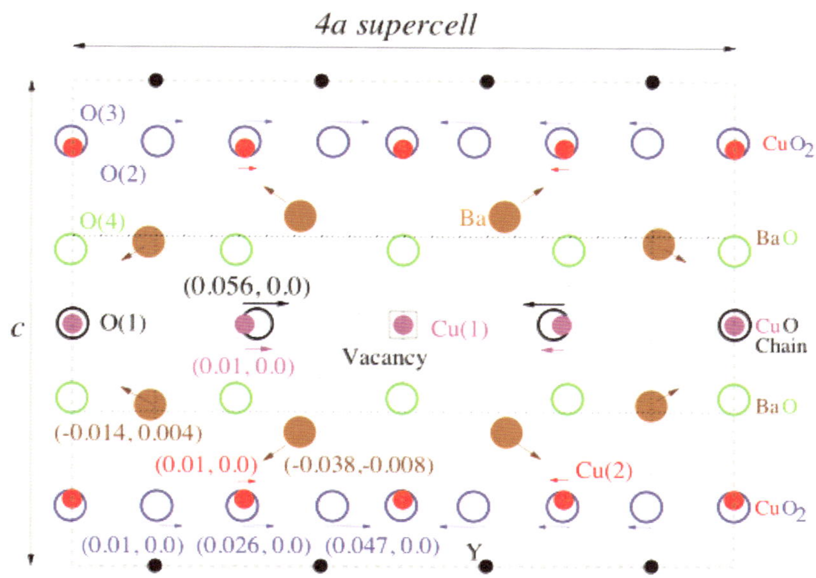

Figure 20.7. *An ideal atomic displacement (arrows) pattern at low temperature. All the atoms have been projected on the ac plane. Note that primary diplacements (δu's) are along the* **a** *axis, that is, along the shorter Cu-Cu bond direction. δu's, (δ_a, δ_c) in units of a and c, of respective atoms are given in parentheses. δu's of all other atoms are related by mirror symmetry. Adapted from Islam et al. [6].*

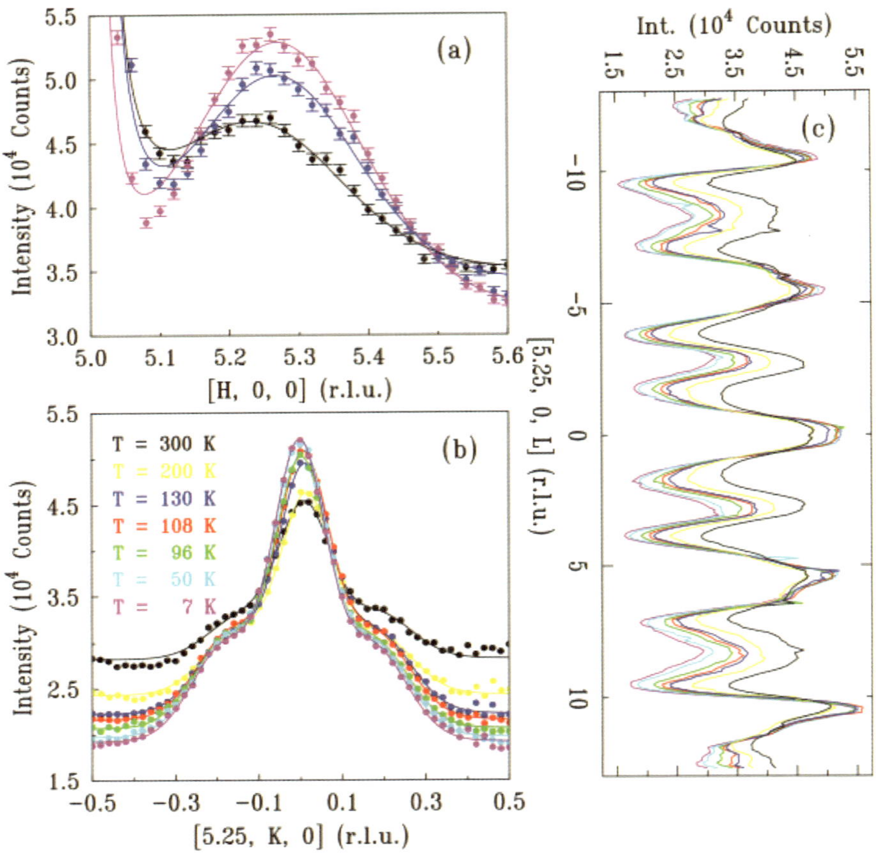

Figure 20.8. *Ortho-IV superstructure peak and its temperature dependence. (a) Scans along* **a*** *corresponding to a Cu-O-Cu bond direction in the real space at several representative temperatures. (b) K scans (scans along* **b*** *parallel to the CuO chain direction) through the peak at the same temperatures. (c) Intensity modulation along* **c*** *at the corresponding temperatures. Adapted from Islam et al. [6].*

same temperatures are shown in the bottom panel. It is clear from these scans that as the temperature is increased the intensity of the **q** peak decreases relative to the TDS. From the data one can infer that the integrated intensity (area under the peak corrected for the TDS tail emanating from the (5,0,0) Bragg point) at low temperature is twice that at room temperature. *L* scans also show a reduction of maximum amplitude of the intensity oscillations. However, in order to get quantitative behavior as a function of temperature, we need to take the underlying "bowtie"-shaped Huang scattering and TDS into account. This is better accomplished via [5.25,*K*,0] scans shown in the lower panel (Figure 20.8). The data for the [5.25,*K*,0] scans are well represented by a combination of three Gaussian line profiles, one for the central satellite peak and two for the Huang scattering lobes and a constant term to account for the background, which is predominantly made up of TDS. Since the peak widths and positions do not change with temperature, only the peak heights of the Gaussians and the constant term (i.e., four parameters all together) were needed to fit the entire data.

Integrated intensity and maximum amplitude of L scans consistently show gradual diminishment of the diffuse peak on warming, as presented in Figure 20.9. Fourier amplitudes, which are a measure of strength of interplanar correlations, also weaken with increasing temperature as the ordering is destroyed. Since diffusive motion of chain oxygens practically freezes below ~250 K, the growth of Ortho-IV regions in size or number seems unlikely. Note that the rate of decrease is faster than that expected from thermal effects of DWFs of the average lattice [5,6]. Given that atomic displacements (Figure 20.7) are clearly anharmonic in these imperfectly ordered nanoscale regions, it appears that enhanced elastic softening of the lattice takes place within these regions, providing a means for a faster decay of intensity with temperature.

2.5 *Fermi Surface-Induced Effects*

In YBCO, one-dimensional metallic CuO chains are interleaved with two-dimensional CuO_2 bilayers that not only act as charge reservoirs for the CuO_2 planes but also provide the grounds for the interplay of electronic properties of systems with different dimensionality in close proximity to each other. It has been known [26–28] for some time that the chains exert profound influence on both normal and superconducting states. Local density approximation (LDA) calculations [29] indicate Fermi surface (FS) sections that are open electron "ridges" orthogonal to **b*** (i.e., normal to the CuO chain direction), derived predominantly from electronic states on the CuO chains, which have been confirmed by positron annihilation [30] and photoemission [31] measurements. Intriguingly, these sections of the

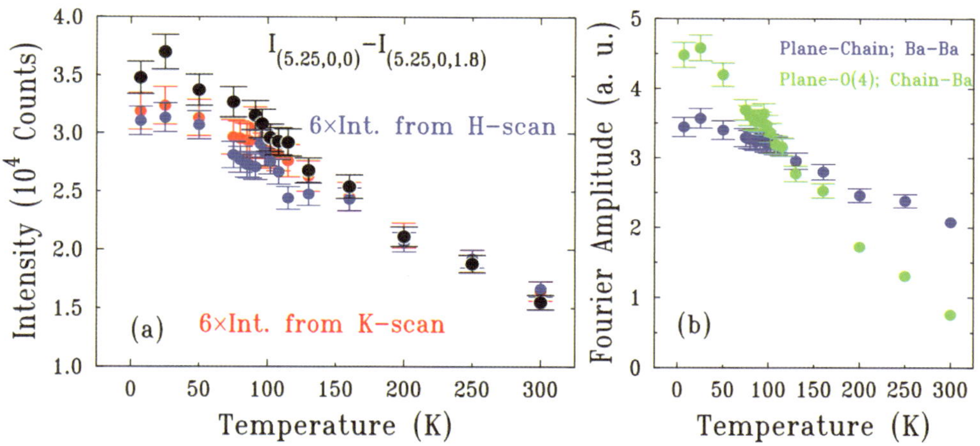

Figure 20.9. *Temperature dependence of Ortho-IV peak. (a) Integrated intensity, peak intensity, and the maximum oscillation amplitude obtained from various scans shown in Figure 20.8. (b) Temperature dependence of the Fourier amplitudes derived from L scans. Adapted from Islam et al. [6].*

FS associated with the one-dimensional CuO chain subsystem are in a near nesting condition, which can lead to an FS-induced instability, such as a charge density wave (CDW), at an incommensurate wave vector $\mathbf{q}_{IC} = (0,2k_F \approx 0.22,0)$ (in reduced lattice units; $\mathbf{k}_F$ is the Fermi wave vector). Although NMR/NQR measurements [32], STM studies [33,34], and inelastic neutron scattering investigations [35] have reported on the existence of some form of electronic instability originating from CuO-chain derived states, a consensus among different groups has not been reached to date.

On the other hand, there is a clear X-ray scattering evidence of lattice modulations with a propagation vector $\mathbf{q}_1 = (0,\sim 0.21,0)$ along the CuO chain direction observed at optimal doping. In reciprocal lattice scans, they appear as shoulders at $\mathbf{q} \pm \mathbf{q}_1$ with respect to Bragg points, as shown in Figure 20.10. Note that **K** scans through different **q** peaks show a definite coupling between the intensities of the shoulder peaks with the center Ortho-IV peaks, both having the same correlation lengths and obeying the same parity relations [6]. The parity relations impose stringent phase correlations among atomic displacements as those in Ortho-IV model (Figure 20.7).

Secondly, detailed temperature dependence measurements revealed a covariance of intensities of **q** satellite and $\mathbf{q}_1$ shoulder with temperature. Their ratio is essentially a constant within the whole temperature range (7–300 K), as illustrated in the bottom panel of Figure 20.11. The fact that the intensity of the shoulders obeys the same temperature dependence as the center **q** peak indicates that they are strongly related. As discussed above (see Islam et al. [6] also), the temperature dependence of the **q** peaks distinguishes the Ortho-IV phase islands from the average lattice, which has rapidly varying DWFs. Therefore, the $\mathbf{q}_1$ modulation must coexist with Ortho-IV islands and does not form

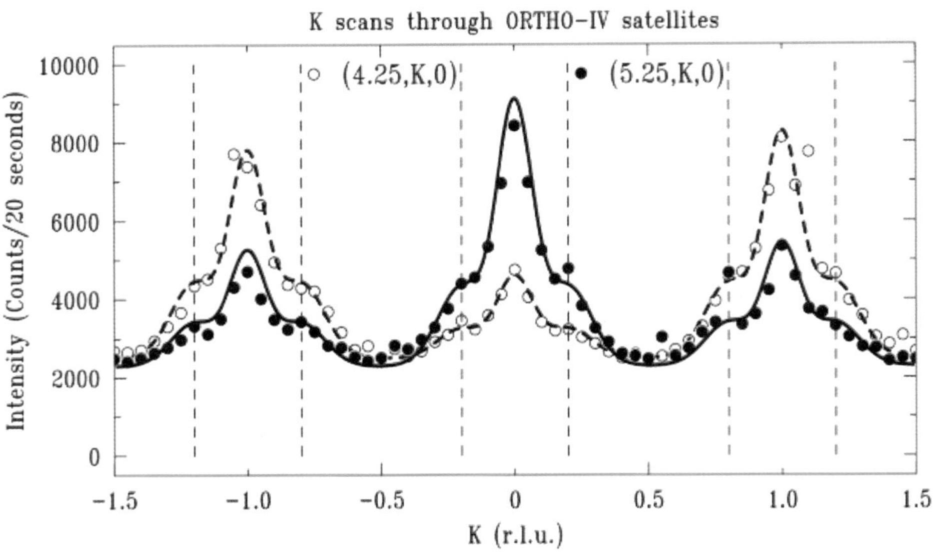

Figure 20.10. *Various reciprocal lattice scans along* **b*** *(i.e., K scans) showing shoulder structures on either side of the central Ortho-IV satellite peaks. Vertical dashed lines indicate the position of the shoulders, which are displaced by ±0.2 from the respective central satellite. Note that the relative intensity of the shoulders follows the systematics of the central satellite with respect to the parity of Miller indices of the parent Bragg point. Data were taken at ~7 K.*

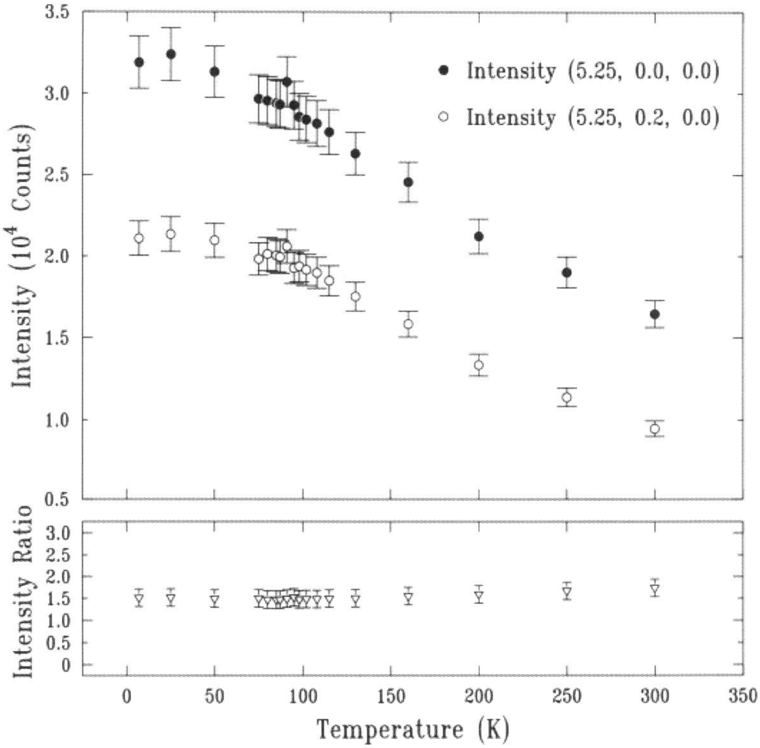

Figure 20.11. *Top: The intensity of* **q** *satellite and that of the* $\mathbf{q}_1$ *shoulders as a function of temperature. Bottom: Intensity ratio remains constant as a function of temperature.*

homogeneously throughout the whole crystal. If lattice modulations with wave vector $\mathbf{q}_1$ are present throughout the whole crystal, they should be observed by X-ray scattering as peaks (or rods) centered at $(m,n,L) \pm \mathbf{q}_1$ (m and n are integers) in the H-K plane of the reciprocal space, depending on their interchain correlations. A thorough investigation in the vicinity of $(m,n,L) \pm \mathbf{q}_1$ for different integer m and n values showed no indication of such modulations along **b***.

Our experimental observations therefore indicate the existence of a **b**-direction lattice modulation with $\mathbf{q}_1 = (0,\delta \approx 0.21,0)$ within the Ortho-IV phase islands in optimally doped YBCO. $\mathbf{q}_1$ is practically the same as the nesting wave vector $\mathbf{q}_{IC}$, suggesting that this modulation is induced by the FS. In our view, nested "ridges" (or "sheets") of the FS simply induce a $\mathbf{q}_1$ amplitude modulation of the Ortho-IV displacement pattern with a concomitant presence of a CDW on the CuO chains. Qualitatively, given that the intensity of the pair of shoulders is symmetric about the central peak and that the $\mathbf{q}_1$ peaks appear strongly near the **a*** axis, atomic displacements associated with the $\mathbf{q}_1$ modulation must be transverse (i.e., perpendicular) to the chains, implying an unusual form of coupling to the lattice. Since the Ortho-IV displacement pattern is a *static* relaxation of the lattice [15], we expect the $\mathbf{q}_1$ modulation to be static as well.

An alternative possibility to a transverse-displacement modulation and a concomitant CDW in YBCO is the idea of short-range ordered correlated microdomain structures within a disordered matrix, which has been discussed [36] in the study of order-disorder transitions in binary alloys such as Cu-Au and was used subsequently in investigations [37] of disordered alloys. In the context of the Ortho-IV phase, any one of the four permutations of <1110> will nucleate as antiphase (AP) domains with periodicity along the **b** axis determined by $\mathbf{q}_1$, giving rise to symmetric shoulders of $\mathbf{q}$, along the K axis, without any amplitude modulations of the Ortho-IV displacement pattern [6]. Structure factor calculations based on various combinations, however, do not yield relative intensities of $\mathbf{q}$ and $\mathbf{q}_1$ satellites consistent with the observations (Figure 20.10). Although a more subtle arrangement of AP domains cannot be completely ruled out, a simple transverse modulation of a monodomain Ortho-IV pattern adequately accounts for the data [38].

3 Summary

In summary, YBCO compounds are intrinsically inhomogeneous due to the formation of short-range ordered superstructures characterized by a single wave vector of the form $\mathbf{q} = (q_x,0,0)$, as revealed by systematic X-ray diffuse scattering experiments [5–9,20,39]. The modulation vector $\mathbf{q}$ and its range of formation as a function of oxygen stoichiometry are consistent with the theoretical phase diagram of the Ortho phases determined by de Fontaine et al. within the framework of an anisotropic Ising model [8]. Although the Ortho phases form well above room temperature due to kinetic limitations of oxygen mobility, they organize into "nanodomains" (or "nanopatches") with anisotropic correlation lengths of ~3–6 unit cells in extent for the higher order phases. Earlier aging experiments observing the effects of thermal treatment on superconducting transitions have inferred the presence of oxygen vacancy ordering phenomena as well [19]. Recent X-ray diffraction work [39] on YBCO did not find any changes in a 4-unit-cell superstructure at optimal doping when oxygen stoichiometry was kept fixed but the charge-carrier concentration was changed independently. Finally, most remarkable signature of the presence of the Ortho phases, however, is the pattern of correlated atomic displacements of Ba, Cu, and O atoms that give rise to X-ray scattering effects discussed here. Experimentally deduced patterns of atomic displacements are consistent with the theoretical model obtained via first-principles calculations [15]. So it is compelling to conclude that the superstructures characterized by the wave vector $\mathbf{q}$ in YBCO originate from oxygen vacancy ordering throughout the phase diagram. Furthermore, "bowtie"-shaped HDS indicate characteristic strain on longer length scales at all compositions regardless of the average structure.

At higher doping levels for which a coherent Fermi surface is observed in the normal state, effects associated with FS nesting of CuO-chain derived states manifest. In particular, an additional modulation along the chain direction, namely, $\mathbf{q}_1$, coexists with Ortho phases and form inhomogeneously in the lattice. It appears to be associated with an additional transverse-displacement modulation of the Ortho-phase pattern (Figure 20.7) as observed in the optimally doped YBCO [38].

The presence of inhomogeneous O-ordered Ortho phases and associated lattice strain introduces intrinsic length scales ranging from a few unit cells to several nanometers as well as local anisotropy, both of which can have profound and fundamental influence on electronic properties of YBCO compounds. In the context of the related BSCCO family of superconductors, it was discussed how lattice distortions due to O dopants spatially modulate the LDOS and SC gap. On similar grounds, comparable effects on electronic inhomogeneties in YBCO may be expected. Indeed, theoretical work [40] on YBCO also suggests how lattice deformations can suppress the SC order parameter and modulate LDOS. Given the fact that YBCO is intrinsically inhomogeneous throughout the phase diagram, the treatment of superconductivity in terms of a single homogeneous phase in YBCO, and perhaps in related SC cuprates, seems unrealistic.

4 Acknowledgement

High-quality single-crystal samples of YBCO were grown and characterized by Dr. B. W. Veal and Dr. U. Welp of Argonne National Laboratory. Use of the Advanced Photon Source is supported by the U.S. Department of Energy (DOE), Office of Science, under contract no. DE-AC02-06CH11357. The work at University of California, San Diego, was supported by the DOE through Grant No. DE-FG02-03ER46084. At Houston support comes from the State of Texas through the Texas Center for Superconductivity at the University of Houston (TcSUH).

References

1. K. McElroy, J. Lee, J. A. Slejak, D.-H. Lee, H. Eisaki, S. Uchida, and J. C. Davis, *Science* 309 (2005): 1048.
2. T. S. Nunner, B. M. Andersen, A. Melikyan, and P. J. Hirschfeld, *Phys. Rev. Lett.* 95 (2005): 177003.
3. J. P. Castellan, B. D. Gaulin, H. A. Dabkowska, A. Nabialek, G. Gu, X. Liu, and Z. Islam, *Phys. Rev. B* 73 (2006): 174505; M. Izquierdo, S. Megtert, J. P. Albouy, J. Avila, M. A. Valbuena, G. Gu, J. S. Abell, G. Yang, M. C. Asensio, and R. Comes, *Phys. Rev. B* 74 (2006): 054512.
4. S. H. Pan, J. P. O'Neal, R. L. Badzey, C. Chamon, H. Ding, J. R. Engelbrecht, Z. Wang, H. Eisaki, S. Uchida, A. K. Gupta, et al., *Science* 413 (2001): 282.
5. Z. Islam, S. K. Sinha, D. Haskel, J. C. Lang, G. Srajer, B. Veal, D. R. Haeffner, and H. A. Mook, *Phys. Rev. B* 66 (2002): 92501.
6. Z. Islam, X. Liu, S. K. Sinha, J. C. Lang, S. C. Moss, D. Haskel, G. Srajer, P. Wochner, D. R. Lee, D. R. Haeffner, and U. Welp, *Phys. Rev. Lett.* 93 (2004): 157008.
7. J. D. Jorgensen, M. A. Beno, D. G. Hinks, L. Soderholm, K. J. Volin, R. L. Hitterman, J. D. Grace, I. K. Schuller, C. U. Segre, K. Zhang, and M. S. Kleefisch, *Phys. Rev. B* 36 (1987): 3608; C. Chaillot, M. A. Alario-Franco, J. J. Capponi, J. Chenavas, J. L. Hodeau, and M. Marezio, ibid., (1987): 7118; R. Beyers, B. T. Ahn, G. Gorman, V. Y. Lee, S. S. P. Parkin, M. L. Ramirez, K. P. Roche, J. E. Vazquez, T. M. Gür, and R. A. Huggins, *Nature* 340 (1989): 619; D. J. Werder, C. H. Chen, R. J. Cava, and B. Batlogg, *Phys. Rev. B* 37 (1987): 2317; V. Plakhty, A. Stratilatov, Yu. Chenenkov, V. Fedorov, S. K. Sinha, C. K. Loong, B. Gaulin, M. Vlasov, and S. Moshkin, *Solid State Commun.* 84 (1992): 639; F. Yakhou, J.-Y. Henry, P. Burlet, V. P. Plakhty, M. Vlasov, S. Moshkin, Physica C 333 (2000): 146.
8. D. de Fontaine, G. Ceder, and M. Asta, *Nature* 343 (1990): 544; M. Asta, D. de Fontaine, G. Ceder, E. Salomons, and M. Kraichtman, J. *Less. Common Met.* 168 (1991): 39; D. de Fontaine, M. Asta, G. Ceder, R. McCormack, and G. Van Tendeloo, *Euro. Phys. Lett.* 19 (1992): 229; G. Ceder, M. Asta, W. C. Carter, M. Kraichtman, D. de Fontaine, M. E. Mann, and M. Sluiter, *Phys. Rev. B* 41 (1990): 8698.
9. N. H. Andersen, M. von Zimmermann, T. Frello, M. Käll, D. Mønster, P.-A. Lindgård, J. Madsen, T. Niemöller, H. F. Poulsen, O. Schmidt, et al., *Physica C* 317-318 (1999): 259.
10. X. Xiong, P. Wochner, S. C. Moss, Y. Cao, K. Koga, and M. Fujita, *Phys. Rev. Lett.* 76 (1996): 2997.
11. Y. S. Lee, R. J. Birgeneau, M. A. Kastner, Y. Endoh, S. Wakimoto, K. Yamada, R. W. Erwin, S.-H. Lee, and G. Shirane, *Phys. Rev. B* 60 (1999): 3643.
12. G. Chabot-Couture, J. N. Hancock, L. Lu, Z. Islam, and M. Greven, unpublished x-ray scattering work on Hg-cuprate superconductors at the APS.
13. D. de Fontaine, L. T. Willie, and S. C. Moss, *Phys. Rev. B* 36 (1987): 5709.
14. X. Jiang, P. Wochner, S. C. Moss, and P. Zschack, *Phys. Rev. Lett.* 67 (1991): 2167.
15. D. de Fontaine, V. Ozolins, Z. Islam, and S. C. Moss, *Phys. Rev. B* 71 (2005): 212504.

16. J. Zaanen and O. Gunnarsson, *Phys. Rev. B* 40 (1989): 7391; V. J. Emery, S. A. Kivelson, and J. M. Tranquada, *Proc. Natl. Acad. Sci.* 96 (1999): 8814; S. A. Kivelson, E. Fradkin, and V. J. Emery, *Nature* 393 (1998): 550; K. Machida, *Physica C* (Amsterdam) 158 (1989): 192; V. J. Emery, S. A. Kivelson, and O. Zachar, *Phys. Rev. B* 56 (1997): 6120; C. Nayak and F. Wilczek, *Phys. Rev. Lett.* 78 (1997): 2465; B. G. Levi, *Phys. Today* 51 (1998): 19; J. Orenstein and A. J. Millis, *Science* 288 (2000): 468.
17. H. A. Mook, P. Dai, and F. Dǒgan, *Phys. Rev. Lett.* 88 (2002): 97004; P. Dai, H. A. Mook, R. D. Hunt, F. Dǒgan, *Phys. Rev. B* 63 (2001): 54525; H. A. Mook and F. Dǒgan, *Nature* 401 (1999): 145; H. A. Mook, P. Dai, S. M. Hayden, G. Aeppli, T. G. Perrings, and F. Dǒgan, *Nature* 395 (1998): 580.
18. R. J. McQueeney, J. L. Sarrao, P. G. Pagliuso, P. W. Stephens, and R. Osborn, *Phys. Rev. Lett.* 87 (2001): 77001; R. J. McQueeney, Y. Petrov, T. Egami, M. Yethiraj, G. Shirane, and Y. Endoh, *Phys. Rev. Lett.* 82 (1999): 628; T. Egami, J.-H. Chung, R. J. McQueeney, M. Yethiraj, H. A. Mook, C. Frost, Y. Petrov, F. Dogan, Y. Inamura, M. Arai, S. Tajima, and Y. Endoh, *Physica B* 316-317 (2002): 62; L. Pintschovius, W. Reichardt, M. Kläser, T. Wolf, and H. V. Löhneysen, *Phys. Rev. Lett.* 89 (2002): 37001; A. Lanzara, P. V. Bogdanov, X. J. Zhou, S. A. Kellar, D. L. Feng, E. D. Lu, T. Yoshida, H. Eisaki, A. Fujimori, K. Kishio, et al., *Nature* 412 (2001): 510.
19. B. W. Veal, A. P. Paulikas, H. You, H. Shi, Y. Fang, and J. W. Downey, *Phys. Rev. B* 42 (1990): 6305; S. J. Rothman, J. L. Routbort, and J. E. Baker, *Phys. Rev. B* 40 (1989): 8852.
20. M. V. Zimmermann, J. R. Schneider, T. Frello, N. H. Andersen, J. Madsen, M. Käll, H. F. Poulsen, R. Liang, P. Dosanjh, and W. N. Hardy, *Phys. Rev. B* 68 (2003): 104515.
21. W. A. Wooster, *Diffuse X-ray Reflections from Crystals* (Mineola, NY: Dover, 1997).
22. Z.-X. Cai, Y. Zhu, and D. O. Welch, *Phys. Rev. B* 46 (1992): 11014.
23. M. A. Krivoglaz, *Theory of X-ray and Thermal-Neutron Scattering by Real Crystals* (New York: Plenum, 1969).
24. Z. Islam, unpublished x-ray diffuse scattering studies on composition, temperature, and magnetic field dependence of inhomogeneous phases in the entire phase diagram of YBCO .
25. R. W. James, *Optical Principles of the Diffraction of X-rays*, (New York: Oxford University Press, 1954), 565.
26. T. A. Friedmann, M. W. Rabin, J. Giapintzakis, J. P. Rice, and D. M. Ginsberg, *Phys. Rev. B* 42 (1990): 6217.
27. D. N. Basov, R. Liang, D. A. Bonn, W. N. Hardy, B. Dabrowski, M. Quijada, D. B. Tanner, J. P. Rice, D. M. Ginsberg, and T. Timusk, *Phys. Rev. Lett.* 74 (1995): 598.
28. R. Gagnon, S. Pu, B. Ellman, and L. Taillefer. *Phys. Rev. Lett.* 78 (1997): 1976.
29. W. E.Pickett, R. E. Cohen, and H. Krakauer, *Phys. Rev. B* 42 (1990): 8764; W. E.Pickett, H. Krakauer, R. E. Cohen, and D. J. Singh, Science 255 (1992): 46.
30. H. Haghighi, J. H. Kaiser, S. Rayner, R. N. West, J. Z. Liu, R. Shelton, R. H. Howell, F. Solal, and M. J. Fluss, *Phys. Rev. Lett.* 67 (1991): 382.
31. M. C. Schabel, C.-H. Park, A. Matsuura, Z.-X. Shen, D. A. Bonn, R. Liang, and W. N. Hardy, *Phys. Rev. B.* 57 (1998): 6107.
32. S. Krämer and M. Mehring, *Phys. Rev. Lett.* 83 (1999): 396; B. Grévin, Y. Berthier, and G. Collin, *Phys. Rev. Lett.* 85 (2000): 1310; Z. Yamani, B. W. Statt, W. A. MacFarlane, R. Liang, D. A. Bonn, and W. N. Hardy, *Phys. Rev. B* 73 (2006): 212506.
33. H. L. Edwards, D. J. Derro, A. L. Barr, J. T. Markert, and A. L. de Lozanne, *Phys. Rev. Lett.* 75 (1995): 1387.
34. M. Maki, T. Nishizaki, K. Shibata, and N. Kobayashi *Phys. Rev. B* 72 (2005): 024536.
35. H. A. Mook, P. Dai, K. Salama, D. Lee, F. Dǒgan, G. Aeppli, A. T. Boothroyd, and M. E. Mostoller, *Phys. Rev. Lett.* 77 (1996): 370.
36. S. C. Moss, *Phys. Rev. Lett.* 22 (1969): 1108.
37. S. Hashimoto, *Acta. Cryst. A* 30 (1974): 792; ibid., 37 (1981): 511; ibid., 39 (1983): 524.
38. X. Liu, Z. Islam, S. K. Sinha, S. C. Moss, R. J. McQueeney, J. C. Lang, and U. Welp, Phys. Rev. B 78, 134526 (2008).
39. J. Strempfer, I. Zegkinoglou, U. Rütt, M. V. Zimmermann, C. Bernhard, C. T. Lin, Th. Wolf, and B. Keimer, *Phys. Rev. Lett.* 93 (2004): 157007.
40. J.-X. Zhu, K. H. Ahn, Z. Nussinov, T. Lookman, A. V. Balatsky, and A. R. Bishop, *Phys. Rev. Lett.* 91 (2003): 057004.

Part V

Final Remarks

21

Dynamical Bragg and Diffuse Scattering Effects and Implications for Diffractometry in the Twenty-First Century

V. B. Molodkin[1]

M. V. Kovalchuk[2]

A. P. Shpak[1]

S. I. Olikhovskii[1]

Ye. M. Kyslovskyy[1]

A. I. Nizkova[1]

E. G. Len[1]

T. P. Vladimirova[1]

E. S. Skakunova[1]

V. V. Molodkin[1]

G. E. Ice[3]

R. I. Barabash[3]

I. M. Karnaukhov[4]

1. *Institute for Metal Physics, National Academy of Sciences of Ukraine, Kyiv, Ukraine*

2. *Institute of Crystallography, Russian Academy of Sciences, Moscow, Russia*

3. *Oak Ridge National Laboratory, Oak Ridge, TN 37831, USA*

4. *National Scientific Centre "Kharkiv Institute of Physics and Technology," National Academy of Sciences of Ukraine, Kharkiv, Ukraine*

Abstract

We review dynamical diffraction from defective crystals with an emphasis on the role of dynamical effects in diffuse scattering. The principles of a statistical-dynamical theory of X-ray scattering in crystals with homogeneously distributed microdefects of arbitrary types and sizes provides a framework for the remaining chapter. The advantage of this theory is the ability to reduce the predicted scattering to explicit analytical expressions for both coherent and diffuse scattering intensities with direct connections to defect characteristics. These expressions highlight new opportunities to characterize defects. The general theory is specialized to examples of imperfect multilayered and inhomogeneous crystal structures. Dynamical recurrence relations for the amplitudes of coherent waves in a multilayered structure with defects are derived and expressions for the diffuse component of the reflection coefficient of such structures are obtained, which account for the dynamical redistribution of intensities in the transmitted and diffracted coherent waves in each layer. The expressions self-consistently account for both the diffuse scattering contribution to the diffracted intensity and for extinction due to diffuse scattering. Analytic expressions for coherent and diffuse scattering intensities measured by double-crystal diffractometers (DCDs) and triple-crystal diffractometers (TCDs) are derived from the statistical dynamical theory. These expressions account for instrumental factors and take into account the influence on diffuse scattering from imperfections in the monochromator and analyzer crystals. Two example applications are presented: (1) an analysis of structural imperfections in silicon crystals grown by Czochralski- and float-zone methods and (2) a characterization of multilayered semiconductor structures with a single quantum well InGaAs/GaAs.

1 Introduction

Theories for X-ray and neutron structural characterization of real crystals can be categorized into four generations. The first generation, or *kinematic* scattering theory, assumes single scattering from crystals without microdefects [1,2]. The second generation, or *dynamical* scattering theory, assumes multiple scattering from perfect crystals [1,3,4]. An approximate dynamical diffraction theory for crystals with smooth and macroscopically homogeneous strain fields, as developed by Takagi [5,6], Taupin [7], and others [8–13], can also be ascribed to the second structural characterization generation. In this approach, the second derivatives of wave fields in the wave equations are neglected. As a result, this theory is inappropriate for applications with high concentrations of microdefects or to nanosystem applications in which the criterion of smoothness of the scattering potential at distances on the order of the extinction length is not satisfied.

Complementary to the first- and second-generation theories are third- and fourth-generation theories that explicitly introduce statistically distributed point defects and micro defects. The statistical-kinematic (third generation) approach was pioneered by Krivoglaz [14], and the statistical-dynamical (fourth generation) approach was pioneered by Molodkin [15]. In Krivoglaz's kinematic theory, the nonperiodical scattering potential depends on random variables characterizing the distribution of defects and is represented by two sums. The first sum determines an average potential, which when averaged over the entire crystal lattice retains periodicity for a random (homogeneous) distribution of defects. The second sum contains the fluctuating part of the potential, that is, its deviation from the average potential. The periodical part of the potential depends on statistical characteristics of the defects mainly through Krivoglaz's factor, named by him as the static Debye-Waller factor, and is associated with Bragg scattering. The fluctuating part of the potential is associated with the diffuse scattering, whose intensity distribution in reciprocal space is expressed through defect characteristics by the method of Krivoglaz's fluctuation waves of the defect concentration.

When dynamical diffraction takes place, the Bragg and diffuse components of the diffraction intensity depend on both the periodic and fluctuating parts of the potential, which are due to multiple scattering processes, as shown for the first time in works by Molodkin and Tikhonova and others [16–23]. As a consequence, dynamical Bragg scattering is determined not by an average potential, as in Krivoglaz's kinematic theory, but by a renormalized effective periodic potential due to rescattering on the fluctuating part of the sample potential. This effective potential is non-Hermitian and nonlocal and is similar to the coherent potential known from the electron theory of ideal crystals or to the optical potential from nuclear theory. It can also differ substantially from a potential formed by a simple statistical average over the defects, which is its zeroth approximation with respect to the small dynamical parameter of the perturbation theory. The key result of the statistical-dynamical theory is the appearance of an extinction factor due to diffuse scattering. This basic result was recognized for the first time in a work by Molodkin and Tikhonova [16], where the relationship to defect characteristics and physics of the extinction factor were also established. Some time later, a similar extinction factor was introduced independently by Dederichs [24,25] in the framework of an optical potential method that is limited to Bragg scattering. The attenuation of Bragg waves due to diffuse scattering, which is described by the Molodkin-Tikhonova-Dederichs factor, is in general substantially more essential and sensitive to defect characteristics than that due to the Krivoglaz-Debye-Waller factor. Thus, unlike its kinematic limit, Bragg scattering with significant multiple scattering is not less sensitive to defect characteristics than diffuse scattering. On the other hand, it has been shown [16–23] that multiple rescattering of the diffuse scattering waves by the periodic part of the potential transforms these waves into Bloch waves. The predicted [16–23] effects of anomalous transmission and extinction of the diffuse scattering intensity have been verified in numerous experimental investigations [26–29].

Statistical-dynamical scattering theory has been applied to a number of specific problems. These problems include characterization of crystals with randomly distributed small Coulomb-type defects with weak distortion fields, characterization of microdefects with sizes commensurate with an extinction length [30,31], and characterization of crystals with a superposition of an elastic macroscopic strain field and a fluctuating strain field caused by randomly distributed microdefects [32,33]. Other theories, which have been developed *de facto* for mosaic crystals (Kato, Holý, Chukhovskii, and others [34–39]), have been applied for practical analysis in many cases; however, they are sometimes ill suited for the characterization of randomly distributed defects. The statistical-dynamical approach with its **direct analytical relations between the coherent and diffuse scattering intensities and defect characteristics** is much more sensitive to random defects (see, e.g., experimental verifications of the previous statistical dynamical theories by Olekhnovich et al. [40] and Schneider et al. [41]).

In addition to the above mentioned effects, which describe the distinguishing features of defect influences on the diffraction patterns in kinematic and dynamical theories, there are completely new effects and phenomena in the dynamical diffraction theory of imperfect crystals, which are connected with the fundamental distinction in dependencies of Bragg and diffuse components on conditions of dynamical diffraction [42]. In this connection, the contribution of diffuse components has appeared depending on the conditions of dynamical diffraction. These new effects offer approaches to defect characterization, which are fundamentally absent in the kinematic theory. These effects and phenomena are summarized below:

Dynamical Effect I. The relative contribution of the diffuse scattering intensity component [42] is found to be *thickness dependent* due to a substantial difference between the dynamical extinction factors for Bragg scattering and for diffuse scattering. As a result, even for extremely weak distortions in single crystals, if the crystal thickness is on the order of the radiation absorption depth, then there is an anomalously large increase in the diffuse scattering intensity component, and the diffuse scattering component can dominate.

This means, for example, that even when the static Debye-Waller factor exponent is small (e.g., <1%) and the diffuse scattering predicted by kinematic theory is less than 2%, the diffuse scattering predicted by dynamical theory can be two orders of magnitude larger. Of course, an increase in the static Debye-Waller factor exponent leads to further increases in the contribution of the diffuse component and enhances the effect of its anomalous growth with the thickness increase.

This effect challenges the prevailing intuition that only a negligibly small contribution of the diffuse component exists for such systems and establishes qualitatively new effects in dynamical diffraction in single crystals. This also impacts the classification of crystal defects with respect to their influence on the diffraction pattern.

Dynamical Effect II. Dynamical diffraction from imperfect crystals (as in perfect crystals) violates the conservation law of total (coherent plus diffuse) integrated reflection power (TIRP) of crystals with defects [42]. In particular, total integrated reflectivity is no longer independent on crystal distortions as observed in the kinematic scattering approximation. This behavior provides for the sensitivity of thickness, spectral, azimuth, or strain TIRP dependencies to defect characteristics. These dependencies can be expressly measured and interpreted by using statistical-dynamical formulas.

Dynamical Effect III. There is a significant change in the influence of defects on the diffraction pattern depending on the diffraction geometry (i.e., when making the transition from Laue to Bragg geometries or from thin to thick crystals [42]). This behavior allows for more robust investigations through independent experiments on the same crystal.

Dynamical Effect IV. There is an asymmetry in the azimuth TIRP dependencies for crystals with large defects [43]. It should be noted that the new effects listed above are related to diffuse scattering and are caused by multiple scattering processes on both periodic and fluctuating potential parts, that is, by the dynamical character of Bragg and diffuse components. For this reason, they are explicitly describable neither by the dynamical theory of perfect crystals, in which diffuse scattering is absent, nor by the kinematic theory for crystals with defects, where multiple scattering effects are not taken into account.

Based on the theoretical results above, the experimental foundation of fourth-generation crystallography—that is, diffuse dynamical combined diffractometry (DDCD)—has been developed through joint efforts by scientists from Ukraine, Russia, and the United States, namely, by A. P. Shpak with collaborators in Ukraine, M. V. Kovalchuk with collaborators in Russia, and G. E. Ice with collaborators in the United States, and has been protected by patents from Ukraine [44–46]. Fourth-generation experimental methods include dynamical differential [47], differential-integral [48,49], and integral [50, 51] triple-crystal diffractometry methods; the method of total rocking curves [52–54]; TIRP methods [42]; and standing-wave methods [55–58] that account for diffuse scattering.

As described above, the essential new behavior observed in the statistical-dynamical theory is the dependence of the diffuse-to-Bragg component ratios on both the crystal defect characteristics *and* the conditions of the dynamical diffraction. This means that there can be purposeful and controllable substantial changes of the diffuse scattering contribution and their sensitivity to various defect types, even for a crystal with a fixed defect structure, by changing the diffraction geometry (from transmission to reflection) or by changing the approximation of dynamical diffraction (from thin to thick crystal). Changes can also be introduced by altering the experimental techniques (from one-axis to double- and triple-axis diffractometry). The distinctions in selective sensitivity to defects due to the purposeful changes of the diffuse scattering contribution are connected with different signs of their influence on Bragg and diffuse-scattering components of the diffraction intensity and offer a way to extend experimental control by providing independent diffraction experiments to help in multiparametric characterization of materials with several defect types.

Applications include the following:

- Quantitative nondestructive characterization of multiple defect types simultaneously present in imperfect crystals
- Nondestructive layer-by-layer characterization of strains and microdefects in multilayered or inhomogeneous crystal systems
- Integral diffractometry of fast structural transformations (X-ray movies)
- Characterization of defect structures in macroscopically strained crystals and controlled reflectivity of focusing elements for synchrotron radiation by using microdefects

Here we develop the underlying principles of a generalized statistical-dynamical theory of X-ray or neutron scattering from imperfect single crystals with randomly distributed microdefects (section 2) or from imperfect multilayered structures with two-dimensional and three-dimensional defects, also called microdefects for the sake of brevity (section 4).

Analytical expressions for coherent and diffuse scattering intensity components measured by double-crystal diffractometers (DCDs) (sections 3 and 4) and triple-crystal diffractometers (TCDs) (section 5) have been derived from the statistical-dynamical theory, including instrumental factors. The expressions explicitly take into account the influence of diffuse scattering from imperfections in the monochromator and analyzer crystals and have been applied to characterize structural imperfections in silicon crystals grown by Czochralski- and float-zone methods (sections 3.2 and 5.3) and to characterize the structure of InGaAs/GaAs single quantum well (QW) multilayered structures (section 4.3).

2 Dynamical X-ray Diffraction by Imperfect Single Crystals

The classical dynamical scattering theory developed by Ewald [59], Bethe [60], and von Laue [1] for perfect crystals can be generalized for the case of X-ray diffraction in imperfect crystals containing randomly distributed defects (point defects, their clusters, and microdefects) by representing the crystal polarizability as the sum of average and fluctuating parts and by using perturbation theory with these quantities as small parameters [16–23,30–33]. This approach makes substantial use of terms in the dispersion equation and surface, and simplifies the scattering problem in momentum space.

2.1 Basic Equations

The polarizability $\chi(\mathbf{r})$ of an imperfect crystal with defects is a nonperiodic function of space coordinates and can be represented as a Fourier integral over all momentum space. After performing a

Fourier transformation of Maxwell's equations, we obtain at $|\chi| << 1$ the set of basic equations in momentum space:

$$\left(K^2 - k^2\right)\mathbf{D}_{\mathbf{k}} - \sum_{\mathbf{G}}\sum_{\mathbf{q}} \chi_{\mathbf{G+q}} \mathbf{k} \times \mathbf{k} \times \mathbf{D}_{\mathbf{k-G-q}} = 0 . \tag{21.1}$$

Here $K = 2\pi / \lambda$ is the modulus of the wave vector **K** of the incident plane wave, λ is the X-ray wavelength, $\mathbf{D}_{\mathbf{q}}$ and $\chi_{\mathbf{q}}$ are Fourier components of electric induction and crystal polarizability, respectively, **G** and **q** run over all reciprocal lattice vectors, and N is the number of discrete values in the first Brillouin zone (N is the number of crystal lattice points). This set consists of an infinite number of linear algebraic equations for an infinite number of unknown amplitudes of coherent and diffusely scattered plane waves in the crystal.

The set of basic Equation 21.1 is valid only for a homogeneous random distribution of microdefects. Such a distribution is equivalent to averaging over the crystal lattice points and averaging over the ensemble of defects, including the translation invariance of the ensemble-averaged polarizability:

$$\left\langle \chi\left(\mathbf{r} + \mathbf{R}_s^0\right)\right\rangle = \left\langle \chi(\mathbf{r})\right\rangle = \sum_{\mathbf{G}} \chi_{\mathbf{G}} \exp(-i\mathbf{G}\mathbf{r}), \tag{21.2}$$

The corner brackets denote the ensemble averaging, **r** is a space coordinate, $\mathbf{R}_s^0 = \left\langle \mathbf{R}_s \right\rangle$ are the average lattice vectors (the subscript s runs over N lattice points), and **G**s are 2π times the corresponding reciprocal lattice vectors. The Fourier component of polarizability in Equation 21.2 is equal to $\chi_{\mathbf{G}} = \chi_{\mathbf{G}}^0 E$ if changes of structure factors of the unit cells due to static distortions can be neglected [14]. Here the value $\chi_{\mathbf{G}}^0$ is referenced to the perfect crystal and the static Debye-Waller factor is defined as $E = \exp(-L_{\mathbf{G}}) = \left\langle \exp - i\mathbf{G}\mathbf{u}(\mathbf{R}_s)\right\rangle$, where the static displacement field $\mathbf{u}\left(\mathbf{R}_s\right) = \mathbf{R}_s - \mathbf{R}_s^0$ can be described at small concentrations of randomly distributed defects $c << 1$ as the superposition of static displacements of matrix atoms **U**(**r**), which are caused by a single defect [14]:

$$\mathbf{u}(\mathbf{R}_s) = \sum_t (c_t - c)\mathbf{U}(\mathbf{R}_s - \mathbf{R}_t) . \tag{21.3}$$

The random occupation numbers, c_t, in Equation 21.3 are equal to 1 if the lattice point with a radius-vector $\mathbf{R}_t^0$ is occupied by the center of a microdefect and 0 if not. The averaging in Equation 21.2 is performed just over these random numbers, for which $\left\langle c_t \right\rangle = c$ is valid.

Thus the polarizability of the crystal with randomly distributed microdefects can be represented as the sum of average and fluctuating (random) parts; that is,

$$\chi(\mathbf{r}) = \left\langle \chi(\mathbf{r})\right\rangle + \delta\chi(\mathbf{r}) . \tag{21.4}$$

The average polarizability $\langle\chi(\mathbf{r})\rangle$ in Equation 21.4 is described by Equation 21.2, and the fluctuating part by definition is equal to $\delta\chi(\mathbf{r})=\chi(\mathbf{r})-\langle\chi(\mathbf{r})\rangle$

or

$$\delta\chi(\mathbf{r})=\sum_{\mathbf{G}}\sum_{\mathbf{q}\neq 0}\delta\chi_{\mathbf{G+q}}\,e^{-i(\mathbf{G+q})\mathbf{r}}, \tag{21.5}$$

where $\delta\chi_{\mathbf{G+q}}=\chi_{\mathbf{G+q}}-\chi_{\mathbf{G}}\delta_{0,\mathbf{q}}$ with $\delta_{0,\mathbf{q}}=1$ if $\mathbf{q}=0$ and 0 if $\mathbf{q}\neq 0$.

Similarly, the wave field in the crystal can be subdivided into average and fluctuating parts,

$$\mathbf{D}(\mathbf{r})=\langle\mathbf{D}(\mathbf{r})\rangle+\delta\mathbf{D}(\mathbf{r}), \tag{21.6}$$

which correspond to coherent and diffuse scattering components:

$$\langle\mathbf{D}(\mathbf{r})\rangle=e^{-i\mathbf{K}_0\mathbf{r}}\sum_{\mathbf{G}}\mathbf{D}_{\mathbf{G}}\,e^{-i\mathbf{G}\mathbf{r}}, \tag{21.7}$$

$$\delta\mathbf{D}(\mathbf{r})=e^{-i\mathbf{K}_0\mathbf{r}}\sum_{\mathbf{G}}\sum_{\mathbf{q}\neq 0}\mathbf{D}_{\mathbf{G+q}}\,e^{-i(\mathbf{G+q})\mathbf{r}}. \tag{21.8}$$

The constant wave vector $\mathbf{K}_0$ of the transmitted coherent plane wave in the crystal has been introduced for the sake of convenience.

In the two-beam case of diffraction, only two so-called strong Bragg (coherent) waves are excited in the crystal, and consequently, in Equation 21.1 as well as in Equations 21.7 and 21.8, only two amplitudes of coherent waves $(\mathbf{D}_{\mathbf{G}})$ and corresponding $2N$ amplitudes of diffuse scattering waves $(\mathbf{D}_{\mathbf{G+q}})$, where $\mathbf{G}=0$ and $\mathbf{H}$, should be retained, whereas the rest of amplitudes of weak Bragg and diffuse scattering waves can be neglected. The infinite set (1) can be represented, then, as two sets, namely, one for two amplitudes of coherent transmitted (D_0) and diffracted ($D_{\mathbf{H}}$) plane waves with wave vectors $\mathbf{k}=\mathbf{K}_0$ and $\mathbf{k}=\mathbf{K}_{\mathbf{H}}=\mathbf{K}_0+\mathbf{H}$, respectively,

$$\begin{cases}(-2\varepsilon_0+\chi_0)D_0+CE\chi_{-\mathbf{H}}D_{\mathbf{H}}=-\sum_{\mathbf{q}}(\delta\chi_{\mathbf{q}}D_{-\mathbf{q}}+C\delta\chi_{-\mathbf{H+q}}D_{\mathbf{H-q}}),\\ CE\chi_{\mathbf{H}}D_0+(-2\varepsilon_{\mathbf{H}}+\chi_0)D_{\mathbf{H}}=-\sum_{\mathbf{q}}(C\delta\chi_{\mathbf{H+q}}D_{-\mathbf{q}}+\delta\chi_{\mathbf{q}}D_{\mathbf{H-q}}),\end{cases} \tag{21.9}$$

and another one for $2N$ amplitudes of transmitted ($D_{\mathbf{q}}$) and diffracted ($D_{\mathbf{H+q}}$) diffuse scattering waves with wave vectors $\mathbf{k} = \mathbf{K}_{0\mathbf{q}} = \mathbf{K}_0 + \mathbf{q}$ and $\mathbf{k} = \mathbf{K}_{\mathbf{Hq}} = \mathbf{K}_{\mathbf{H}} + \mathbf{q}$, respectively,

$$\begin{cases} (-2\varepsilon_{0\mathbf{q}} + \chi_0) D_{\mathbf{q}} + CE\chi_{-\mathbf{H}} D_{\mathbf{H+q}} = -(\delta\chi_{\mathbf{q}} D_0 + C\delta\chi_{-\mathbf{H+q}} D_{\mathbf{H}}) \\ \qquad - \sum\limits_{\mathbf{q}' \neq \mathbf{q}} \left(\delta\chi_{\mathbf{q}'} D_{\mathbf{q}-\mathbf{q}'} + C\delta\chi_{-\mathbf{H}+\mathbf{q}'} D_{\mathbf{H}+\mathbf{q}-\mathbf{q}'} \right), \\ CE\chi_{\mathbf{H}} D_{\mathbf{q}} + (-2\varepsilon_{\mathbf{Hq}} + \chi_0) D_{\mathbf{H+q}} = -(C\delta\chi_{\mathbf{H+q}} D_0 + \delta\chi_{\mathbf{q}} D_{\mathbf{H}}) \\ \qquad - \sum\limits_{\mathbf{q}' \neq \mathbf{q}} \left(C\delta\chi_{\mathbf{H}+\mathbf{q}'} D_{\mathbf{q}-\mathbf{q}'} + \delta\chi_{\mathbf{q}'} D_{\mathbf{H}+\mathbf{q}-\mathbf{q}'} \right), \end{cases} \tag{21.10}$$

where the polarization factor is $C = 1$ or $\cos(2\theta_B)$ for σ and π polarization states, respectively; θ_B is the Bragg angle; and the excitation errors for strong Bragg waves in Equation 21.9 and for diffuse scattering waves in Equation 21.10, respectively, are defined as

$$\varepsilon_0 = \frac{K_0 - K}{K}, \quad \varepsilon_{\mathbf{H}} = \frac{K_{\mathbf{H}} - K}{K}, \quad \varepsilon_{0\mathbf{q}} = \frac{K_{0\mathbf{q}} - K}{K}, \quad \varepsilon_{\mathbf{Hq}} = \frac{K_{\mathbf{Hq}} - K}{K}. \tag{21.11}$$

To describe the attenuation of the coherent waves due to diffuse scattering, one should substitute the amplitudes of diffuse scattering waves from Equation 21.10 into the right-hand side of Equation 21.9. Neglecting small terms in the framework of perturbation theory, we obtain the set of basic equations for strong Bragg waves:

$$\begin{cases} \left(-2\varepsilon_0 + \chi_0 + \Delta\chi_{00}\right) D_0 + \left(CE\chi_{-\mathbf{H}} + \Delta\chi_{0\mathbf{H}}\right) D_{\mathbf{H}} = 0, \\ \left(CE\chi_{\mathbf{H}} + \Delta\chi_{\mathbf{H}0}\right) D_0 + \left(-2\varepsilon_{\mathbf{H}} + \chi_0 + \Delta\chi_{\mathbf{HH}}\right) D_{\mathbf{H}} = 0. \end{cases} \tag{21.12}$$

Here $\Delta\chi_{\mathbf{GG}'}$ are the dispersion corrections due to diffuse scattering and superscript 0 at Fourier components of the perfect crystal polarizability $\chi^0_{\pm\mathbf{H}}$ has been dropped.

It should be noted that the sets of coupled Equations 21.9 and 21.10 can be solved by iteration [30] if the following inequalities hold:

$$\left|\Delta\chi_{\mathbf{GG}'}\right| << E\left|\chi_{\pm\mathbf{H}}\right|, \tag{21.13}$$

where $\mathbf{G}$ and $\mathbf{G}'$ = 0 or $\mathbf{H}$. In the case when inequalities

$$\left|\langle\delta\chi_{\mathbf{q}}\delta\chi_{-\mathbf{q}}\rangle\right| << \left|\langle\delta\chi_{\mathbf{q}}\delta\chi_{-\mathbf{q}\pm\mathbf{H}}\rangle\right| << \left|\langle\delta\chi_{\mathbf{q}\pm\mathbf{H}}\delta\chi_{-\mathbf{q}\pm\mathbf{H}}\rangle\right| \tag{21.14}$$

are valid, which takes place at $q << H$ for typical microdefects at sufficiently small $\mathbf{q}$ values [14], the dispersion corrections to the wave vectors of coherent waves, which describe the attenuation of coherent waves due to diffuse scattering and simultaneously are renormalizing contributions to Fourier components of the crystal polarizability in Equation 21.12, satisfy four integral equations $(\mathbf{G},\mathbf{G}'=0,\mathbf{H})$:

$$\Delta\chi_{\mathbf{GG}'} = sC^2 \sum_{\mathbf{q}} \frac{a_{\mathbf{GG}'}(\mathbf{q}) + \Delta\chi_{\mathbf{G}'\mathbf{G}}(\mathbf{q})}{d(\mathbf{q})} S_{\mathbf{GG}'}(-\mathbf{q}), \tag{21.15}$$

$$d(\mathbf{q}) = \left[-2\varepsilon_{0\mathbf{q}} + \chi_0 + \Delta\chi_{00}(\mathbf{q})\right]\left[-2\varepsilon_{\mathbf{Hq}} + \chi_0 + \Delta\chi_{\mathbf{HH}}(\mathbf{q})\right]$$

$$-\left[CE\chi_{-\mathbf{H}} + \Delta\chi_{0\mathbf{H}}(\mathbf{q})\right]\left[CE\chi_{\mathbf{H}} + \Delta\chi_{\mathbf{H}0}(\mathbf{q})\right], \tag{21.16}$$

$$a_{\mathbf{GG}}(\mathbf{q}) = -2\varepsilon_{\mathbf{Gq}} + \chi_0,$$

$$a_{\mathbf{GG}'}(\mathbf{q}) = CE\chi_{\mathbf{G}'-\mathbf{G}} \text{ at } \mathbf{G}' \neq \mathbf{G}.$$

Here $s=1$ at $\mathbf{G}'\neq\mathbf{G}$, and $s=-1$ at $\mathbf{G}'=\mathbf{G}$. Four similar integral equations hold for the four dispersion corrections to the wave vectors of the diffuse scattering waves, $\Delta\chi'_{\mathbf{GG}'}(\mathbf{q})$. The correlation functions in Equation 21.15 are determined by ensemble averaging of the quadratic combinations of fluctuating Fourier components of polarizability as

$$S_{\mathbf{GG}'}(\mathbf{q}) = \left\langle \delta\chi_{\mathbf{q}-\mathbf{H}+2\mathbf{G}}\delta\chi_{-\mathbf{q}+\mathbf{H}-2\mathbf{G}'} \right\rangle. \tag{21.17}$$

These functions have been introduced instead of the corresponding fluctuating products of these components, which is justified for sufficiently large concentrations of defects.

After solving Equation 21.11 by iterations and performing integration over $\mathbf{q}$, one obtains the following expression for the dispersion corrections to the wave vectors of coherent waves due to diffuse scattering ($\delta,\ \tau = 1,\ 2$):

$$\Delta\chi^{\delta}_{\mathbf{GG}'}(\Delta\theta) = P^{\delta}_{\mathbf{GG}'}(\Delta\theta) - i\mu^{\delta}_{\mathbf{GG}'}(\Delta\theta)/K, \tag{21.18}$$

and, similarly, for the dispersion corrections to the wave vectors of diffuse scattering waves due to n-multiple diffuse-scattering processes [30,31]:

$$\Delta\chi'^{\delta\tau}_{\mathbf{GG}'}(\Delta\theta,\Delta\theta') = P'^{\delta\tau}_{\mathbf{GG}'}(\Delta\theta,\Delta\theta') - i\mu'^{\delta\tau}_{\mathbf{GG}'}(\Delta\theta,\Delta\theta') / K \cdot \tag{21.19}$$

Here real and imaginary parts in Equations 21.18 and 21.19 are connected via dispersion relations [30,32] and are of the same order of magnitude.

2.2 Coherent X-ray Scattering

The coherent wave field in a crystal with a homogeneous distribution of defects can be represented in the two-beam approximation as the sum of transmitted and diffracted waves for each polarization state (σ and π):

$$D(\mathbf{r}) = D_{\mathrm{T}}(\mathbf{r}) + D_{\mathrm{S}}(\mathbf{r}), \tag{21.20}$$

$$D_{\mathrm{T}}(\mathbf{r}) = \sum_{\delta} D_0^{\delta}\, \mathrm{e}^{-i\mathbf{K}_0^{\delta}\mathbf{r}},$$

$$D_{\mathrm{S}}(\mathbf{r}) = \sum_{\delta} D_{\mathbf{H}}^{\delta}\, \mathrm{e}^{-i\mathbf{K}_{\mathbf{H}}^{\delta}\mathbf{r}}, \tag{21.21}$$

where $\delta = 1,2$. The amplitudes of transmitted (D_0^{δ}) and diffracted ($D_{\mathbf{H}}^{\delta}$) coherent plane waves in Equation 21.21 can be found from Equation 21.12 after imposing appropriate boundary conditions. In the case of Bragg diffraction geometry one obtains

$$D_0^{\delta} = (-1)^{\delta} E_0 \frac{B_{\delta'}}{B_1 - B_2},$$

$$D_{\mathbf{H}}^{\delta} = c^{\delta} D_0^{\delta},$$

$$\delta' \neq \delta, \tag{21.22}$$

$$B_{\delta} = c^{\delta} e^{-iK\Delta_{\delta} t},$$

$$c^{\delta} = -\frac{-2\gamma_0\Delta_{\delta} + \chi_0 + \Delta\chi_{00}^{\delta}}{CE\chi_{-\mathbf{H}} + \Delta\chi_{0\mathbf{H}}^{\delta}}, \tag{21.23}$$

where E_0 is the amplitude of the incident plane wave in a vacuum and t is the crystal thickness. Amplitudes of coherent wave fields are related with characteristics of crystal defect structure by the static Debye-Waller factor, E, and dispersion corrections due to diffuse scattering $\Delta\chi^{\delta}_{\mathbf{GG}'}$ ($\mathbf{G}$ and $\mathbf{G}' = 0, \mathbf{H}$). Accommodations of the wave vectors of strong Bragg waves $\mathbf{K}^{\delta}_{\mathbf{H}} = \mathbf{K}^{\delta}_{0} + \mathbf{H}$ and $\mathbf{K}^{\delta}_{0} = \mathbf{K} + K\Delta_{\delta}\mathbf{n}$ in Equation 21.21 are described by the following expression:

$$\Delta_\delta = \frac{1}{2\gamma_0}\left(\chi_0 + \Delta\chi^{\delta}_{00}\right) + \frac{\lambda}{2\Lambda}\left[y + (-1)^{\delta}\sqrt{y^2 - 1}\right]. \tag{21.24}$$

The normalized angular deviation of the crystal is defined in Equation 21.24 as

$$y = (\alpha - \alpha_0)\sqrt{b} / \sigma,$$

$$\sigma^2 = \left(CE\chi_{\mathbf{H}} + \Delta\chi_{\mathbf{H}0}\right)\left(CE\chi_{-\mathbf{H}} + \Delta\chi_{0\mathbf{H}}\right), \tag{21.25}$$

$$\alpha = (\mathbf{K} + \mathbf{H})\Delta\mathbf{H} / K^2 = -\Delta\theta \ \sin(2\theta_B),$$

$$2\alpha_0 = \chi_0 + \Delta\chi_{\mathbf{HH}} + \left(\chi_0 + \Delta\chi_{00}\right) / b, \tag{21.26}$$

where $\Lambda = \lambda\sqrt{\gamma_0|\gamma_{\mathbf{H}}|} / \sigma$ is the extinction length, $\Delta\theta$ is an angular deviation of the crystal from its exact Bragg reflection position, $b = \gamma_0|\gamma_{\mathbf{H}}|^{-1}$ is the parameter of diffraction asymmetry, $\gamma_0 = \sin(\theta_B - \psi)$ and $\gamma_{\mathbf{H}} = -\sin(\theta_B + \psi)$ are direction cosines with respect to the inner normal $\mathbf{n}$ to the entrance crystal surface for wave vectors of transmitted and diffracted waves, respectively, and ψ is the angle between crystal surface and reflecting planes.

The amplitude of diffracted plane wave in a vacuum $E_{\mathbf{H}}$, which is generated by the superposition of diffracted coherent plane waves in the crystal $D_S(\mathbf{r})$, is obtained from the boundary condition for this wave field on the crystal entrance surface,

$$D_S(\mathbf{r}) = E_{\mathbf{H}} \exp\left(-i\mathbf{K}'_{\mathbf{H}}\mathbf{r}\right)\Big|_{z=0}, \tag{21.27}$$

where $\mathbf{K}'_{\mathbf{H}} = \mathbf{K}^{\delta}_{\mathbf{H}} - K\varepsilon^{\delta}_{\mathbf{H}}\gamma^{-1}_{\mathbf{H}}\mathbf{n}$, $\varepsilon^{\delta}_{\mathbf{H}} = \varepsilon^{\delta}_{0}\gamma_{\mathbf{H}}/\gamma_0 + \alpha$, $\varepsilon^{\delta}_{0} = \gamma_0\Delta_\delta$, and $K'_{\mathbf{H}} = K$. The amplitude reflectivity of the crystalline plate is described then by the expression

$$r(\Delta\theta) = (\zeta b)^{-1/2}\frac{E_{\mathbf{H}}}{E_0} = (\zeta b)^{1/2}\frac{e^{iK\Delta_1 t} - e^{iK\Delta_2 t}}{c^2\, e^{iK\Delta_1 t} - c^1\, e^{iK\Delta_2 t}}, \tag{21.28}$$

where $\zeta = (CE\chi_{\mathbf{H}} + \Delta\chi_{\mathbf{H}0})(CE\chi_{-\mathbf{H}} + \Delta\chi_{0\mathbf{H}})^{-1}$. In the limiting case of the thick crystal—that is, at μ_0 $t >> 1$, where μ_0 is the linear coefficient of photoelectric absorption—Equation 21.28 can be simplified to the form

$$r(\Delta\theta) = X_0 = (\zeta\ b)^{-1/2}\left(y - s\sqrt{y^2 - 1}\right), \tag{21.29}$$

where $s = \operatorname{sgn} y_r$ and $y_r = \operatorname{Re}\ y$.

Thus the coherent component of the differential crystal reflectivity is described by the expression

$$R_{\mathrm{B}}(\mathbf{k}) = R_{\mathrm{coh}}(\Delta\theta)\ \delta(\Delta\theta' - b\Delta\theta)\ \delta(\varphi' - \varphi), \tag{21.30}$$

$$R_{\mathrm{coh}}(\Delta\theta) = \left| r(\Delta\theta) \right|^2, \tag{21.31}$$

where $\delta(x)$ is the Dirac δ function, $\Delta\theta$ and $\Delta\theta'$ are angular deviations of the wave vectors of incident and scattered coherent plane waves in the diffraction plane (**K,H**) from corresponding exact Bragg directions, and φ and φ' are angular deviations of the same vectors in a vertical plane.

2.3 Absorption Due to Diffuse Scattering

The coefficients of absorption due to diffuse scattering can be calculated in the first approximation of the dynamical perturbation theory developed by Molodkin et al. [30] as

$$\mu^{\delta}_{\mathbf{GG}'}(\Delta\theta) = C^2 K^2 \frac{V}{8\pi^2} \sum_{\tau=1,2} (-1)^{\tau} \operatorname{Re} \int_{K'=K^{\delta\tau}_{0\mathbf{q}}} dS_{\mathbf{K}'} f_{\mathbf{GG}'}(\mathbf{q}_{\delta\tau}) S_{\mathbf{GG}'}(-\mathbf{q}_{\delta\tau}), \tag{21.32}$$

where V is the crystal volume, the integration at fixed δ is performed over two sheets of the dispersion surface for wave vectors $\mathbf{K}^{\delta t}_{0\mathbf{q}}$ of diffuse scattering plane waves, and the following notation was used:

$$f_{\mathbf{GG}'}(\mathbf{q}_{\delta\tau}) = a^{\tau}_{\mathbf{GG}'}\left[4\gamma_{\mathbf{H}}\left(\Delta'^{\,0}_{\,2} - \Delta'^{\,0}_{\,1}\right)\right]^{-1}, \tag{21.33}$$

$$a^{\tau}_{\mathbf{GG}'} = -2\varepsilon^{0\tau}_{\mathbf{Gq}} + \chi_0, \quad a^{\tau}_{0\mathbf{H}} = CE\chi_{\mathbf{H}}, \quad a^{\tau}_{\mathbf{H}0} = CE\chi_{-\mathbf{H}}.$$

The wave vectors of diffuse scattering waves are complex due to complex momentum transfers, $\mathbf{q}_{\delta\tau}$, from wave vectors of strong Bragg waves, $\mathbf{K}_0^{\delta}$,

$$\mathbf{K}_{0\mathbf{q}}^{\delta\tau}=\mathbf{K}_0^{\delta}+\mathbf{q}_{\delta\tau},\quad \mathbf{q}_{\delta\tau}=\mathbf{k}+K\left(\Delta'^{\,0}_{\tau}-\Delta^{\,0}_{\delta}\right)\mathbf{n}\,, \tag{21.34}$$

where $\mathbf{k}=\mathbf{K}'-\mathbf{K}-\mathbf{H}$ is the deviation of the wave vector $\mathbf{K}'$ of a diffusely scattering wave in a vacuum from the reciprocal lattice point H. The accommodations of wave vectors of diffuse scattering waves in Equation 21.34 are described by the expression

$$\Delta'^{\,0}_{\tau}=\frac{\chi_0}{2\gamma_0}+\frac{\lambda}{2\Lambda_0}\left[y_0'+(-1)^{\tau}\sqrt{y_0'^2-1}\right], \tag{21.35}$$

$$y_0'=\left(\alpha'-\alpha_0^0\right)\sqrt{b}\,/\,\sigma_0,\qquad \alpha'=-\Delta\theta'\sin\left(2\theta_B\right),$$

where $\alpha_0^0=\chi_0\left(1+b^{-1}\right)/2$ and $\sigma_0=C^2E^2\chi_{\mathbf{H}}\chi_{-\mathbf{H}}$.

As the behavior of the correlation functions in Equation 21.32 is weakly dependent on different δ and τ indices, one can approximately put

$$S_{\mathbf{GG}'}(\mathbf{q}_{\delta\tau})\approx S(\mathbf{q})=\mathrm{Re}\left\langle\delta\chi_{\mathbf{q}-\mathbf{H}+2\mathbf{G}}\delta\chi_{-\mathbf{q}+\mathbf{H}-2\mathbf{G}'}\right\rangle, \tag{21.36}$$

where $\mathbf{q}=\mathbf{k}+i\mu_i\mathbf{n}$, and μ_i is the limiting value of an interference absorption coefficient $\mu_i^{\delta\tau}=K\,\mathrm{Im}\left(\Delta'^{\,0}_{\tau}-\Delta^{0}_{\delta}\right)$ at $|\Delta\theta|$ and $|\Delta\theta'|\to\infty$. Then we can perform the approximate summation in Equation 21.32 accounting for Equations 21.33 to 21.35 and can replace the integration over sheets of the dispersion surface by the integration over the plane tangent to the Ewald sphere near the reciprocal lattice point, considered as

$$\mu_{00}(\Delta\theta)\approx b\frac{C^2V}{4\lambda^2}\int d\mathbf{k}'S(\mathbf{q}), \tag{21.37}$$

$$\mu_{\mathbf{HH}}(\Delta\theta)\approx b^{-1}\mu_{00}(\Delta\theta),\quad \mu_{0\mathbf{H}}(\Delta\theta)\approx\mu_{\mathbf{H}0}(\Delta\theta)\approx 0\,. \tag{21.38}$$

To calculate the correlation function $S(\mathbf{q})$, when *several* types of microdefects are simultaneously present in the crystal, one can use the method of fluctuation waves of the defect concentration [14]. Using this method at a small concentration of α type defects, $c_{\alpha}<<1$, one can write the approximate expression in the Huang scattering region, $q << k_{m\alpha} = 1/R_{eff}^{\alpha}$, where R_{eff}^{α} is an effective microdefect radius specified below for various microdefect types:

$$\delta\chi_{\mathbf{H+q}} \approx iE\chi_{\mathbf{H}} \sum_{\alpha} \left(\mathbf{HU}_{\mathbf{q}\alpha}\right) c_{\mathbf{q}\alpha}. \quad (21.39)$$

Fourier components of the displacement field from a single defect and of the fluctuating field of the concentration of α type defects in Equation 21.39 are defined as

$$\mathbf{U}_{\mathbf{q}\alpha} = \frac{1}{v_c} \int d\mathbf{r} \mathbf{U}_{\alpha}(\mathbf{r}) e^{i\mathbf{qr}}, \qquad c_{\mathbf{q}\alpha} = \frac{1}{N} \sum_{t} (c_{t\alpha} - c_{\alpha}) e^{i\mathbf{qR}_t}, \quad (21.40)$$

where v_c is the unit cell volume. If only pair correlations in defect distributions exist, the relation holds [14] as

$$\left\langle c_{\mathbf{q}\alpha} c^{*}_{\mathbf{q}\alpha'} \right\rangle = \frac{1}{N} \left[\delta_{\alpha\alpha'} c_{\alpha} (1 - c_{\alpha}) + \sum_{\rho \neq 0} \varepsilon_{\alpha\alpha'}(\rho) e^{i\mathbf{q}\rho} \right], \quad (21.41)$$

where $\delta_{\alpha\alpha'}$ is Kronecker's symbol and $\varepsilon_{\alpha\alpha'}(\rho)$ are pair correlation parameters with $\rho = \mathbf{R}_s - \mathbf{R}_{s'}$. If correlations between defect positions are absent, then the function $S(\mathbf{q})$ can be written as

$$S(\mathbf{q}) = \sum_{\alpha} S_{\alpha}(\mathbf{q}), \qquad S_{\alpha}(\mathbf{q}) = \frac{c_{\alpha}}{N} E^2 \chi_{\mathbf{H}} \chi_{-\mathbf{H}} F_{\alpha}^{\mathrm{H}}(\mathbf{q}), \quad (21.42)$$

$$F_{\alpha}^{\mathrm{H}}(\mathbf{q}) = \left(\mathbf{HU}_{\mathbf{q}\alpha}\right)\left(\mathbf{HU}_{-\mathbf{q}\alpha}\right) \cdot \quad (21.43)$$

The superscript H in Equations 21.42 and 21.43 denotes that these equations are valid only in the Huang scattering region, that is, at $k << k_{m\alpha}$. When considering momentum transfers in the Stokes-Wilson scattering region, that is, the region of asymptotic diffuse scattering at $k >> k_{m\alpha}$, one has to multiply Equation 21.43 by $k_{m\alpha}/q$ and also replace the function $F_{\alpha}^{\mathrm{H}}(\mathbf{q})$ in Equation 21.42 by

$$F_{\alpha}^{\mathrm{SW}}(\mathbf{q}) = F_{\alpha}^{\mathrm{H}}(\mathbf{q}) k_{m\alpha}^2 / q^2. \quad (21.44)$$

The expressions derived for the absorption coefficients due to diffuse scattering in Equations 21.37 and 21.38 allow correct calculations of angular dependencies of absorption effects because of diffuse scattering for Bragg and diffuse scattering waves in crystals with microdefects of various types and with arbitrary sizes. Expressions for the absorption coefficients of diffuse scattering waves in Equation 21.19 can be derived along similar lines. Real parts of the dispersion corrections due to diffuse scattering in Equations 21.18 and 21.19 are of the same order of magnitude as the imaginary parts and can be calculated through the absorption coefficients in Equations 21.37 and 21.38 by using known dispersion relations [33].

2.4 Dynamical Diffuse Scattering Intensity

Diffuse scattering waves are created due to the scattering of strong Bragg (coherent) waves and the rescattering of diffuse scattering waves (due to multiple diffuse scattering processes) by the fluctuation field of static displacements of crystal atoms, which are caused by randomly distributed defects. Similar to coherent waves, they also form a dynamical wave field in the crystal due to the scattering by the periodical part of the crystal's polarizability. In the two-beam diffraction case, the amplitudes of diffuse scattering plane waves $D_{\mathbf{q}}$ and $D_{\mathbf{H+q}}$, which form the diffuse quasi-Bloch waves, satisfy the set of inhomogeneous equations in Equation 21.10. The solving this set of equations by iteration [30,31] enables one to account for any multiple diffuse-scattering processes. The amplitude of a diffuse quasi-Bloch wave created in the crystal by the δth quasi-Bloch coherent wave can then be found from the set of equations

$$\begin{cases}(-2\varepsilon_{0\mathbf{q}}^{\delta}+\chi_0+\Delta\chi_{00}^{\prime\delta})D_{\mathbf{q}}+(CE\chi_{-\mathbf{H}}+\Delta\chi_{0\mathbf{H}}^{\prime\delta})D_{\mathbf{H+q}}=-(\delta\chi_{\mathbf{q}}D_0^{\delta}+C\delta\chi_{-\mathbf{H+q}}D_{\mathbf{H}}^{\delta}),\\(CE\chi_{\mathbf{H}}+\Delta\chi_{\mathbf{H}0}^{\prime\delta})D_{\mathbf{q}}+(-2\varepsilon_{\mathbf{Hq}}^{\delta}+\chi_0+\Delta\chi_{\mathbf{HH}}^{\prime\delta})D_{\mathbf{H+q}}=-(C\delta\chi_{\mathbf{H+q}}D_0^{\delta}+\delta\chi_{\mathbf{q}}D_{\mathbf{H}}^{\delta}),\end{cases} \quad (21.45)$$

where $\varepsilon_{0\mathbf{q}}^{\delta}$ and $\varepsilon_{\mathbf{Hq}}^{\delta}$ are the excitation errors of diffuse scattering waves with the wave vectors of coherent plane waves $\mathbf{K}_0^{\delta}$ and $\mathbf{K}_{\mathbf{H}}^{\delta}$, which belong to δth sheet of the dispersion surface, and $\Delta\chi_{\mathbf{GG}'}^{\prime\delta}$ are the dispersion corrections accounting for multiple diffuse scattering processes.

The solvability condition for the homogeneous set of equations, which corresponds to the inhomogeneous Equation 21.45, requires the equality $d_{\delta}(\mathbf{q})=0$, where the determinant $d_{\delta}(\mathbf{q})$ is described by Equation 21.16. This equality gives the dispersion equation for the wave vector $\mathbf{K}_{0\mathbf{q}}^{\delta}$, which can be solved by using the relations between the excitation errors $\varepsilon_{0\mathbf{q}}^{\delta}$ and $\varepsilon_{\mathbf{Hq}}^{\delta}$. Its solution, $\varepsilon_{0\mathbf{q}}^{\delta\tau}=\gamma_0\Delta_{\tau}'$ (τ = 1 and 2), gives the accommodation of the wave vector $\mathbf{K}_{0\mathbf{q}}^{\delta\tau}=\mathbf{K}+K\Delta_{\tau}'\mathbf{n}$,

$$\Delta_{\tau}'=\frac{1}{2\gamma_0}\left(\chi_0+\Delta\chi_{00}^{\prime\delta}\right)+\frac{\lambda}{2\Lambda'}\left[y'+(-1)^{\tau}\sqrt{y'^2-1}\right], \quad (21.46)$$

$$y'=(\alpha'-\alpha_0')\sqrt{b}/\sigma',\ \Lambda'=\lambda|\gamma_{\mathbf{H}}|\sqrt{b}/\sigma',$$

$$2\alpha_0' = \chi_0 + \Delta\chi_{\mathbf{HH}}'^{\delta} + b^{-1}\left(\chi_0 + \Delta\chi_{00}'^{\delta}\right),$$

$$\sigma'^2 = \left(CE\chi_{\mathbf{H}} + \Delta\chi_{\mathbf{H}0}'^{\delta}\right)\left(CE\chi_{-\mathbf{H}} + \Delta\chi_{0\mathbf{H}}'^{\delta}\right),$$

where indices δ on quantities y', Λ', and so on have been dropped for brevity.

To determine the far-field diffuse scattering intensity in a given direction $\mathbf{K}'$, the amplitudes of diffuse scattering waves at the crystal surface, which can be found from Equation 21.46 after imposing proper boundary conditions and which can be considered as sources of divergent spherical waves, should be recalculated into the amplitudes of these waves. Later ones have the asymptotic form

$$E_{\mathbf{G}}^{\inf}(\mathbf{r}) = f_{\mathbf{G}}(\mathbf{K}',\mathbf{K})\frac{e^{-iK'r}}{r}, \quad \mathbf{K}' = K\frac{\mathbf{r}}{r}, \quad r \to \infty, \tag{21.47}$$

where $f_{\mathbf{G}}$ $(\mathbf{K}',\mathbf{K})$ is the scattering amplitude in the direction $\mathbf{K}' = \mathbf{K} + \mathbf{G} + \mathbf{k}$, where $\mathbf{G} = 0$ or $\mathbf{H}$ for transmitted and diffracted waves, respectively. The final expression for the dynamical diffuse scattering amplitude in the diffraction direction has the form [31]

$$f_{\mathbf{H}}(\mathbf{K}',\mathbf{K}) = \sum_{\delta}\sum_{\mathbf{G}} D_{\mathbf{G}}^{\delta}(\mathbf{K}) F_{\mathbf{HG}}^{\delta}(\mathbf{K}',\mathbf{K}), \tag{21.48}$$

where $\delta = 1$ and 2, $\mathbf{G} = 0$ and $\mathbf{H}$, and partial diffuse scattering amplitudes are described by the expression

$$F_{\mathbf{HG}}^{\delta}(\mathbf{K}',\mathbf{K}) = \frac{VK^2}{4\pi\left(B_1' - B_2'\right)}\sum_{\tau}(-1)^{\tau} M_{\mathbf{HG}}^{\delta\tau} e^{-iK\Delta_{\tau}'t}. \tag{21.49}$$

In Equation 21.49, the following notation was used:

$$B_{\tau}' = c'^{\tau} e^{-iK\Delta_{\tau}'t}, \quad c'^{\tau} = -\frac{-2\gamma_0\Delta_{\tau}' + \chi_0 + \Delta\chi_{00}'^{\delta}}{CE\chi_{-\mathbf{H}} + \Delta\chi_{0\mathbf{H}}'^{\delta}}, \quad \tau = 1, 2, \tag{21.50}$$

$$M_{\mathbf{HG}}^{\delta\tau} = c'^{\tau} M_{0\mathbf{G}}^{\delta\tau}, \quad M_{0\mathbf{G}}^{\delta\tau} = \zeta' X_{0\mathbf{G}}(\mathbf{q}_{\delta\tau})/c'^{\tau} + X_{\mathbf{HG}}(\mathbf{q}_{\delta\tau}), \tag{21.51}$$

$$X_{\mathbf{GG}'}(\mathbf{q}) = C_{\mathbf{GG}'}\delta\chi_{\mathbf{G}-\mathbf{G}'+\mathbf{q}}, \quad \zeta' = \left(CE\chi_{\mathbf{H}} + \Delta\chi_{\mathbf{H}0}^{\prime\tau}\right)\cdot\left(CE\chi_{-\mathbf{H}} + \Delta\chi_{0\mathbf{H}}^{\prime\tau}\right)^{-1}, \qquad (21.52)$$

where $C_{\mathbf{GG}'} = 1$ at $\mathbf{G}' = \mathbf{G}$, and $C_{\mathbf{GG}'} = C$ at $\mathbf{G}' \neq \mathbf{G}$.

To calculate the diffuse component of the differential transmission or reflection power of the crystal, one has to average the absolute square of the diffuse scattering amplitude from Equation 21.48 over a random distribution of defects and divide it by the incident X-ray intensity

$$R_{\mathrm{D}}(\mathbf{k}) = \frac{\left\langle \left| f_{\mathrm{H}}(\mathbf{K}',\mathbf{K}) \right|^2 \right\rangle}{\gamma_0 S \left| E_0 \right|^2}, \qquad (21.53)$$

where S is the entrance crystal surface and E_0 is the incident plane wave amplitude.

Equations 21.48 to 21.53 enable the derivation of an explicit analytic expression for the diffuse scattering component of reflectivity from an imperfect crystal with an arbitrary thickness, which includes the complex interactions of coherent and diffuse scattering wave fields, but will have an unfortunately complicated form. However, very often in practical applications the crystal thickness is on the same order or significantly exceeds the absorption length, and consequently, the approximation of a semi-infinite crystal can be used, which provides an appreciable simplification of the resulting formula.

3 The Generalized Method of "Integral" Diffuse Scattering

The generalized method of "integral" diffuse scattering, unlike the well-known kinematic method of "integral" diffuse scattering [61,62], is based on the joint treatment of coherent and diffuse components of rocking curves measured by DCD [52–54,63–65] and gives a self-consistent description of rocking curves measured from an imperfect single crystal over all angles, including the total reflection region. This method was applied to determine characteristics of dislocation loops in silicon single crystals grown by the float-zone method (FZ Si) [54] as well as more complicated defect structures consisting of oxygen precipitates and dislocation loops in isothermally annealed Czochralski-grown silicon samples (Cz Si) [52,53,63–65].

3.1 "Integral" Diffuse Scattering Intensity

When the approximation of a semi-infinite crystal is valid, that is, at $\mu_0 t >> 1$, only one quasi-Bloch wave remains in the wave fields of both strong Bragg and diffuse scattering waves. It follows, then, that (approximately) from Equations 21.48 to 21.53,

$$R_{\mathrm{D}}(\mathbf{k}) \approx \frac{1}{\gamma_0 S}\left(\frac{CVK^2}{4\pi}\right)^2 F_{\mathrm{dyn}} \left\langle \left| \delta\chi_{\mathbf{H}+\mathbf{q}} \right|^2 \right\rangle, \qquad (21.54)$$

where $F_{\mathrm{dyn}}(\Delta\theta)=\left|1-c^{\delta}\right|^2$. The diffuse reflectivity in Equation 21.54 can be rewritten through the correlation function according to Equations 21.17 to 21.19 by using relations in Equations 21.20 to 21.22 as

$$R_{\mathrm{D}}(\mathbf{k})\cong\frac{1}{\gamma_0 S}\left(\frac{CVK^2}{4\pi}\right)^2 F_{\mathrm{dyn}}(\Delta\theta)S(\mathbf{q})p(t), \tag{21.55}$$

$$p(t)=\left(1-e^{-2\mu_i t}\right)/\left(2\mu_i t\right)\approx 1/\left(2\mu_i t\right), \tag{21.56}$$

where $\mathbf{q}=\mathbf{k}+i\mu_i\mathbf{n}$. The interference absorption coefficient $\mu_i = K\,\mathrm{Im}\left(\Delta'_{\tau}-\Delta_{\delta}\right)$ in the complex momentum transfer $\mathbf{q}$ can be estimated as $\mu_i \approx (1+b)\mu_0/(2\gamma_0)$ at $|y|,\ |y'|>>1$ and $\mu_i \sim \pi/\Lambda$ at $|y|,\ |y'|\leq 1$; that is, it describes the extinction effect for diffuse scattering waves. The factor F_{dyn} describes the angular modulation of diffuse scattering intensity, which is caused by the dynamical interference of strong Bragg waves.

The integration of Equation 21.55 over the Ewald sphere ($K'=K$) near the reciprocal lattice point H accounting for Equations 21.37 and 21.38 gives the "integral" diffuse component of the reflection coefficient measured by DCD with a widely open detector window,

$$R_{\mathrm{diff}}(\Delta\theta)=\int_{K'=K} R_{\mathrm{D}}(\mathbf{k})\,\mathrm{d}\Omega_{\mathbf{K}'}=F_{\mathrm{dyn}}(\Delta\theta)\mu_{\mathbf{HH}}(\Delta\theta)p(t)\,t/\gamma_0, \tag{21.57}$$

where $\mathrm{d}\Omega_{\mathbf{K}'}$ is a solid angle in $\mathbf{K}'$ direction and $\mathrm{d}S_{\mathbf{K}'}=K^2\mathrm{d}\Omega_{\mathbf{K}'}$ is an element of surface area in the reciprocal space. It should be emphasized here that diffuse scattering intensity integrated over exit angles at fixed orientation $\Delta\theta$ of the incident wave vector $\mathbf{K}$ appears to be proportional to the coefficient of absorption of coherent waves due to diffuse scattering in Equation 21.37.

To obtain an analytic expression for the "integral" diffuse reflectivity in Equation 21.57 one should specify microdefect types and corresponding Fourier components of static displacement fields. For two types of microdefects—namely, spherically symmetric clusters and randomly oriented prismatic dislocation loops—the function $F_{\alpha}^{\mathrm{H}}(\mathbf{q})$ in the Huang scattering region can be represented in a unified form as [15]

$$F_{\alpha}^{\mathrm{H}}(\mathbf{q})=H^2\left(B_1^{\alpha}+B_2^{\alpha}\left|\frac{\mathbf{H}_0\mathbf{q}}{q}\right|^2\right)\frac{1}{|\mathbf{q}|^2}, \tag{21.58}$$

where $\mathbf{H}_0 = \mathbf{H} / H$ is the unit vector. For the simplist model of spherical clusters ($\alpha = 1$), one should put in Equation 21.58

$$B_1^1 = 0, \quad B_2^1 = \left(4\pi A_{\mathrm{C}} / v_c\right)^2. \tag{21.59}$$

Here the strength of the cluster is defined as

$$A_{\mathrm{C}} = \Gamma \varepsilon R_{\mathrm{C}}^3, \quad \Gamma = \frac{1}{3}(1+\nu)/(1-\nu), \tag{21.60}$$

where ε is the strain at the cluster interface, R_{C} is the cluster radius, and ν is the Poisson ratio. In more complicated models of clusters with lower symmetry (e.g., platelike, ellipsoidal, or disc-shaped new phase particles lying in {100} planes of a cubic crystal), one can put approximately in Equations 21.59 and 21.60 $B_1^1 \approx B_2^1$ and $A_{\mathrm{C}} = 3\Gamma\varepsilon V_{\mathrm{C}} / (4\pi)$, respectively, where V_{C} is the cluster volume [63].

For a model with randomly oriented prismatic dislocation loops ($\alpha = 2$), one should use Equation 21.58 after averaging over all the continuous orientations of Burgers vectors:

$$B_1^2 = \frac{4}{15}\left(\pi \left|\mathbf{b}\right| R_{\mathrm{L}}^2 / v_c\right)^2 \; B_2^2 = \beta\, B_1^2, \; \beta = \frac{1}{4}\left(3\nu^2 + 6\nu - 1\right)\left(1-\nu\right)^{-2}, \tag{21.61}$$

where $\mathbf{b}$ is the Burgers vector and R_{L} is the radius of the dislocation loop.

For each type of microdefect, the proper value of the boundary between Huang and Stokes-Wilson scattering regions $k_{\mathrm{m}\alpha} = 1 / R_{\mathrm{eff}}^{\alpha}$ should also be chosen. By making effective radii equal $R_{\mathrm{eff}}^1 = \sqrt{H\left|A_{\mathrm{C}}\right|E}$ for clusters and $R_{\mathrm{eff}}^2 = R_{\mathrm{L}}\sqrt{H\left|\mathbf{b}\right|E}$ for dislocation loops, one can provide an accurate description of asymptotic diffuse scattering intensity with respect to both momentum transfer $\mathbf{k}$ and such parameters as H, $\mathbf{b}$, and A_{C} [14]. Due to this choice of the boundary radius $k_{\mathrm{m}\alpha}$, there is also a physically reasonable matching of Equations 21.43 and 21.44 at $k = k_{\mathrm{m}\alpha}$.

In general, the diffuse scattering intensity from microdefects is localized near reciprocal lattice points, where it is, as a rule, significantly larger in comparison with the contribution of thermal diffuse scattering (TDS) in these regions, and for this reason TDS can be neglected in the analysis of measured diffraction patterns. However, at high reflection indices, the TDS intensity can be comparable with the diffuse scattering intensity from microdefects in some angular regions and therefore should be taken into account to provide the correct quantitative interpretation of measured diffraction profiles [54].

In the case of TDS ($\alpha = 3$), the Fourier component of the field of dynamic atom displacements from an oscillating atom in a single atom crystal can be considered in a harmonic approximation [14] according to which the function $F_3^{\mathrm{H}}(\mathbf{q})$ has the form

$$F_3^{\mathrm{H}}(\mathbf{q}) = \frac{\hbar H^2}{6m} \sum_{\rho=1}^{3} \frac{2n_{\mathbf{q}\rho}+1}{\omega_{\mathbf{q}\rho}}, \tag{21.62}$$

$$n_{\mathbf{q}\rho} = \left[\exp\left(\hbar\,\omega_{\mathbf{q}\rho}/k_{\mathrm{B}}T\right)-1\right]^{-1}, \tag{21.63}$$

where $\hbar$ is Planck's constant, m is an atom mass, $\omega_{\mathbf{q}\rho}$ is the phonon frequency, and the summation is carried out over phonon polarizations. When considering TDS intensity in a close vicinity of a reciprocal lattice point, the relation holds $\hbar\omega_{\mathbf{q}\rho}/k_{\mathrm{B}}T << 1$, and therefore one can put $n_{\mathbf{q}\rho} \approx k_{\mathrm{B}}T/\hbar\omega_{\mathbf{q}\rho}$ in Equation 21.63 and, consequently, Equation 21.62 can be rewritten as

$$F_3^{\mathrm{H}}(\mathbf{q}) \approx \frac{H^2}{m}\frac{k_{\mathrm{B}}T}{\omega_{\mathbf{q}}^2}. \tag{21.64}$$

The phonon frequency in Equation 21.64 can be calculated as $\omega_{\mathbf{q}} = v_{\mathrm{S}}q$, where v_{S} is the phonon velocity, which can be calculated by using elastic constants and crystal density according to the relationship $v_{\mathrm{S}} = \left[1/3\left(C_{11}+2C_{22}+4C_{44}\right)/\rho\right]^{\frac{1}{2}}$ [66]. Equation 21.64 can thus be considered then as a particular case of Equation 21.58 with coefficients $B_1^3 = k_{\mathrm{B}}T/(mv_{\mathrm{S}}^2)$ and $B_2^3 = 0$.

The derivation of the analytical expression for the "integral" diffuse component of crystal reflectivity [57] is reduced to the integration in Equation 21.37 for the coefficient of absorption due to diffuse scattering, which gives, in the case of the simultaneous presence of several microdefect types without mutual correlations [52,53,63,64],

$$\mu_{\mathbf{HH}}(\Delta\theta) = \sum_{\alpha} \mu_{\mathrm{ds}}^{\alpha}(k_0). \tag{21.65}$$

The absorption coefficient due to diffuse scattering for microdefects of α type in Equation 21.65 is described by the following expressions:

$$\mu_{\mathrm{ds}}^{\alpha}(k_0) = c_{\alpha}C^2E^2m_0J^{\alpha}(k_0), \tag{21.66}$$

$$m_0 = \frac{\pi v_c}{4}\left(H\left|\chi_{r\mathbf{H}}\right|/\lambda\right)^2, \tag{21.67}$$

$$J^{\alpha}(k_0)=\frac{1}{\pi}\int d\mathbf{k}'F_{\alpha}(\mathbf{q})\,. \tag{21.68}$$

The integration in Equation 21.68 can be performed after decomposing the momentum transfer $\mathbf{k}$ into the components parallel ($\mathbf{k}_0$) and perpendicular ($\mathbf{k}'$) to the wave vector $\mathbf{K}'$ (see Figure 21.1). Then relations $\mathbf{k} = \mathbf{k}_0 + \mathbf{k}'$ and $|\mathbf{q}|^2 = k'^2 + k_0^2 + \mu_i^2$ hold, and the integration can be performed in a polar coordinate system chosen in the plane tangent to the Ewald sphere near the reciprocal lattice

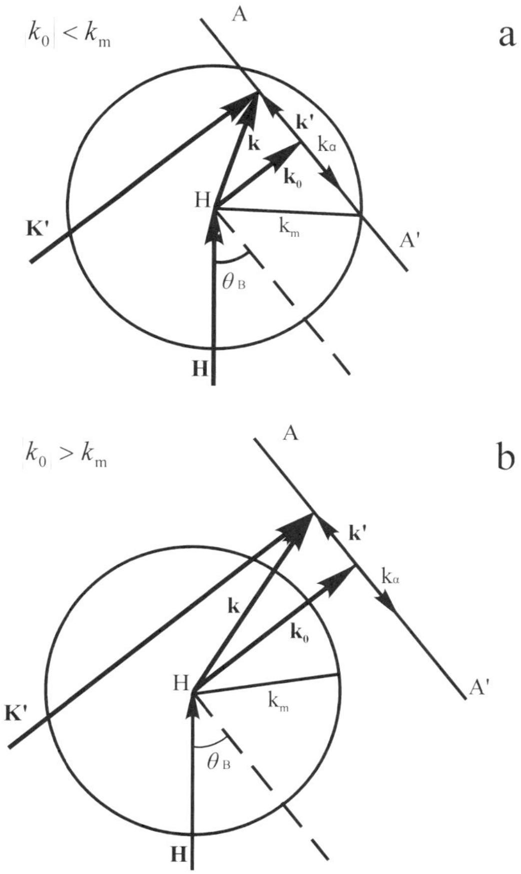

Figure 21.1. *Schematic for the integration in a momentum space for two diffuse scattering intensity cases in which the AA′ line represents the intersection of a coherent scattering plane* ($\mathbf{K},\mathbf{H}$) *with an integration plane tangent to an Ewald sphere. (a)* $|\mathbf{k}_0| < k_m$*, the integration interval, includes both Huang (*$k < k_m$*) and Stokes-Wilson (*$k > k_m$*) scattering regions. (b)* $|\mathbf{k}_0| > k_m$*, the integration interval, lies only in the Stokes-Wilson scattering region.*

point H. When integrating over k' in Equation 21.68, the expression in Equation 21.58 for $F_{\alpha}^{\mathrm{H}}(\mathbf{q})$ is used in the Huang scattering region ($0 \le k'^2 + k_0^2 \le k_{\mathrm{m}}^2$), and this expression multiplied by $k_{\mathrm{m}}^2 / |\mathbf{q}|^2$ is substituted as $F_{\alpha}^{\mathrm{SW}}(\mathbf{q})$ into Equation 21.68 in the Stokes-Wilson scattering region ($k'^2 + k_0^2 \ge k_{\mathrm{m}}^2$). Also, the dependence of the interference absorption coefficient μ_{i} on exit angles $\Delta\theta'$ is neglected because its influence is smoothed due to integration and thus μ_{i} can be replaced by its limiting value μ at $|y'| >> 1$.

Then the integration in Equation 21.68 can easily be performed, and by accounting for the angular aperture of the detector $\Delta\theta_a$, we obtain [54]

$$J^{\alpha}(k_0) = \begin{cases} J_{\mathrm{H}}^{\alpha}(k_0) + J_{\mathrm{H-SW}}^{\alpha}(k_0) + J_{\mathrm{H}}^{*\alpha}(k_0), & \text{at} \quad |k_0| \le k_{\mathrm{m}\alpha}; \\ J_{\mathrm{SW}}^{\alpha}(k_0), & \text{at} \quad |k_0| \ge k_{\mathrm{m}\alpha}, \end{cases} \tag{21.69}$$

where $k_0 = K\Delta\theta \sin(2\theta_{\mathrm{B}})$. The angular dependence of the "integral" diffuse scattering intensity in the Huang scattering region is described by symmetric and antisymmetric ($J_{\mathrm{H}}^{*\alpha}$) components in Equation 21.69 as follows:

$$\begin{aligned} J_{\mathrm{H}}^{\alpha}(k_0) &= \frac{1}{2\pi} \int_0^{2\pi} d\varphi \int_0^{k_{\mathrm{m}\alpha}^2 - k_0^2} dk'^2 F_{\alpha}^{\mathrm{H}}(\mathbf{q}) = \\ &= b_2 \ln \frac{k_{\mathrm{m}\alpha}^2 + \mu^2}{k_0^2 + \mu^2} + \left(b_3 k_0^2 + b_4 \mu^2\right) \left(\frac{1}{k_{\mathrm{m}\alpha}^2 + \mu^2} - \frac{1}{k_0^2 + \mu^2} \right) \end{aligned} \tag{21.70}$$

$$\begin{aligned} J_{\mathrm{H-SW}}^{\alpha}(k_0) &= \frac{1}{2\pi} \int_0^{2\pi} d\varphi \int_{k_{\mathrm{m}\alpha}^2 - k_0^2}^{k_a^2} dk'^2 F_{\alpha}^{\mathrm{SW}}(\mathbf{q}) = \frac{k_{\mathrm{m}\alpha}^2 \left(k_a^2 + k_0^2 - k_{\mathrm{m}\alpha}^2\right)}{\left(k_{\mathrm{m}\alpha}^2 + \mu^2\right)\left(k_a^2 + k_0^2 + \mu^2\right)} \times \\ &\times \left(b_2 + \frac{1}{2} \frac{\left(b_3 k_0^2 + b_4 \mu^2\right)\left[\left(k_{\mathrm{m}\alpha}^2 + \mu^2\right)^2 - \left(k_a^2 + k_0^2 + \mu^2\right)^2\right]}{\left(k_{\mathrm{m}\alpha}^2 + \mu^2\right)\left(k_a^2 + k_0^2 + \mu^2\right)\left(k_a^2 + k_0^2 - k_{\mathrm{m}\alpha}^2\right)} \right), \end{aligned} \tag{21.71}$$

$$J_{\mathrm{H}}^{*\alpha}(k_0) = b_1 \left(\sqrt{k_{\mathrm{m}\alpha}^2 + \mu^2} - \sqrt{k_0^2 + \mu^2} \right), \tag{21.72}$$

where $k_a = K\Delta\theta_a$ was supposed to be always larger in comparison with $k_{m\alpha}$. In the Stokes-Wilson scattering region, the diffuse scattering intensity profile is described by the expression

$$J_{\mathrm{SW}}^{\alpha}(k_0) = \frac{1}{2\pi}\int_0^{2\pi} d\varphi \int_0^{k_a^2} dk'^2 F_{\alpha}^{\mathrm{SW}}(\mathbf{q}) = \frac{k_{m\alpha}^2 k_a^2}{\left(k_0^2+\mu^2\right)\left(k_a^2+k_0^2+\mu^2\right)} \times \left\{ b_2 + \frac{1}{2}\frac{\left(b_3k_0^2+b_4\mu^2\right)\left[\left(k_0^2+\mu^2\right)^2-\left(k_a^2+k_0^2+\mu^2\right)^2\right]}{k_a^2\left(k_0^2+\mu^2\right)\left(k_a^2+k_0^2+\mu^2\right)} \right\} \tag{21.73}$$

The coefficients $b_i\left(i=\overline{1,4}\right)$ in Equations 21.70 to 21.73 are connected to microdefects characteristics by the relations

$$b_1 = \frac{4L_{\mathbf{H}}^{\alpha}\sqrt{B_{2\alpha}}}{c_{\alpha}H}, \qquad b_2 = B_{1\alpha} + \frac{1}{2}B_{2\alpha}\cos^2\theta_{\mathrm{B}},$$

$$b_3 = B_{2\alpha}\left(\frac{1}{2}\cos^2\theta_{\mathrm{B}} - \sin^2\theta_{\mathrm{B}}\right), \qquad b_4 = B_{2\alpha}\left(\frac{1}{2}\cos^2\theta_{\mathrm{B}} - \cos^2\psi\right). \tag{21.74}$$

If the inequality $k_a >> k_{m\alpha}$ holds, then $k_a \to \infty$ in Equations 21.71 and 21.73, which simplifies them to the following form [63]:

$$J_{\mathrm{H-SW}}^{\alpha}(k_0) = \frac{k_{m\alpha}^2}{k_{m\alpha}^2+\mu^2}\left(b_2 - \frac{1}{2}\frac{b_3k_0^2+b_4\mu^2}{k_{m\alpha}^2+\mu^2}\right), \tag{21.75}$$

$$J_{\mathrm{SW}}^{\alpha}(k_0) = \frac{k_{m\alpha}^2}{k_0^2+\mu^2}\left(b_2 - \frac{1}{2}\frac{b_3k_0^2+b_4\mu^2}{k_0^2+\mu^2}\right). \tag{21.76}$$

Thus, in the semi-infinite crystal approximation ($\mu_0 t >> 1$), we obtain the following expression for the diffuse component of rocking curves measured by DCD with a widely open detector window:

$$R_{\mathrm{diff}}(\Delta\theta) \cong F_{\mathrm{dyn}}(\Delta\theta)\frac{\mu_{\mathbf{HH}}(\Delta\theta)}{2\gamma_0\mu(\Delta\theta)}. \tag{21.77}$$

Unlike the well-known kinematical expression for the diffuse scattering intensity integrated over an Ewald sphere [61,62], where a cutoff parameter $k_c \propto 1/\Lambda$ is introduced artificially to remove the divergence at $k \to 0$, the absorption coefficient μ, which has the order of magnitude of k_c within the total reflection range, appears as a cutoff parameter in Equations 21.70 to 21.73 as a natural consequence of dynamical theory.

Also, the restriction on the maximum allowable effective radius of microdefects, R_{eff}, is absent in the derived formulas, thus providing the ability to quantitatively describe in explicit analytical terms the Huang diffuse scattering intensity distribution in the total reflection range even when the effective microdefect radii R_{eff} exceeds the extinction length Λ. The importance of accounting for the detector entrance aperture should also be mentioned. This is essential for a correct quantitative characterization of crystal imperfections, especially when effective radii of microdefects are so small that the relation $k_{m\alpha} \geq k_a$ holds.

3.2 The Treatment of Total Rocking Curves from Single Crystals with Microdefects

The problem of a statistical treatment of rocking curves from crystals with a wide spread of defect radii or with several types of defects is connected with an increasing number of unknown defect characteristics that should be determined. This problem can be overcome to a large extent if effective radii of different defect types do not coincide. In this case, the various types of defects produce contributions to the diffuse scattering component of the rocking curves with distinct dependences in different angular intervals due to mutual displacement of boundaries between Huang and Stokes-Wilson scattering regions. The uncertainties in defect characterization are also largely decreased due to the self-consistent description of coherent and diffuse scattering components of rocking curves because the behavior of the central part of the coherent component of rocking curves is very sensitive to resulting values of the static Debye-Waller factor and the coefficient of absorption due to diffuse scattering. Obviously, such a complex approach to the treatment of measured rocking curves implies the need to account for instrumental factors, for example, by performing a convolution of the intrinsic rocking curves of the sample under investigation with reflection coefficients of the monochromator crystals in the high-resolution DCD [54,63].

The statistical treatment of rocking curves measured from the silicon sample grown by the float-zone method (FZ Si) was carried out in Vladimirova et al. [54] independently for 111 (Figure 21.2) and 333 reflections of characteristic CuK_α radiation. It was assumed in the defect model that only perfect dislocation loops of the interstitial type are present in the sample crystal [67–69]. However, all the attempts to describe the measured rocking curves over the entire angular range were unsuccessful with only a single-loop radius. The most probable cause of this difficulty is a wide distribution of loop radii, which leads to corresponding different diffuse scattering intensity dependencies in different angular ranges.

The good agreement between experimental and calculated rocking curves over the whole measured angular ranges was achieved only after introducing three different loop populations with small, intermediate, and large sizes, respectively. The final refinement of all the varied characteristics of dislocation loops with a weighted reliability factor near 10% has given their values, shown in Figure 21.3. It should be emphasized especially that the set of dislocation loop characteristics obtained provides a

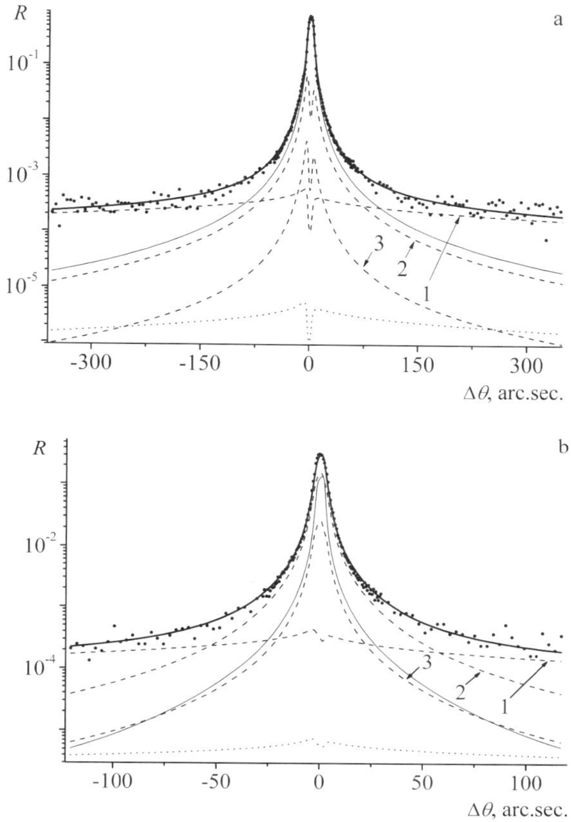

Figure 21.2. *Rocking curves of an FZ Si sample for (a) 111 and (b) 333 reflections of CuK 1 radiation. The calculated total rocking curves and their coherent components are shown by thick and thin solid lines, respectively. Contributions to diffuse scattering components of rocking curves from dislocation loops with different sizes are shown by numerated broken lines, and the thermal diffuse scattering components are shown by dotted lines.*

good fit of the calculated to measured RCs for both 111 and 333 reflections simultaneously. This characterization of defects has been achieved due to the qualitatively different behavior of the diffuse scattering intensity profiles in different angular intervals for dislocation loops with significantly different sizes. This shifts the corresponding boundaries between Huang and Stokes-Wilson scattering regions.

For the detailed quantitative characterization of microdefects in Czochralski-grown silicon crystal (Cz Si), the appropriate model of the defect structure should be chosen. It is well known that in the Cz Si oxygen precipitates grow during anneals at elevated temperatures [70]. Because of the high stress at the interface between the silicon matrix and new phase particles, SiO_x is released via emission of interstitial silicon atoms, which agglomerate into dislocation loops and stacking faults [71–74]. For this reason, in the microdefect model, which has been applied in Kyslovskyy et al. [65] to treat the rocking curves measured from a Cz Si sample for 111 and 333 reflections (see Figure 21.4), the simultaneous

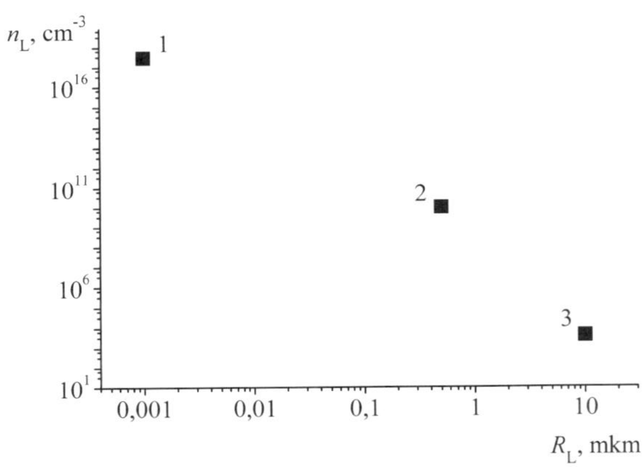

Figure 21.3. *Size distribution of dislocation loops in an FZ Si sample.*

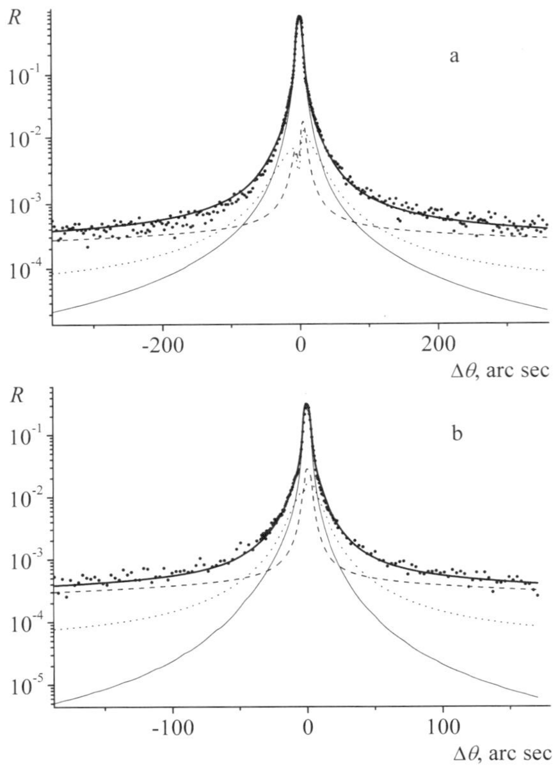

Figure 21.4. *The fitted rocking curves for (a) 111 and (b) 333 reflections of CuK_1 radiation from the sample Cz Si annealed at 750°C for 50 hours. The coherent component is shown by a thin solid line, and diffuse scattering components from precipitates and dislocation loops are shown by broken and dotted lines, respectively.*

presence of two types of microdefects was supposed, namely, disc-shaped oxygen precipitates [75] with a large radius R_C and circular dislocation loops $\langle 110 \rangle$ with a radius R_L, both randomly distributed without mutual correlation.

The results of X-ray diffractometric characterization of Cz Si samples annealed at 750°C during various time intervals are shown in Figure 21.5. It should be noted first that in this case the ability to determine the characteristics of two types of microdefects has been realized due to the significant shift of the boundary between the Huang and Stokes-Wilson scattering region for precipitates from that for dislocation loops (see Figure 21.4). These boundaries were well defined on rocking curves for each sample under investigation. As a result, each of the measured rocking curves has been fitted with a weighted reliability factor less than 10%.

As can be seen from Figure 21.5a, the approximated dependency (dashed line) of the experimental radius R_C of ellipsoidal precipitates on the annealing time is close to the known power law $t^{1/2}$ [76–78]. Simultaneously, the approximated number density n_C of precipitates (dotted line) decreases linearly with annealing time. This behavior of oxygen precipitate concentrations during various thermal anneals of silicon has often been observed in experimental investigations [79] and can be explained by Ostwald ripening of precipitates. A similar dependency of the radius R_L of dislocation loops (dashed line) versus annealing time (Figure 21.5b) approaches approximately the power law $t^{3/4}$ [78], whereas the number density n_L of loops (dotted line) decreases linearly with annealing time, and this dependency can also be caused by coarsening of larger dislocation loops at the cost of smaller ones.

The dependencies determined for microdefect characteristics versus annealing time support predictions of simplified models describing growth laws of oxygen precipitates [76–78] and dislocation loops [78] at sufficiently long annealing times, but significant discrepancies are observed at short times. Obviously a more complete explanation of the observed time dependence of both microdefect sizes and, especially, densities can be achieved by using a more sophisticated microdefect model that describes the evolution of such a complex interacting cluster system including sets of coupled kinetic equations for size distribution functions of microdefects (see, e.g., [80–85]). The rigorous dynamical

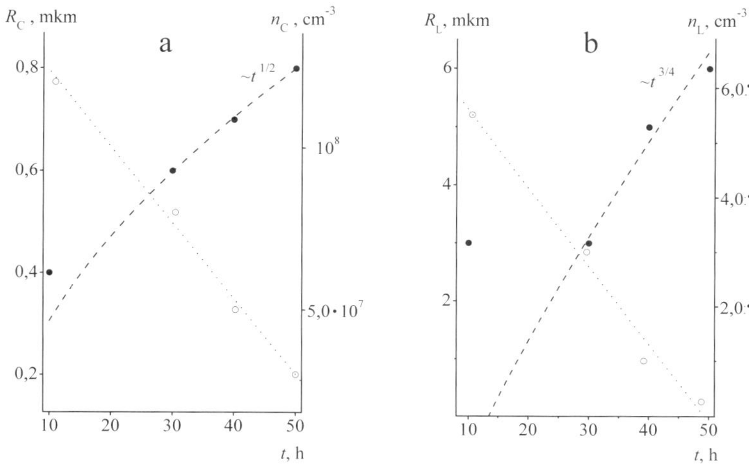

Figure 21.5. *Experimental dependencies of radii and number densities (black and white markers, respectively) of (a) oxygen precipitates and (b) dislocation loops versus annealing time.*

theory of X-ray diffraction developed for single crystals with complicated defect structures can provide a theoretical basis for the quantitative diffraction characterization of microdefect distributions in silicon under various technological treatments.

Overall, the important role of a self-consistent treatment of the absorption due to diffuse scattering in the coherent rocking curve components should be emphasized for reliable characterization in both FZ Si and Cz Si samples. It is also important to account for contributions of TDS and to account for the influence of the instrumental function of DCD to avoid systematical errors when determining microdefect characteristics by rocking curve analysis.

4 Dynamical Theory of Diffuse Scattering by Imperfect Multilayered Structures

The analysis of X-ray diffraction profiles from multilayered structures, as a rule, makes use of the theoretical diffraction models for perfect crystals [86–89]. Imperfections in such structures are typically assumed only at layer interfaces and are taken into account by introducing amorphization factors [90,91]. Imperfections in crystalline layers are sometimes simulated by mosaic blocks [92]. Both approaches to defect modeling do not include a range of possible imperfections in real multilayered crystal structures. Also, the semikinematical distorted-wave Born approximation (DWBA) method for describing diffuse scattering intensities from small-volume defects in layers [93] cannot properly account for diffuse scattering effects introduced by microdefects in thick substrates. For this reason, fit quality to measured rocking curves with theoretical diffraction profiles is poor, which causes systematic errors in estimates of strain and layer chemical compositions. Of course, it is practically impossible with these approaches to obtain detailed information on the characteristics of defect structure in multilayered systems.

X-ray diffraction characterization of imperfect multilayered structures can be improved using the dynamical theory of X-ray diffraction by imperfect crystals with randomly distributed Coulomb-type defects, namely, two-dimensional (dislocation loops, stacking faults) and three-dimensional defects (clusters, precipitates, inclusions, etc.) or, shortly, microdefects [94–96].

4.1 Coherent X-ray Scattering by Multilayered Structures

For a multilayer crystal system with M layers on a substrate with randomly distributed microdefects in each layer and in the substrate, the reflectivity amplitude and transmissivity of the jth layer can be written as [95,96]

$$r_j = \left(\zeta_j b_j\right)^{-1/2} \frac{D_{\mathbf{H}j}}{D_{0j}} = \left(\zeta_j b_j\right)^{1/2} \frac{e^{iK\Delta_1^j d_j} - e^{iK\Delta_2^j d_j}}{c_j^2 e^{iK\Delta_1^j d_j} - c_j^1 e^{iK\Delta_2^j d_j}}, \tag{21.78}$$

$$t_j = \frac{D_{0,j-1}}{D_{0j}} = \frac{c_j^2 - c_j^1}{c_j^2 e^{iK\Delta_1^j d_j} - c_j^1 e^{iK\Delta_2^j d_j}}, \tag{21.79}$$

where $\zeta = \left(CE\chi_{\mathbf{H}} + \Delta\chi_{\mathbf{H}0}\right)\left(CE\chi_{-\mathbf{H}} + \Delta\chi_{0\mathbf{H}}\right)^{-1}$, $j = \overline{1, M}$, M is the number of layers, and d_j is the thickness of the jth layer.

To determine the reflectivity amplitude of the total multilayered system $R = \left(\zeta b\right)^{-1/2} E_{\mathbf{H}}/E_0$, it is necessary to find amplitudes D^{δ}_{0M} and $D^{\delta}_{\mathbf{H}M}$ of a wave field in Mth layer, which are connected to the amplitude of the wave reflected in $(M-1)$ th layer, and so on. From new boundary conditions for a wave field in an arbitrary jth layer, which take into account all waves entering into this layer from its both sides, the set of equations for the determination of amplitudes D^{δ}_{0j} is

$$\begin{cases} D^{1}_{0j}\,\mathrm{e}^{-iK\Delta^{j}_{1}z_{j}} + D^{2}_{0j}\,\mathrm{e}^{-iK\Delta^{j}_{2}z_{j}} = D_{0j}, \\ D^{1}_{\mathbf{H}j}\,\mathrm{e}^{-iK\Delta^{j}_{1}z_{j-1}} + D^{2}_{\mathbf{H}j}\,\mathrm{e}^{-iK\Delta^{j}_{2}z_{j-1}} = D_{\mathbf{H},j-1}, \end{cases} \tag{21.80}$$

where $D^{\delta}_{\mathbf{H}j} = c^{\delta}_{j} D^{\delta}_{0j}$. From the set in Equation 21.80, we obtain

$$D^{\delta}_{0j} = \frac{D_{0j}c^{\delta'}_{j} - D_{\mathbf{H},j-1}\mathrm{e}^{iK\Delta^{j}_{\delta'}d_j}}{c^{\delta'}_{j}\mathrm{e}^{iK\Delta^{j}_{\delta}d_j} - c^{\delta}_{j}\mathrm{e}^{iK\Delta^{j}_{\delta'}d_j}}\,\mathrm{e}^{iK\Delta^{j}_{\delta}z_{j-1}}. \tag{21.81}$$

The amplitude of the diffracted wave, which has emerged from the jth layer, can be determined from the boundary condition $D^{j}_{S}(\mathbf{r}) = D_{\mathbf{H}j}\,\mathrm{e}^{-i\mathbf{K}'_{\mathbf{H}}\mathbf{r}}\Big|_{z=z_j}$ according to the equation

$$D_{\mathbf{H}j} = D^{1}_{\mathbf{H}j}\mathrm{e}^{-iK\Delta^{j}_{1}z_{j}} + D^{2}_{\mathbf{H}j}\mathrm{e}^{-iK\Delta^{j}_{2}z_{j}}. \tag{21.82}$$

From Equation 21.82, the recurrence relation between amplitude reflectivities of adjacent layers can be determined by

$$R_j = \frac{r_j + R_{j-1}\left(t^2_j e_j - r^2_j\right)}{1 - r_j R_{j-1}}, \tag{21.83}$$

where $R_j = \left(\zeta_j b_j\right)^{-1/2} D_{\mathbf{H}j}/D_{0j}$.

Equation 21.83 is similar to the well-known recurrence relation for perfect multilayered structures with the exception of a phase factor $e_j = \exp\left[iK\left(\Delta^{j}_{1} + \Delta^{j}_{2}\right)d_j\right]$, which is obtained by calculating the infinite geometrical progression arising from the summation of an infinite number of reflections of coherent waves in each layer with diffraction vectors $\mathbf{H}_j$ and $-\mathbf{H}_j$ [97]. The phase factor e_j in Equation 21.83, which arises after applying boundary conditions at the multilayer interfaces, can be set approximately equal to unity in a wide angular range $|y| < \Lambda_j / d_j$. However, it should be emphasized that Equation 21.83 has been derived by consistently using the concept of the dynamical wave field in the imperfect crystal and includes not only the effective Fourier components of the crystal polarizability, which are modified by the static Debye-Waller factor and the complex additions due to diffuse scattering, but also the accommodations of wave vectors with complex dispersion corrections describing the attenuation of coherent waves due to diffuse scattering.

4.2 Diffuse Scattering in Multilayered Structures

If small defects are distributed randomly and independently in each layer of the multilayered structure, then the diffuse scattering amplitude for the *j*th layer can be represented in accordance with Equation 21.48 as

$$f_{\mathbf{H}j}\left(\mathbf{K}'_j, \mathbf{K}_j\right) = \sum_{\delta}\sum_{\mathbf{G}j} D^{\delta}_{\mathbf{G}j} F^{\delta}_{\mathbf{H}j\mathbf{G}j}\, e^{-\gamma_j}, \tag{21.84}$$

$$\gamma_j = \sum_{i=j+1}^{M} \mu_i d_i / (2|\gamma_{\mathbf{H}}|), \qquad \mu_i = \mu^i_0 + \mu^i_{\mathrm{ds}}. \tag{21.85}$$

Here $\delta = 1$ and 2, $\mathbf{G}_j = 0$ and $\mathbf{H}_j$, $F^{\delta}_{\mathbf{H}j\mathbf{G}j}$ are partial amplitudes of scattering of strong Bragg waves into diffuse scattering waves in the diffraction direction, $\mathbf{K}_j + \mathbf{H}_j$, μ^j_0 is the photoelectric absorption coefficient in the *j*th layer, and μ^j_{ds} is the absorption coefficient due to diffuse scattering in the *j*th layer. The exponential factor in Equation 21.84 takes into account the absorption of diffuse scattering waves during their transmission through layers above the *j*th layer. The diffuse component of differential reflectivity of the multilayered system can be represented then as

$$R_{\mathrm{D}}(\mathbf{k}) = \sum_{j=0}^{M} \left\langle \left| f_{\mathbf{H}j} \right|^2 \right\rangle_j \Big/ (\gamma_0 S |E_0|^2), \tag{21.86}$$

where $\langle \ldots \rangle_j$ means averaging over the random distribution of defects in the *j*th layer. After integration of Equation 21.86 over the exit solid angle of diffuse scattering waves, the diffuse component of rocking curves measured by DCD with a wide-open detector window takes the form [95,96]

$$R_{\mathrm{diff}}(\Delta\theta) = \sum_{j=0}^{M} F^j_{\mathrm{abs}} F^j_{\mathrm{dyn}} R^j_{\mathrm{diff}}, \quad F^j_{\mathrm{abs}} = e^{-2\gamma_j}, \tag{21.87}$$

where the reflectivities of thick substrate (j = 0) and thin layers ($j=\overline{1,M}$) are calculated as

$$R_{\rm diff}^{0}=\mu_{\rm ds}^{0}(\Delta\theta)\Big/\Big[(1+b)\mu_{0}^{0}\Big],\qquad R_{\rm diff}^{j}=\mu_{\rm ds}^{j}(\Delta\theta)d_{j}\Big/\gamma_{0}\,, \tag{21.88}$$

and the factor $F_{\rm abs}^{j}$ describes photoelectric and diffuse scattering absorption in layers lying above the jth layer. The influence of redistribution of X-ray intensity between transmitted and diffracted coherent waves in the jth layer is described by dynamical factors

$$F_{\rm dyn}^{0}=\left|1-c_{0}^{\delta}\right|^{2},\qquad F_{\rm dyn}^{j}=\left|\frac{c_{j}^{\delta'}}{c_{j}^{1}-c_{j}^{2}}\left(1-c_{j}^{\delta}\right)\right|^{2},\qquad j=\overline{1,M}\,, \tag{21.89}$$

where the index δ is equal to 1 or 2 in dependence on the sign of $y_{jr}={\rm Re}\,y_{j}$ and designates the number of the wave field existing in the thin layer at $d_{j}/\Lambda_{jr}<<1$ ($\Lambda_{jr}={\rm Re}\Lambda_{j}$) or in the thick substrate at $\mu_{0}d_{0}>>1$.

4.3 Structural Characterization of Inhomogeneous Crystals by Double-Crystal Diffractometer

Quantitative analysis and reliable interpretation of rocking curves measured from multilayered structures require both adequate theoretical models of diffraction *and* realistic models of structural imperfections in these structures. In Molodkin et al. [96], the multilayered sample structure $In_xGa_{1-x}As/GaAs$ with a single QW (x = 0.37) was grown on a GaAs (001) substrate by molecular-beam epitaxy (MBE) (see scheme in Figure 21.6). Randomly distributed dislocation loops were assumed in the substrate. For the InGaAs QW layer and the adjoining barrier layer, the existence of a nonrectangular profile in the In concentration due to In segregation was also assumed.

In the first step of the rocking curve analysis, nominal chemical compositions and layer thicknesses were refined *without* taking into account the influence of defects. This approach resulted in significant qualitative and quantitative discrepancies between calculated and measured rocking curves (Figure 21.6, solid line 1) and justifies the need to include diffuse scattering effects in the measured rocking curves of defective multilayered structures.

In the next stage of the rocking curve treatment, the radii and concentrations of dislocation loops in the substrate were varied to fit the calculated and experimental rocking curves in various angular ranges. These loops introduce an additional average strain $\varepsilon_{\rm L}=\pi\,|\,\mathbf{b}\,|\,R_{\rm L}^{2}n_{\rm L}\,/\,3$ in the substrate, which leads to a corresponding shift of the central rocking curve peak position. Also, a diffuse scattering contribution to the rocking curve is produced by dislocation loops and modification of amplitudes of coherent waves in the substrate due to the static Debye-Waller factor and coefficients of absorption due to diffuse scattering. Simultaneously, the value of chemical composition of buffer layers ($Al_{0.3}Ga_{0.7}As$)

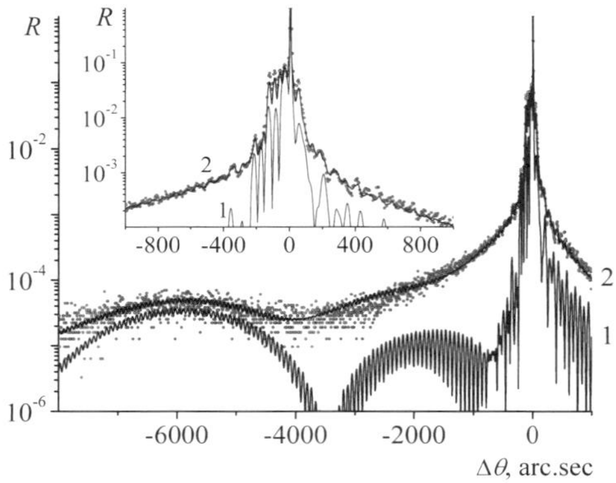

Scheme of $In_xGa_{1-x}As$ multilayered structure with QW

GaAs	10 nm
$Al_{0.3}Ga_{0.7}As$	100 nm
GaAs	150 nm
$In_xGa_{1-x}As$ QW	6.8 nm
GaAs	150 nm
$Al_{0.3}Ga_{0.7}As$	300 nm
GaAs substrate	

Figure 21.6. *Experimental rocking curves of a $In_xGa_{1-x}As$ multilayered structure and corresponding theoretical rocking curves calculated with and without taking into account diffuse scattering effects from defects (solid lines 1 and 2, respectively). The central part of the fitted rocking curve is shown in the insertion.*

and the thicknesses of barrier (GaAs) and buffer layers have been refined, in particular, due to the very high sensitivity of the fine structure of rocking curves (on the high θ side of the substrate peak) to these parameters. When varying QW thickness, the height of the peak from this layer and also the influence on shift and transformation of peaks set from GaAs layers were the basic criteria in fitting.

In the final stage of the rocking curve analysis, the best fit of calculated and experimental rocking curves is achieved (Figure 21.6, solid line 2) by choosing the most appropriate In segregation profile in the QW and adjoining barrier layer. This profile was calculated according to the empirical expressions [98]

$$x_{\mathrm{QW}}(z) = x_0\left(1-\rho^n\right), \qquad x_{\mathrm{B}}(z) = x_0\left(1-\rho^N\right)\rho^{n-N}, \tag{21.90}$$

where x_0 is the nominal In concentration in the layer, N is the number of monolayers in QW, n is the current monolayer number, and ρ is the segregation parameter determined to be $\rho \approx 0.5$.

As illustrated above, a dynamical analysis of the rocking curve from a multilayered structure with a single InGaAs QW extracts more detailed information about the layer structure and defects than can be obtained by using the classical diffraction model for perfect multilayers. The new diagnostic opportunities that result from a careful analysis of diffuse scattering effects from defects allow for correct calculations of both coherent and diffuse components of the measured rocking curve, and the resulting models are sensitive to defect characteristics.

A similar theoretical approach has been used to investigate the structural parameters and defect characteristics in as-grown [99] and ion-implanted [100] gadolinium gallium garnet ($Gd_3Ga_5O_{12}$) crystals. In these cases the inhomogeneous subsurface crystal layers were subdivided into thin laminas with constant strain and a random distribution of defects was assumed for each lamina. Again, in these investigations, an account for diffuse scattering effects from dislocation loops and clusters in the crystal bulk played an important role in the structural characterization.

5 Dynamical Diffraction Profiles Measured by Triple-Crystal Diffractometer

Triple-crystal X-ray diffractometry provides the most complete and precise characterization of defects in bulk crystals and strain parameters in the disturbed surface layer (see, e.g., [101–112]). A traditional analysis of TCD measurements is usually restricted to an analysis of diffuse scattering intensity in those regions where the coherent component of the diffraction intensity can be neglected. This approach can lead to systematic errors when determining the characteristics of imperfections.

In addition, there are experimental challenges associated with the influence of instrumental factors; the reflectivities of monochromator and analyzer crystals are sometimes simulated by Gaussians [113,114] to decrease calculation time and to account partially for imperfections in these crystals. It is clear, however, that these reflectivities will be substantially modified in the presence of diffuse scattering from crystal imperfections, and such a simplified description can significantly distort the results of TCD characterization of real samples. For this reason, it is necessary to consider both the instrumental factors (horizontal and vertical divergence and dispersion of X-rays) and peculiarities of the defect structure in the sample crystal, as well as in monochromator and analyzer crystals, to accurately characterize defects in sample crystals.

Analytical expressions for X-ray diffraction profiles measured by TCD from single crystals with homogeneously distributed defects of Coulomb type that account for instrumental factors have been derived in Molodkin et al. [47], where the consideration has been restricted to the Huang scattering region in the diffuse scattering component of reflectivity of the investigated crystal. The model was generalized some time later to include the integral-differential method of TCD [48,49] and to account for asymptotic diffuse scattering [49,115].

A generalization of the dynamical theory of TCD has been carried out (see [116] and [117]). The influence of disturbed surface layers in each of three crystals on their reflectivities and the contribution of TDS were taken into account. The theoretical results were used for the analysis of the TCD profiles measured from Cz Si [116,117] and FZ Si [118] single crystals.

5.1 Coherent Component

In the case of the nondispersive (n,–n,n) geometry, the coherent component of the X-ray intensity measured by TCD can be represented in the form [116]

$$I_{\text{coh}}\left(\Delta\theta,\Delta\theta'\right)=I_0\int_{-\infty}^{\infty} dx\; R_{\text{M}}^{n_{\text{M}}}\left(b_{\text{M}}^{-1}[b_{\text{S}}^{-1}(x-\Delta\theta)-\Delta\theta]\right)\quad R_{\text{coh}}\left(b_{\text{S}}^{-1}(x-\Delta\theta)\right)R_{\text{A}}^{n_{\text{A}}}\left(x-\Delta\theta'\right), \quad (21.91)$$

where R_{M} and R_{A} are reflection coefficients of monochromator and analyzer crystals; R_{coh} is the coherent component of reflectivity (from Equation 21.31) of the crystal sample under investigation; $\Delta\theta$ and $\Delta\theta'$ are angular deviations of sample and analyzer crystals, respectively; n_{M} and n_{A} are numbers of reflections at the channel-cut grooved monochromator and analyzer crystals; and b_M and b_S are asymmetry parameters of monochromator and sample crystals, respectively.

The influence of disturbed subsurface layers can be taken into account in the general case by considering all the coherent components of crystal reflectivities in Equation 21.91. This approach is similar to the approach in the previous section for multilayered systems. In the particular case of sufficiently thin layers, their contribution to the coherent component of crystal reflectivity can be represented in explicit analytical terms using the kinematic approximation [116].

If structural defects are present in monochromator and analyzer crystals, then their reflection coefficients also consist of coherent and diffuse components similar to those in Equations 21.30 and 21.53 for the investigated crystal:

$$R_{\mathrm{M,A}}(\mathbf{k}) = R_{\mathrm{B}}^{\mathrm{M,A}}(\mathbf{k}) + R_{\mathrm{D}}^{\mathrm{M,A}}(\mathbf{k}). \tag{21.92}$$

Such a modification of reflection coefficients for monochromator and analyzer crystals requires additional integrations over solid angles $d\Omega_{\mathbf{K}}$ and $d\Omega_{\mathbf{K}'}$ in Equation 21.91. These angles correspond to deviations of wave vectors of diffuse scattering waves in the monochromator and analyzer crystals. After the integrations, the reflection coefficients for monochromator and analyzer crystals in Equation 21.91 take the form

$$R_{\mathrm{M,A}}(\Delta\theta) = R_{\mathrm{coh}}^{\mathrm{M,A}}(\Delta\theta) + R_{\mathrm{diff}}^{\mathrm{M,A}}(\Delta\theta), \tag{21.93}$$

where the coherent components $R_{\mathrm{coh}}^{\mathrm{M}}(\Delta\theta)$ and $R_{\mathrm{coh}}^{\mathrm{A}}(\Delta\theta)$ in general are described by Equation 21.31, whereas the diffuse components are represented by the following expressions:

$$\begin{aligned} R_{\mathrm{diff}}^{\mathrm{M}}(\Delta\theta) &= \int d\Omega_{\mathbf{K}} R_{\mathrm{D}}^{\mathrm{M}}(\mathbf{k}) = \int d(\Delta\theta)\, r_{\mathrm{diff}}^{\mathrm{M}}(\text{æ}), \\ R_{\mathrm{diff}}^{\mathrm{A}}(\Delta\theta) &= \int d\Omega_{\mathbf{K}'} R_{\mathrm{D}}^{\mathrm{A}}(\mathbf{k}) = \int d(\Delta\theta')\, r_{\mathrm{diff}}^{\mathrm{A}}(\text{æ}). \end{aligned} \tag{21.94}$$

Here the two-dimensional vector in a momentum space is defined as $\text{æ} = k_x\mathbf{e}_x + k_z\mathbf{e}_z$, where $\mathbf{e}_x$ and $\mathbf{e}_z \equiv \mathbf{n}$ are orthogonal unit vectors in the scattering plane, $\mathbf{n}$ is the normal to an entrance crystal surface, and the diffuse components of crystal reflectivities integrated over the vertical divergence are defined as

$$r_{\mathrm{diff}}^{\mathrm{M}}(\text{æ}) = \frac{1}{K}\int dk_y R_{\mathrm{D}}^{\mathrm{M}}(\mathbf{k}), \quad r_{\mathrm{diff}}^{\mathrm{A}}(\text{æ}) = \frac{1}{K}\int dk_y R_{\mathrm{D}}^{\mathrm{A}}(\mathbf{k}). \tag{21.95}$$

As can be seen from Figure 21.7, the presence of diffuse scattering from defects in the monochromator crystal leads to a rapid increase of the pseudo-peak height in comparison to the main peak. This effect is often observed in TCD profiles [102,110,112,119] but explained only by the influence of TDS in the monochromator crystal [120]. However, the observed variety of height differences between the main and pseudo-peaks can only be explained by accounting for diffuse scattering

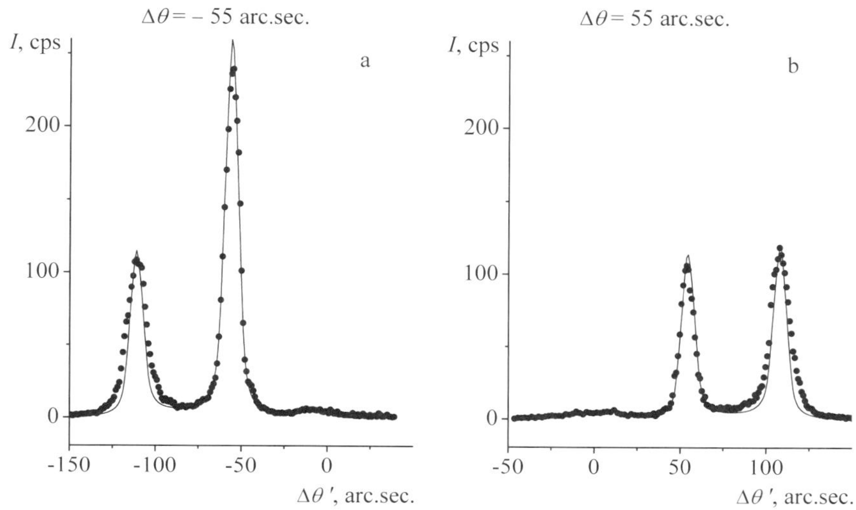

Figure 21.7. *Diffraction profiles measured by TCD from an FZ Si sample at its (a) negative and (b) positive deviations from exact reflecting position.*

from microdefects in monochromator crystal [116,118]. In particular, the asymmetric behavior of the pseudo-peak height at mirrored diffraction conditions of the investigated crystal (see Figure 21.7) can be explained only by the contribution of an antisymmetric part of the diffuse scattering intensity from microdefects in the monochromator crystal [118]. It should be noted also that the presence of microdefects in the analyzer crystal has a smaller influence on the shape of TCD profiles and only leads to some broadening of coherent peaks.

In addition to diffuse scattering from microdefects, weak elastic strains in surface layers of the monochromator crystal can influence the shape of TCD profiles [116]. These strains in surface layers can lead to a wide spectrum of variations in the heights of coherent peaks at different angular deviations of the investigated crystal, but the basic feature of this influence is to further increase the height of the pseudo-peak in comparison to the main peak. The presence of strain in the surface layer of the analyzer results in some broadening of both coherent peaks and their small additional asymmetry, similar to the presence of microdefects.

5.2 Diffuse Component

The diffuse component of the X-ray intensity measured by TCD can be represented in the form [117]

$$I_{\mathrm{diff}}\left(\Delta\theta,\Delta\theta'\right)=I_0\int_{-\infty}^{\infty}dx\;R_{\mathrm{M}}^{n_{\mathrm{M}}}(x)\int_{-\infty}^{\infty}dx'\;r_{\mathrm{diff}}(\text{æ})\;R_{\mathrm{A}}^{n_{\mathrm{A}}}\left(x'-\Delta\theta'\right). \qquad (21.96)$$

Here the diffuse component of the differential reflection coefficient in Equation 21.53 integrated over a vertical divergence is defined as

$$r_{\text{diff}}\left(\text{æ}\right)=\frac{1}{K}\int dk_y R_{\text{D}}\left(\mathbf{k}\right). \tag{21.97}$$

Two components, k_x and k_z, of the vector $\mathbf{k} = (k_x, k_y, k_z)$ lie in the plane of coherent scattering (**K**,**H**). The component k_z is directed along the normal **n**, and components k_x and k_y lie in the crystal surface. The components k_x and k_z are connected by simple relationships with angular deviations of the wave vectors of incident (x) and scattered (x') plane waves from their exact Bragg directions in the scattering plane. In the case of the symmetric Bragg diffraction geometry, these relationships have the form

$$\begin{aligned} k_x &= K(2\Delta\theta + x - x')\sin\theta_{\text{B}}, \\ k_z &= -K(x + x')\cos\theta_{\text{B}}, \end{aligned} \tag{21.98}$$

and for an asymmetric Bragg diffraction geometry, Equation 21.98 should be replaced by

$$\begin{aligned} k_x &= -K(x + x')\sin(\theta_{\text{B}} + \psi) + 2K(\Delta\theta + x)\sin\theta_{\text{B}}\cos\psi, \\ k_z &= -K(x + x')\cos(\theta_{\text{B}} + \psi) - 2K(\Delta\theta + x)\sin\theta_{\text{B}}\sin\psi. \end{aligned} \tag{21.99}$$

If the half width of the function r_{diff} in Equation 21.96 is much larger than half widths of reflection coefficients R_{M} and R_{A}, these functions can approximately be replaced by δ functions, and then one can write

$$I_{\text{diff}}\left(\Delta\theta,\ \Delta\theta'\right) \approx I_0 R_{i\text{M}} R_{i\text{A}} r_{\text{diff}}\left(k_x, k_z\right), \tag{21.100}$$

where $R_{i\text{M}}$ and $R_{i\text{A}}$ are integrated reflectivities of the monochromator and analyzer crystals and components k_x and k_z are described by the following expressions:

$$\begin{aligned} k_x &= -K\Delta\theta'\sin(\theta_{\text{B}} + \psi) + 2K\,\Delta\theta\sin\theta_{\text{B}}\cos\psi, \\ k_z &= -K\Delta\theta'\cos(\theta_{\text{B}} + \psi) - 2K\,\Delta\theta\sin\theta_{\text{B}}\sin\psi. \end{aligned} \tag{21.101}$$

After integrating in Equation 21.97, where $\mathbf{k} = \text{æ} + k_y\mathbf{e}_y$ ($\mathbf{e}_y$ is a normal to the scattering plane), we obtain in the Huang scattering region ($\text{æ} \le k_{\text{m}}$):

$$r_{\text{diff}}\left(\text{æ}\right) = M\left[A_{\text{H}}\left(\text{æ}\right) + A_{\text{S-W}}\left(\text{æ}\right) + A_{\text{a}}\left(\text{æ}\right)\right], \tag{21.102}$$

$$A_{\mathrm{H}}(\text{æ}) = \frac{K}{\pi\sqrt{\text{æ}^2+\mu^2}}(2\beta_1 + a_{\text{æ}}\beta_2)\operatorname{arctg}\frac{\sqrt{k_{\mathrm{m}}^2-\text{æ}^2}}{\sqrt{\text{æ}^2+\mu^2}}, \quad a_{\text{æ}} = \frac{(\mathbf{H}_0\text{æ})^2}{\text{æ}^2+\mu^2}, \tag{21.103}$$

$$A_{\mathrm{S-W}}(\text{æ}) = \frac{K}{\pi\sqrt{\text{æ}^2+\mu^2}}\left(\beta_1 + \frac{3}{4}a_{\text{æ}}\beta_2\right)\left[\left(\frac{\pi}{2} - \operatorname{arctg}\frac{\sqrt{k_{\mathrm{m}}^2-\text{æ}^2}}{\sqrt{\text{æ}^2+\mu^2}}\right)\frac{k_{\mathrm{m}}^2}{\text{æ}^2+\mu^2} - \frac{\sqrt{k_{\mathrm{m}}^2-\text{æ}^2}}{\sqrt{\text{æ}^2+\mu^2}}\right] + \frac{K}{2\pi}a_{\text{æ}}\beta_2\frac{\sqrt{k_{\mathrm{m}}^2-\text{æ}^2}}{\sqrt{\text{æ}^2+\mu^2}}, \tag{21.104}$$

$$A_{\mathrm{a}}(\text{æ}) = \frac{K}{\pi H}\beta_3\left[\ln\frac{\sqrt{k_{\mathrm{m}}^2+\mu^2}+\sqrt{k_{\mathrm{m}}^2-\text{æ}^2}}{\sqrt{k_{\mathrm{m}}^2+\mu^2}-\sqrt{k_{\mathrm{m}}^2-\text{æ}^2}} + \frac{2k_{\mathrm{m}}^2}{\text{æ}^2+\mu^2}\left(1-\frac{\sqrt{k_{\mathrm{m}}^2-\text{æ}^2}}{\sqrt{k_{\mathrm{m}}^2+\mu^2}}\right)\right], \tag{21.105}$$

where $M = cm_0C^2E^2(2\gamma_0\mu)^{-1}$, $\beta_i = B_i/B$, and $i = \overline{1,3}$.

Similarly, in the Stokes-Wilson scattering region ($\text{æ} \geq \mathrm{k_m}$) we obtain

$$r_{\mathrm{diff}}(\text{æ}) = MB_{\mathrm{S-W}}(\text{æ}), \qquad B_{\mathrm{S-W}}(\text{æ}) = \frac{Kk_{\mathrm{m}}^2}{2(\text{æ}^2+\mu^2)^{3/2}}\left(B_1 + \frac{3}{2}a_{\text{æ}}B_2\right). \tag{21.106}$$

If several types of microdefects are present in a crystal simultaneously, Equations 21.102 and 21.106 should simply be replaced by the sum of corresponding expressions for each type of defect.

The interference absorption coefficient μ in Equations 21.103 to 21.106 acts as the cut parameter and removes the divergence at $\text{æ} \to 0$. This allows for a correct description of the diffuse scattering component to TCD profiles in the coherent peak region, where this contribution is rapidly decreasing. In the analysis of diffuse scattering peaks on TCD profiles, as a rule, the condition $\text{æ} >> \mu$ is satisfied and consequently this parameter can be neglected.

The above approach describes the complex diffraction patterns measured by TCD from single crystals with complicated defect structures and explains the asymmetrical behavior of coherent peaks on TCD profiles by taking into account the influence of microdefects and disturbed layers in the monochromator, analyzer, and sample crystals.

5.3 Analysis of Experimental Triple-Crystal Diffractometer Profiles

The theoretical model developed for analysis of full TCD profiles has allowed quantitative characterization of complicated defect structures in silicon single crystals at a qualitatively new level of sensitivity and reliability [116,117]. Monochromator and analyzer crystals used in TCD measurements in

Kislovskii et al. [118] were identical to the investigated sample; that is, they have been cut from the central part of the same FZ Si plate. TCD measurements were carried out in a nondispersive scheme (n,–n,n) with symmetric 111 reflections in Bragg diffraction geometry on all three crystals by using $CuK_{\alpha 1}$ radiation from the conventional X-ray tube. The TCD profiles were measured at fixed sample deviation $\Delta\theta$ by rotating the analyzer crystal ($\Delta\theta'$) with a stepper motor. The corresponding diffraction profiles are shown in Figure 21.7.

Very low heights of the diffuse scattering peaks compared with coherent ones indicate the high perfection of the crystal under investigation. The fit of diffuse scattering peaks on these profiles (Figure 21.8) has been performed assuming the same defect model as for the analysis of rocking curves in Vladimirova et al. [54]. It should be noted that both diffuse scattering peaks in Figure 21.8 have been fitted simultaneously and in the whole measured angular range, including the regions of coherent peaks. The characteristics of dislocation loops, which have been determined for all three FZ Si crystals of the TCD scheme simultaneously, coincide well with those of larger dislocation loops in Vladimirova et al. [54]. However, the diffuse scattering intensity from smaller dislocation loops was too low in the angular range measured by TCD, and these defects could not be detected and characterized. In view of this circumstance, the new combined DCD + TCD method proposed in Shpak et al. [121] seems to be helpful for the quantitative characterization of complicated defect structures with wide spreads of microdefect sizes.

The analysis of TCD profiles measured from the Cz Si sample, which has been annealed at 750°C for 50 hours, was performed by Olikhovskii et al. [116,117] (see Figure 21.9). The diffraction profiles were taken at different deviation angles of the investigated sample, and their joint treatment determined the characteristics of two types of microdefects, namely, oxygen precipitates and dislocation loops, which correlate well with results of DCD measurements [65].

The thorough analysis of diffraction profiles using the dynamical diffraction model determined not only the characteristics of oxygen precipitates and dislocation loops in the crystal bulk but also the strain parameters in the subsurface layer. This strain is caused by "mirror-image forces" from micro-

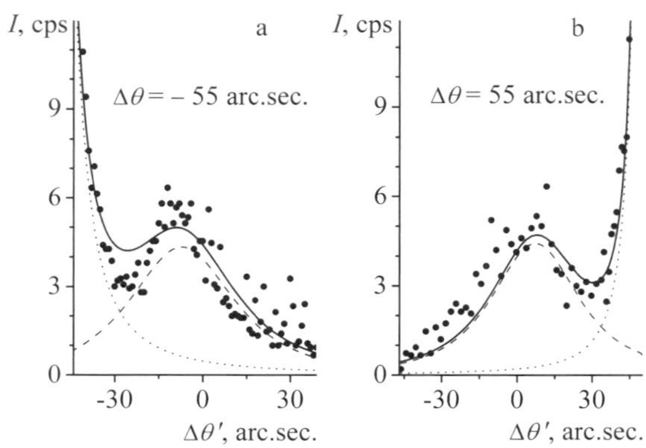

Figure 21.8. *Fitted diffuse scattering peaks on diffraction profiles measured from a FZ Si sample by TCD for the Si 111 reflection of CuK_1 radiation (dotted and dashed lines describe the coherent and diffuse scattering components of the diffracted intensity, respectively).*

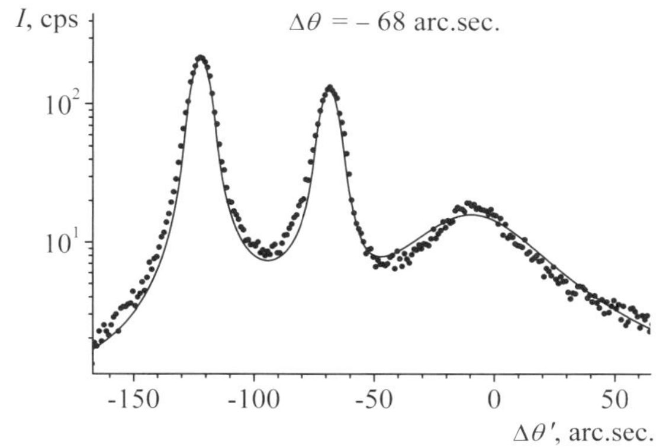

Figure 21.9. *Fitted total diffraction profile measured by TCD from the Cz Si sample for the Si 111 reflection of CuK$_1$ radiation.*

defects [21] and was described by the exponential law $\varepsilon_\perp = \varepsilon_{0\perp} \exp(-z/t_0)$, where z is the crystal depth, and the parameters that fit were t_0 = 7 nm and $\varepsilon_{0\perp} = 10^{-4}$. The quantitative description of the asymmetrical pseudo-peak heights in TCD profiles also determines the characteristics of dislocation loops in the monochromator crystal.

Thus the theoretical model developed for a description of full TCD profiles enables quantitative characterization of complicated defect structures in silicon single crystals at a qualitatively new level of completeness, uniqueness, and sensitivity. The thorough analysis of total diffraction profiles measured by TCD allows for the reliable determination of the role of diffuse scattering in the monochromator crystal in the formation of coherent peaks observed in TCD profiles.

6 Conclusions

This review summarizes the key principles of the generalized statistical dynamical theory of X-ray scattering in imperfect single crystals with randomly distributed microdefects. The expressions derived for the dispersion corrections to the wave vectors of coherent and diffuse scattered waves take into account, for the first time, multiple scattering of diffuse scattering waves from both average and fluctuating parts of the crystal polarizability, that is, for dynamical effects in diffuse scattering and for multiple diffuse scattering processes themselves. This provides an opportunity to correctly model the diffraction pattern from crystals with large microdefects with dimensions similar to the extinction length, including in the total reflection range. The main advantage of the statistical dynamical theory developed is the derivation of explicit analytical expressions for coherent and diffuse scattering intensities and their direct relations with defect characteristics. This extends quantitative characterization of defect structures to real crystals with multiple defect types and improves the reliability of the characterization.

The dynamical expressions for coherent and diffuse scattering components of the crystal reflectivity integrated over exit angles form the theoretical foundation of the generalized method of "integral" diffuse scattering. This method treats the measured rocking curves quantitatively over the whole angle range, including the total reflection range, and determines microdefect characteristics without restrictions on the defect sizes.

The extension of statistical dynamical theory to imperfect multilayered crystal structures has also been described. The recurrence relations for amplitudes of coherent waves in a multilayered structure with two-dimensional and three-dimensional defects take into account their extinction due to diffuse scattering in layers and in the substrate. The expression for the diffuse scattering component of the reflection coefficient of such structures accounts for the dynamical redistribution of transmitted and diffracted coherent waves in each layer.

Analytical expressions for coherent and diffuse scattering intensity components measured by DCD and TCD have been derived from statistical dynamical theory, taking into account instrumental factors. The obtained expressions take into account the influence of diffuse scattering effects from imperfections in monochromator and analyzer crystals as well. It should be emphasized that a self-consistent description of the diffracted intensity components and a careful treatment of instrumental factors are essential for reliable quantitative characterization of complicated defect structures in real crystals, especially with large defect sizes, by using the measured full diffraction profiles or reciprocal lattice maps.

It has also been shown that different diffraction conditions can lead to substantial changes of the ratio of diffuse to Bragg component contributions to the scattering pattern. Thus the character of the influence of defects on the general scattering pattern is changed due to opposite signs of the influence of defects on the coherent and diffuse components.

The analytic formulas derived have been applied to quantitatively characterize structural imperfections in silicon crystals grown by Czochralski- and float-zone methods as well as in multilayered semiconductor structures with a single QW by using the diffraction profiles measured by DCD and TCD. Overall, the analyzed diffraction profiles account for diffuse scattering effects, and the statistical dynamical theory appears essential for a correct interpretation and full description of the observed diffraction patterns. In general, the derived self-consistent dynamical expressions for the coherent and diffuse scattering components of both differential and integrated reflectivities of imperfect single crystals and multilayered structures, which contain randomly distributed microdefects, and the method of DDCD, which has been developed on this basis, can be used to perform quantitative characterization of various defect structures [42], particularly, with several microdefect types.

Finally, it should be mentioned in this conclusion that in cases where dynamical effects do not take place, that is, where the scattering is kinematic, the use of the dynamical formulas remains useful. For example, the dynamical formulas automatically give a correct description of the kinematic scattering pattern even when dynamical effects are small. Second, if the dynamical approach is used, it is not necessary to analyze and justify its applicability for each specific experiment, which is required for the kinematic theory.

Acknowledgment

This research was supported by the Science and Technology Center in Ukraine (Project #3085).

References

1. M. Von Laue, *Röntgenstrahlinterferezen* (Leipzig: Akademishe Verlagsges, 1948).

2. C. Hammond, *The Basics of Crystallography and Diffraction*, 2nd ed. (London: Oxford University Press, 2001).
3. R. W. James, "The dynamical theory of X-ray diffraction," *Solid State Phys.* 15 (1963): 55.
4. B. W. Batterman and H. Cole, *Rev. Mod. Phys.* 36 (1964): 681.
5. S. Takagi, *Acta Crystallogr.* 15 (1962): 1131.
6. S. Takagi, *J. Phys. Soc. Jpn.* 26 (1969): 1239.
7. D. Taupin, *Bull. Soc. Franc. Miner. Crist.* 87, no. 2 (1964): 469.
8. P. Penning and D. Polder, *Philips Res. Repts.* 16, no. 2 (1961): 419.
9. P. Penning, *Philips Res. Repts.* Suppl., 21, no. 5 (1966): 1.
10. N. Kato, *Acta Crystallogr.* 16 (1963): 276.
11. N. Kato, *Z. Naturforsch. A.* 28 (1973): 604.
12. L. V. Azaroff, R. Kaplow, N. Kato, R. J. Weiss, A. J. C. Wilson, and R. A. Young, *X-ray Diffraction* (New York: McGraw-Hill, 1974).
13. A. Authier, *Dynamical Theory of X-ray Diffraction* (London: Oxford University Press, 2001).
14. M. Krivoglaz A., *X-ray and Neutron Diffraction in Nonideal Crystals* (Berlin: Springer, 1996).
15. L. I. Datsenko, V. B. Molodkin, and M. E. Osinovskii, *Dynamical Scattering of X-rays by Real Crystals* (Kiev: Naukova Dumka, 1988).
16. V. B. Molodkin and E. A. Tikhonova, *Fiz. Met. i Metalloved.* 24 (1967): 385.
17. V. B. Molodkin, *Fiz. Met. i Metalloved.* 25 (1968): 410.
18. V. B. Molodkin, *Fiz. Met. i Metalloved.* 27 (1969): 582.
19. V. B. Molodkin, *Metallofiz.* 2, no. 1 (1980): 3.
20. V. B. Molodkin, *Phys. Metals* 3 (1981): 615.
21. V. B. Molodkin, S. I. Olikhovskii, and M. E. Osinovskii, *Phys. Metals* 5 (1984): 1.
22. V. B. Molodkin, S. I. Olikhovskii, and M. E. Osinovskii, *Phys. Metals* 5 (1985): 847.
23. V. V. Kochelab, V. B. Molodkin, S. I. Olikhovskii, and M. E. Osinovskii, *Phys. Stat. Solidi A* 108, no. 1 (1988): 67.
24. P. H. Dederichs, *Phys. Stat. Solidi B* 23 (1967): 377.
25. P. H. Dederichs, *Solid State Phys.* 27 (1972): 135.
26. V. V. Ratnikov, E. K. Kovev, and L. M. Sorokin, *Fiz. Tverd. Tela* 26 (1984): 2155.
27. V. V. Ratnikov and L. M. Sorokin, *Fiz. Tverd. Tela* 26 (1984): 3445.
28. V. V. Ratnikov and R. N. Kyutt, *Zhurn. Techn. Fiz.* 55 (1985): 391.
29. R. N. Kyutt and V. V. Ratnikov, *Metallofiz.* 7, no. 1 (1985): 36 [Engl. translation: *Phys. Metals* 7 (1987): 58].
30. V. B. Molodkin, S. I. Olikhovskii, E. N. Kislovskii, E. G. Len, and E. V. Pervak, *Phys. Stat. Solidi B* 227 (2001): 429.
31. S. I. Olikhovskii, V. B. Molodkin, E. N. Kislovskii, E. G. Len, and E. V. Pervak, *Phys. Stat. Solidi B* 231 (2002): 199.
32. A. N. Kostyuk, V. B. Molodkin, and S. I. Olikhovskii, *Phys. Stat. Solidi B* 178 (1993): 45.
33. V. B. Molodkin, S. I. Olikhovskii, and A. N. Kostyuk, *Phys. Stat. Solidi B* 183 (1994): 59.
34. M. Kato, *Acta Crystallogr. A* 36 (1980): 763; ibid., 36 (1980): 770; ibid., 47 (1991): 1.
35. V. Holý and K. T. Gabrielyan, *Phys. Stat. Solidi B* 140 (1987): 39.
36. V. Holý and J. Kuběna, *Phys. Stat. Solidi B* 151 (1989): 23; ibid., 155 (1989): 339.
37. M. Al Haddad and P. Becker, *Acta Crystallogr. A* 44 (1988): 262; P. Becker and M. Al Haddad, ibid., 45 (1989): 333; ibid., 46 (1990): 123; ibid., 48 (1992): 121.
38. A. M. Polyakov, F. N. Chukhovskii, and D. I. Piskunov, *Zhurn. Exp. Theor. Fiz.* 99 (1991): 589 [Engl. translation: *Sov. Phys. JETP* 72 (1991): 330].
39. J. P. Guigay and F. N. Chukhovskii, *Acta Crystallogr. A* 48 (1992): 819; ibid., 51 (1995): 288.
40. N. M. Olekhnovich, A. L. Karpey, A. I. Olekhnovich, and L. D. Puzenkova, *Acta Crystallogr. A* 39 (1983): 116.
41. J. R. Schneider, R. Bouchard, H. A. Graf, and H. Nagasava, *Acta Crystallogr. A* 48 (1992): 804.
42. V. B. Molodkin, A. I. Nizkova, A. P. Shpak, V. Ph. Machulin, V. P. Kladko, I. V. Prokopenko, R. N. Kyutt, E. N. Kislovskii, S. I. Olikhovskii, E. V. Pervak, I. M. Fodchuk, et al., *Diffractometry of Nanosized Defects and Heterolayers of Crystals* (Kiev: Akademperiodyka, 2005).
43. V. B. Molodkin, S. V. Dmitriev, E. V. Pervak, A. A. Belotskaya, A. I. Nizkova, and A. P. Melnik, *Metallofiz. Noveishie Tekhnol.* 28 (2006): 1077.

44. V. G. Bar'yakhtar, V. V. Nemoshkalenko, V. B. Molodkin, A. P. Shpak, R. M. Kutt, M. V. Kovalchuk, S. I. Olikhovskii, G. I. Nizkova, Ye. M. Kislovskii, T. A. Grischenko et al., Patent of Ukraine no. 14831A, М К И6 G 01 no. 23/ 20, Bulletin no. 2 (1997).
45. V. G. Bar'yakhtar, V. V. Nemoshkalenko, V. B. Molodkin, A. P. Shpak, J. Chikawa, K. Kohra, R. M. Kutt, M. V. Kovalchuk, S. I. Olikhovskii, G. I. Nizkova, et al., Patent of Ukraine no. 17146A, М К И6 G 01 no. 23/ 20, Bulletin no. 3. (1997).
46. A. M. Gureev, V. G. Bar'yakhtar, M. A. Gureev, M. T. Kogut, V. P. Krivitskii, G. I. Nizkova, S. I. Olikhovskii, M. E. Osinovskii, O. V. Polinur, D. I. Bidnik, et al., Patent of Ukraine no. 19220, М К И6 G 01 no. 23/ 20, Bulletin no. 6. (1997).
47. V. B. Molodkin, S. I. Olikhovskii, M. E. Osinovskii, V. V. Kochelab, A. Yu. Kazimirov, M. V. Kovalchuk, F. N. Chukhovskii, *Metallofiz.* 6, no. 3 (1984): 7.
48. V. V. Nemoshkalenko, V. B. Molodkin, S. I. Olikhovskii, E. N. Kislovskii, T. A. Grischenko, M. T. Kogut, M. V. Kovalchuk, S. I. Haratyan, and G. V. Grin', *Metallofiz.* 15, no. 11 (1993): 53.
49. V. B. Molodkin, V. V. Nemoshkalenko, S. I. Olikhovskii, E. N. Kislovskii, O. V. Reshetnyk, T. P. Vladimirova, V. P. Krivitsky, V. F. Machulin, I. V. Prokopenko, G. E. Ice et al., *Metallofiz. Noveishie Tekhnol.* 20, no. 11 (1998): 29.
50. V. V. Nemoshkalenko, V. B. Molodkin, E. N. Kislovskii, S. I. Olikhovskii, T. A. Grischenko, M. T. Kogut, and E. V. Pervak, *Metallofiz. Noveishie Tekhnol.* 22, no. 2 (2000): 42.
51. V. V. Nemoshkalenko, V. B. Molodkin, E. N. Kislovskii, S. I. Olikhovskii, M. T. Kogut, L. M. Sheludchenko, and E. V. Pervak, *Metallofiz. Noveishie Tekhnol.* 22, no. 2 (2000): 51.
52. V. B. Molodkin, S. I. Olikhovskii, E. N. Kislovskii, V. P. Krivitsky, E. G. Len, E. V. Pervak, G. E. Ice, and B. C. Larson, *J. Phys. D : Appl. Phys.* 34 (2001): A82.
53. E. N. Kislovskii, S. I. Olikhovskii, V. B. Molodkin, V. V. Nemoshkalenko, V. P. Krivitsky, E. G. Len, E. V. Pervak, G. E. Ice, and B. C. Larson, *Phys. Stat. Solidi B* 231 (2002): 213.
54. T. P. Vladimirova, R. F. Seredenko, V. B. Molodkin, S. I. Olikhovskii, and E. N. Kislovskii, *Metallofiz. Noveishie Tekhnol.* 29 (2007): 711.
55. M. V. Kovalchuk, and V. Kohn, *Uspekhi Fizicheskikh Nauk* 149 (1986): 69.
56. M. V. Kovalchuk and E. H. Mukhamedjanov, *Fiz. Tverd. Tela* 25 (1983): 3532.
57. M. J. Bedzyk, G. Materlik, and M. V. Kovalchuk, *Phys. Rev. B.* 30 (1984): 2453.
58. S. I. Zheludeva, S. Lagomarsino, N. N. Novikova, M. V. Kovalchuk, and F. Scarinci, *Thin Solid Films* 193 (1990): 395.
59. P. P. Ewald, *Ann. Physik* 49 (1916): 1; ibid., 49 (1916): 117; ibid., 54 (1917): 519; *Z. Krist.* 97 (1937): 1.
60. H. Bethe, *Ann. Physik* 87 (1928): 55.
61. P. H. Dederichs, *Phys. Rev. B* 4 (1971): 1041.
62. B. C. Larson, *J. Appl. Cryst.* 8 (1975): 150.
63. S. Y. Olikhovsky, Ye. M. Kyslovskyy, V. B. Molodkin, E. G. Len, T. P. Vladimirova, and O. V. Reshetnyk, *Metallofiz. Nov. Techn.* 22, no. 6 (2000): 3.
64. I. V. Prokopenko, E. N. Kislovskii, S. I. Olikhovskii, V. M. Tkach, P. M. Lytvyn, and T. P. Vladimirova, *Semiconductor Physics, Quantum Electronics & Optoelectronics* 3, no. 3 (2000): 275.
65. Ye. M. Kyslovskyy, T. P. Vladimirova, S. I. Olikhovskii, V. B. Molodkin, E. V. Kochelab, and R. F. Seredenko, *Phys. Stat. Solidi A* 204 (2007): 2591.
66. C. Kittel, *Introduction to Solid-State Physics* (New York: John Wiley & Sons, 1986).
67. H. Föll, V. Gösele, and B. O. Kolbesen, *J. Cryst. Growth* 52 (1981): 907.
68. P. J. Roksnoer and M. M. B. van der Boom, *J. Cryst. Growth* 53 (1981): 563.
69. P. J. Roksnoer, *J. Cryst. Growth* 68 (1984): 596.
70. A. Borghesi, B. Pivac, A. Sassella, and A. Stella, *J. Appl. Phys.* 77 (1995): 4169.
71. T. Y. Tan and W. K. Tice, *Phil. Mag.* 34 (1976): 615.
72. D. M. Maher, A. Staudinger, and J. R. Patel, *J. Appl. Phys.* 47 (1976): 3813.
73. J. R. Patel, K. A. Jackson, and H. Reiss, *J. Appl. Phys.* 48 (1977): 5279.
74. W. Patrick, E. Hearn, W. Westdorp, and A. Bohg, *J. Appl. Phys.* 50 (1979): 7156.
75. S. M. Hu, *Appl. Phys. Lett.* 48 (1986): 115.
76. F. S. Ham *J. Appl. Phys.* 30 (1959): 1518.
77. K. Sueoka, N. Ikeda, T. Yamamoto, and S. Kobayashi, *J. Appl. Phys.* 74 (1993): 5437.

78. J. R. Patel, K. A. Jackson, and H. Reiss, *J. Appl. Phys.* 48 (1977): 5279.
79. K. F. Kelton, R. Falster, D. Gambaro, M. Olmo, M. Comara, and P. F. Wei, *J. Appl. Phys.* 85 (1999): 8097.
80. S. Senkader, J. Esfandyari, and G. J. Hobler, *Appl. Phys.* 78 (1995): 6469.
81. B. G. Ko and K. D. Kwack, *J. Appl. Phys.* 85 (1999): 2100.
82. S. I. Olikhovskii, M. M. Belova, and E. V. Kochelab, *Uspehi Fiziki Metallov* 7, no. 3 (2006): 135.
83. M. M. Belova, M. N. Moskalkov, A. E. Rudich, S. I. Olikhovskii, and E. V. Kochelab, *Metallofiz. Noveishie Tekhnol.* 29 (2007): 427.
84. S. I. Olikhovskii, M. M. Belova, M. N. Moskalkov, Yu. A. Belov, A. E. Rudich, and E. V. Kochelab, *Metallofiz. Noveishie Tekhnol.* 29 (2007): 649.
85. M. M. Belova, M. N. Moskalkov, Yu. A. Belov, A. E. Rudich, S. I. Olikhovskii, and E. V. Kochelab, *Metallofiz. Noveishie Tekhnol.* 29 (2007): 727.
86. V. S. Speriosu, and T. Vreeland, *J. Appl. Phys.* 56 (1984): 1591.
87. M. A. G. Halliwell, M. H. Lyons, and M. J. Hill, *J. Cryst. Growth* 68 (1984): 523.
88. W. J. Bartels, J. Hornstra, and D. J. Lobeek, *Acta Crystallogr. A* 42 (1986): 539.
89. L. Tapfer and K. Ploog, *Phys. Rev. B* 33 (1986): 5565.
90. V. Holý, J. Kuběna, and K. Ploog, *Phys. Stat. Solidi B* 162 (1990): 347.
91. A. M. Afanas'ev and R. M. Imamov, *Crystallogr. Rep.* 48 (2003): 728.
92. P. F. Fewster, *J. Appl. Crystallogr.* 25 (1992): 714.
93. V. Holý, U. Pietsch, and T. Baumbach, *High-Resolution X- Ray Scattering from Thin Films and Multilayers* (Berlin Heideiberg: Springer Verlag, 1999).
94. S. Y. Olikhovskyy, V. B. Molodkin, Ye. M. Kysloskyy, I. M. Fodchuk, O. S. Skakunova, and K. V. Pervak, *Metallofiz. Noveishie Tekhnol.* 27 (2005): 197.
95. A. P. Shpak, V. B. Molodkin, S. Y. Olikhovskyy, Ye. M. Kysloskyy, I. M. Fodchuk, O. S. Skakunova, and K. V. Pervak, *Metallofiz. Noveishie Tekhnol.* 27 (2005): 721.
96. V. B. Molodkin, S. I. Olikhovskii, E. N. Kislovskii, I. M. Fodchuk, E. S. Skakunova, E. V. Pervak, and V. V. Molodkin, *Phys. Stat. Solidi A* 204 (2007): 2606.
97. D. M. Vardanyan, H. M. Manoukyan, and H. M. Petrosyan, *Acta Crystallogr. A* 41 (1985): 212.
98. S. Martini, A. A. Quivy, M. J. da Silva, T. E. Lamas, E. C. F. da Silva, J. R. Leite, and E. Abramof, *J. Appl. Phys.* 94 (2003): 7050.
99. Ye. M. Kysloskyy, S. Y. Olikhovskyy, V. B. Molodkin, B. K. Ostafiychuk, T. P. Vladimirova, E. G. Len, and O. V. Reshetnyk, *Metallofiz. Noveishie Tekhnol.* 27 (2005): 217.
100. S. Y. Olikhovskyy, V. B. Molodkin, Ye. M. Kysloskyy, E. G. Len, T. P. Vladimirova, O. S. Skakunova, and B. K. Ostafiychuk, *Metallofiz. Noveishie Tekhnol.* 27 (2005): 653.
101. B. C. Larson and W. G. Schmatz, *Phys. Rev. B* 10 (1974): 2307.
102. A. Iida and K. Kohra, *Phys. Stat. Solidi A* 51 (1979): 533.
103. A. A. Lomov, P. Zaumseil, and U. Winter, *Acta Crystallogr. A* 41 (1985): 223.
104. V. Holý and J. Kuběna, *Phys. Stat. Solidi B* 170 (1992): 9.
105. L. A. Charniy, A. N. Morozov, V. T. Bublik, K. D. Scherbachev, I. V. Stepantsova, and V. M. Kaganer, *J. Cryst. Growth* 118 (1992): 163.
106. V. T. Bublik and N. M. Zotov, *Crystallogr. Rep.* 44 (1999): 635.
107. V. T. Bublik and K. D. Scherbachev, *Poverkhnost'. Rentgenovskie, Sinkhrotronnye i Neitronnye Issledovaniya* 2 (2001): 87.
108. K. D. Scherbachev and V. T. Bublik, *Poverkhnost'. Rentgenovskie, Sinkhrotronnye i Neitronnye Issledovaniya* 9 (2000): 53.
109. P. Zaumseil, U. Winter, F. Cembali, M. Servidori, and Z. Sourek, *Phys. Stat. Solidi A* 100 (1987): 95.
110. P. Zaumseil and U. Winter, *Phys. Stat. Solidi A* 120 (1990): 67.
111. M. Servidori, R. Angelucci, F. Cembali, P. Negrini, S. Solmi, P. Zaumseil, and U. Winter, *J. Appl. Phys.* 61 (1987): 1834.
112. A. M. Afanas'ev, P. A. Aleksandrov, and R. M. Imamov, *X-ray Diffraction Diagnostics of Submicronic Layers* (Moscow: Nauka, 1989).
113. R. A. Cowley, *Acta Crystallogr. A* 43 (1987): 825.
114. E. L. Garstein, M. Mandelbrot, and D. Mogilyanski, *J. Phys. D* 31 (2001): A57.

115. V. B. Molodkin, M. Ando, E. N. Kislovskii, S. I. Olikhovskii, T. P. Vladimirova, O. V. Reshetnyk, E. G. Len, E. A. Evgrafova, and E. V. Pervak, *Metallofiz. Noveishie Tekhnol.* 24 (2002): 541.
116. S. I. Olikhovskii, V. B. Molodkin, E. N. Kislovskii, O. V. Reshetnyk, T. P. Vladimirova, E. G. Len, G. E. Ice, R. Barabash, R. Köhler, and D. O. Grigoriev, *Metallofiz. Noveishie Tekhnol.* 27 (2005): 947.
117. S. I. Olikhovskii, V. B. Molodkin, E. N. Kislovskii, O. V. Reshetnyk, T. P. Vladimirova, G. E. Ice, R. Barabash, Köhler, R., and D. O. Grigoriev, *Metallofiz. Noveishie Tekhnol.* 27 (2005): 1251.
118. E. N. Kislovskii, O. V. Reshetnyk, T. P. Vladimirova, V. B. Molodkin, S. I. Olikhovskii, B. V. Sheludchenko, R. F. Seredenko, and O. S. Skakunova, *Metallofiz. Noveishie Tekhnol.* 29 (2007): 701.
119. A. Y. Kazimirov, M. V. Kovalchuk, and V. G. Kon, *Metallofiz.* 9, no. 4 (1987): 54.
120. A. Y. Kazimirov, M. V. Kovalchuk, and V. G. Kon, *Kristallografiya* 32 (1987): 1360.
121. A. P. Shpak, V. B. Molodkin, S. I. Olikhovskii, Ye. M. Kyslovskyy, O. V. Reshetnyk, T. P. Vladimirova, E. G. Len, V. M. Venger, and S. V. Dmitriev, *Phys. Stat. Solidi A* 204 (2007): 2651.

INDEX

A

B

C

D

E

F

G

O

P

Q

R

S

Author E-mail Index

Abe, Hiroshi <ab@nda.ac.jp>

Asahi, R. <rasahi@mosk.tytlabs.co.jp>

Barabash, Rozaliya <barabashr@ornl.gov>

Bassler, Kevin <bassler@uh.edu>

Bauer, Guenther <guenther.bauer@jku.at>

Boesecke, Peter <boesecke@esrf.fr>

Castro-Colin, Miguel <mccolin@utep.edu>

Caha, Ondrej, <caha@monoceros.physics.muni.cz>

Chiang, T.C. <chiang@mrl.uiuc.edu>

de Boissieu, Marc <Marc.de-Boissieu@simap.inpg.fr>

Donner, Wolfgang <wdonner@tu-darmstadt.de>

Egami, Takeshi <egami@utk.edu>

Francoual, Sonia <sfranco@lanl.gov>

Goossens, Darren <goossens@rsc.anu.edu.au>

Gao, Carrie Y. gaoy1@ornl.gov

van Gastel, Raoul, <R.vanGastel@tnw.utwente.nl>

Holy, Vaclav <holy@mag.mff.cuni.cz>

Hong, Hawoong hong@aps.anl.gov

Ice, Gene <icege@ornl.gov>

Islam, Zahirul <zahir@aps.anl.gov>

Kaminski, Daniel, <daniel_kaminski3@wp.pl>

Karnaukhov I. M. <karnaukhov@kipt.kharkov.ua>

Keffer, D. J. <dkeffer@utk.edu>

Kosenkov, Vladimir <kosenkovv@vinf.ru>

Kostorz, Gernot <kostorz@emeritus.ethz.ch>

Kovalchuk, M. V. <koval@ns.crys.ras.ru>

Kyslovskyy, Ye. M. <yekis@imp.kiev.ua>

Larson, Bennett <larsonbc @ornl.gov>

Len, Evgen <len@imp.kiev.ua>

Levine, Lyle <lyle.levine@nist.gov>

Li, Jianhua <Jianhua.Li@Rigaku.com>

Liu, Xuerong <xliu@physics.ucsd.edu>

Long, Gabrielle <gglong@aps.anl.gov>

Metzger, Harmut <metzger@esrf.fr>

Mizutani, Uichiro <riken-mizutani@mosk.tytlabs.co.jp>

Molodkin, V.B. <molodkn@imp.kiev.ua>

Moss, Simon <smoss@uh.edu>

Nemanich, Robert, <robert_nemanich@ncsu.edu>

Nicholson, Donald <dmn@ornl.gov>

Nizkova, A.I. <nizkova@imp.kiev.ua>

Olikhovskii, S. I. <olikhovsky@meta.ua>

Poelsema, Bene, <B.Poelsema@tnw.utwente.nl>

Puzyrev, Y.S. <yevgeniy.s.puzyrev@vanderbilt.edu>

Sato, H. <hsato@auecc.aichi-edu.ac.jp>

Schönfeld, Bernd <bernd.schoenfeld@mat.ethz.ch>

Sinha, Sunil <ssinha@physics.ucsd.edu>

Silantiev, P.P. <fae@niiar.ru>

Shülli, Tobias <schulli@esrf.fr>

Skakunova, E. S <sunnydale@ukr.net>

Shpak, A. P. <metall@imp.kiev.ua>

Stangl, Julian <julian.stangl@jku.at;>

Steurer, Walter <steurer@mat.ethz.ch>;

Takeuchi, T. <takeuchi@nuap.nagoya-u.ac.jp>

Turchi, Patrice <turchi1@llnl.gov>

Vladimirova, T.P.< tvlad@imp.kiev.ua>

Vlieg, Elias <E.Vlieg@science.ru.nl>

Weber, Thomas <thomas.weber@mat.ethz.ch>

Welberry, Richard <welberry@rsc.anu.edu.au>

Xu, Ruqing <ruqingxu@uiuc.edu>

Contributors

Hiroshi Abe, Department of Materials Science and Engineering, National Defense Academy, Yokosuka 239-8686, Japan

R. Asahi, Toyota Central Research Laboratories, LTD, Nagakute, AICHI, 480-1192, Japan

Rozaliya I. Barabash, Oak Ridge National Laboratory, Oak Ridge, TN 37831, USA, and Department of Materials Science and Engineering, University of Tennessee, Knoxville, TN 37996, USA

Kevin E. Bassler, Department of Physics, University of Houston, 617 Science and Research Blvd. I, Houston, TX 77204-5005, USA

Günther Bauer, Institute of Semiconductor Physics, Johannes Kepler University Linz, A-4040 Linz, Austria

Marc de Boissieu, Sciences de l'Ingéniérie, des Matériaux et des Procédés (SIMaP), Grenoble INP, CNRS, UJF, BP 75 38402 St. Martin d'Hères Cedex, France

Peter Bösecke, E. Synchrotron Radiation Facility, BP 220; 38043 Grenoble CEDEX 9, France

Ondrej Caha, Institute of Condensed Matter Physics, Masaryk University, Kotlářská 2, 611 37 Brno, Czech Republic

Miguel Castro-Colin, Department of Physics, University of Texas El Paso, El Paso, TX 79968, USA

T.-C. Chiang, Department of Physics, University of Illinois at Urbana-Champaign, 1110 West Green Street, Urbana, IL 61801, USA

Wolfgang Donner, Technische Universität Darmstadt, Fachbereich Material und Geowissenschaften, Strukturforschung Petersenstrasse 23, 64287 Darmstadt, Germany

Takeshi Egami, Joint Institute for Neutron Sciences, Department of Materials Science and Engineering, Department of Physics and Astronomy, University of Tennessee, Knoxville, TN 37996, USA, and Materials Science and Technology Division, Oak Ridge National Laboratory, Oak Ridge, TN, 37831, USA

Sonia Francoual, National High Magnetic Field Laboratory, Los Alamos National Laboratory, Los Alamos, NM 87545, USA

Carrie Y. Gao, Department of Chemical Biomechanical Engineering, University of Tennessee, Knoxville, TN 37996, USA

Raoul van Gastel, University of Twente, Solid State Physics, MESA+ Institute for Nanotechnology, PO Box 217, 7500 AE Enschede, The Netherlands

Darren Goossens, Research School of Chemistry and Department of Physics, Australian National University, Canberra, ACT 0200, Australia

Václav Holý, Faculty of Mathematics and Physics, Charles University, Prague, Czech Republic

Hawoong Hong, Advanced Photon Source, Argonne National Laboratory, 9700 S. Cass Avenue, Argonne, IL 60439, USA

Gene E. Ice, Oak Ridge National Laboratory, Oak Ridge, TN 37831, USA

Zahirul Islam, X-ray Science Division, Advanced Photon Source, Argonne National Laboratory, 9700 S. Cass Avenue, Argonne, IL 60439, USA

Daniel Kaminski, IMM Solid State Chemistry, Radboud University Nijmegen Toernooiveld 1, 6525 ED Nijmegen, The Netherlands

Ivan M. Karnaukhov, National Scientific Centre "Kharkiv Institute of Physics and Technology," National Academy of Sciences of Ukraine, Kharkiv, Ukraine

David J. Keffer, Department of Chemical Biomechanical Engineering, University of Tennessee, Knoxville, TN 37996, USA

Vladimir M. Kosenkov, Research Institute of Atomic Reaktors, Dimitrovgrad, Russia

Gernot Kostorz, Professor emeritus, Department of Physics, ETH Zurich, 8093 Zurich, Switzerland

Mikhail V. Kovalchuk, Institute of Crystallography, Russian Academy of Sciences, Moscow, Russia

Yevgen M. Kyslovskyy, Institute for Metal Physics, National Academy of Sciences of Ukraine, Kyiv, Ukraine

Bennett C. Larson, Materials Science and Technology Division, Oak Ridge National Laboratory, Oak Ridge, TN 37831, USA

Evgen G. Len, Institute for Metal Physics, National Academy of Sciences of Ukraine, Kyiv, Ukraine

Lyle E. Levine, Materials Science and Engineering Laboratory, National Institute of Standards and Technology, Gaithersburg, MD 20899, USA

Jianhua Li, Rigaku Americas Corporation, 9009 New Trails Drive, The Woodlands, TX 77381, USA

Xuerong Liu, Physics Department, University of California, San Diego, 9500 Gilman Drive, La Jolla, CA 92093, USA

Gabrielle G. Long, X-ray Science Division, Advanced Photon Source, Building 401, Argonne National Laboratory, 9700 South Cass Avenue, Argonne, IL 60439 USA

Harmut T. Metzger, E. Synchrotron Radiation Facility, BP 220; 38043 Grenoble CEDEX 9, France

Uichiro Mizutani, Toyota Physical & Chemical Research Institute, Nagakute, AICHI, 480-1192, Japan

Vadim B. Molodkin, Institute for Metal Physics, National Academy of Sciences of Ukraine, Kyiv, Ukraine

Vitaliy V. Molodkin, Institute for Metal Physics, National Academy of Sciences of Ukraine, Kyiv, Ukraine

Simon C. Moss, Department of Physics, University of Houston, Houston, TX 77204, USA

Robert Nemanich, Department of Physics, North Carolina State University, Raleigh, NC, 27695, USA

Don M. C. Nicholson, Oak Ridge National Laboratory, Oak Ridge, TN 37831, USA

Anna I. Nizkova, Institute for Metal Physics, National Academy of Sciences of Ukraine, Kyiv, Ukraine

Stepan I. Olikhovskii, Institute for Metal Physics, National Academy of Sciences of Ukraine, Kyiv, Ukraine

Bene Poelsema, University of Twente, Solid State Physics, MESA+ Institute for Nanotechnology, PO Box 217, 7500 AE Enschede, The Netherlands

Yevgeniy S. Puzyrev, Department of Physics and Astronomy, Vanderbilt University 6301 Stevenson Center, Nashville, TN 37235, USA

H. Sato, Aichi University of Education, Kariya, Aichi-ken, 448-8542, Japan

Bernd Schönfeld, Laboratory of Metal Physics and Technology, Department of Materials, ETH Zurich, 8093 Zurich, Switzerland

Anatoliy P. Shpak, Institute for Metal Physics, National Academy of Sciences of Ukraine, Kyiv, Ukraine

Tobias Shülli, E. Synchrotron Radiation Facility, BP 220; 38043 Grenoble CEDEX 9, France

Pavel P. Silantiev, Research Institute of Atomic Reaktors, Dimitrovgrad, Russia

Sunil K. Sinha, Physics Department, University of California, San Diego, 9500 Gilman Drive, La Jolla, CA 92093, USA

Elena S. Skakunova, Institute for Metal Physics, National Academy of Sciences of Ukraine, Kyiv, Ukraine

Julian Stangl, Institute of Semiconductor Physics, Johannes Kepler University Linz, A-4040 Linz, Austria

Walter Steurer, Laboratory of Crystallography, Department of Materials, Wolfgang-Pauli-Strasse 10, CH-8093 Zurich, Switzerland

T. Takeuchi, Ecotopia Science Institute, Nagoya University, Furo-cho, Chikusa-ku, Nagoya 464-8603, Japan

Patrice E. A. Turchi, Lawrence Livermore National Laboratory, P. O. Box 808, Livermore CA 94551, USA

Tetiana P. Vladimirova, Institute for Metal Physics, National Academy of Sciences of Ukraine, Kyiv, Ukraine

Elias Vlieg, IMM Solid State Chemistry, Radboud University Nijmegen, Toernooiveld 1, 6525 ED Nijmegen, The Netherlands

Thomas Weber, Laboratory of Crystallography, Department of Materials, ETH Zurich, Wolfgang-Pauli-Strasse 10, CH-8093 Zurich, Switzerland

T. R. Welberry, Research School of Chemistry, Australian National University, Canberra, ACT 0200 Australia

Ruqing Xu, Department of Physics, University of Illinois at Urbana-Champaign, 1110 West Green Street, Urbana, IL 61801, USA